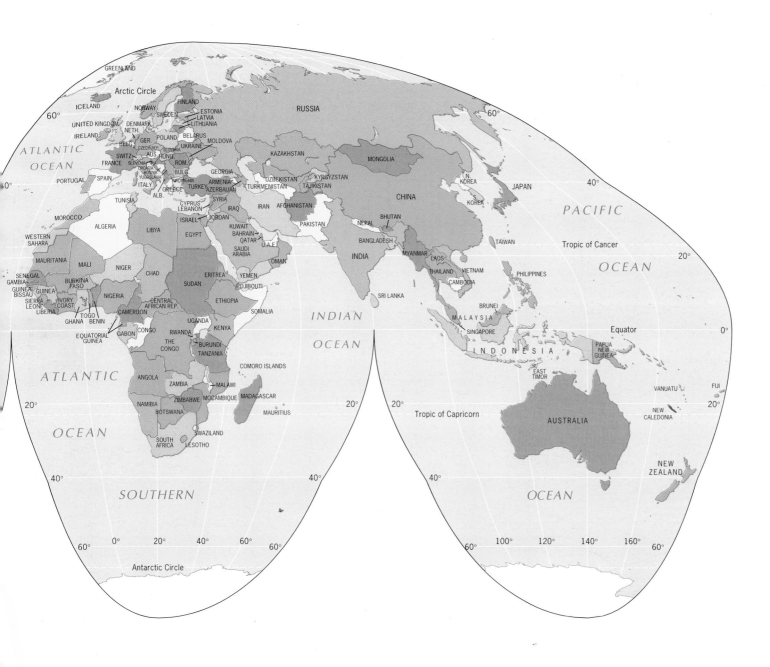

Geography links the study of human societies and their natural-environmental settings through a revealing, spatial approach, aided in this book by a unique series of thematic maps. From glaciation to globalization, desertification to devolution, Mexico to Madagascar, *Concepts and Regions in Geography, 2e* puts processes and places in spatial context. The focus shifts seamlessly from the global to the regional to the local, yielding insights into the forces that shape the human mosaic of our planet.

Inside these covers lies an information highway to geographic literacy. The textbook and Student Web site will help you to understand geographic concepts, prepare for examinations, and ultimately improve your grade. To create an individualized learning strategy that combines use of text and media according to your specific strengths, complete the **Learning Styles Survey** on the Concepts and Regions Web site found at www.wiley.com/college/deblij.

As you work through the text, special icons direct you to learning resources available on the GeoDiscoveries Web site. Each icon will lead you to additional content, activities and quizzing.

STUDY TOOLS

1 **A**s you read the textbook, you will notice that each chapter begins with a list of key **Concepts, Ideas, and Terms**. These are noted by numbers in the margins that correspond to the introduction of each item in the text. The Web site provides simulated flashcards and multiple-choice quizzes to help you study and memorize their definitions.

Student Web Site. This comprehensive on-line resource will contain chapter-based self-quizzes and extensive links to Web material providing real-world examples and additional research tools.

At the end of each chapter of this book you will find boxed features which map, in detail, the chapter-specific resources available on the *Concepts and Regions in Geography, 2e* Web site, found at **http://www.wiley.com/college/deblij**.

Please let us know if you have any comments or questions by sending an e-mail to: **ConceptsAndRegions@wiley.com.**

DISCOVERY TOOLS

The **Photo Gallery** provides numerous photographs, most taken by H.J. de Blij during his continuing geographic fieldwork. The senior author's field notes are provided to help you view each photograph through the eyes of a professional geographer. Locator maps add further spatial context.

More to Explore essays fall into two categories: **Systematic Essays** and **Issues in Geography**. Regional geography allows us to integrate concepts and principles from the disciplines major subfields to create an overall image of our divided world. These constitute topical or systematic geography, and are described in the Systematic Essays that supplement each chapter. Issues in Geography essays explore interesting and current topics that are specific to each of the world's realms.

Each **GeoDiscoveries** module uses videos, animations, quizzing, and critical thinking exercises to help you learn core content and improve your grade. These modules challenge you to form spatial questions and apply the tools of geography to find answers.

The **Expanded Regional Coverage** of every world region is available on the Student Web site. Although *Concepts and Regions in Geography, 2e* is the most compact world regional geography textbook available, the Web site has all of the regional coverage available in H. J. de Blij and Peter Muller's best-selling textbook, *Geography: Realms, Regions and Concepts, 11e*.

eGrade Plus

www.wiley.com/college/deblij

Based on the Activities You Do Every Day

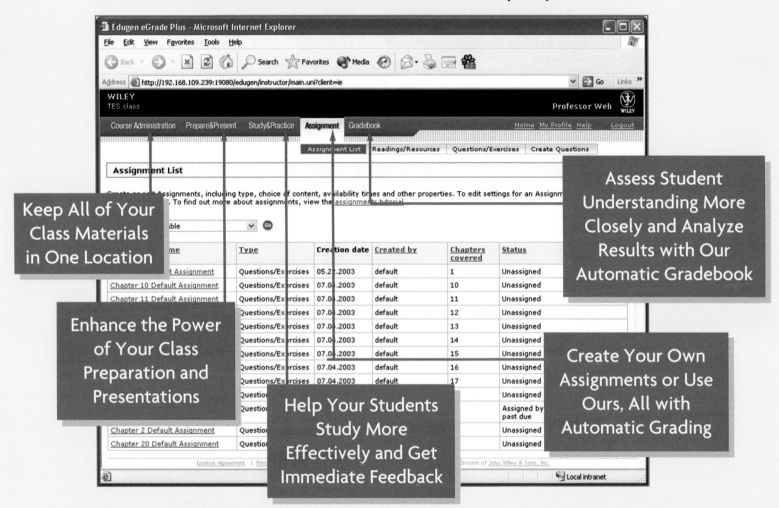

Keep All of Your Class Materials in One Location

Enhance the Power of Your Class Preparation and Presentations

Help Your Students Study More Effectively and Get Immediate Feedback

Assess Student Understanding More Closely and Analyze Results with Our Automatic Gradebook

Create Your Own Assignments or Use Ours, All with Automatic Grading

All the content and tools you need, all in one location, in an easy-to-use browser format.

Choose the resources you need, or rely on the arrangement supplied by us.

Now, many of Wiley's textbooks are available with eGrade Plus, a powerful online tool that provides a completely integrated suite of teaching and learning resources in one easy-to-use website. eGrade Plus integrates Wiley's world-renowned content with media, including a multimedia version of the text, PowerPoint slides, and more. Upon adoption of eGrade Plus, you can begin to customize your course with the resources shown here.

See for yourself! Go to www.wiley.com/college/egradeplus for an online demonstration of this powerful new software.

Instructors,
eGrade Plus Allows You to:

Keep All of Your Class Materials in One Location

Administer Your Course:
eGrade Plus can easily be integrated with another course management system, gradebook, or other resources you are using in your class, providing you with the flexibility to build your course, your way.

Enhance the Power of Your Class Preparation and Presentations

Prepare & Present:
Create class presentations using a wealth of Wiley-provided resources—such as an **online version of the textbook, PowerPoint slides, animations, video clips,** and **additional photographic resources**—making your preparation time more efficient. You may easily adapt, customize, and add to this content to meet the needs of your course.

Create Your Own Assignments or Use Ours, All with Automatic Grading

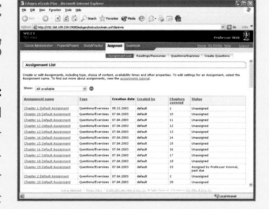

Create Assignments:
Automate the assigning and grading of homework or quizzes by using Wiley-provided **question banks**, or by writing your own. Student results will be automatically graded and recorded in your gradebook. eGrade Plus can link homework problems to the relevant section of the online text, providing students with context-sensitive help.

Assess Student Understanding More Closely

Track Student Progress:
Keep track of your students' progress via an **instructor's gradebook**, which allows you to analyze individual and overall class results to determine their progress and level of understanding.

Students,
eGrade Plus Allows You to:

Study More Effectively

Get Immediate Feedback When You Practice on Your Own

A **"Study & Practice"** area links directly to text content, allowing you to review the text while you study and complete homework assignments. Additional resources can include **virtual tours of regions covered, animations, video clips, expanded regional coverage,** and **issues in geography highlights.**

Complete Assignments / Get Help with Problem Solving

An **Assignment** area keeps all your assigned work in one location, making it easy for you to stay "on task." You will have access to a variety of **interactive resources,** as well as other materials for building your confidence and understanding. In addition, many homework problems contain a **link** to the relevant section of the **multimedia book,** providing you with context-sensitive help that allows you to master key concepts.

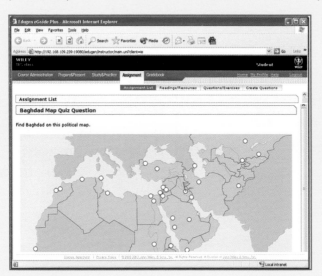

Keep Track of How You're Doing

A **Personal Gradebook** allows you to view your results from past assignments at any time.

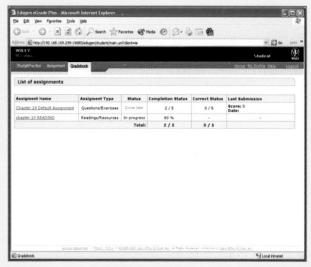

Concepts and Regions in
GEOGRAPHY

Second Edition

Concepts and Regions in
GEOGRAPHY
Second Edition

H. J. de Blij
Michigan State University

Peter O. Muller
University of Miami

With Media Integration by
Eugene J. Palka
United States Military Academy

WILEY

JOHN WILEY & SONS, INC.

About the cover:
Global game in a globalizing country: practicing for the future. South Africa will be the first African country to host the World Cup competition in 2010.

ACQUISITIONS EDITOR:	Jerry Correa
MARKETING MANAGER:	Frank Lyman
SENIOR PRODUCTION EDITOR:	Kelly Tavares
SENIOR DESIGNER:	Karin Kincheloe
SENIOR PHOTO EDITOR:	Jennifer MacMillan
ILLUSTRATION EDITOR:	Sigmund Malinowski
TEXT DESIGN:	Lee Goldstein
COVER DESIGN:	David Levy
FRONT/BACK COVER PHOTO:	Tim Flach/Taxi/Getty Images
ELECTRONIC ILLUSTRATIONS:	Mapping Specialists

This book was set in Times Roman by GGS Book Services, Atlantic Highlands and printed and bound by Von Hoffman Press.

ISBN 0-471-64991-0

Printed in the United States of America

10 9 8 7 6 5 4 3 2 1

All Opener Photographs by H. J. de Blij

Preface

This is a new and substantially revised Second Edition of a book that represents a new era in educational technologies. We wrote in the Preface to the First Edition that *Concepts and Regions in Geography* was "drawn from [the] full-length text, *Geography: Realms, Regions, and Concepts*." This new edition retains those foundations, but it is to a far greater extent a different and independent book. It is much more than a short version of *Regions*: it contains text and emphases not found in the larger book because of the demands of its dimensions and the pace of change in the world today.

An important structural change involves the repositioning of Chapters 6 and 7 from the First Edition. Chapter 7 (Subsaharan Africa) now follows South America and is renumbered 6; Chapter 6 (North Africa/Southwest Asia [NASWA]) is renumbered 7, a more logical sequence. The regional discussion in Chapter 6 has also been altered, beginning now with Southern Africa and concluding with West Africa and the Transition Zone to the Muslim world.

Beside the addition of several photos to each chapter, the most obvious visual modification is the addition of a number of detailed shaded-relief maps to replace the atlas maps that open each chapter. In response to suggestions from readers we have increased our emphasis on environmental relationships, which could be done more effectively using terrain representation (see, for example, the opening map in the Introduction and the enhanced physical map of Europe, Fig. 1-4). Again at the recommendation of a colleague, we have renumbered the maps in the opening chapter with a G (Global) rather than the potentially confusing I (Introduction). Topical maps are updated to mid-2004.

The numerous new and extensively revised maps in this edition reflect the content changes we have introduced. These maps were drawn specifically for *Concepts 2e*. In Chapter 1 (Europe) we have brought the expansion of the European Union to its current stage and added a map of the Northeastern Mediterranean, the theater of significant developments. In Chapter 2 (Russia) a new map of climates supports the discussion of environmental prospects and challenges. Chapter 3 (North America) is likewise enhanced by a new physical map. In Chapter 4 (Middle America) the extensively revised narrative required new maps of physical geography and the regional geography of Mexico. We also expanded the discussion of Mexico's Amerindian minorities as well as the Greater Antilles and augmented these with maps from *Regions 11*. A new physical map in Chapter 5 (South America) as well as improved cartographic coverage of the Southern Cone support extensive and detailed revisions of the text for each of the regions of this fast-changing geographic realm.

Chapter 6 (Subsaharan Africa) has undergone major revision as well as structural reorganization. A map of Africa's ethnic mosaic facilitates the new narrative. The focus on South Africa is strengthened by a new map that replaces the pre- and post-liberation provincial maps in the First Edition, representing a process that has run its course. Because it is now possible to transit from Subsaharan Africa to "NASWA", we have transferred maps of the African Transition Zone and the Horn of Africa to this chapter from Chapter 7. This chapter (7), which deals with some of the world's most intractable political-geographical problems, has an exhaustively revised map of Iraq and accompanying text, new realities (including the demarcation phenomenon of the Israeli "security fence"), revisions in the Afghanistan map, and the addition of a new relief map.

In Chapter 8 (South Asia) a new physical map incorporates the monsoon phenomenon illustrated separately in the earlier edition, and text changes focus on Pakistan, Nepal, Bangladesh, and Sri Lanka. Chapter 9 (East Asia) did not require significant intervention in this edition, but Chapter 10 (Southeast Asia) has a new relief map as well as new text and cartography outlining the maritime and resource implications of the independence of East Timor. Chapter 11 (the Austral Realm) has a new relief map and was updated in its economic geography sections, and for Chapter 12 (Pacific Realm) we recast the physical map for clarity and improved text linkage.

In terms of text revision, new and substituted material, noted by chapter, includes national debt and globalization (G), supranationalism and Muslim integration (1), regional economic development (3), economic and social crises (4), politico-geographical and democratic setbacks (5), poverty and international neglect (6), culture conflict, foreign intervention, and energy resources (7), nationalism and insurgency (8), periphery problems (9), maritime issues (10), relative location (11), and environmental hazards (12). Linking geography's perspectives to the problems and issues of the world today remains a key objective of this book.

PEDAGOGY

We continue to devise ways to help students learn important geographic concepts and ideas, and to make sense of our complex and rapidly changing world. Continuing special features from *Regions* include the following:

Concepts, Ideas, and Terms. Each chapter begins with a boxed sequential listing of the key geographic concepts, ideas, and terms that appear in the pages that follow. These are noted by numbers in the margins (e.g. 1) that correspond to the introduction of each item in the text.

Two-Part Chapter Organization. To help the reader to logically organize the material within chapters, we have broken the regional chapters into two distinct parts: first, "Defining the Realm" includes the general physiographic, historical, and human-geographic background common to the realm, and the second section, "Regions of the Realm," presents each of the distinctive regions within the realm (denoted by the symbol ▶).

List of Regions. Also on the chapter-opening page, a list of the regions within the particular realm provides a preview and helps to organize the chapter. For ease of identification, the semi-circular symbol (shown at the end of the previous paragraph) that denotes the regions list here also appears beside each region heading in the chapter.

Major Geographic Qualities. Near the beginning of each realm chapter, we list, in boxed format, the major geographic qualities that best summarize that portion of the Earth's surface.

Appendices and Glossary. At the end of the book, the reader will find three sections that enrich and/or supplement the main text: (1) *Appendix A*, a guide to Using the Maps; (2) *Appendix B*, an overview of Career Opportunities in Geography; and (3) an exten-sive *Glossary*. The general index follows. A geographical index or *gazetteer* of the place names contained in our maps now appears in the book's Web site. The book's references, and pronunciation guide are also contained on the Web site.

GeoDiscoveries. This robust media tool has been integrated into the *Concepts 2e* Web site. It contains an Interactive Globe that allows students to explore and understand the world by changing the face of this three-dimensional globe using 5 distinct textures. It also contains several quizzes per chapter that test student understanding of map features. The regional material contains *Presentations* that use videos, animations, and other resources to focus on key concepts from the chapter; *Interactivities* that engage students in concept-based exercises; and *Assessment* self-tests that allow students to measure their comprehension of the concept being explored. There is extensive expanded coverage from each realm as well, including text, illustrations, and maps.

Web site. Additional resources for students include annotated web links, web quizzes with feedback, links to webcams and live radio from around the globe, blank outline maps of each region, and a learning-style survey that provides students' feedback on their preferred method of learning and how the book and media pedagogy teach to these styles. Additional resources for instructors include the Test Bank, Guide to Virtual Field Trips, Virtual Field Trips, Lesson Outlines, Concepts-Ideas-Terms, and Using Geographic Qualities.

ANCILLARIES

A broad spectrum of print and electronic ancillaries are available to accompany *Concepts and Regions in Geography*, 2nd edition. Additional information, including prices and ISBNs for ordering, can be obtained by contacting John Wiley & Sons.

Data Sources

For all matters geographical, of course, we consult *The Annals of the Association of American Geographers, The Professional Geographer, The Geographical Review, The Journal of Geography*, and many other academic journals published regularly in North America—plus an array of similar periodicals published in English-speaking countries from Scotland to New Zealand.

As with every new edition of this book, all quantitative information was updated to the year of publication and checked rigorously. Hundreds of other modifications were made, many in response to readers' and reviewers' comments. The stream of new spellings of geographic names continues, and we pride ourselves in being a reliable source for current and correct usage.

The statistical data that constitute Table G-1 (pp. 24–29) are derived from numerous sources. As users of such data are aware, considerable inconsistency marks the reportage by various agencies, and it is often necessary to make informed decisions on contradictory information. For example, some sources still do not reflect the rapidly declining rates of population increase or life expectancies in AIDS-stricken African countries. Others list demographic averages without accounting for differences between males and females in this regard.

In formulating Table G-1 we have used among our sources the United Nations, the Population Reference Bureau, the World Bank, the Encyclopaedia Britannica *Books of the Year*, the *Economist* Intelligence Unit, the *Statesman's Year-Book*, and the *The New York Times Almanac*.

The urban population figures—which also entail major problems of reliability and comparability—are mainly drawn from the most recent database published by the United Nations' Population Division. For cities of less than 750,000, we developed our own estimates from a variety of other sources. At any rate, the urban population figures used here are estimates for 2005 and they represent *metropolitan-area totals* unless otherwise specified.

For Sale to the Student

Student Study Guide. Text co-author Peter O. Muller and his geographer daughter, Elizabeth Muller Hames, have written a popular Study Guide to accompany the book that is packed with useful study and review tools. For each chapter in the textbook, the Study Guide gives students and faculty access to chapter objectives, content questions-and-answers, outline maps of each realm, sample tests, and more.

Goode's Atlas **from Rand McNally.** We are delighted to be able to offer the 21st edition of the *Goode's Atlas* at a deeply-discounted price when shrink-wrapped with the text. Economies of scale allow us to provide this at a net price that is close to our cost. Our partnership with Rand McNally and the widely-popular *Goode's Atlas* is an arrangement that is exclusive to John Wiley & Sons.

For Instructors

PowerPoint Slides. Available for this edition, these electronic files outline the main concepts of each chapter in *Concepts 2nd Edition* in a highly visual manner. These presentations are available on the Instructor's Web Site and can be uploaded to presentation programs such as PowerPoint, or to any major word processing program.

Instructor's Manual. Distributed on-line to instructors via a secure, password-protected Instructor's Web Site, the *Instructor's Manual* by Wendy Shaw of Southern Illinois University, Edwardsville, provides outlines, descriptions, and key terms to help professors organize the concepts in the book for classroom use.

Test Bank. Prepared by long-term Test Bank author Ira Sheskin of the University of Miami, the *Test Bank* contains over 3000 test items including multiple-choice, fill-in, matching, and essay questions. It is distributed via the secure Instructor's

Web Site as electronic files, which can be saved into all major word processing programs.

Computerized Test Bank. An easy-to-use program that can be used to create and customize exams.

Student Web Site. This comprehensive on-line resource contains chapter-based self-quizzes and extensive links to Web material providing real-world examples and additional research tools.

Course Management. On-line course management assets are available to accompany the Second Edition of *Concepts*. This includes content in WebCT, Blackboard, and Wiley's EGrade+.

Other Resources for the Classroom

Overhead Transparencies. The book's maps and diagrams are available in their entirety for transparency in beautifully rendered, four-color format.

Instructor's Web site. This rich resource contains animations, videos, PowerPoint presentations, the Test Bank, and the Instructor's Manual. Organized by chapter, the instructor's Web site has a tested, intuitive interface that allows for easy file management and presentation-building.

ACKNOWLEDGMENTS

Over the third of a century since the publication of the First Edition of *Geography: Realms, Regions, and Concepts* (and joined most recently by the appearance of *Concepts*), we have been fortunate to receive advice and assistance from literally hundreds of people. One of the rewards associated with the publication of books of this kind is the steady stream of correspondence and other feedback they generate. Geographers, economists, political scientists, education specialists, and others have written us, often

with fascinating enclosures. We make it a point to respond personally to every such letter, and our editors have communicated with many of our correspondents as well. Moreover, we have considered every suggestion made—and many who wrote or transmitted their reactions through other channels will see their recommendations in print in this edition.

STUDENT RESPONSE

A major part of the correspondence we receive comes from student readers. We would like to take this opportunity to extend our deep appreciation to the several million students around the world who have studied from the first 12 editions of our texts. In particular, we thank the students from more than 100 different colleges across the United States who took the time to send us their opinions.

Students told us they found the maps and graphics attractive and functional. We have not only enhanced the map program with exhaustive updating, but have added a number of new maps to this Second Edition as well as making significant changes in many others.

Generally, students have told us that they found the pedagogical devices quite useful. We have kept the study aids the students cited as effective: a boxed list of each chapter's key concepts, ideas, and terms (now numbered for quick reference in both the box and text margins); a box summarizing each realm's major geographic qualities; and an extensive Glossary. Many additional study aids can be found on the book's Web site.

FACULTY FEEDBACK

Faculty members from a large number of North American colleges and universities continue to supply us with vital feedback and much-appreciated advice. Our publishers arranged several feedback sessions, and we are most grateful to the following professors for showing us where the written text could be strengthened and made more precise:

RANDY BERTOLAS, Wayne State College

JONATHAN C. COMER, Oklahoma State University

MICHAEL CORNEBISE, University of Tennessee

FIONA M. DAVIDSON, University of Arkansas

MEL DROUBAY, University of West Florida

DAVID J. KEELING, Western Kentucky University

MOHAMEDEN OULD-MEY, Indiana State University

THOMAS W. PARADIS, Northern Arizona University

JAMES W. PENN, JR., University of Florida

JOHN D. REILLY, University of Florida

THOMAS C. SCHAFER, Fort Hays State University

In addition, several faculty colleagues from around the world assisted us with earlier editions, and their contributions continue to grace the pages of this book. Among them are:

JAMES P. ALLEN, California State University, Northridge

STEPHEN S. BIRDSALL, University of North Carolina

J. DOUGLAS EYRE, University of North Carolina

FANG YONG-MING, Shanghai, China

EDWARD J. FERNALD, Florida State University

RAY HENKEL, Arizona State University

RICHARD C. JONES, University of Texas at San Antonio

GIL LATZ, Portland State University (Oregon)

IAN MACLACHLAN, University of Lethbridge (Alberta)

MELINDA S. MEADE, University of North Carolina

HENRY N. MICHAEL, Temple University (Pennsylvania)

CLIFTON W. PANNELL, University of Georgia

J. R. VICTOR PRESCOTT, University of Melbourne (Victoria)

JOHN D. STEPHENS, University of Washington

CANUTE VANDER MEER, University of Vermont

We also received input from a much wider circle of academic geographers. The list that follows is merely representative of a group of colleagues across North America to whom we are grateful for taking the time to share their thoughts and opinions with us:

MEL AAMODT, California State University—Stanislaus

R. GABRYS ALEXSON, University of Wisconsin—Superior

NIGEL ALLAN, University of California—Davis

JAMES P. ALLEN, California State University, Northridge

JOHN L. ALLEN, University of Connecticut

JERRY R. ASCHERMANN, Missouri Western State College

JOSEPH M. ASHLEY, Montana State University

THEODORE P. AUFDEMBERGE, Concordia College (Michigan)

EDWARD BABIN, University of South Carolina—Spartanburg

MARVIN W. BAKER, University of Oklahoma

THOMAS F. BAUCOM, Jacksonville State University (Alabama)

GOURI BANERJEE, Boston University (Massachusetts)

J. HENRY BARTON, Thiel College (Pennsylvania)

STEVEN BASS, Paradise Valley Community College (Arizona)

KLAUS J. BAYR, University of New Hampshire—Manchester

JAMES BELL, Linn Benton Community College (Oregon)

WILLIAM H. BERENTSEN, University of Connecticut

ROYAL BERGLEE, Indiana State University

RIVA BERLEANT-SCHILLER, University of Connecticut

THOMAS BITNER, University of Wisconsin

WARREN BLAND, California State University—Northridge

DAVIS BLEVINS, Huntington College (Alabama)

S. BO JUNG, Bellevue College (Nebraska)

MARTHA BONTE, Clinton Community College (Idaho)

GEORGE R. BOTJER, University of Tampa (Florida)

R. LYNN BRADLEY, Belleville Area College (Illinois)

KEN BREHOB, Elmhurst, Illinois

JAMES A. BREY, University of Wisconsin—Fox Valley

ROBERT BRINSON, Santa Fe Community College (Florida)

REUBEN H. BROOKS, Tennessee State University

LARRY BROWN, Ohio State University

LAWRENCE A. BROWN, Troy State—Dothan (Alabama)

ROBERT N. BROWN, Delta State University (Mississippi)

STANLEY D. BRUNN, University of Kentucky

RANDALL L. BUCHMAN, Defiance College (Ohio)

DIANN CASTEEL, Tusculum College (Tennessee)

JOHN E. COFFMAN, University of Houston (Texas)

DAWYNE COLE, Grand Rapids Baptist College (Michigan)

JONATHAN C. COMER, Oklahoma State University

BARBARA CONNELLY, Westchester Community College (New York)

WILLIS M. CONOVER, University of Scranton (Pennsylvania)

OMAR CONRAD, Maple Woods Community College (Missouri)

BARBARA CRAGG, Aquinas College (Michigan)

GEORGES G. CRAVINS, University of Wisconsin

ELLEN K. CROMLEY, University of Connecticut

JOHN A. CROSS, University of Wisconsin—Oshkosh

WILLIAM CURRAN, South Suburban (Illinois)

ARMANDO DA SILVA, Towson State University (Maryland)

DAVID D. DANIELS, Central Missouri State University

RUDOLPH L. DANIELS, Morningside College (Iowa)

SATISH K. DAVGUN, Bemidji State University (Minnesota)

JAMES DAVIS, Illinois College

JAMES L. DAVIS, Western Kentucky University

KEITH DEBBAGE, University of North Carolina—Greensboro

MOLLY DEBYSINGH, California State University, Long Beach

DENNIS K. DEDRICK, Georgetown College (Kentucky)

STANFORD DEMARS, Rhode Island College

THOMAS DIMICELLI, William Paterson College (New Jersey)

D.F. DOEPPERS, University of Wisconsin—Madison

ANN DOOLEN, Lincoln College (Illinois)

STEVEN DRIEVER, University of Missouri—Kansas City

WILLIAM ROBERT DRUEN, Western Kentucky University

ALASDAIR DRYSDALE, University of New Hampshire

KEITH A. DUCOTE, Cabrillo Community College (California)

WALTER N. DUFFET, University of Arizona

CHRISTINA DUNPHY, Champlain College (Vermont)

ANTHONY DZIK, Shawnee State University (Kansas)

DENNIS EDGELL, Firelands BGSU (Ohio)

JAMES H. EDMONSON, Union University (Tennessee)

M.H. EDNEY, State University of New York—Binghamton

HAROLD M. ELLIOTT, Weber State University (Utah)

JAMES ELSNES, Western State College

DINO FIABANE, Community College of Philadelphia (Pennsylvania)

G.A. FINCHUM, Milligan College (Tennessee)

IRA FOGEL, Foothill College (California)

ROBERT G. FOOTE, Wayne State College (Nebraska)

G.S. FREEDOM, McNeese State University (Louisiana)

RONALD FORESTA, University of Tennessee

OWEN FURUSETH, University of North Carolina—Charlotte

RICHARD FUSCH, Ohio Wesleyan University

GARY GAILE, University of Colorado—Boulder

EVELYN GALLEGOS, Eastern Michigan University & Schoolcraft College

JERRY GERLACH, Winona State University (Minnesota)

LORNE E. GLAIM, Pacific Union College (California)

SHARLEEN GONZALEZ, Baker College (Michigan)

DANIEL B. GOOD, Georgia Southern University

GARY C. GOODWIN, Suffolk Community College (New York)

S. GOPAL, Boston University (Massachusetts)

ROBERT GOULD, Morehead State University (Kentucky)

GORDON GRANT, Texas A&M University

DONALD GREEN, Baylor University (Texas)

GARY M. GREEN, University of North Alabama

MARK GREER, Laramie County Community College (Wyoming)

STANLEY C. GREEN, Laredo State University (Texas)

W. GREGORY HAGER, Northwestern Connecticut Community College

RUTH F. HALE, University of Wisconsin—River Falls

JOHN W. HALL, Louisiana State University—Shreveport

PETER L. HALVORSON, University of Connecticut

MERVIN HANSON, Willmar Community College (Minnesota)

ROBERT J. HARTIG, Fort Valley State College (Georgia)

JAMES G. HEIDT, University of Wisconsin Center—Sheboygan

CATHERINE HELGELAND, University of Wisconsin—Manitowoc

NORMA HENDRIX, East Arkansas Community College

JAMES HERTZLER, Goshen College (Indiana)

JOHN HICKEY, Inver Hills Community College (Minnesota)

THOMAS HIGGINS, San Jacinto College (Texas)

EUGENE HILL, Westminster College (Missouri)

LOUISE HILL, University of South Carolina—Spartanburg

MIRIAM HELEN HILL, Indiana University Southeast

SUZY HILL, University of South Carolina—Spartanburg

ROBERT HILT, Pittsburg State University (Kansas)

SOPHIA HINSHALWOOD, Montclair State University (New Jersey)

PRISCILLA HOLLAND, University of North Alabama

ROBERT K. HOLZ, University of Texas—Austin

R. HOSTETLER, Fresno City College (California)

LLOYD E. HUDMAN, Brigham Young University (Utah)

JANIS W. HUMBLE, University of Kentucky

WILLIAM IMPERATORE, Appalachian State University (North Carolina)

RICHARD JACKSON, Brigham Young University (Utah)

MARY JACOB, Mount Holyoke College (Massachusetts)

GREGORY JEANE, Samford University (Alabama)

SCOTT JEFFREY, Catonsville Community College (Maryland)

JERZY JEMIOLO, Ball State University (Indiana)

SHARON JOHNSON, Marymount College (New York)

SARA MAYFIELD, San Jacinto College, Central (California)

DAVID JOHNSON, University of Southwestern Louisiana

JEFFREY JONES, University of Kentucky

MARCUS E. JONES, Claflin College (South Carolina)

MOHAMMAD S. KAMIAR, Florida Community College, Jacksonville

MATTI E. KAUPS, University of Minnesota—Duluth

COLLEEN KEEN, Gustavus Adolphus College (Minnesota)

GORDON F. KELLS, Mott Community College

SUSANNE KIBLER-HACKER, Unity College (Maine)

JAMES W. KING, University of Utah

JOHN C. KINWORTHY, Concordia College (Nebraska)

ALBERT KITCHEN, Paine College

TED KLIMASEWSKI, Jacksonville State University (Alabama)

ROBERT D. KLINGENSMITH, Ohio State University—Newark

LAWRENCE M. KNOPP, JR., University of Minnesota—Duluth

TERRILL J. KRAMER, University of Nevada

ARTHUR J. KRIM, Cambridge, Massachusetts

ELROY LANG, El Camino Community College (California)

CHRISTOPHER LANT, Southern Illinois University—Carbondale

A.J. LARSON, University of Illinois—Chicago

LARRY LEAGUE, Dickinson State University (North Dakota)

DAVID R. LEE, Florida Atlantic University

JOE LEEPER, Humboldt State University (California)

YECHIEL M. LEHAVY, Atlantic Community College (New Jersey)

JOHN C. LEWIS, Northeast Louisiana University

CAEDMON S. LIBURD, University of Alaska—Anchorage

T. LIGIBEL, Eastern Michigan University

Z.L. LIPCHINSKY, Berea College (Kentucky)

ALLAN L. LIPPERT, Manatee Community College (Florida)

JOHN H. LITCHER, Wake Forest University (North Carolina)

LI LIU, Stephen F. Austin State University (Texas)

WILLIAM R. LIVINGSTON, Baker College (Michigan)

CYNTHIA LONGSTREET, Ohio State University

TOM LOVE, Linfield College (Oregon)

K.J. LOWREY, Miami University (Ohio)

ROBIN R. LYONS, University of Hawai'i—Leeward Community College

SUSAN M. MACEY, Southwest Texas State University

CHRISTIANE MAINZER, Oxnard College (California)

HARLEY I. MANNER, University of Guam

JAMES T. MARKLEY, Lord Fairfax Community College (Virginia)

SISTER MAY LENORE MARTIN, Saint Mary College (Kansas)

GARY MANSON, Michigan State University

KENT MATHEWSON, Louisiana State University

DICK MAYER, Maui Community College (Hawai'i)

DEAN R. MAYHEW, Maine Maritime Academy

J.P. MCFADDEN, Orange Coast College (California)

BERNARD MCGONIGLE, Community College of Philadelphia (Pennsylvania)

PAUL D. MEARTZ, Mayville State University (North Dakota)

DALTON W. MILLER, JR., Mississippi State University

RAOUL MILLER, University of Minnesota, Duluth

INES MIYARES, Hunter College, CUNY (New York)

BOB MONAHAN, Western Carolina University

KEITH MONTGOMERY, University of Wisconsin—Marathon

JOHN MORTON, Benedict College (South Carolina)

ANNE MOSHER, Syracuse University (New York)

BARRY MOWELL, Broward Community College (Florida)

ROBERT R. MYERS, West Georgia College

YASER M. NAJJAR, Framingham State College (Massachusetts)

JEFFREY W. NEFF, Western Carolina University

DAVID NEMETH, University of Toledo (Ohio)

RAYMOND O'BRIEN, Bucks County Community College (Pennsylvania)

JOHN ODLAND, Indiana University

JOSEPH R. OPPONG, University of North Texas

RICHARD OUTWATER, California State University, Long Beach

PATRICK O'SULLIVAN, Florida State University

BIMAL K. PAUL, Kansas State University

JAMES PENN, Southeastern Louisiana University

PAUL PHILLIPS, Fort Hays State University (Kansas)

MICHAEL PHOENIX, ESRI (California)

JERRY PITZL, Macalester College (Minnesota)

BILLIE E. POOL, Holmes Community College (Mississippi)

VINTON M. PRINCE, Wilmington College (North Carolina)

RHONDA REAGAN, Blinn College (Texas)

DANNY I. REAMS, Southeast Community College (Nebraska)

JIM RECK, Golden West College (California)

ROGER REEDE, Southwest State University (Minnesota)

JOHN RESSLER, Central Washington University

JOHN B. RICHARDS, Southern Oregon State College

DAVID C. RICHARDSON, Evangel College (Missouri)

SUSAN ROBERTS, University of Kentucky

WOLF RODER, University of Cincinnati

JAMES ROGERS, University of Central Oklahoma

PAUL A. ROLLINSON, AICP, Southwest Missouri State University

JAMES C. ROSE, Tompkins/Cortland Community College (New York)

THOMAS E. ROSS, Pembroke State University (North Carolina)

THOMAS A. RUMNEY, State University of New York—Plattsburgh

GEORGE H. RUSSELL, University of Connecticut

RAJAGOPAL RYALI, Auburn University at Montgomery (Alabama)

PERRY RYAN, Mott Community College

ADENA SCHUTZBERG, Middlesex Community College (Massachusetts)

SIDNEY R. SHERTER, Long Island University (New York)

NANDA SHRESTHA, Florida A&M University

WILLIAM R. SIDDALL, Kansas State University

DAVID SILVA, Bee County College (Texas)

DEBRA STRAUSSFOGEL, University of New Hampshire

MORRIS SIMON, Stillman College (Alabama)

KENN E. SINCLAIR, Holyoke Community College (Massachusetts)

ROBERT SINCLAIR, Wayne State University (Michigan)

EVERETT G. SMITH, JR., University of Oregon

RICHARD V. SMITH, Miami University (Ohio)

CAROLYN D. SPATTA, California State University—Hayward

M.R. SPONBERG, Laredo Junior College (Texas)

DONALD L. STAHL, Towson State University (Maryland)

ELAINE STEINBERG, Central Florida Community College

D.J. STEPHENSON, Ohio University Eastern

HERSCHEL STERN, Mira Costa College (California)

REED F. STEWART, Bridgewater State College (Massachusetts)

NOEL L. STIRRAT, College of Lake County (Illinois)

GEORGE STOOPS, Mankato State University (Minnesota)

JOSEPH P. STOLTMAN, Western Michigan University

PHILIP SUCKLING, University of Northern Iowa

CHRISTOPHER SUTTON, Western Illinois University

T. L. TARLOS, Orange Coast College (California)

MICHAEL THEDE, North Iowa Area Community College

DERRICK J. THOM, Utah State University

CURTIS THOMSON, University of Idaho

S. TOOPS, Miami University (Ohio)

ROGER T. TRINDELL, Mansfield University of Pennsylvania

DAN TURBEVILLE, East Oregon State College

NORMAN TYLER, Eastern Michigan University

GEORGE VAN OTTEN, Northern Arizona University

C.S. VERMA, Weber State College (Utah)

GRAHAM T. WALKER, Metropolitan State College of Denver

DEBORAH WALLIN, Skagit Valley College (Washington)

MIKE WALTERS, Henderson Community College (Kentucky)

J.L. WATKINS, Midwestern State University (Texas)

P. GARY WHITE, Western Carolina University (North Carolina)

W.R. WHITE, Western Oregon University

GARY WHITTON, Fairbanks, Alaska

GENE C. WILKEN, Colorado State University

STEPHEN A. WILLIAMS, Methodist College

P. WILLIAMS, Baldwin-Wallace College

MORTON D. WINSBERG, Florida State University

ROGER WINSOR, Appalachian State University (North Carolina)

WILLIAM A. WITHINGTON, University of Kentucky

A. WOLF, Appalachian State University, N.C.

JOSEPH WOOD, University of Southern Maine

RICHARD WOOD, Seminole Junior College (Florida)

GEORGE I. WOODALL, Winthrop College (North Carolina)

STEPHEN E. WRIGHT, James Madison University (Virginia)

LEON YACHER, Southern Connecticut State University

DONALD J. ZEIGLER, Old Dominion University (Virginia)

In assembling this newest edition, we are indebted to the following people for advising us on a number of matters:

THOMAS L. BELL, University of Tennessee

KATHLEEN BRADEN, Seattle Pacific University

JESUS CAÑAS, Research Deparment, Federal Reserve Bank of Dallas, El Paso Branch

STUART E. CORBRIDGE, University of Miami/London School of Economics

WILLIAM V. DAVIDSON, Louisiana State University

JAMES D. FITZSIMMONS, U.S. Bureau
of the Census (D.C.)

GARY A. FULLER, University of Hawai'i

RICHARD J. GRANT, University of Miami

MARGARET M. GRIPSHOVER, University
of Tennessee

TRUMAN A. HARTSHORN, Georgia State
University

PHILIP L. KEATING, Indiana University

DAVID LEY, University of British Columbia

RICHARD LISICHENKO, Fort Hays State
University (Kansas)

GLEN M. MACDONALD, University
of California, Los Angeles

IAN MACLACHLAN, University of Lethbridge
(Alberta)

DALTON MILLER, Mississippi State University

ANNE MOSHER, Syracuse University

VALIANT C. NORMAN, Lexington Community
College (Kentucky)

PAI YUNG-FENG, New York City

EUGENE J. PALKA, U.S. Military Academy
(New York)

JOSEPH L. SCARPACI, JR., Virginia Tech

ROLF STERNBERG, Montclair State University
(New Jersey)

COLLEEN J. WATKINS, Linfield College
(Oregon)

BARBARA A. WEIGHTMAN, California State
University, Fullerton

KRISTOPHER D. WHITE, University
of Connecticut

For assistance with the map of North American indigenous peoples, we are greatly indebted to Jack Weatherford, Professor of Anthropology at Macalester College (Minnesota); Henry T. Wright, Professor and Curator of Anthropology at the University of Michigan; and George E. Stuart, President of the Center for Maya Research (North Carolina), who also assisted with the Altun Ha site in Belize. The map of Russia's federal regions could not have been compiled without the invaluable help of David B. Miller, Senior Edit Cartographer at the National Geographic Society, and Leo Dillon of the Russia Desk of the U.S. Department of State. And special thanks, too, go to Charles Pirtle, Professor of Geography at Georgetown University's School of Foreign Service for his exhaustive review of the entire book, and to Charles Fahrer of the Department of Geography at the University of South Carolina for his suggestions on Chapter 7.

We also record our appreciation to those geographers who ensured the quality of this book's ancillary products: Ira M. Sheskin (University of Miami) prepared the *Test Bank* and manipulated a large body of demographic data to derive the tabular display in Table G-1; and Elizabeth Muller Hames (M.A. in Geography, University of Miami) co-authored the *Study Guide*, re-compiled the Geographical Index found on the Web site, and handled the data preparation for Table G-1. At the University of Miami's Department of Geography and Regional Studies, Peter Muller is most grateful for the advice he received from faculty colleagues Stuart Corbridge, Douglas Fuller, Richard Grant, Daniel Griffith, Jan Nijman, Rinku Roy Chowdhury, and Ira Sheskin as well as GIS Lab Manager Chris Hanson.

PERSONAL APPRECIATION

We are privileged to work with a team of professionals at John Wiley & Sons that is unsurpassed in the college textbook publishing industry. As authors we are acutely aware of these talents on a daily basis during the crucial production stage, especially the outstanding coordination and leadership skills of Production Manager Kelly Tavares, Production Editor Sarah Wolfman-Robichaud, Illustration Editor Sigmund Malinowski, Photo Editor Jennifer MacMillan, and Production Assistants Rebecca Rothaug and Carmen Hernandez. Others who played a leading role in this process were Senior Designer Karin Kincheloe, Copy Editor Betty Pessagno, Photo Director Marge Graham, Photo Manager Hilary Newman, and Don Larson and Emily Hardy of Mapping Specialists, Ltd. in Madison, Wisconsin. We appreciated the leadership of former Geography Editor Ryan Flahive, who was the prime mover in launching *Concepts*, a venture enthusiastically supported by current Geography Editor Jerry Correa. Both were ably assisted by Tom Kulesa, Christine Cordek, and Lindsay Lovier. We also thank our College Marketing Manager Clay Stone for his many efforts on behalf of our books. Beyond this immediate circle, we acknowledge the support and encouragement we have received over the years from others at Wiley including Publisher Anne Smith and Vice-President for Production Ann Berlin.

Finally, and most of all, we thank our wives, Bonnie and Nancy, for once again seeing us through the challenging schedule of our latest collaboration.

H. J. de Blij
Boca Grande, Florida

Peter O. Muller
Coral Gables, Florida

July 21, 2004

Brief Contents

Contents

Chapter 10

SOUTHEAST ASIA   339

Chapter 11

THE AUSTRAL REALM   367

Concepts and Regions in
GEOGRAPHY

Second Edition

World Regional Geography: Global Perspectives

CONCEPTS, IDEAS, AND TERMS

1 Geographic realm
2 Spatial perspective
3 Taxonomy
4 Transition zone
5 Geographic change
6 New World Order
7 Regional concept
8 Regional boundaries
9 Location
10 Absolute location
11 Relative location
12 Formal region
13 Spatial system

14 Hinterland
15 Functional region
16 Scale
17 Natural landscape
18 Physical geography
19 Continental drift
20 Tectonic plate
21 Subduction
22 Pacific Ring of Fire
23 Climate
24 Desertification
25 Glaciation
26 Ice age

27 Interglaciation
28 Hydrologic cycle
29 Climatic region
30 Physiography
31 Culture
32 Regional character
33 Cultural landscape
34 Central business district (CBD)
35 Ethnicity
36 Population distribution
37 Population density
38 Urbanization
39 State

40 European state model
41 Development
42 Economic geography
43 Core area
44 Periphery
45 Regional disparity
46 Advantage
47 Neocolonialism
48 Globalization
49 Regional geography
50 Systematic geography

*W*HAT A TIME this is to be studying geography! The world is undergoing a historic transformation, one of those momentous political, economic, and social upheavals about which we have read in history books. It is happening today, and as in the past, new alliances are forming, old unions are fracturing, novel ideas are spreading, older notions are fading. We hear from our political leaders talk about the emergence of a "New World Order," but such pronouncements are premature. A future world order is indeed in the offing, but it is too early to tell what it will be like. Of this we can be sure: the new world map will look quite different from the old. Our task is to understand the ongoing changes, to make sense of the new directions our world is taking. Geography is our most powerful ally in this mission.

GEOGRAPHIC PERSPECTIVES

In this book, we take a penetrating look at the geographic framework of the contemporary world, the grand design that is the product of thousands of years of human achievement and failure, movement and

1

STUDY TOOLS

3D GLOBE
AREA AND DEMOGRAPHIC DATA

FLASHCARDS
MAP QUIZZES

AUDIO PRONUNCIATION GUIDE
CHAPTER QUIZ

ANNOTATED WEBLINKS
LONELY PLANET WEBLINKS

stagnation, revolution and stability, interaction and isolation. Ours is an interconnected world of travel and trade, tourism and television, a global village—but the village still has neighborhoods. Their names are Europe, South America, Southeast Asia, and others familiar to all of us. We call such global **1** neighborhoods **geographic realms**, and when we subject these realms to geographic scrutiny, we find that each has its own identity and distinctiveness.

Geographers study the location and distribution of features on the Earth's surface. These features may be the landmarks of human occupation, the properties of the natural environment, or both: one of the most interesting themes in geography is the relationship between natural environments and human societies, which is why the first map in this introduc-

tory chapter summarizes the prominent natural (physical) features of the continents we inhabit (Fig. G-1). Geographers investigate the reasons for these **2** distributions. Their approach is guided by a **spatial perspective**. Just as historians focus on chronology, geographers concentrate on space and place. The spatial structure of cities, the layout of farms and fields, the networks of transportation, the system of rivers, the pattern of climate—all these and more go into the study of a geographic realm. As you will discover, geography is full of spatial terms: area, distance, direction, clustering, proximity, accessibility, isolation, and many others which we will encounter in the pages that follow.

In this book we use the geographic perspective and geography's spatial terminology to investigate

the world's great geographic realms. We will find that each of these realms possesses a special combination of cultural, organizational, and environmental properties. These characteristic qualities are imprinted on the landscape, giving each realm its own traditional attributes and social settings. As we come to understand the human and natural makeup of those geographic realms, we learn not only *where* they are located (always a crucial question in geography, and often the answer is not as simple as it may appear), but also *why they are located where they are*, how they are constituted, and what their future is likely to be in our changing world.

It is not enough, however, to study the world from a geographic viewpoint without learning something about the discipline of geography itself. Beginning

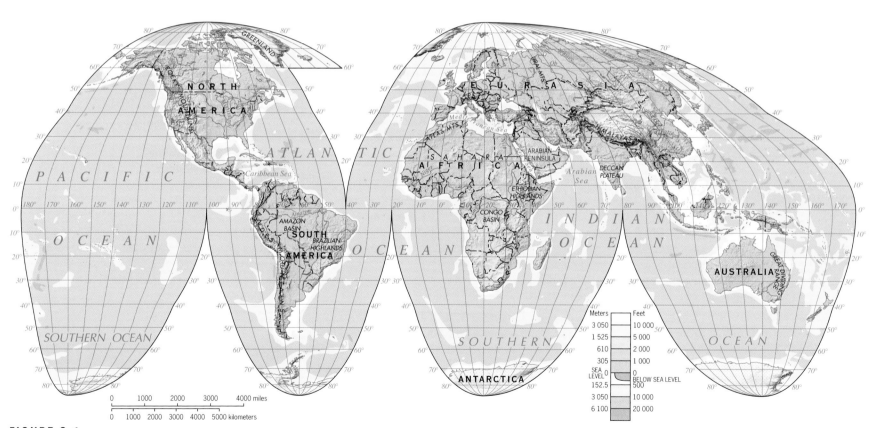

FIGURE G-1

in this introductory chapter, we introduce the fundamental ideas and concepts that make modern geography what it is. Not only will these geographic ideas enhance your awareness of the many dimensions of our complex, multicultural, interconnected world, but also many of them will remain useful to you long after you have closed this book. Welcome to geography . . . realms, regions, *and* concepts.

REALMS AND REGIONS

Geographers, like other scholars, seek to establish order from the countless data that confront them. Biologists have established a system of classification, **3** or **taxonomy**, to categorize the many millions of plants and animals into a hierarchical system of seven ranks. In descending order, we humans belong to the animal *kingdom*, the *phylum* (division) named chordata, the *class* of mammals, the *order* of primates, the *family* of hominids, the *genus* designated *Homo*, and the *species* known as *Homo sapiens*. Geologists classify the Earth's rocks into three major (and many subsidiary) categories and then fit these categories into a complicated geologic time scale that spans hundreds of millions of years. Historians define eras, ages, and periods to conceptualize the sequence of the events they study.

Geography, too, employs systems of classification. When geographers deal with urban problems, for instance, they use a classification scheme based on the sizes and functions of the places involved. Some of the terms in this classification are part of our everyday language: megalopolis, metropolis, city, town, village, hamlet.

In regional geography, which is the focus of this book, our challenge is different. We, too, need a hierarchical framework for the areas of the world we study, from the largest to the smallest. But our classification scheme is horizontal, not vertical. It is *spatial*. Our equivalent of the biologists' overarching kingdoms (of plants and animals) is the Earth's natural partitioning into landmasses and oceans. The next level is the division of the inhabited landmasses into geographic realms based on human as well as natural properties. The great global realms, in turn, divide into regions.

The Criteria for Realms

In any classification system, criteria are the key. Not all animals are mammals; the criteria for inclusion in that biological class are more specific and restrictive. A dolphin may look and act like a fish, but both anatomically and functionally dolphins belong to the class of mammals.

Geographic realms are based on sets of spatial criteria. First, they are the largest units into which the inhabited world can be divided. The criteria on which such a broad regionalization is based include both physical (that is, natural) and human (or social) yardsticks. South America is a geographic realm, for example, because physically it is a continent and culturally it is dominated by a set of social norms. The realm called South Asia, on the other hand, lies on a Eurasian landmass shared by several other geographic realms; high mountains, wide deserts, and dense forests combine with a distinctive social fabric to create this well-defined realm centered on India.

Second, geographic realms are the result of the interaction of human societies and natural environments, a *functional* interaction revealed by farms, mines, fishing ports, transport routes, dams, bridges, villages, and countless other features that mark the landscape. According to this criterion, Antarctica is a continent but not a geographic realm.

Third, geographic realms must represent the most comprehensive and encompassing definition of the great clusters of humankind in the world today. China lies at the heart of such a cluster, as does India. Africa constitutes a geographic realm from the southern margin of the Sahara (an Arabic word for "desert") to the Cape of Good Hope and from its Atlantic to its Indian Ocean shores.

Figure G-2 displays the 12 world geographic realms based on the these criteria. As we will show in more detail later, waters, deserts, and mountains as well as cultural and political shifts mark the borders of these realms. We will discuss the position of these boundaries as we examine each realm. For the moment, keep in mind the following:

4 *Where geographic realms meet,* **transition zones**, *not sharp boundaries, mark their contacts.*

We need only remind ourselves of the border zone between the geographic realm in which most of us live, North America, and the adjacent realm of Middle America. The line on Figure G-2 coincides with the boundary between Mexico and the United States, crosses the Gulf of Mexico, and then separates Florida from Cuba and the Bahamas. But Hispanic influences are strong in North America north of this boundary, and the U.S. economic influence is strong south of it. The line, therefore, represents an ever-changing zone of regional interaction. Again, there are many ties between South Florida and the Bahamas, but the Bahamas resemble a Caribbean more than a North American society.

In Africa, the transition zone from Subsaharan to North Africa is so wide and well defined that we have put it on the world map; elsewhere, transition zones tend to be narrower and less easily represented. In these early years of the twenty-first century, such countries as Belarus (between Europe and Russia) and Kazakhstan (between Russia and Muslim Southwest Asia) lie in inter-realm transition zones. Remember, over much (though not all) of their length, borders between realms are zones of regional change.

5 *Geographic realms* **change** *over time.*

Had we drawn Figure G-2 before Columbus made his voyages (1492–1504), the map would have looked different: Amerindian states and peoples would have determined the boundaries in the Americas; Australia and New Guinea would have

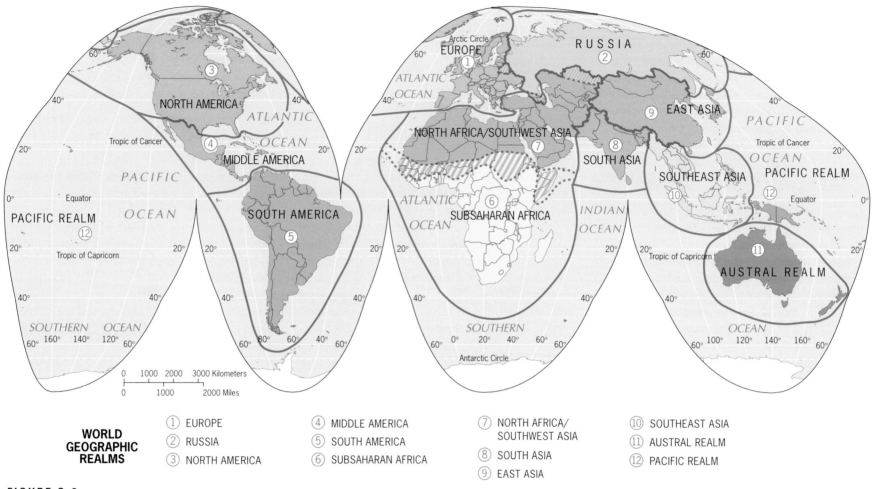

FIGURE G-2

① EUROPE
② RUSSIA
③ NORTH AMERICA

④ MIDDLE AMERICA
⑤ SOUTH AMERICA
⑥ SUBSAHARAN AFRICA

⑦ NORTH AFRICA/ SOUTHWEST ASIA
⑧ SOUTH ASIA
⑨ EAST ASIA

⑩ SOUTHEAST ASIA
⑪ AUSTRAL REALM
⑫ PACIFIC REALM

constituted one realm, and New Zealand would have been part of the Pacific Realm. The colonization, Europeanization, and Westernization of the world changed that map dramatically. During the four decades after World War II relatively little change took place, but since 1985 far-reaching re-alignments have again been occurring.

6 As we try to envisage what a **New World Order** will look like on the map, note that the 12 geographic realms can be divided into two groups: (1) those dominated by one major political entity, in terms of territory or population or both (North America/United States, Middle America/Mexico, South America/Brazil, South Asia/India, East Asia/ China, Southeast Asia/Indonesia as well as Russia and Australia), and (2) those that contain many countries but no dominant state (Europe, North Africa/ Southwest Asia, Subsaharan Africa, and the Pacific Realm). For several decades two major powers, the United States and the former Soviet Union, dominated the world and competed for global influence. Today, the United States is dominant in what has become a unipolar world. What lies ahead? Will China and/or some other power challenge U.S. supremacy? Is our map a prelude to a multipolar world? We will address such questions in the pages that follow.

The Criteria for Regions

The spatial division of the world into geographic realms establishes a broad global framework, but for our purposes a more refined level of spatial classification is needed. This brings us to an important **7** organizing concept in geography: the **regional concept**. To continue the analogy with biological

FROM THE FIELDNOTES

"The Atlantic-coast city of Bergen, Norway displayed the Norse cultural landscape more comprehensively, it seemed, than any other Norwegian city, even Oslo. The high-relief site of Bergen creates great vistas, but also long shadows; windows are large to let in maximum light. Red-tiled roofs are pitched steeply to enhance runoff and inhibit snow accumulation; streets are narrow and houses clustered, conserving warmth. . . . The coastal village of Mengkabong on the Borneo coast of the South China Sea represents a cultural landscape seen all along the island's shores, a stilt village of the Bajau, a fishing people. Houses and canoes are built of wood as they have been for centuries. But we could see some evidence of modernization: windows filling wall openings, water piped in from a nearby well."

taxonomy, we now go from phylum to order. To establish regions within geographic realms, we need more specific criteria.

Let us use the North American realm to demonstrate the regional idea. When we refer to a part of the United States or Canada (e.g., the South, the Midwest, or the Prairie Provinces), we employ a regional concept—not scientifically but as part of everyday communication. We reveal our *perception* of local or distant space as well as our mental image of the region we are describing.

But what exactly is the Midwest? How would you draw this region on the North American map? Regions are easy to imagine and describe, but they can be difficult to outline on a map. One way to define the Midwest is to use the borders of States: certain States are part of this region, others are not. You could also use agriculture as the chief criterion: the Midwest is where corn and/or soybeans occupy a certain percentage of the farmland. Each method results in a different delimitation; a Midwest based on States is different from a Midwest based on farm production. Therein lies an important principle: regions are scientific devices that allow us to make spatial generalizations, and they are based on artificial criteria we establish to help us construct them. If you were studying the geography behind politics, then a Midwest region defined by State boundaries would make sense. If you were studying agricul-

tural distributions, you would need a different definition.

Given these different dimensions of the same region, we can identify properties that all regions have in common. To begin with, all regions have *area*. This observation would seem obvious, but there is more to this idea than meets the eye. Regions may be intellectual constructs, but they are not abstractions: they exist in the real world, and they occupy space on the Earth's surface.

It follows that regions have *boundaries*. Occasionally, nature itself draws sharp dividing lines, for example, along the crest of a mountain range or the ⬛ 8 margin of a forest. More often, **regional boundaries** are not self-evident, and we must determine

FROM THE FIELDNOTES

"Walking the hilly countryside anywhere in Indonesia leaves you in no doubt about the properties of an *Af* climate (p. 11). The sweltering sun, the hot, humid air, the daily afternoon rains, and the lack of relief even when the sun goes down—the still atmosphere lies like a heavy blanket on the countryside to make this a challenging field trip. Deep, fertile soils form rapidly here on the volcanic rocks, and the entire landscape is green, most of it now draped by rice fields and terraced paddies. The farmer whose paddies I photographed here told me that his land produces three crops per year. Consider this: the island of Jawa, about the size of Louisiana, has a population of more than 125 million—its growth made possible by this combination of equatorial circumstances. . . . Alaska, almost a dozen times as large as Jawa, has a population under three-quarters of a million. Here climates range from *Cfc* to *E* (see p. 13), soils are thin and take thousands of years to develop, and the air is arctic. We sailed slowly into Glacier Bay, in awe of the spectacular, unspoiled scenery, and turned into a bay filled with ice floes, some of them serving as rafts for sleeping seals. Calving (breaking up) into the bay was the Grand Pacific Glacier with evidence of its recent recession all around. Less than 300 years ago, all of Glacier Bay was filled with ice; today you can sail miles to the Johns Hopkins (shown here) and Margerie tidewater glaciers' current outer edges. Global warming in action and a reminder that any map of climate is a still picture of a changing world."

them using criteria that we establish for that purpose. For example, to define a citrus-growing agricultural region, we may decide that only areas where more than 50 percent of all farmland stands under citrus trees qualify to be part of that region.

9 All regions also have **location**. Often the name of a region contains a locational clue, as in Amazon Basin or Indochina (a region of Southeast Asia lying between India and China). Geographers refer **10** to the **absolute location** of a place or region by providing the latitudinal and longitudinal extent of the region with respect to the Earth's grid coordinates. A far more practical measure is a **11** region's **relative location**, that is, its location with reference to other regions. Again, the names of some regions reveal aspects of their relative locations, as in *Eastern* Europe and *Equatorial* Africa.

Many regions are marked by a certain *homogeneity* or sameness. Homogeneity may lie in a region's human (cultural) properties, its physical (natural) characteristics, or both. Siberia, a vast region of northeastern Russia, is marked by a sparse human population that resides in widely scattered, small settlements of similar form, frigid climates, extensive areas of permafrost (permanently frozen subsoil), and cold-adapted vegetation. This dominant uniformity makes it one of Russia's natural and cultural regions, extending from the Ural Mountains in the west to the Pacific Ocean in the east. When regions display a measurable and often visible internal **12** homogeneity, they are called **formal regions**. But not all formal regions are visibly uniform. For example, a region may be delimited by the area in which, say, 90 percent of the people or more speak a particular language. This cannot be seen in the landscape, but the region is a reality, and we can use this criterion to draw its boundaries accurately. It, too, is a formal region.

Other regions are marked *not* by their internal sameness, but by their functional integration—that is, the way they work. These regions are defined as **13** **spatial systems** and are formed by the areal extent of the activities that define them. Take the case of a large city with its surrounding zone of suburbs, urban-fringe countryside, satellite towns, and farms. The city supplies goods and services to this encircling zone, and it buys farm products and other commodities from it. The city is the heart, the *core* of this region, and we call the surrounding zone of interaction the city's **14** **hinterland**. But the city's influence wanes on the outer periphery of that hinterland, and there lies the boundary of the functional region of **15** which the city is the focus. A **functional region**, therefore, is forged by a structured, urban-centered system of interaction. It has a core and a periphery. As we shall see, core-periphery contrasts in some parts of the world are becoming strong enough to endanger the stability of countries.

All human-geographic regions are *interconnected*, being linked to other regions. We know that the borders of geographic realms sometimes take on the character of transition zones, and so do neighboring regions. Trade, migration, education, television, computer linkages, and other interactions blur regional boundaries. These are just some of the links in the fast-growing interdependence among the world's peoples, and they reduce the differences that still divide us. Understanding these differences will lessen them further.

REGIONS AT SCALE

In this book we examine the geography of the world by dividing it into regions and using regional concepts as we proceed. These concepts form the building blocks from which the regional framework emerges.

Regions come in many sizes. The Russian formal region of Siberia is huge. By comparison, a functional region formed by a city and its hinterland is small. It is always possible to identify smaller regions within larger ones. And this brings us to the geographic **16** concept of **scale**.

The map is the geographer's strongest ally. It does for geography what taxonomic (and other) classification systems do for the other sciences. Maps display enormous amounts of information. They can suggest spatial relationships, they pose questions for

EFFECT OF SCALE

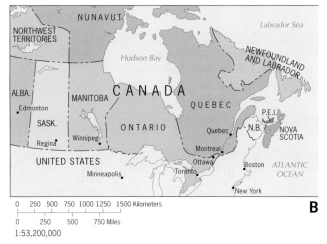

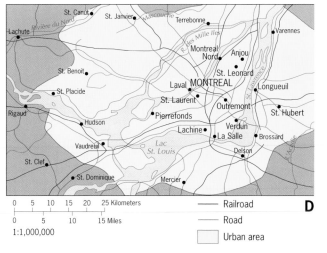

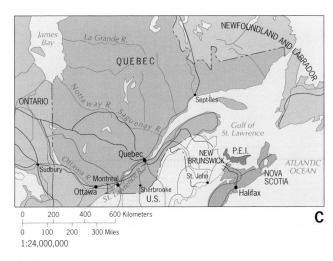

FIGURE G-3

researchers, and they often suggest where the answers may lie. Maps represent the surface of the Earth at various levels of generalization. A map's scale is the ratio of the distance between two locations on a map and the actual distance between those two locations on the Earth's surface. That ratio is expressed as a fraction.

In Figure G-3, note how that fraction grows larger as the level of detail increases—and the area displayed shrinks. Map A shows almost all of North America at the small scale of 1:103,000,000. The scale of map B is almost twice as large but shows only the northeast quadrant of the realm. At 1:24,000,000 (map C) only the Canadian province of Quebec can mostly be shown; at 1:1,000,000 (map D) you can see details of the urban area of Montreal at a scale about 100 times as large as map A.

In a book that surveys the major regions of each world realm, we will obviously have to operate at relatively small scales. Occasionally, we will enlarge our scale to focus on an issue in somewhat greater detail, but mostly our view will be macroscopic and general.

THE PHYSICAL SETTING

This book focuses on the geographic realms and regions produced by human activity over thousands of years. But we should never forget the natural environments in which all this activity took place because we can still recognize the role of these environments in how people make their living. Certain areas of the world, for example, presented opportunities for plant and animal domestication that other areas did not. The people who happened to live in those favored areas learned to grow wheat, rice, or root crops, and to domesticate oxen, goats, or llamas. We can still discern those early "patterns of

opportunity" on the map in the twenty-first century. From such opportunities came adaptation and invention, and thus arose villages, towns, cities, and states. But people living in more difficult environments found it much harder to achieve this organization. Take tropical Africa. There, the human communities could not domesticate any of the many species of wildlife, from gazelles and zebras to giraffes and buffalo. Wild animals were a threat, not an opportunity. Eventually humans domesticated only one African animal, the guinea fowl. Early African peoples faced environmental disadvantages that persist today. The modern map carries many imprints of the past.

The landmasses of Planet Earth present a jumble **17** of **natural landscapes** ranging from rugged mountain chains to smooth coastal plains (Fig. G-1). The map of natural (physical) landscapes reminds us that the names of the continents and their mountain backbones are closely linked: North America

and its Rocky Mountains, South America and the Andes, Eurasia with its Alps (in the west) and Himalayas (toward the east), Australia and its Great Dividing Range. There is, however, a prominent exception: Africa, the Earth's second-largest landmass, has peripheral mountains in the north (the Atlas) and south (the Cape Ranges), but no linear chain from coast to coast. We will soon discover why.

In his book *Guns, Germs and Steel*, UCLA geographer Jared Diamond argues that the orientation of these continental "axes" has had "enormous, sometimes tragic, consequences [that] affected the rate of spread of crops and livestock, and possibly also of writing, wheels, and other inventions. That basic feature of geography thereby contributed heavily to the very different experiences of Native Americans, Africans, and Eurasians in the last 500 years." Inevitably it also affected the evolution of the regional map we are set to investigate.

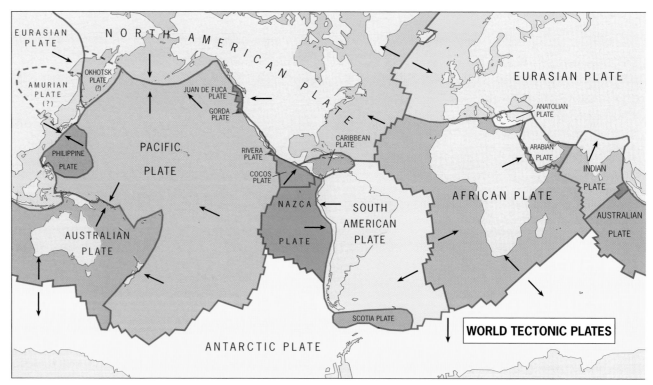

FIGURE G-4

And it affected not only its evolution, but also its future. Mountain ranges formed barriers to movement but, as Diamond stresses, also channeled the spread of agricultural and technical innovations. Rugged mountains obstructed contact but also protected peoples and cultures against their enemies (today Taliban, Al-Qaeda, and Chechnyan fugitives are using remote and rugged mountain refuges for such purposes). As we study each of the world's geographic realms, we will find that physical landscapes continue to play significant roles in this modern world. River basins in Asia still contain several of the planet's largest population concentrations: the advantages of fertile soils and ample water that first enabled clustered human settlement now sustain hundreds of millions in crowded South and East Asia. But river basins in Africa and South America support no such numbers because the combination of natural and historical factors is different. That is one reason why the study of world regional geography is important: it puts the human map in environmental as well as regional perspective.

Continents and Plates

18 The study of natural landscapes is part of **physical geography**. Nearly a century ago a physical geographer, Alfred Wegener, used spatial analysis to explain the details he saw on maps like Figure G-1. He studied the outlines of the continents and, marshalling vast evidence, theorized that the Earth's landmasses are pieces of a giant supercontinent that existed hundreds of millions of years ago. He named that postulated supercontinent *Pangaea*, and Africa lay at the heart of it. When North and South America split away from Eurasia and Africa, the Atlantic Ocean opened—but you can still see how the opposite coastlines fit together.

19 Wegener's hypothesis of **continental drift** seemed to explain much of what Figure G-1 shows, including the global distribution of major mountain ranges. As South America "drifted" westward, away from Africa, the rocks of its leading edge crumpled like folds of an accordion, creating the Andes Mountains. On the opposite side of Africa, Australia moved eastward, so that its major mountain chain now lies near its eastern margin. Africa itself moved little, which is why no major linear mountain range formed there.

Wegener's hypothesis, based on spatial evidence, pointed the way for the research of others. Eventually geologists, building on the geographic evidence, formulated the notion of plate tectonics. The continental landmasses, they reasoned, rest **20** upon great slabs of crust called **tectonic plates**. These plates are mobile, moving slowly a few centimeters (an inch or so) a year, propelled by heat-driven convection cells in the molten rock below.

Where these giant plates collide, the plate that consists of lighter (less heavy) rock rides up over the weightier plate, which is pushed downward **21** in a process called **subduction**. Such gigantic collisions create the mountain chains that girdle the Earth, causing earthquakes and volcanic eruptions in the process. In Figure G-4, note that the Pacific Ocean is encircled by a nearly continuous **22** plate-collision zone called the **Pacific Ring of Fire**. Disastrous earthquakes, volcanic eruptions, landslides, and other threats are facts of daily life here (Fig. G-5).

The complex and varied natural landscapes of the continents result not only from geologic forces but also from processes of rock *weathering* and *erosion*. Rocks vary greatly in their resistance to these processes, resulting in highly diverse and often spectacular scenery. Rivers, glaciers, coastal waves, even the wind all sculpt the surface, creating a mosaic of natural landscapes that is unique to the Solar System.

Climate

It would be impossible to discuss the world's geographic realms and regions without reference to their **23** **climates**. Many of humanity's

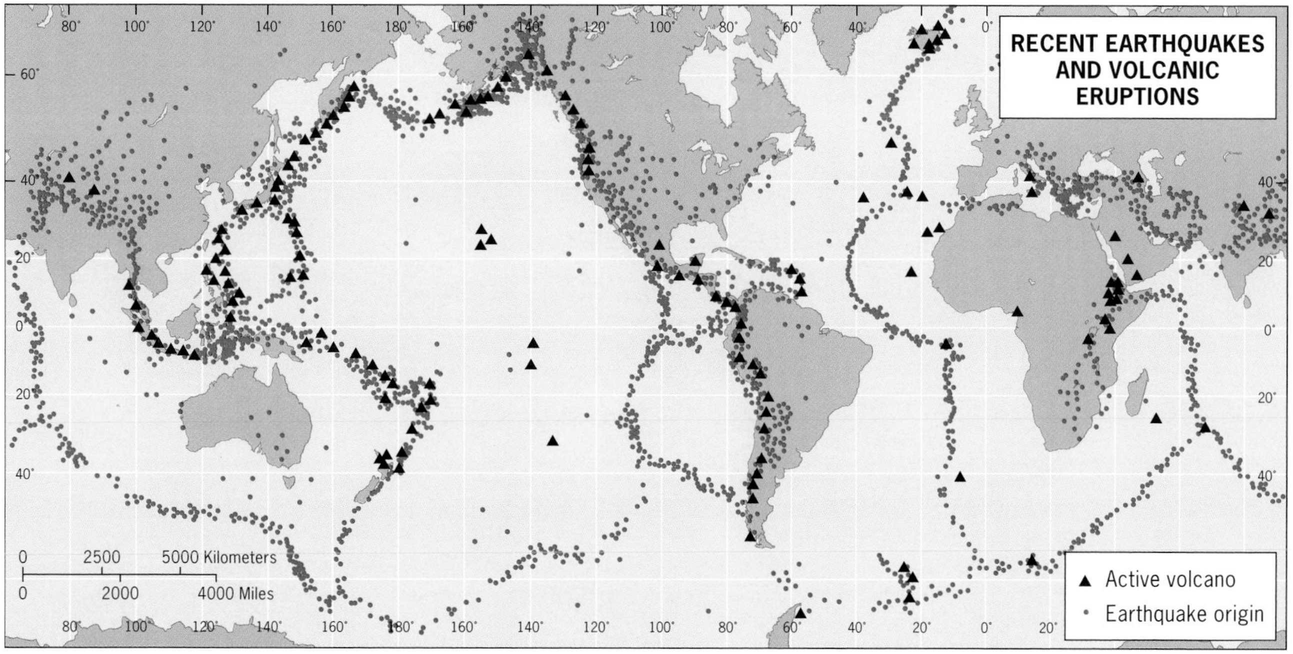

RECENT EARTHQUAKES AND VOLCANIC ERUPTIONS

▲ Active volcano
• Earthquake origin

0   2500   5000 Kilometers
0   2000   4000 Miles

FIGURE G-5

inventions were motivated by the need to overcome problems posed by climate: clothing against cold, irrigation against drought, air conditioning against heat. When we think of some distant region, the first question we are likely to ask is "What is its climate?"

A more appropriate question would be "What is its climate *today*?" Global climates change, sometimes rapidly. Archeological evidence tells us that farms, villages, and towns that sprang up in well-watered, fertile areas of Southwest Asia were 24 overtaken by **desertification**. Dependable rainfall cycles gave way to intermittent droughts, river levels dropped, wells failed. Eventually the expanding desert claimed what had once been thriving farmlands and bustling settlements. Some settlements were buried under advancing dunes.

Over the longer term, too, climate is subject to wide swings. Little more than 12,000 years ago, an eyeblink in geologic history, giant ice sheets covered virtually all of Canada and much of the U.S. Midwest (Fig. G-6) as far south as the Ohio River. Most of Europe was covered from the Alps to Scandinavia. This was only the latest 25 phase of **glaciation**, a period of lowered temperatures, ice surges, and dropping sea levels that has prevailed over the past 12 million years. It all started with a cooling trend, perhaps as long as 20 million years 26 ago, the beginning of a global **ice age**. Eventually this gradual cooling gave way to massive advances of ice sheets and valley glaciers, marking the beginning of what geologists call the Pleistocene epoch less than 3 million years ago. Ever since, frigid glaciations have been followed by 27 relatively warm spells called **interglaciations**. During the current interglaciation, named the *Holocene*

and already over 10,000 years in duration, human civilizations have evolved from the smallest and simplest communities to the populous countries and megacities of today.

Climatic Regions

Climatic variability also occurs in the short term, and we should not be surprised that climates continue to change. Unless we prepare for it, these dynamics can produce social and political dislocation. Thus we need to understand the climatic environments of the world's geographic realms and regions. Our technological advances notwithstanding, climate still plays a key role in the lives

and livelihoods of billions of people. A perturbation in the system—such as an El Niño event or a drought—can have widespread repercussions for economies and even politics.

It is no easier to generalize about climates than it is about natural landscapes. A regional climate is the average of countless weather observations among which measures of precipitation and temperature are key. Also taken into account are seasonal variations and extremes. In terms of precipitation, the Earth displays enormous spatial variation (Fig. G-7). Equatorial and tropical areas tend to be well watered 28 because the **hydrologic cycle**—the system that carries evaporated ocean water (leaving the salt be-

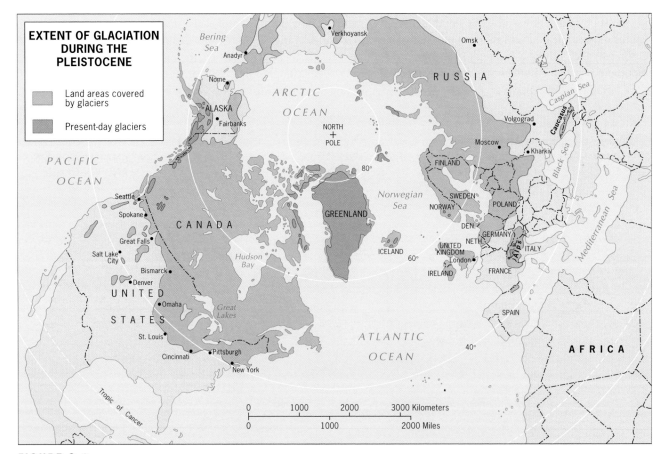

FIGURE G-6

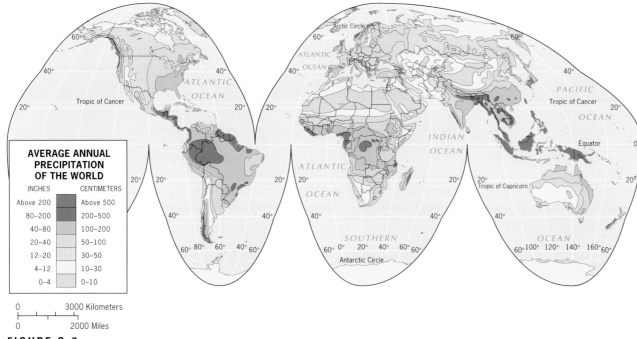

**AVERAGE ANNUAL
PRECIPITATION
OF THE WORLD**

INCHES	CENTIMETERS
Above 200	Above 500
80–200	200–500
40–80	100–200
20–40	50–100
12–20	30–50
4–12	10–30
0–4	0–10

0   3000 Kilometers

0   2000 Miles

FIGURE G-7

weather in each color-coded region is relatively standard. If, for example, you are familiar with the weather in the large area mapped as *Cfa* in the southeastern United States, you will feel at home in Uruguay (South America), Kwazulu-Natal (South Africa), New South Wales (Australia), and Fujian Province (China). Let us look at the world's climatic regions in some detail.

Humid Equatorial (*A*) Climates

The humid equatorial, or tropical, climates are characterized by high temperatures all year and by heavy precipitation. In the *Af* subtype, the rainfall arrives in substantial amounts every month; but in the *Am* areas, the arrival of the annual wet *monsoon* (the Arabic word for "season" [see p. 269]) marks a sudden enormous increase in precipitation. The *Af* subtype is named after the vegetation that develops there—the tropical rainforest. The *Am* subtype, prevailing in part of peninsular India, in a coastal area of West Africa, and in sections of Southeast Asia, is appropriately referred to as the monsoon climate. A third tropical climate, the savanna (*Aw*), has a wider daily and annual temperature range and a more strongly seasonal distribution of rainfall. As Figure G-7 indicates, savanna rainfall totals tend to be lower than those in the rainforest zone, and savanna seasonality is often expressed in a "double maximum." Each year produces two periods of increased rainfall separated by pronounced dry spells. In many savanna zones, inhabitants refer to the "long rains" and the "short rains" to identify those seasons; a persistent problem is the unpredictability of the rain's arrival. Savanna soils are not among the most fertile, and when the rains fail hunger looms. Savanna regions are far more densely peopled than rainforest areas, and millions of residents of the savanna subsist on what they cultivate.

hind) over the landmasses where it falls as rain or snow—is most efficient in the high heat and humidity of the low latitudes. But in Africa this equatorial bounty soon gives way to subtropical aridity. From over 80 inches (200 cm) of rain in The Congo near the equator, the annual total drops to 4 inches (10 cm) or less in the Sahara to the north and the Kalahari-Namib to the south. In higher latitudes, only certain favored areas (notably Western Europe) receive ample precipitation.

Average temperatures decline from equatorial toward polar latitudes, but the gradation is irregular, affected by elevation (mountains have a cooling effect) and location relative to coastlines (interiors of continents are hotter in summer, cooler in winter). The size, shape, and terrain of the landmasses complicate the map of average temperatures. Ocean currents also influence the pattern. Warm waters offshore, combined with onshore

winds, give most of Europe a warmth that is not experienced at similar latitudes elsewhere.

How can we forge regional coherence out of such a jumble of data? The effort to create a relatively simple, small-scale world map of climates has continued for nearly a century. Figure G-8 is based on the system Wladimir Köppen devised and Rudolf Geiger then modified. This system, which has the advantage of simplicity, is represented by a set of letter symbols. The first (capital) letter is the critical one: the *A* climates are humid and tropical, the *B* climates are arid, the *C* climates are mild and humid, the *D* climates show increasing extremes of seasonal heat and cold, and the *E* climates reflect the frigid conditions at and near the poles.

Figure G-8 merits your attention because familiarity with it will help you understand much of what follows in this book. The map has practical utility, 29 too. Although it depicts **climatic regions**, daily

WORLD CLIMATES
After Köppen–Geiger

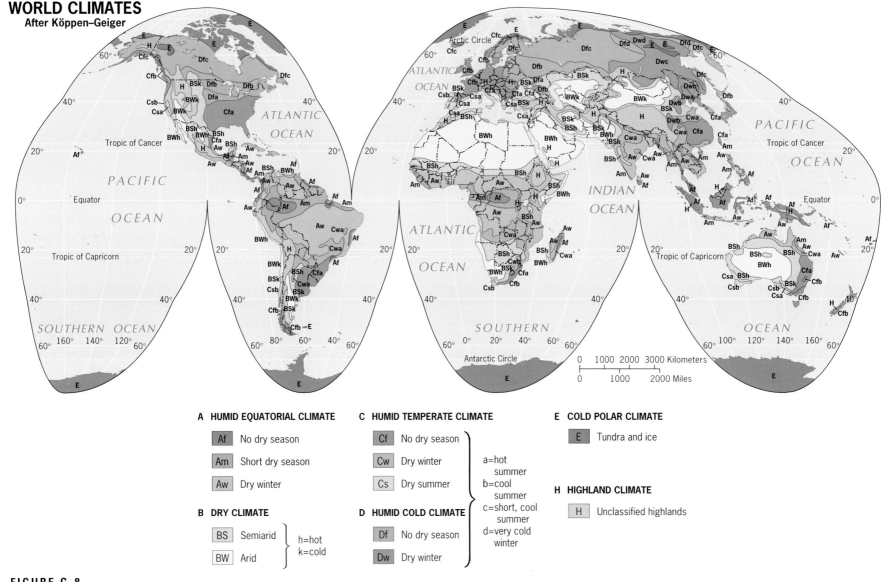

FIGURE G-8

A HUMID EQUATORIAL CLIMATE

Af	No dry season
Am	Short dry season
Aw	Dry winter

B DRY CLIMATE

BS	Semiarid	h=hot
BW	Arid	k=cold

C HUMID TEMPERATE CLIMATE

Cf	No dry season	
Cw	Dry winter	a=hot summer
Cs	Dry summer	b=cool summer

D HUMID COLD CLIMATE

Df	No dry season	c=short, cool summer
Dw	Dry winter	d=very cold winter

E COLD POLAR CLIMATE

E	Tundra and ice

H HIGHLAND CLIMATE

H	Unclassified highlands

Rainfall variability is their principal environmental problem.

Dry (*B*) Climates

Dry climates occur in both lower and higher latitudes. The difference between the *BW* (true desert) and the moister *BS* (semiarid steppe) varies but may be taken to lie at about 10 inches (25 cm) of annual precipitation. Parts of the central Sahara in North Africa receive less than 4 inches (10 cm) of rainfall. Most of the world's arid areas have an enormous daily temperature range, especially in subtropical deserts. In the Sahara, there are recorded instances of a maximum daytime shade temperature of over 120°F (49°C) followed by a nighttime low of 48°F (9°C). Soils in these arid areas tend to be thin and poorly developed; soil scientists have an appropriate name for them—aridisols.

Humid Temperate (*C*) Climates

As the map shows, almost all these mid-latitude climate areas lie just beyond the Tropics of Cancer and Capricorn (23½° North and South latitude, respectively). This is the prevailing climate in the southeastern United States from Kentucky to central Florida, on North America's west coast, in most of Europe and the Mediterranean, in southern Brazil and northern Argentina, in coastal South Africa, in eastern Australia, and in eastern China and southern Japan. None of these areas suffers climatic extremes or severity, but the winters can be cold, especially away from water bodies that moderate temperatures. These areas lie midway between the winterless equatorial climates and the summerless polar zones. Fertile and productive soils have developed under this regime, as we will note in our discussion of the North American and European realms.

The humid temperate climates range from moist, as along the densely forested coasts of Oregon, Washington, and British Columbia, to relatively dry, as in the so-called Mediterranean (dry-summer) areas that include not only coastal southern Europe and northwestern Africa but also the southwestern tips of Australia and Africa, central Chile, and Southern California. In these Mediterranean environments, the scrubby, moisture-preserving vegetation creates a natural landscape different from that of richly green Western Europe.

Humid Cold (*D*) Climates

The humid cold (or "snow") climates may be called the continental climates, for they seem to develop in the interior of large landmasses, as in the heart of Eurasia or North America. No equivalent land areas at similar latitudes exist in the Southern Hemisphere; consequently, no *D* climates occur there.

Great annual temperature ranges mark these humid continental climates, and cold winters and relatively cool summers are the rule. In a *Dfa* climate, for instance, the warmest summer month (July) may average as high as 70°F (21°C), but the coldest month (January) might average only 12°F

(−11°C). Total precipitation, much of it snow, is not high, ranging from over 30 inches (75 cm) to a steppe-like 10 inches (25 cm). Compensating for this paucity of precipitation are cool temperatures that inhibit the loss of moisture from evaporation and evapotranspiration (moisture loss to the atmosphere from soils and plants).

Some of the world's most productive soils lie in areas under humid cold climates, including the U.S. Midwest, parts of southern Russia and Ukraine, and Northeast China. The winter dormancy (when all water is frozen) and the accumulation of plant debris during the fall balance the soil-forming and enriching processes. The soil differentiates into well-defined, nutrient-rich layers, and substantial organic humus accumulates. Even where the annual precipitation is light, this environment sustains extensive coniferous forests.

Cold Polar (*E*) and Highland (*H*) Climates

Cold polar (*E*) climates are differentiated into true icecap conditions, where permanent ice and snow keep vegetation from gaining a foothold, and the tundra, which may have average temperatures above freezing up to four months of the year. Like rainforest, savanna, and steppe, the term *tundra* is vegetative as well as climatic, and the boundary between the *D* and *E* climates in Figure G-8 corresponds closely to that between the northern coniferous forests and the tundra.

Finally, the *H* climates—unclassified highlands mapped in gray (Fig. G-8)—resemble the *E* climates. High elevations and the complex topography of major mountain systems often produce near-Arctic climates above the tree line, even in the lowest latitudes such as the equatorial section of the high Andes of South America.

Let us not forget an important qualification concerning Figure G-8: this is a still-picture of a changing scene, a single frame from an ongoing film. Climate is still changing, and less than a century from now climatologists are likely to be modifying the climate maps to reflect new data. Who knows: we may have to redraw even those familiar

coastlines. Environmental change is a never-ending challenge.

REGIONS AND CULTURES

Whenever we explore a geographic realm or region, we should assess the physical stage that forms its base—its total physical geography or **30 physiography**. Still, the realms and regions we discuss in the chapters that follow are defined primarily by human-geographic criteria, and one criterion, culture, is especially significant. Therefore, we should carefully consider the culture concept in a regional context.

31 When anthropologists define **culture**, they tend to concentrate on abstractions: learning, knowledge and its transmission, and behavior. Geographers are most interested in how culture and the patterns of behavior associated with it are imprinted on the landscape. Thus we will examine how members of a society perceive and exploit their available resources, how they maximize the opportunities and adapt to the limitations of their natural environment, and how they organize their portion of the Earth. Some human works remain etched on the Earth's surface for a long time: the Egyptian pyramids, the Great Wall of China, and Roman roads and bridges are still in place millennia after they were built. Over time, regions take on dominant **32** qualities that collectively constitute a **regional character**, a personality, a distinct atmosphere. Regional character is a crucial criterion in ascertaining how we divide the human world into major geographic realms and regions.

Cultural Landscapes

As we noted, geographers are particularly concerned with the impress of culture on the Earth's physical surface. Culture gives visible character to a region in many ways. Often a single scene in a photograph or picture can reveal, in general terms, where the photo was taken. The architecture, forms of transportation, goods being carried, and clothing of the people (all these are part of culture) help us guess the region in the photo. We can do this because the people of any culture are active agents of change; they transform the land they occupy by building structures, creating lines of transport and communication, parceling out fields, and tilling the soil (among countless other activities).

The composite of human imprints on the Earth's 33 surface is called the **cultural landscape**, a term that came into general use in geography during the 1920s. Carl Ortwin Sauer (a professor of geography at the University of California, Berkeley) developed a school of cultural geography that focused on the concept of cultural landscape. In a paper titled "Recent Developments in Cultural Geography" (1927), Sauer proposed his most straightforward (though, for an obvious reason, much-criticized) definition of the cultural landscape: *the forms superimposed on the physical landscape by the activities of man.* Such forms result from cultural processes—causal forces that shape cultural patterns—that unfold over a long time and involve the cumulative influences of successive occupants.

When it comes to mapping a cultural landscape, its material, tangible properties are the easiest to observe and record. Take, for instance, the urban "townscape"—a prominent element of the overall cultural landscape—and compare a major U.S. city with, say, a major Japanese city. Photographs of these two metropolitan scenes would reveal the differences quickly, of course, but so would maps. The American city with its square-grid layout of the 34 **central business district (CBD)** and its widely dispersed, now heavily urbanized suburbs contrasts sharply with the clustered, space-conserving Japanese metropolis. In a rural example, the spatially lavish subdivision and ownership patterns of American farmland look unmistakably different from the traditional African countryside, with its irregular, often tiny patches of land surrounding a village. Still, the totality of a cultural landscape can never be captured by a photograph or map because the personality of a region involves more than its prevailing spatial organization. One must also include the region's visual appearance, its noises and odors, the shared experiences of its inhabitants, and even their pace of life.

Culture and Ethnicity

Language, religion, and other cultural traditions often are durable and persistent. But culture is not 35 necessarily based on **ethnicity**, and as we study the world's human-geographic regions we should be aware that peoples of different ethnic stocks can achieve a common cultural landscape, while people of the same ethnic background can be divided along cultural lines.

The recent events in the former Eastern European country of Yugoslavia are a case in point. As the old Yugoslavia broke up in the early 1990s, its component parts were first affected and then engulfed by what was often described as "ethnic" conflict. When the crisis reached Bosnia in the heart of Yugoslavia, three groups fought a bitter civil war. These groups were identified as Bosnian Muslims, Serbs, and Croats—but in fact all were ethnic Slavs (Yugoslavia means "Land of the South Slavs"). What distanced them from one another, and what kept the conflict going, was cultural tradition, not ethnicity. The Bosnian Muslims and the Serbs had developed different communities in different parts of the former Yugoslavia. The Muslims were descendants of Slavs who, a century ago or more, had been converted by the Turks (who once ruled here) from Christianity to Islam. Each of these groups feared domination by the others, and so South Slav turned against South Slav in what was, in truth, a culture-based conflict.

Even though the post–Cold War world is still young, it has already witnessed numerous intra-regional conflicts; in later chapters, we will explain some of these conflicts in geographic context. Not all of them have ethnicity at their roots. Culture is a great unifier; it can also be a powerful divider.

REALMS OF POPULATION

Earlier we noted that population numbers by themselves do not define geographic realms or regions. Population distributions, and the functioning society that gives them common ground, are more significant criteria. That is why we can identify one geographic realm (the Austral) with barely more than 20 million people and another (East Asia) with 1.5 billion inhabitants. Neither population numbers nor territorial size alone can delimit a geographic realm. Nevertheless, the map of world population distribution (Fig. G-9) suggests the relative location of several of the world's geographic realms, based on the strong clustering of population in certain areas. Before we examine these clusters in some detail, remember that the Earth's human population now totals over 6 billion—6 thousand million people confined to the land-masses that constitute less than 30 percent of our planet's surface, much of which is arid desert, rugged mountain terrain, or frigid tundra. (Remember that Fig. G-9 is a turn-of-the-millennium still-picture of an ever-changing scene: the rapid growth of humankind continues.) After thousands of years of slow growth, world population during the nineteenth and twentieth centuries expanded at an increasing rate. That rate has recently been slowing down somewhat, but consider this: it took about 17 centuries after the birth of Christ for the world to add 250 million people to its numbers; now we are adding 250 million about every *four years*.

Demographic (population-related) issues will arise repeatedly in later chapters as we survey the world's most crowded realms. As we will note, there are indications that the explosive population growth of the twentieth century will slow down during the twenty-first. For the moment, we will confine ourselves to the overall global situation. 36 Let us compare the map of world **population distribution** to the demographic data shown in Table G-1 (pp. 24–29). This table provides information about the total populations in each geographic realm as well as their individual countries. Also, if you compare Figure G-9 to the maps of terrain (Fig. G-1) and climate (Fig. G-8), you will note that some of the largest population concentrations lie in the fertile basins of major rivers

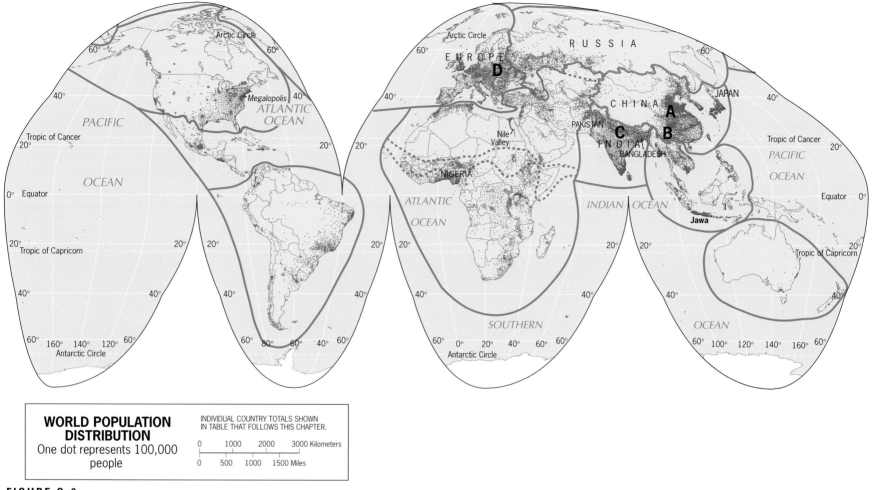

FIGURE G-9

(China's Huang He and Chang Jiang, India's Ganges). We live in a modern world, but old ways of life still prescribe where hundreds of millions on this Earth live.

Major Population Clusters

The world's greatest population cluster, *East Asia*, lies centered on China and includes the Pacific-facing Asian coastal zone from the Korean Peninsula to Vietnam. The map indicates that the number of 37 people per unit area—the **population density**—tends to decline from the coastal zone toward the interior. Note, however, the ribbon-like extensions marked *A* and *B* (Fig. G-9). As the map of world natural landscapes (Fig. G-1) confirms, these are populations concentrated in the valleys of China's major rivers, the Huang He (Yellow) and the Chang Jiang (Yangzi), respectively. Here in East Asia, most people are farmers, not city dwellers. There are great cities in China (such as Beijing and Shanghai), but their total population still is far outnumbered by the farmers—those who live and work on the land and whose crops of rice and wheat feed not only themselves but also the people in those cities.

The *South Asia* population cluster lies centered on India and includes the populous neighboring countries of Bangladesh and Pakistan. This huge agglomeration of humanity focuses on the broad plain of the lower Ganges River (*C* in Fig. G-9). It is nearly as large as that of East Asia and at present growth rates will overtake East Asia in less than 20 years. As in East Asia, most people are farmers, but in South Asia pressure on the land is even greater, and farming is less efficient. As we note in Chapter 8, the population issue looms large in this realm.

The third-ranking population cluster, *Europe*, also lies on the world's biggest landmass but at the opposite end from China. The European cluster, including western Russia, counts over 700 million inhabitants, which puts it in a class with the two larger Eurasian concentrations—but there the similarity ends. In Europe, the key to the linear, east-west orientation of the axis of population (*D* in Fig. G-9) is not a fertile river basin but a zone of raw materials for industry. Europe is among the world's most highly urbanized and industrialized realms, its human agglomeration sustained by factories and offices rather than paddies and pastures.

The three world population concentrations just discussed (East Asia, South Asia, and Europe) account for more than 3.6 billion of the world's 6.4 billion people. No other cluster comes close to these numbers. The next-ranking cluster, *Eastern North America*, is only about one-quarter the size of the smallest of the Eurasian concentrations. As in Europe, the population in this area is concentrated in major metropolitan complexes; the rural areas are now relatively sparsely settled. Geographic realms and 38 regions, therefore, display varying levels of **urbanization**, the percentage of the total population living in cities and towns. And some regions are urbanizing much more rapidly than others, a phenomenon we will explain as we examine the world's realms.

REALMS, REGIONS, AND STATES

Our analysis of the world's regional geography requires data, and it is crucial to know the origin of these data. Unfortunately, we do not have a uniformly-sized grid to superimpose over the globe: we must depend on the world's 190-plus countries to report vital information. Irregular as the boundary framework shown on the world map (see Front Endpaper or Fig. G-10) may be, it is all we have. Fortunately, all large and populous countries are

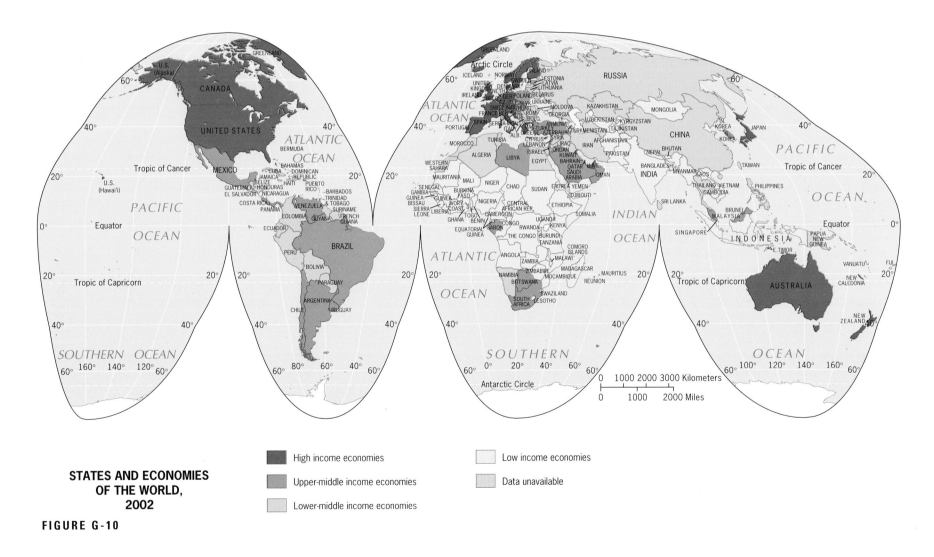

STATES AND ECONOMIES OF THE WORLD, 2002

High income economies

Upper-middle income economies

Lower-middle income economies

Low income economies

Data unavailable

FIGURE G-10

subdivided into provinces, States, or other internal entities, and their governments provide information on each of these subdivisions when they conduct their census.

Although we often refer to the political entities on the world map as *countries*, the appropriate geographic term for them is *states*. To avoid confusion with subdivisions of the same name, we capitalize internal States (as in the preceding paragraph). Thus we refer to the State of California and the State of Queensland (Australia), but to the state of Japan and the state of Argentina.

39 The **state** as a political, social, and economic entity has been developing for thousands of years, ever since agricultural surpluses made possible the growth of large and powerful towns that could command hinterlands and control peoples far beyond their walls. But the modern state is a relatively recent phenomenon. The boundary framework we see on the world political map today substantially came about during the nineteenth century, and the independence of dozens of former colonies (which made them states as well) occurred during the twentieth **40** century. Today, the **European state model**—a clearly and legally defined territory inhabited by a population governed from a capital city by a representative government—prevails in the aftermath of the collapse of colonial and communist empires.

As Figures G-2 and G-10 suggest, geographic realms are mostly assemblages of states, and the borders between realms frequently coincide with the boundaries between countries—for example, between North America and Middle America along the U.S.-Mexico boundary. But a realm boundary can also cut *across* a state, as does the one between Subsaharan Africa and the Muslim-dominated realm of North Africa/Southwest Asia. Here the boundary takes on the properties of a wide transition zone, but it still divides states such as Nigeria, Chad, and Sudan. The transformation of the margins of the former Soviet Union is creating similar cross-country transitions. Newly independent states such as Belarus (between Europe and Russia) and Kazakhstan (between Russia and Muslim Southwest Asia) lie in zones of regional change.

Most often, however, geographic realms consist of groups of states whose boundaries also mark the limits of the realms. Look at Southeast Asia, for instance. Its northern border coincides with the political boundary that separates China (a realm practically unto itself) from Vietnam, Laos, and Myanmar (Burma). The boundary between Myanmar and Bangladesh (which is part of the South Asian realm) defines its western border. Here, the state boundary framework helps delimit geographic realms.

The global boundary framework is even more useful in delimiting regions *within* geographic realms. We shall discuss regional divisions every time we introduce a geographic realm, but an example is appropriate here. In the Middle American realm, we recognize four regions. Two of these lie on the mainland: Mexico, the giant of the realm, and Central America, which consists of the seven comparatively small countries located between Mexico and the Panama-Colombia border (which is the boundary with the South American realm). Central America often is misdefined in news reports; the correct regional definition is based on the politico-geographical framework.

To our earlier criteria of physiography, population distribution, and cultural geography, we now add political geography as a determinant of world-scale geographic regions. In doing so, we should be aware that the global boundary framework continues to change and that boundaries are created (as in 1993 between the Czech Republic and Slovakia) as well as eliminated (e.g., between former West and East Germany in 1990). But the overall system, much of it resulting from colonial and imperial expansionism, has endured, despite the predictions of some geographers that the "boundaries of imperialism" would be replaced by newly negotiated ones in the postcolonial period.

Toward the end of this book, we will discuss a recent, ominous development in boundary-making: the extension of boundaries onto and into the oceans and seas. This process has been consuming the last of the Earth's open frontiers, with uncertain consequences.

PATTERNS OF DEVELOPMENT

Finally, we turn to economic geography for criteria that allow us to place states, regions, and realms in regional **41** groupings based on their level of **development**. **42** **Economic geography** focuses on spatial aspects of how people make a living; thus it deals with patterns of production, distribution, and consumption of goods and services. As with all else in this world, these patterns exhibit considerable variation. Individual states report their imports and exports, farm and factory production, and many other economic data to the United Nations and other international agencies. From such information we can determine the comparative economic well-being of the world's countries.

The World Bank, one of the agencies that monitor economic conditions, classifies countries into four groups: (1) high-income, (2) upper-middle-income, (3) lower-middle-income, and (4) low-income countries. These groupings display regional clustering when mapped (Fig. G-10). The higher-income economies are concentrated in Europe, North America, and along the western Pacific Rim, most notably in Japan and Australia. The low-income countries dominate in Africa and parts of Asia.

As Figure G-10 shows, certain geographic-realm boundaries can be discerned on this economic map, including the one between North America and Middle America and between Australia and Southeast Asia. Within several geographic realms, you can see some distinct regional boundaries: between Western and Eastern Europe, between Brazil and Andean South America (the region to the west), between the oil-rich Arabian Peninsula and the countries of the Middle East to its northwest, and between Mongolia and neighboring parts of China.

Figure G-10 reveals the level of economic development among the world's countries—based, we should remind ourselves, on the political framework as the reporting grid. Until recently, economists and economic geographers used this basis to divide the world into developed countries (DCs) and underdeveloped countries (UDCs). But that distinction has lost much of its relevance, for reasons Figure G-10 cannot show.

Spatial contrasts in well-being have been attributed to many factors, including climate (as a conditioner of human capacity), cultural heritage (e.g., receptivity or resistance to change and innovation), colonial exploitation, and, more recently, neocolonialism. Obviously, the distribution of accessible natural resources and the relative location of countries play their parts. Certain areas and peoples have had more opportunities for interaction and exchange than others. Also, richer countries perpetuate their advantages by erecting high tariffs (tolls) against the products of poorer countries and by subsidizing their own producers (such as farmers) when they compete on world markets. Europe, the United States, and Japan do this even as they advocate "free trade" and "open markets," giving more than $300 billion to their farmers and thus depriving poor-country farmers of the opportunity to compete on the global "level playing field" they claim to envision. Such unfair policies have much to do with the pattern you see in Figure G-10.

Cores and Peripheries

For as long as states have existed, they have been **43** organized around **core areas**, the foci of human activity that function as the leading locales of authority and productivity. In the modern state, the core area is a *heartland*, where the largest population cluster, major cities, the capital, most efficient transport networks, firms and factories, and other leading assets of the country are concentrated.

If states have core areas, then they also have **44** **peripheries**. From the earliest states onward, the core held the advantage. Through its authority and organizational capacity, the government in the core could extract taxes and tribute from those in the weaker periphery. Today, even in democratic countries, outlying States and provinces, through their representatives, complain of neglect and disadvantage. A state's historic core area holds the advantage.

Indeed, modern times have worsened, not improved, this core-periphery dichotomy. Not only are there core (*have*)-periphery (*have-not*) relationships in virtually all countries of the world, but core-periphery systems now encompass the entire globe. Today the whole of Western Europe, North America, and Japan function as macroscale, global core areas. Power, money, influence, and decision-making (and enforcing) organizations are concentrated here. Here the terms of an economic system that perpetuates the core's primacy are dictated.

Even as the core countries ensure that the prices of commodities from periphery sources remain low and competition from countries on the margin is suppressed, their wealth and opportunity attract skilled and professional people from afar. How many thousands of doctors, engineers, and teachers from India are working in England and the United States? The core's impact on the periphery takes many forms.

The countries of the periphery, therefore, face severe problems. They are pawns in a global economic game whose rules they cannot change. Their internal problems are intensified by the aggressive involvement of core-area interests. The way they use resources (such as whether to produce food for local consumption or to produce export crops) is strongly affected by foreign interference. They suffer far more from environmental degradation, overpopulation, and mismanagement than core-area countries do. Their inherited disadvantages have grown, not lessened, over time. The widening gaps that result between enriching cores and persistently impoverished peripheries threaten the future of the world. Better than "national" economic statistics, an index **45** representing **regional disparity** would give us a true picture of the economic and social conditions in geographic realms, their component regions, and individual countries.

It is therefore no longer appropriate to divide the world geographically into developed and underdeveloped realms. East Asia is dominated by lower-income economies, but it includes Japan, one of the world's richest states. Europe is one of the world's highest-income realms, but not in Albania, Moldova, or Ukraine.

Not long ago, countries also were categorized as being part of a First World (capitalist), Second World (socialist/communist), Third World (underdeveloped), Fourth World (severely underdeveloped), or even Fifth World (poorest). This classification, too, has lost most of its significance in the twenty-first century. The most meaningful distinction we can **46** make is based on the notion of **advantage**: some countries have it and many others do not. Such advantage takes many forms: geographic location (to be on the coast is generally better than to be landlocked), raw materials, government, political stability, productive skills, and much more. As the gap between the advantaged and the disadvantaged states widens (and it *is* widening), the stability of the political world is at risk, and with it the advantages of the "haves" over the "have nots."

As we proceed with our investigation of the world's geographic realms, the concept of development will be examined in a regional context and from several viewpoints. Figure G-10 reflects events that began long before the Industrial Revolution of the late eighteenth and nineteenth centuries: Europe, even by the middle of the eighteenth century, had laid the foundations for its colonial expansion. The Industrial Revolution then magnified Europe's demands for raw materials, and its manufactures increased its imperial control. Western countries thereby gained an enormous head start, while colonial dependencies continued to supply raw materials and to consume Western industrial products. Thus was born a system of international exchange and capital flow that changed little when the colonial period ended. Developing countries, well aware of their predicament, accused the developed world of perpetuating its **47** long-term advantage through **neocolonialism**— the entrenchment of the old system under a new guise.

Symptoms of Underdevelopment

Although it is therefore no longer appropriate to divide the world into developed and underdeveloped

FROM THE FIELDNOTES

"February 1, 2003. A long-held hope came true today: thanks to a Brazilian intermediary I was allowed to enter and spend a day in two of Rio de Janeiro's hillslope *favelas*, an eight-hour walk through one into the other. Here live millions of the city's poor, in areas often ruled by drug lords and their gangs, with minimal or no public services, amid squalor and stench, in discomfort and danger. And yet life in the older *favelas* has become more comfortable as shacks are replaced by more permanent structures, electricity is sometimes available, water supply, however haphazard, is improved, and an informal economy brings goods and services to the residents. I stood in the doorway of a resident's single-room dwelling for this overview of an urban landscape in transition: satellite-television disks symbolize the change going on here. The often blue cisterns catch rainwater; walls are made of rough brick and roofs of corrugated iron or asbestos sheeting. No roads or automobile access, so people walk to the nearest road at the bottom of the hill. Locals told me of their hope that they will some day have legal rights to the space they occupy. During his campaign for president of Brazil, Lula da Silva suggested that long-term inhabitants should be awarded title, and in 2003 his government approved the notion. It will be complicated: as the photo shows, people live quite literally on top of one another, and mapping the chaos will not be simple (but will be made possible with Geographic Information Systems). This would allow the government to tax residents, but it would also allow residents to obtain loans based on the value of their *favela* properties, and bring millions of Brazilians into the formal economy. The hardships I saw on this excursion were often dreadful, but you could sense the hope for and anticipation of a better future."

geographic realms, make no mistake: there still *are* regions within realms, states within regions, and provinces within states that suffer from underdevelopment. As we will discover during our global journey, underdevelopment takes many forms, displays many symptoms, and has many causes. To grasp some of the symptoms, see Table G-1 and compare, say, Bangladesh or Moçambique with Japan or Canada. Population growth rates in the periphery are higher, life expectancies shorter, urbanization levels lower, incomes smaller. As we will find, the poorer countries suffer from high infant and child mortality rates, poor overall health and sanitation, inefficient farming, insufficient diets, overcrowded cities, and many other ills. Many countries in the periphery are trapped in a global economic system in which the export of unfinished or partially finished raw materials is their only source of outside income.

The Specter of Debt

All of us are aware of the dangers of going too deeply into debt. Borrow too much money, and payments of interest and principal may leave too little for routine needs such as food and clothing, not to mention repairs and replacement of equipment. Individuals, families, villages, and cities must devise budgets and balance incomes against expenditures.

So it is with countries. National governments, like individuals and families, must sometimes borrow to make ends meet. If a drought curtails domestic food production, a government may borrow against future oil or ore exports to pay for urgently needed staples. Should domestic harvests fail in the following year as well, further borrowing will be necessary. This means that the government's future budgets will be burdened by ever-higher "debt retirement" payments, limiting its ability to spend on schools, roads, and clinics. Should the decision be made to raise taxes to help make the payments, the result may be a slowdown in the economy. The cycle of debt is hard to escape.

All too often it is not just a food emergency or some other unavoidable problem, but a government's mismanagement that leads to excessive debt. In the postcolonial period, many governments of former colonies sought shortcuts to boost their economies, building dams, factories, and ports that did not justify their huge investments. Of course the governments and corporations of the former colonial powers were all too willing to do the building, making large profits and even lending the new governments money to build still more.

Soon, many of the civilian governments that took over from the colonial powers, weakened by such errors and losing the confidence of their citizens, fell prey to military takeovers. And the military dictatorships that replaced them needed weapons to maintain their hold on power—weapons readily sold for credit by the arms manufacturers in the wealthy core countries. During the Cold War, dictatorial regimes in the periphery became close allies of both superpowers, and proxy wars were fought in Asia, Africa, and the Americas. By the time the Cold War ended and representative government began to return to some (though not all) of these countries, they were so deeply in debt to the rich core countries that they had no prospect of recovery.

External debt is not confined to the countries in the periphery, and even the richer countries (including the United States) have national debts. What matters is a country's ability to service its debt and

FROM THE FIELDNOTES

"It was an equatorial day here today, in hot and humid Singapore. Walked the three miles to the Sultan Mosque this morning and observed the activity arising from the Friday prayers, then sat in the shade of the large ficus tree on the corner of Arab Street and watched the busy pedestrian and vehicular traffic. An Indian man saw me taking notes and sat down beside me. For some time he said nothing, then asked whether I was writing a novel. 'If only I could!' I answered, and explained that I was noting locations and impressions for photos and text in a geography book. He said that he was Hindu but liked living in this area because some of what he called 'Old Singapore' still survived. 'No problems between Hindus and Muslims here,' he said. 'The police maintain strict security here around the Mosques, especially after September 2001. The government tolerates all religions, but it doesn't tolerate religious conflict.' He offered to show me the Hindu shrine where he and his family joined others from Singapore's Hindu community. As we walked down Arab Street, he pointed skyward. 'That's the future,' he said. 'The symbol of globalization. Do you mention it in your book? It's in the papers every day, supposedly a good thing for Singapore. Well, maybe. But a lot of history has been lost to make space for what you see here.' I took the photo, and promised to mention his point."

still have the capacity to pay for its other needs. South Korea, for example, currently has the highest per-capita debt in the world, but its per-capita *gross national product* (*GNP*—the total value of all goods and services produced by the citizens of a country, within or outside of its boundaries, during a calendar year) is more than three times as high. Nicaragua, a Middle American Cold War victim, is much worse off: its per-capita debt is nearly *four times* its per-capita GNP. When a destructive hurricane struck Nicaragua in 1998, the country was left totally dependent on outside help. It had no reserves nor, given its debts, did it have the ability to borrow.

Several Middle American countries suffer from heavy debt burdens, but the realm most severely afflicted in this respect is Subsaharan Africa. As we note in Chapter 6, postcolonial tropical Africa suffers from a combination of conditions and circumstances that mires most of its countries in debt-ridden poverty.

Globalization

If a geographic concept can arouse strong passions, **48** **globalization** is it. To leading economists, politicians, and businesspeople this is the best of all worlds: the march of international capitalism, open markets, and free trade. To millions of poor farmers and powerless citizens in debt-mired African and Asian countries, it represents a system that will keep them in poverty and subservience forever. Economic geographers can prove that global economic integration allows the overall economies of poorer countries to grow faster: compare their international trade to their national income, and you will find that the GNP of those that engage in more international trade (and thus are more "globalized") rises, while the GNP of those with less international trade actually declines. The problem is that those poorest countries contain more than 2 billion people, and their prospects in this globalizing world are worsening, not rising. In countries like the Philippines, Kenya, and Nicaragua, income per person has shrunk, and there globalization is seen as a culprit, not a cure.

Globalization in the economic sphere is proceeding under the auspices of the World Trade Organization (WTO), of which the United States is the leading architect. To join, countries must agree to open their economies to foreign trade and investment and to adhere to a set of economic rules discussed annually at ministerial meetings. By 2004, the WTO had approximately 150 member-states, all expecting benefits from their participation. But the driving forces of globalization, including European countries and Japan, did not always do their part when it came to giving the poorer countries the chances the WTO seemed to promise. The case of the Philippines is often cited: joining the WTO, its government assumed, would open world markets to Filipino farm products, which can be brought to those markets cheaply because farm workers' wages in the Philippines are very low. Economic planners forecast a huge increase in farm jobs, a slow rise in wages, and a major expansion in produce exports. But instead they found themselves competing against American and European farmers who receive subsidies toward the production as well as the export of their products—and losing out. Meanwhile, low-priced, subsidized American corn appeared on Filipino markets. As a result, the

Philippine economy lost several hundred thousand farm jobs, wages went down, and WTO membership had the effect of severely damaging its agricultural sector. Not surprisingly, the notion of globalization is not popular among rural Filipinos.

This story has parallels among banana and sugar growers in the Caribbean, cotton and sugar producers in Africa, and fruit and vegetable farmers in Mexico. Economic globalization has not been the two-way street most WTO members, especially the poorer countries, had hoped. Nor is rich-country protectionism confined to agriculture. When cheap foreign steel began to threaten what remains of the U.S. steel industry in 2001, the American government erected tariffs to guard domestic producers against unwanted competition.

When meetings of economic ministers take place in major cities, they attract thousands of vocal and sometimes violent opponents who demonstrate and occasionally riot to express their opposition. This happens because globalization is seen as more than an economic strategy to perpetuate the advantages of rich countries: it is also viewed as a cultural threat. Globalization constitutes Americanization, these protesters say, eroding local traditions, endangering moral standards, and menacing the social fabric. Such views are not confined to the disadvantaged, poorer countries still losing out in our globalizing world: in France the leader of a so-called Peasant Federation became a national hero after his sympathizers destroyed a McDonald's restaurant in what he described as an act of cultural self-defense. McDonald's restaurants, he argued, promote the consumption of junk food over better, more traditional fare. Elsewhere the target is violent American movies or the lyrics of certain American popular music.

Nevertheless, the process of globalization—in cultural as well as economic spheres—continues to change our world, for better and for worse. (In the United States, opponents of globalization bemoan the loss of American jobs to foreign countries where labor is cheaper.) In theory, globalization breaks down barriers to international trade, stimu-

lates commerce, brings jobs to remote places, and promotes social, cultural, political, and other kinds of exchanges. High-tech workers in India are employed by computer firms based in California. Japanese cars are assembled in Thailand. American shoes are made in China. McDonald's restaurants spread standards of service and hygiene as well as familiar (and standard) menus from Tokyo to Tel Aviv. If wages and standards of employment are lower in newly involved countries than in the global core, the gap is shrinking, employment conditions are more open to scrutiny, and global openness is a byproduct of globalization.

We should be aware that the current globalization process is a revolution—but also that it is not the first of its kind. The first "globalization revolution" occurred during the nineteenth and early twentieth centuries, when Europe's colonial expansion spread ideas, inventions, products, and habits around the world. Colonialism transformed the world as the European powers built cities, transport networks, dams, irrigation systems, power plants, and other facilities, often with devastating impact on local traditions, cultures, and economies. From goods to games (soap to soccer) people in much of the world started doing similar things. The largest of all colonial empires, that of Britain, made English a worldwide language, a key element in the current, second globalization process.

The current globalization is even more revolutionary than the colonial phase because it is driven by more modern, higher-speed communications. When the British colonists planned the construction of their ornate Victorian government and public buildings in (then) Bombay, the architectural drawings had to be prepared in London and sent by boat to India. When the Chinese government in the 1980s decided to create a Manhattan-like commercial district on the riverfront in Shanghai, the plans were drawn in the United States, Japan, and Western Europe and transmitted to Shanghai via the Internet. One container ship carrying products from China to the American market hauls more cargo than a hundred colonial-era boats.

And, as the pages that follow will show frequently, the world's national political boundaries are becoming increasingly porous. Economic alliances enable manufacturers to send raw materials and finished products across borders that once inhibited such exchanges. Groups of countries forge economic unions whose acronyms (NAFTA, Mercosur) stand for freer trade. The ultimate goal of the World Trade Organization is to lower the remaining trade barriers the world over, boosting not just regional commerce but also global trade.

As with all revolutions, the overall consequences of the ongoing globalization process are uncertain. Critics underscore that one of its outcomes is a growing gap between rich and poor, a polarization of wealth that will destabilize the world. Core-periphery contrasts are intensified, not lessened, by globalization as the poor in peripheral societies are exploited by core-based corporations. Proponents argue that, as with the Industrial Revolution, it will take time for the benefits to spread—but that globalization's ultimate effects will be advantageous to all.

Indeed, the world is functionally shrinking, and we will find evidence for this throughout the book. But the "global village" still retains its distinctive neighborhoods, and two revolutionary globalizations have failed to erase their particular properties. In the chapters that follow we use the vehicle of geography to visit and investigate them.

REGIONAL FRAMEWORK AND GEOGRAPHIC PERSPECTIVE

At the beginning of this chapter, we outlined a map of the great geographic realms of the world (Fig. G-2). We then addressed the task of dividing these realms into regions, and we used criteria ranging from physical geography to economic geography. The result is Figure G-11.

On this map, note that we display not only the great geographic realms but also the regions into which they subdivide. The numbers in the legend reveal the order in which the realms and regions are discussed, starting with Europe (1) and ending with the Pacific Realm (12).

As this introductory chapter demonstrates, our world regional survey is no mere description of places and areas. We have combined the study of realms and regions with a look at geography's ideas and concepts—the notions, generalizations, and basic theories that make the discipline what it is. We continue this method in the chapters ahead so that we will become better acquainted with the world and with geography. By now you are aware that geography is a wide-ranging, multifaceted dis-

cipline. It is often described as a social science, but that is only half the story: in fact, geography straddles the divide between the social and the physical (natural) sciences. Many of the ideas and concepts you will encounter have to do with the multiple interactions between human societies and natural environments.

49 **Regional geography** allows us to view the world in an all-encompassing way. As we have seen, regional geography borrows information from many sources to create an overall image of our divided world. Those sources are not random. They form **50** topical or **systematic geography**. Research in

the systematic fields of geography makes our world-scale generalizations possible. As Figure G-12 shows, these systematic fields relate closely to those of other disciplines. Cultural geography, for example, is allied with anthropology; it is the spatial perspective that distinguishes cultural geography. Economic geography focuses on the spatial dimensions of economic activity; political geography concentrates on the spatial imprints of political behavior. Other systematic fields include historical, medical, behavioral, environmental, and coastal geography. We will also draw on information from biogeography, marine geography,

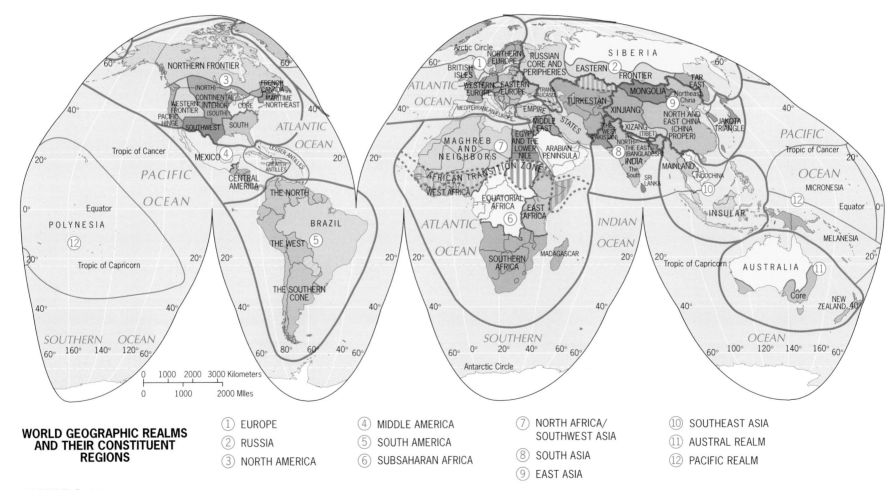

WORLD GEOGRAPHIC REALMS AND THEIR CONSTITUENT REGIONS

① EUROPE
② RUSSIA
③ NORTH AMERICA

④ MIDDLE AMERICA
⑤ SOUTH AMERICA
⑥ SUBSAHARAN AFRICA

⑦ NORTH AFRICA/ SOUTHWEST ASIA
⑧ SOUTH ASIA
⑨ EAST ASIA

⑩ SOUTHEAST ASIA
⑪ AUSTRAL REALM
⑫ PACIFIC REALM

FIGURE G-11

THE RELATIONSHIP BETWEEN REGIONAL AND SYSTEMIC GEOGRAPHY

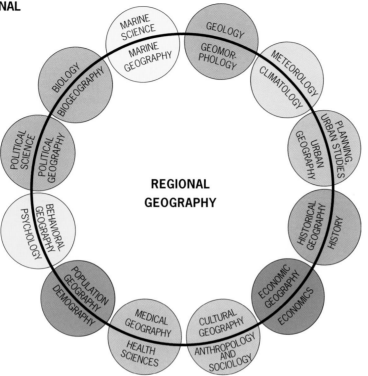

REGIONAL GEOGRAPHY

MARINE SCIENCE — MARINE GEOGRAPHY
GEOLOGY
GEOMOR-PHOLOGY
METEOROLOGY CLIMATOLOGY
PLANNING, URBAN STUDIES — URBAN GEOGRAPHY
HISTORICAL GEOGRAPHY — HISTORY
ECONOMIC GEOGRAPHY — ECONOMICS
CULTURAL GEOGRAPHY — ANTHROPOLOGY AND SOCIOLOGY
MEDICAL GEOGRAPHY — HEALTH SCIENCES
POPULATION GEOGRAPHY — DEMOGRAPHY
BEHAVIORAL GEOGRAPHY — PSYCHOLOGY
POLITICAL SCIENCE — POLITICAL GEOGRAPHY
BIOLOGY — BIOGEOGRAPHY

FIGURE G-12

population geography, and climatology (as we did earlier in this chapter).

These systematic fields of geography are so named because their approach is global, not regional. Take the geographic study of cities, urban geography. Urbanization is a worldwide process, and urban geographers can identify certain human activities that all cities in the world exhibit in one form or another. But cities also display regional properties. The model Japanese city is quite distinct from, say, the African city. Regional geography, therefore, borrows from the systematic field of urban geography, but it injects this regional perspective.

In the following chapters we call upon these systematic fields to give us a better understanding of the world's realms and regions. As a result, you will gain insights into the discipline of geography as well as the regions we investigate. This will prove that geography is a relevant and practical discipline when it comes to comprehending, and coping with, our fast-changing world.

DISCOVERY TOOLS www.wiley.com/college/deblij

The *Concepts and Regions* Website, featuring *GeoDiscoveries*, offers many additional resources to enhance your understanding and experience of this chapter. Be sure to explore the following:

Photo Galleries

South Island, New Zealand
Mauna Kea, Hawai'i
Central Valley, California
Kashgar, China
Bergen, Norway
Mengkabong, Malaysia
Livingstonia, Malawi

Kyongju, South Korea
Vancouver, Canada
Cochin, India
Myanmar
Jawa, Indonesia
Rio de Janeiro, Brazil
Singapore
Seoul, South Korea
Tasman Sea, New Zealand

More To Explore

Systematic Essay: Core-Periphery Regional Relationships

Issues in Geography: Core-Periphery Regional Relationships; What Do Geographers Do?

Table G-1
AREA AND DEMOGRAPHIC DATA FOR THE WORLD'S STATES

	Land Area (sq mi)	Population 2004 (Millions)	Population 2025 (Millions)	Population Density Arithmetic	Population Density Physiologic	Birth Rate	Death Rate	Natural Increase %	Doubling Time (years)	Infant Mortality Per 1000 (births)	Life Expectancy Male (years)	Life Expectancy Female (years)	Percent Urban Pop	Per Capita GNP ($US)
World	**51,789,601**	**6,365.1**	**7,859.0**	**123**	**1,286**	**21**	**9**	**1.2**	**58**	**54**	**65**	**69**	**47**	**$7,140**
Europe	**2,284,860**	**583.2**	**588.7**	**255**	**901**	**10**	**11**	**−0.1**		**8**	**70**	**78**	**73**	**$16,150**
Albania	11,100	3.2	4.1	286	1,426	17	5	1.2	58	12	72	76	46	$3,600
Austria	32,378	8.2	8.4	253	1,512	9	9	0.0		5	75	81	54	$26,330
Belarus	80,154	9.8	9.4	122	422	9	14	−0.5		9	63	75	70	$7,550
Belgium	11,787	10.3	10.8	876	3,644	11	10	0.1	700	5	75	82	97	$27,470
Bosnia	19,741	3.4	3.6	174	1,243	12	8	0.4	175	7	70	77	54	$6,050
Bulgaria	42,822	7.7	6.6	180	421	9	14	−0.5		13	68	75	69	$5,560
Croatia	21,830	4.3	4.1	196	944	10	12	−0.2		7	70	77	54	$7,960
Cyprus	3,571	0.9	1.0	255	2,104	12	7	0.5	140	5	75	80	66	$20,780
Czech Republic	30,448	10.3	10.3	337	840	9	11	−0.2		4	72	78	77	$13,780
Denmark	16,637	5.4	5.9	325	550	12	11	0.1	700	5	75	79	85	$27,250
Estonia	17,413	1.4	1.2	80	340	9	14	−0.5		9	65	76	69	$9,340
Finland	130,560	5.2	5.3	40	554	11	10	0.1	700	4	74	81	61	$24,570
France	212,934	60.0	64.2	282	856	13	9	0.4	175	5	76	83	74	$24,420
Germany	137,830	82.4	78.1	598	1,851	9	10	−0.1		4	75	81	86	$24,920
Greece	50,950	11.0	10.4	216	1,163	10	10	0.0		6	76	81	59	$16,860
Hungary	35,919	10.0	9.2	280	551	10	13	−0.3		9	67	76	64	$11,990
Iceland	39,768	0.3	0.3	8		15	6	0.9	78	3	78	81	93	$28,710
Ireland	27,135	3.8	4.5	142	1,112	14	8	0.6	117	6	74	79	58	$25,520
Italy	116,320	58.1	57.5	499	1,651	9	9	0.0		5	77	83	90	$23,470
Latvia	24,942	2.3	2.2	91	351	8	14	−0.6		11	65	76	68	$7,070
Liechtenstein	62	0.1	0.04	489		12	7	0.5	140	8			23	
Lithuania	25,174	3.5	3.5	138	398	9	12	−0.3		8	68	78	67	$6,980
Luxembourg	999	0.5	0.6	505	2,100	13	9	0.4	175	5	75	81	88	$45,470
Macedonia	9,927	2.0	2.2	204	861	15	9	0.6	117	12	70	75	59	$5,020
Malta	124	0.4	0.4	3,245		11	8	0.3	233	6	74	80	91	$16,530
Moldova	13,012	4.3	4.5	330	638	9	10	−0.1		18	64	71	46	$2,230
Netherlands	15,768	16.2	17.7	1,029	4,955	13	9	0.4	175	5	76	81	62	$25,850
Norway	125,050	4.5	5.0	36	1,273	13	10	0.3	233	4	76	81	74	$29,630
Poland	124,087	38.6	38.6	311	699	10	10	0.0		8	70	78	62	$9,000
Portugal	35,514	10.4	9.7	294	1,131	12	10	0.2	350	6	73	80	48	$16,990
Romania	92,042	22.3	20.6	242	612	10	12	−0.2		18	67	74	55	$6,360
Serbia-Montenegro	39,448	10.7	10.7	272	1,329	12	11	0.1	700	13	70	75	52	$5,950

	Land Area (sq mi)	Population 2004 (Millions)	Population 2025 (Millions)	Population Density Arithmetic	Population Density Physiologic	Birth Rate	Death Rate	Natural Increase %	Doubling Time (years)	Infant Mortality Per 1000 (births)	Life Expectancy Male (years)	Life Expectancy Female (years)	Percent Urban Pop	Per Capita GNP ($US)
Slovakia	18,923	5.4	5.2	285	937	10	10	0.0		9	69	77	57	$11,040
Slovenia	7,819	2.0	2.0	256	2,137	9	9	0.0		5	72	79	50	$17,310
Spain	195,363	41.4	44.3	212	715	10	9	0.1	700	5	76	83	64	$19,260
Sweden	173,730	8.9	9.1	51	799	10	11	−0.1		3	77	82	84	$23,970
Switzerland	15,942	7.3	7.6	460	4,790	10	8	0.2	350	5	77	83	68	$30,450
Ukraine	233,089	47.5	45.1	204	366	8	15	−0.7		12	62	74	67	$3,700
United Kingdom	94,548	60.3	64.8	638	2,586	11	10	0.1	700	6	75	80	90	$23,550
Russia	**6,592,819**	**143.8**	**130.2**	**22**	**274**	**9**	**16**	**−0.7**		**15**	**59**	**72**	**73**	**$8,010**
Armenia	11,506	3.8	3.7	332	2,059	8	6	0.2	350	16	70	74	67	$2,580
Azerbaijan	33,436	8.3	10.2	249	1,386	14	6	0.8	88	13	69	75	51	$2,740
Georgia	3,571	4.4	3.6	1,232	182	9	9	0.0		18	69	77	56	$2,680
North America	**7,699,508**	**322.2**	**382.0**	**42**	**358**	**14**	**9**	**0.5**	**140**	**6**	**74**	**80**	**75**	**$33,410**
Canada	3,849,670	31.6	36.0	8	164	11	7	0.4	175	5	76	81	78	$27,170
United States	3,717,796	290.9	346.0	78	412	15	9	0.6	117	7	74	80	75	$34,100
Middle America	**1,048,105**	**184.2**	**233.0**	**176**	**1,445**	**26**	**6**	**2.0**	**35**	**31**	**70**	**75**	**66**	
Antigua and Barbuda	170	0.1	0.1	607	2,867	22	6	1.6	44	17	68	73	37	$10,000
Bahamas	5,359	0.3	0.4	57	7,894	18	5	1.3	54	16	70	75	84	$16,400
Barbados	166	0.3	0.3	1,833	4,111	15	8	0.7	100	13	70	76	38	$15,020
Belize	8,865	0.3	0.4	35	1,784	29	6	2.3	30	21	70	74	49	$5,240
Costa Rica	19,730	4.0	5.2	204	3,413	21	4	1.7	41	11	75	79	45	$7,980
Cuba	42,803	11.4	11.8	267	1,111	12	7	0.5	140	6	74	78	75	
Dominica	290	0.1	0.1	350	3,763	16	8	0.8	88	16	71	76	71	
Dominican Rep.	18,815	9.2	12.1	488	2,336	26	5	2.1	33	47	67	71	61	$5,710
El Salvador	8,124	6.9	9.3	850	3,198	30	7	2.3	30	30	67	73	58	$4,410
Grenada	131	0.1	0.1	782	6,828	19	7	1.2	58	14			34	$6,960
Guadeloupe	660	0.5	0.5	774	5,215	17	6	1.1	64	8	73	80	48	
Guatemala	42,042	12.8	19.8	305	2,548	36	7	2.9	24	41	63	69	39	$3,770
Haiti	10,714	7.4	9.6	687	3,471	33	15	1.8	39	80	48	51	35	$1,470
Honduras	43,278	7.1	9.6	163	1,091	33	6	2.7	26	42	64	68	46	$2,400
Jamaica	4,243	2.7	3.3	631	4,555	20	5	1.5	47	24	73	77	50	$3,440
Martinique	425	0.4	0.4	956	12,701	14	6	0.8	88	7	76	82	93	
Mexico	756,052	106.0	131.7	140	1,199	26	5	2.1	33	25	73	78	74	$8,790
Netherlands Antilles	309	0.2	0.2	658	6,774	14	6	0.8	88	12	73	79	70	
Nicaragua	50,193	5.7	8.6	114	1,355	34	5	2.9	24	40	66	71	57	$2,080
Panama	29,158	3.0	3.8	103	1,499	23	4	1.9	37	17	72	77	62	$5,680

	Land Area (sq mi)	Population 2004 (Millions)	Population 2025 (Millions)	Population Density Arithmetic	Population Density Physiologic	Birth Rate	Death Rate	Natural Increase %	Doubling Time (years)	Infant Mortality Per 1000 (births)	Life Expectancy Male (years)	Life Expectancy Female (years)	Percent Urban Pop	Per Capita GNP ($US)
Puerto Rico	3,456	4.0	4.1	1,147	29,137	15	7	0.8	88	11	71	80	71	
Saint Lucia	239	0.2	0.2	857	12,802	18	6	1.2	58	13	69	74	30	$5,400
St. Vincent and the Grenadines	151	0.1	0.1	678	5,121	20	8	1.2	58	22	71	74	44	$5,210
Trinidad and Tobago	1,981	1.3	1.4	664	4,385	14	8	0.6	117	17	68	73	72	$8,220
South America	**6,898,579**	**365.4**	**463.0**	**53**	**1,049**	**22**	**6**	**1.6**	**44**	**29**	**67**	**74**	**79**	**$7,070**
Argentina	1,073,514	37.3	47.2	35	392	19	8	1.1	64	18	70	77	90	$12,050
Bolivia	424,162	9.2	13.2	22	1,100	32	9	2.3	30	61	61	64	64	$2,360
Brazil	3,300,154	178.0	219.0	54	1,090	19	7	1.2	58	33	65	73	81	$7,300
Chile	292,135	16.0	19.5	55	1,105	18	6	1.2	58	12	73	79	86	$9,100
Colombia	439,734	45.2	59.7	103	2,819	22	6	1.6	44	21	68	74	71	$6,060
Ecuador	109,483	13.6	18.5	124	2,117	28	6	2.2	32	30	68	73	61	$2,910
French Guiana	34,749	0.2	0.3	6	613	26	5	2.1	33	17	72	79	79	
Guyana	83,000	0.8	0.7	10	543	24	8	1.6	44	56	59	67	36	$3,670
Paraguay	157,046	6.3	10.1	40	686	31	5	2.6	27	37	69	73	54	$4,450
Peru	496,224	27.7	35.7	56	1,870	26	7	1.9	37	33	66	71	72	$4,660
Suriname	63,039	0.4	0.5	7	687	24	7	1.7	41	26	68	74	69	$3,480
Uruguay	68,498	3.4	3.8	50	728	16	10	0.6	117	14	71	79	92	$8,880
Venezuela	352,143	26.1	34.8	74	1,913	24	5	1.9	37	20	71	77	87	$5,740
Subsaharan Africa	**8,366,638**	**683.8**	**1,025.0**	**82**	**1,360**	**39**	**17**	**2.2**	**32**	**88**	**49**	**50**	**30**	**$1,540**
Angola	481,351	13.4	28.2	28	1,394	48	20	2.8	25	122	44	47	32	$1,180
Benin	43,483	7.0	12.0	161	1,259	41	12	2.9	24	85	53	56	39	$980
Botswana	224,606	1.6	1.2	7	372	31	22	0.9	78	60	39	40	49	$7,170
Burkina Faso	105,792	13.4	21.6	126	974	47	17	3.0	23	105	46	47	15	$970
Burundi	10,745	7.0	12.4	651	1,607	43	21	2.2	32	116	40	41	8	$580
Cameroon	183,568	17.0	24.7	93	3,157	37	12	2.5	28	77	54	56	48	$1,590
Cape Verde Is.	1,556	0.5	0.7	341	3,014	37	7	3.0	23	31	66	72	53	$4,760
Central African Republic	240,533	3.7	4.9	16	519	38	18	2.0	35	98	42	46	39	$1,160
Chad	495,753	9.6	18.2	19	659	49	16	3.3	21	103	49	53	21	$870
Comoros Is.	861	0.6	1.1	746	2,040	47	12	3.5	20	86	54	59	29	$1,590
Congo	132,046	3.4	6.3	26	2,574	44	14	3.0	23	72	49	53	41	$570
Congo, The	905,531	58.7	106.0	65	2,234	46	15	3.1	23	102	47	51	29	$680
Djibouti	8,958	0.7	0.8	81		39	19	2.0	35	117	42	44	83	$690
Equatorial Guinea	10,830	0.5	0.9	49	975	43	17	2.6	27	108	49	53	37	$5,600

	Land Area (sq mi)	Population 2004 (Millions)	Population 2025 (Millions)	Population Density Arithmetic	Population Density Physiologic	Birth Rate	Death Rate	Natural Increase %	Doubling Time (years)	Infant Mortality Per 1000 (births)	Life Expectancy Male (years)	Life Expectancy Female (years)	Percent Urban Pop	Per Capita GNP ($US)
Eritrea	45,045	4.8	8.3	106	1,022	43	12	3.1	23	77	53	58	16	$960
Ethiopia	426,371	71.1	117.6	167	1,535	40	15	2.5	28	97	51	53	15	$660
Gabon	103,347	1.2	1.4	12	1,245	32	16	1.6	44	57	49	51	73	$5,360
Gambia	4,363	1.6	2.7	364	2,262	42	13	2.9	24	82	51	55	37	$1,620
Ghana	92,100	21.1	26.5	229	2,000	32	10	2.2	32	56	56	59	37	$1,910
Guinea	94,927	8.9	14.1	93	4,668	45	18	2.7	26	119	47	48	26	$1,930
Guinea-Bissau	13,946	1.4	2.2	98	1,139	45	20	2.5	28	126	43	46	22	$710
Ivory Coast	124,502	16.8	25.6	135	1,710	36	16	2.0	35	95	44	47	46	$1,500
Kenya	224,081	32.4	33.3	144	2,103	34	14	2.0	35	74	47	49	20	$1,010
Lesotho	11,718	2.3	2.4	195	1,771	33	15	1.8	39	84	50	52	16	$2,590
Liberia	43,000	3.5	6.0	82	9,448	49	17	3.2	22	139	49	52	45	$3,950
Madagascar	226,656	17.6	30.8	78	1,958	34	14	2.0	35	74	47	49	20	$820
Malawi	45,745	11.4	12.8	250	926	46	22	2.4	29	104	37	38	20	$600
Mali	478,838	12.0	21.6	25	1,272	49	19	3.0	23	113	46	48	26	$780
Mauritania	395,954	2.7	5.1	7	683	34	14	2.0	35	74	53	55	55	$1,630
Mauritius	788	1.2	1.4	1,550	3,117	16	7	0.9	78	14	68	75	43	$9,940
Moçambique	309,494	20.4	20.6	66	1,684	43	23	2.0	35	135	38	37	28	$800
Namibia	318,259	1.9	2.0	6	583	35	20	1.5	47	72	44	41	27	$6,410
Niger	489,189	12.4	25.7	25	847	55	20	3.5	20	123	45	46	17	$740
Nigeria	356,668	129.0	198.1	371	1,981	41	17	2.4	29	75	52	52	36	$800
Réunion	969	0.7	0.9	744	4,242	20	5	1.5	47	8	70	79	73	
Rwanda	10,170	7.7	8.0	759	2,320	42	21	2.1	33	107	39	40	5	$930
São Tomé and Principe	371	0.2	0.3	577	35,708	43	8	3.5	20	50	64	67	44	
Senegal	75,954	10.4	16.5	137	1,169	38	12	2.6	27	68	52	55	43	$1,480
Seychelles	174	0.1	0.1	587	25,553	18	7	1.1	64	10	67	73	63	$7,350
Sierra Leone	27,699	5.9	10.6	212	3,028	49	25	2.4	29	153	38	40	37	$480
Somalia	246,201	8.3	14.9	34	1,705	48	19	2.9	24	126	45	48	28	
South Africa	471,444	44.5	35.1	94	943	25	15	1.0	70	45	50	52	54	$9,160
Swaziland	6,703	1.1	1.4	171	1,579	41	20	2.1	33	109	40	41	25	$4,600
Tanzania	364,900	39.2	59.8	108	3,834	40	13	2.7	26	99	51	53	22	$520
Togo	21,927	5.6	7.6	256	703	40	11	2.9	24	80	53	57	31	$1,410
Uganda	93,066	26.2	48.0	282	1,359	48	18	3.0	23	88	42	44	16	$1,210
Zambia	290,583	10.4	14.3	36	518	42	22	2.0	35	95	37	37	38	$750
Zimbabwe	105,873	12.5	10.3	118	1,197	29	20	0.9	78	65	39	36	32	$2,550
North Africa/ SW Asia	**7,459,064**	**530.6**	**730.0**	**71**	**978**	**27**	**8**	**1.9**	**37**	**48**	**65**	**68**	**53**	

	Land Area (sq mi)	Population 2004 (Millions)	Population 2025 (Millions)	Population Density Arithmetic	Population Density Physiologic	Birth Rate	Death Rate	Natural Increase %	Doubling Time (years)	Infant Mortality Per 1000 (births)	Life Expectancy Male (years)	Life Expectancy Female (years)	Percent Urban Pop	Per Capita GNP ($US)
Afghanistan	251,772	29.2	45.9	116	965	43	19	2.4	29	154	46	44	22	
Algeria	919,591	32.5	43.0	35	1,180	23	5	1.8	39	54	68	71	49	$5,040
Bahrain	266	0.7	1.7	2,733	242,284	22	3	1.9	37	9	73	75	87	$14,410
Egypt	386,660	74.1	96.1	192	9,638	27	7	2.0	35	44	65	68	43	$3,670
Iran	630,575	67.1	81.3	106	1,061	17	6	1.1	64	32	68	70	66	$5,910
Iraq	169,236	24.8	41.2	147	1,223	35	10	2.5	28	103	56	59	68	
Israel	8,131	6.8	9.3	836	5,000	21	6	1.5	47	5	76	81	91	$19,330
Jordan	34,444	5.5	8.7	161	4,043	28	5	2.3	30	31	69	71	79	$3,950
Kazakhstan	1,049,151	14.9	14.7	14	121	15	10	0.5	140	20	60	71	56	$5,490
Kuwait	6,880	2.4	3.9	354		32	3	2.9	24	9	74	78	100	$18,690
Kyrgyzstan	76,641	5.1	6.5	67	989	20	7	1.3	54	23	65	72	35	$2,540
Lebanon	4,015	4.4	5.4	1,101	5,263	21	7	1.4	50	33	72	75	88	$4,550
Libya	679,359	5.7	8.3	8	833	28	4	2.4	29	30	73	77	86	
Morocco	172,413	30.8	40.5	179	852	25	6	1.9	37	50	67	71	55	$3,450
Oman	82,031	2.8	5.1	34		33	4	2.9	24	17	72	75	72	
Palestinian Territories	2,417	3.8	7.4	1,554		40	4	3.6	19	26	71	74	57	
Qatar	4,247	0.6	0.8	149	14,717	31	4	2.7	26	12	70	75	91	
Saudi Arabia	829,996	25.4	40.9	31	1,531	35	6	2.9	24	19	71	73	83	$11,390
Sudan	967,494	34.2	49.6	35	745	36	12	2.4	29	82	55	57	27	$1,520
Syria	71,498	18.1	26.5	253	909	31	6	2.5	28	24	70	70	50	$3,340
Tajikistan	55,251	6.5	7.8	117	1,992	19	4	1.5	47	19	66	71	27	$1,090
Tunisia	63,170	10.0	11.6	159	879	17	6	1.1	64	26	70	74	63	$6,070
Turkey	299,158	69.3	85.0	232	729	22	7	1.5	47	35	67	72	66	$7,030
Turkmenistan	188,456	5.8	7.2	31	1,058	19	5	1.4	50	25	63	70	44	$3,800
United Arab Emirates	32,278	3.6	4.5	112		17	2	1.5	47	19	72	77	78	$19,410
Uzbekistan	172,741	26.3	37.2	152	1,826	22	5	1.7	41	20	68	73	38	$2,360
Western Sahara	97,344	0.3	0.4	3		46	17	2.9	24	140	46	48	95	
Yemen	203,849	19.8	39.6	97	3,245	44	11	3.3	21	75	57	61	26	$770
South Asia	**1,732,734**	**1,419.5**	**1,843.0**	**819**	**1,837**	**27**	**9**	**1.8**	**39**	**69**	**62**	**63**	**28**	**$2,213**
Bangladesh	55,598	139.5	177.8	2,510	3,800	30	8	2.2	32	66	59	59	23	$1,590
Bhutan	18,147	0.9	1.4	52	2,598	34	9	2.5	28	61	66	66	16	$1,440
India	1,269,340	1,085.5	1,363.0	855	1,688	26	9	1.7	41	68	62	64	28	$2,340
Maldives	116	0.3	0.5	2,685	31,151	23	4	1.9	37	37	67	66	27	$4,240
Nepal	56,826	24.9	36.1	438	2,770	31	11	2.0	35	64	58	57	11	$1,370

	Land Area (sq mi)	Population 2004 (Millions)	Population 2025 (Millions)	Population Density Arithmetic	Population Density Physiologic	Birth Rate	Death Rate	Natural Increase %	Doubling Time (years)	Infant Mortality Per 1000 (births)	Life Expectancy Male (years)	Life Expectancy Female (years)	Percent Urban Pop	Per Capita GNP ($US)
Pakistan	307,375	149.6	242.1	487	1,862	30	9	2.1	33	86	63	63	33	$1,860
Sri Lanka	25,332	18.7	22.1	740	5,355	18	6	1.2	58	17	70	74	30	$3,460
East Asia	**4,546,050**	**1,530.2**	**1,690.0**	**337**	**3,735**	**13**	**7**	**0.6**	**117**	**31**	**70**	**74**	**44**	**$6,280**
China	3,696,521	1,306.0	1,463.7	353	3,524	13	6	0.7	100	31	70	73	38	$3,920
Japan	145,869	127.4	121.1	873	7,965	8	8	0.0		3	78	85	78	$27,080
Korea, North	46,541	23.6	26.4	506	3,621	18	10	0.8	88	42	62	67	59	
Korea, South	38,324	49.2	50.5	1,283	6,793	13	5	0.8	88	8	72	80	79	$17,300
Mongolia	604,826	2.5	3.3	4	409	23	8	1.5	47	37	61	65	57	$1,760
Taiwan	13,969	22.7	25.3	1,627	6,764	11	6	0.5	140	6	73	78	77	$15,100
Southeast Asia	**1,735,448**	**552.2**	**706.0**	**318**	**2,377**	**22**	**7**	**1.5**	**47**	**41**	**65**	**70**	**36**	**$3,450**
Brunei	2,228	0.4	0.5	186	20,767	22	3	1.9	37	15	71	76	67	$24,630
Cambodia	69,900	12.7	18.4	182	1,439	28	11	1.7	41	95	54	58	16	$1,440
East Timor	5,741	0.8	1.2	143		29	15	1.4	50	135	47	48	8	$480
Indonesia	735,355	224.0	281.9	305	3,176	22	6	1.6	44	46	66	70	39	$2,830
Laos	91,429	5.8	8.6	63	2,153	36	13	2.3	30	104	52	55	17	$1,540
Malaysia	127,317	25.3	35.6	199	6,655	23	4	1.9	37	8	70	75	57	$8,330
Myanmar/Burma	261,228	50.3	60.2	192	1,393	25	12	1.3	54	90	54	59	27	$960
Philippines	115,830	83.9	129.9	724	3,836	29	5	2.4	29	26	65	71	47	$4,220
Singapore	239	4.3	8.0	17,856	1,066,867	12	4	0.8	88	2	76	80	100	$24,910
Thailand	198,116	63.6	72.1	321	948	14	6	0.8	88	20	70	75	31	$6,320
Vietnam	128,066	81.9	104.1	640	3,835	19	5	1.4	50	30	67	70	24	$2,000
Austral Realm	**3,093,340**	**23.9**	**27.8**	**8**		**13**	**7**	**0.6**	**114**	**5**	**77**	**82**	**84**	**$23,894**
Australia	2,988,888	19.9	23.2	7	106	13	7	0.6	117	5	77	82	85	$24,970
New Zealand	104,452	4.0	4.6	38	22	14	7	0.7	100	5	76	81	77	$18,530
Pacific Realm	**213,401**	**8.5**	**12.2**	**40**	**6,928**	**8**	**1**	**0.6**	**110**	**10**	**14**	**14**	**24**	**$628**
Federated States of Micronesia	270	0.1	0.2	389		31	6	2.5	28	45	65	67	27	$3,050
Fiji	7,054	0.9	1.0	132	1,316	25	6	1.9	37	20	65	69	46	$4,480
French Polynesia	1,544	0.2	0.3	134	14,747	21	5	1.6	44	8	69	74	53	
Guam	212	0.2	0.2	982	9,458	24	4	2.0	35	9	75	80	38	$9,010
Marshall Islands	69	0.1	0.1	1,559		42	5	3.7	19	37	66	69	65	$3,080
New Caledonia	7,174	0.2	0.3	29	2,902	21	6	1.5	47	7	70	76	71	$18,870
Papua New Guinea	178,703	5.2	8.0	29		34	11	2.3	30	77	56	58	15	$2,180
Samoa	1,097	0.2	0.2	191	1,003	30	6	2.4	29	25	65	72	21	$5,050
Solomon Islands	11,158	0.5	0.9	48	4,950	41	7	3.4	21	25	67	68	13	$1,710
Vanuatu	4,707	0.2	0.4	45	2,257	36	6	3.0	23	45	67	66	21	$2,960

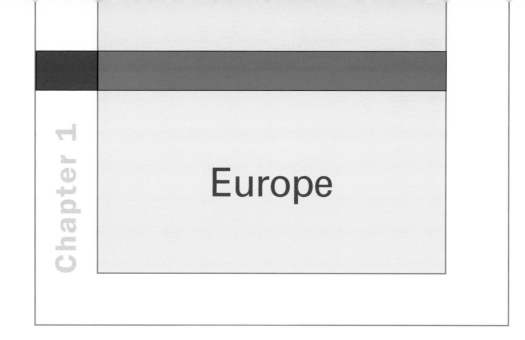

Chapter 1

Europe

CONCEPTS, IDEAS, AND TERMS

1. Land hemisphere
2. Infrastructure
3. Local functional specialization
4. *The Isolated State*
5. Model
6. Industrial Revolution
7. Nation-state
8. Nation
9. Centrifugal forces
10. Centripetal forces
11. Indo-European languages
12. Complementarity
13. Transferability
14. Intervening opportunity
15. Primate city
16. Metropolis
17. Supranationalism
18. Devolution
19. Four Motors of Europe
20. Regional state
21. Site
22. Situation
23. Conurbation
24. Landlocked location
25. Break-of-bulk
26. *Entrepôt*
27. Shatter belt
28. Balkanization
29. Exclave
30. Irredentism

REGIONS

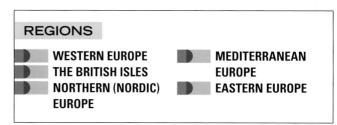

- WESTERN EUROPE
- THE BRITISH ISLES
- NORTHERN (NORDIC) EUROPE
- MEDITERRANEAN EUROPE
- EASTERN EUROPE

*I*T IS APPROPRIATE to begin our investigation of the world's geographic realms in Europe because over the past five centuries Europe and Europeans have influenced and changed the rest of the world more than any other realm or people has done. European empires spanned the globe and transformed societies far and near. European colonialism propelled the first wave of globalization. Millions of Europeans migrated from their homelands to the Old World as well as the New, changing (and sometimes nearly obliterating) traditional communities and creating new societies from Australia to North America. Colonial power and economic incentive combined to impel the movement of millions of imperial subjects from their ancestral homes to distant lands: Africans to the Americas, Indians to Africa, Chinese to Southeast Asia, Malays to South Africa's Cape, Native Americans from east to west. In agriculture, industry, politics, and other spheres, Europe generated revolutions—and then exported those revolutions across the world, thereby consolidating the European advantage.

But throughout much of that 500-year period of European hegemony, Europe also was a cauldron of conflict. Religious, territorial, and political disputes precipitated bitter wars that even spilled over into the colonies. And during the twentieth century, Europe

STUDY TOOLS

3D GLOBE
AREA AND DEMOGRAPHIC DATA

FLASHCARDS
MAP QUIZZES

AUDIO PRONUNCIATION GUIDE
CHAPTER QUIZ

ANNOTATED WEBLINKS
LONELY PLANET WEBLINKS

MAJOR GEOGRAPHIC QUALITIES OF EUROPE

1. The European realm lies on the western extremity of the Eurasian landmass, a locale of maximum efficiency for contact with the rest of the world.

2. Europe's lingering and resurgent world influence results largely from advantages accrued over centuries of global political and economic domination.

3. The European natural environment displays a wide range of topographic, climatic, and soil conditions and is endowed with many industrial resources.

4. Europe is marked by strong internal regional differentiation (cultural as well as physical), exhibits a high degree of functional specialization, and provides multiple exchange opportunities.

5. European economies are dominated by manufacturing, and the level of productivity has been high; levels of development generally decline from west to east.

6. Europe's nation-states emerged from durable power cores that formed the headquarters of world colonial empires. A number of those states are now plagued by internal separatist movements.

7. Europe's rapidly aging population is generally well off, highly urbanized, well educated, and enjoys long life expectancies.

8. A growing number of European countries are experiencing population declines; in many of these countries the natural decrease is partially offset by immigration.

9. Europe has made significant progress toward international economic integration and, to a lesser extent, political coordination.

---

twice plunged the world into war. The terrible, unprecedented toll of World War I (1914–1918) was not enough to stave off World War II (1939–1945), which ended with the first-ever use of nuclear weapons in Japan. In the aftermath of that war, Europe's weakened powers lost most of their colonial possessions and a new rivalry emerged: an ideological Cold War between the communist Soviet Union and the capitalist United States. This Cold War lowered an Iron Curtain across the heart of Europe, leaving most of the east under Soviet control and most of the west in the American camp. Western Europe proved resilient, overcoming the destruction of war and the loss of colonial power to regain economic strength. Meanwhile the Soviet communist experiment failed at home and abroad, and in 1990 the last vestiges of the Iron Curtain were lifted. Since then, a massive effort has been underway to reintegrate and reunify Europe from the Atlantic coast to the Russian border, the key geographic story of this chapter.

DEFINING THE REALM

As Figure 1-1 shows, Europe is a realm of peninsulas and islands on the western margin of the world's largest landmass, Eurasia. It is a realm of 583 million people and 39 countries, but it is territorially quite small. Yet despite its modest proportions it has had—and continues to have—a major impact on world affairs. For many centuries Europe has been a hearth of achievement, innovation, and invention.

The European realm is bounded on the west, north, and south by Atlantic, Arctic, and Mediterranean waters, respectively. But where is Europe's eastern limit? Some scholars place it at the Ural Mountains, deep inside Russia, thereby recognizing a "European" Russia and, presumably, an "Asian" one as well. Our regional definition places Europe's eastern boundary between Russia and its numerous European neighbors to the west. This definition is based on several geographic factors including

FIGURE 1-1

European-Russian contrasts in territorial dimensions, population size, cultural properties, and historic aspects, all discernible on the maps in Chapters 1 and 2.

Europe's peoples have benefited from a large and varied store of raw materials. Whenever the opportunity or need arose, the realm proved to contain what was required. Early on, these requirements included cultivable soils, rich fishing waters, and wild animals that could be domesticated; in addition, extensive forests provided wood for houses and boats. Later, mineral fuels and ores propelled industrialization.

From the balmy shores of the Mediterranean Sea to the icy peaks of the Alps, and from the moist woodlands and moors of the Atlantic fringe to the semiarid prairies north of the Black Sea, Europe presents an almost infinite range of natural environments (Fig. 1-2). Compare Western Europe and eastern North America on Figure G-8 (p. 12) and you will see the moderating influence of the warm ocean current known as the North Atlantic Drift and its onshore windflow.

The European realm is home to peoples of numerous cultural-linguistic stocks, including not only Latins, Germanics, and Slavs but also minorities such as Finns, Magyars (Hungarians), Basques, and Celts. This diversity of ancestries continues to be an asset as well as a liability. It has generated not only interaction and exchange, but also conflict and war.

Europe also has outstanding locational advantages. Its *relative location*, **1** at the heart of the **land hemisphere**, creates maximum efficiency

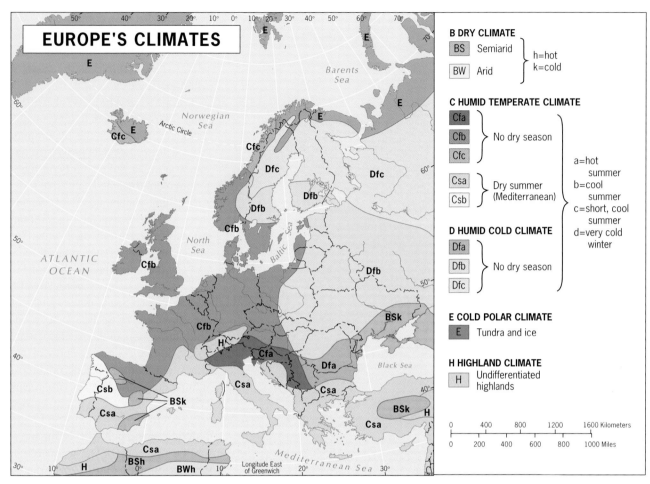

FIGURE 1-2

Mountains, the Western Uplands, and the North European Lowland (Fig. 1-4).

The *Central Uplands* form the heart of Europe. It is a region of hills and low plateaus loaded with raw materials whose farm villages grew into towns and cities when the Industrial Revolution transformed this realm.

The *Alpine Mountains*, a highland region named after the Alps, extend from the Pyrenees on the French-Spanish border to the Balkan Mountains near the Black Sea, and include Italy's Appennines and Eastern Europe's Carpathians.

The *Western Uplands*, geologically older, lower, and more stable than the Alpine Mountains, extend from Scandinavia through western Britain and Ireland to the heart of the Iberian Peninsula in Spain.

The *North European Lowland* extends in a lengthy arc from southeast Britain and central France across Germany and Denmark into Poland and Ukraine, from where it continues well into Russia. Also known as the Great European Plain, this has been an avenue for human migration time after time, so that complex cultural and economic mosaics developed here together with a jigsaw-like political map. As Figure 1-4 shows, many of Europe's major rivers and connecting waterways serve this populous region, where many of Europe's leading cities (London, Paris, Amsterdam, Copenhagen, Berlin, Warsaw) are located.

for contact with much of the rest of the world (Fig. 1-3). A "peninsula of peninsulas," Europe is nowhere far from the ocean and its avenues of seaborne trade and conquest. Hundreds of miles of navigable rivers, augmented by an unmatched system of canals, open the interior of Europe to its neighboring seas and to the waterways of the world.

Also consider the scale of the maps of Europe in this chapter. Europe is a realm of moderate distances and close proximities. Short distances and large cultural differences make for intense interac-tion, the constant circulation of goods and ideas. That has been the hallmark of Europe's geography for over a millennium.

LANDSCAPES AND OPPORTUNITIES

Europe's area may be small, but its landscapes are varied and complex. Regionally, we identify four broad units: the Central Uplands, the southern Alpine

HISTORICAL GEOGRAPHY

Modern Europe was peopled in the wake of the Pleistocene's most recent glacial retreat and global warming—a gradual warming that caused tundra

RELATIVE LOCATION: EUROPE IN THE LAND HEMISPHERE

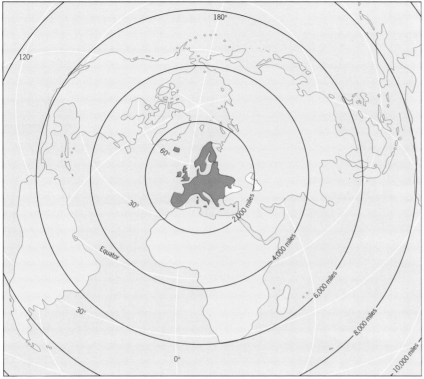

Azimuthal equidistant projection centered on Hamburg, Germany

FIGURE 1-3

to give way to deciduous forest and ice-filled valleys to turn into grassy vales. On Mediterranean shores, Europe witnessed the rise of its first great civilizations, on the islands and peninsulas of Greece and later in what is today Italy.

Ancient Greece lay exposed to influences radiating from the advanced civilizations of Mesopotamia and the Nile Valley, and in their fragmented habitat the Greeks laid the foundations of European civilization. Their achievements in political science, philosophy, the arts, and other spheres have endured for 25 centuries. But the ancient Greeks never managed to unify their domain, and their persistent conflicts proved fatal when the Romans challenged

them from the west. By 147 BC the last of the sovereign Greek intercity leagues (alliances) had fallen to the Roman conquerors.

The center of civilization and power now shifted to Rome in present-day Italy. Borrowing from Greek culture, the Romans created an empire that stretched from Britain to the Persian Gulf and from the Black Sea to Egypt; they made the Mediterranean Sea a Roman lake carrying armies to distant shores and goods to imperial Rome. With an urban population that probably exceeded 1 million, Rome was the first metropolitan-scale urban center in Europe.

The Romans founded numerous other cities throughout their empire and linked them to the

capital through a vast system of highway and water routes, facilitating political control and enabling economic growth in their provinces. It was an **2** unparalleled **infrastructure**, much of which long outlasted the empire itself.

Roman rule brought disparate, isolated peoples into the imperial political and economic sphere. By guiding (and often forcing) these groups to produce particular goods or materials, they launched Europe down a road for which it would become famous: **3** **local functional specialization**. The workers on Elba, a Mediterranean island, mined iron ore. Those near Cartagena in Spain produced silver and lead. Certain farmers were taught irrigation to produce specialty crops. Others raised livestock for meat or wool. The *production of particular goods by particular people in particular places* became and remained a hallmark of the realm.

The Romans also spread their language across the empire, setting the stage for the emergence of the *Romance* languages; they disseminated Christianity; and they established durable systems of education, administration, and commerce. But when their empire collapsed in the fifth century, disorder ensued, and massive migrations soon brought Germanic and Slavic peoples to their present positions on the European stage. Capitalizing on Europe's weakness, the Arab-Berber Moors from North Africa, energized by Islam, conquered most of Iberia and penetrated France. Later the Ottoman Turks invaded Eastern Europe and reached the outskirts of Vienna.

Europe's revival—its *Renaissance*—did not begin until the fifteenth century. After a thousand years of feudal turmoil marking the "Dark" and "Middle" Ages, powerful monarchies began to lay the foundations of modern states. The discovery of continents and riches across the oceans opened a new era of *mercantilism*, the competitive accumulation of wealth chiefly in the form of gold and silver. Best placed for this competition were the kingdoms of Western Europe. Europe was on its way to colonial expansion and world domination.

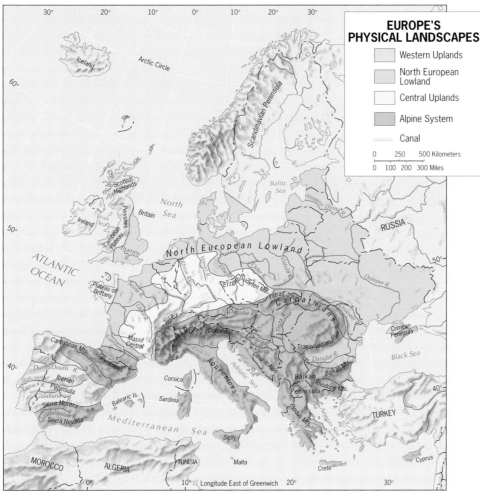

EUROPE'S PHYSICAL LANDSCAPES

- Western Uplands
- North European Lowland
- Central Uplands
- Alpine System
- Canal

0   250   500 Kilometers
0   100 200 300 Miles

FIGURE 1-4

THE REVOLUTIONS OF MODERNIZING EUROPE

Even as Europe's rising powers reached for world domination overseas, they fought with each other in Europe itself. Powerful monarchies and land-owning ("landed") aristocracies had their status and privilege challenged by ever-wealthier merchants and businesspeople. Demands for political recognition grew; cities mushroomed with the development of industries; the markets for farm products burgeoned; and Europe's population, more or less stable at about 100 million since the sixteenth century, began to increase.

The Agrarian Revolution

We know Europe as the focus of the Industrial Revolution, but before this momentous development occurred another revolution was already in progress: the *agrarian revolution*. Port cities and capital cities thrived and expanded, and their growing markets created economic opportunities for farmers. This led to revolutionary changes in land ownership and agricultural methods. Improved farm practices, better equipment, superior storage facilities, and more efficient transport to the urban markets marked a revolution in the countryside. The colonial merchants brought back new crops (the American potato soon became a European staple), and market prices rose, drawing more and more farmers into the economy.

The transformation of Europe's farmlands reshaped its economic geography, producing new patterns of land use and market links. The economic geographer Johann Heinrich von Thünen (1783–1850), himself an estate farmer who had studied these changes for several decades, published his observations in 1826 in a pioneering work entitled **4** ***The Isolated State***, chronicling the geography of Europe's agricultural transformation.

Von Thünen used information from his own **5** farmstead to build what today we call a **model** (an idealized representation of reality that demonstrates its most important properties) of the location of productive activities in Europe's farmlands.

Since a model is an abstraction that must always involve assumptions, von Thünen postulated a self-contained area (hence the "isolation") with a single market center, flat and uninterrupted land without impediments to cultivation or transportation. In such a situation, transport costs would be directly proportional to distance.

Von Thünen's model revealed four zones or rings of land use encircling the market center (Fig. 1-5). Innermost and directly adjacent to the market would lie a zone of intensive farming and dairying, yielding the most perishable and highest-priced products. Immediately beyond lay a zone of forest used for timber and firewood (still a priority in von Thünen's time). Next there would be a ring of field crops, for example, grains or potatoes. A fourth zone would contain pastures and livestock. Beyond lay wildnerness, from where the costs of transport to market would become prohibitive.

In many ways, von Thünen's model was the first analysis in a field that would eventually become

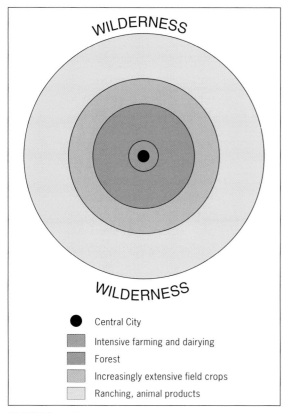

FIGURE 1-5

known as *location theory*. Von Thünen knew, of course, that the real Europe did not present the conditions postulated in his model. But it did demonstrate the economic-geographic forces that shaped the new Europe, which is why it is still being discussed today. More than a century after the publication of *The Isolated State*, geographers Samuel van Valkenburg and Colbert Held produced a map of twentieth-century European agricultural intensity, revealing a striking, ring-like concentricity focused on the vast urbanized area lining the North Sea—now the dominant market for a realmwide, macroscale "Thünian" agricultural system (Fig. 1-6).

The Industrial Revolution

6 The term **Industrial Revolution** suggests that an agrarian Europe was suddenly swept up in whole-sale industrialization that changed the realm in a few decades. In reality, seventeenth- and eighteenth-century Europe had been industrializing in many spheres, long before the chain of events known as the Industrial Revolution began. From the textiles of England and Flanders to the iron farm implements of Saxony (in present-day Germany), from Scandinavian furniture to French linens, Europe had already entered a new era of local functional specialization. It would therefore be more appropriate to call what happened next the period of Europe's *industrial intensification*.

In the 1780s, the Scotsman James Watt and others devised a steam-driven engine, which was soon adopted for numerous industrial uses. At about the same time, coal (converted into carbon-rich coke) was recognized as a vastly superior substitute for charcoal in smelting iron. These momentous innovations had a rapid effect. The power loom revolutionized the weaving industry. Iron smelters, long dependent on Europe's dwindling forests for fuel, could now be concentrated near coalfields. Engines could move locomotives as well as power looms. Ocean shipping entered a new age.

Britain had an enormous advantage, for the Industrial Revolution occurred when British influence reigned worldwide and the significant innovations were achieved in Britain itself. The British controlled the flow of raw materials, they held a monopoly over products that were in global demand, and they alone possessed the skills necessary to make the machines that manufactured the

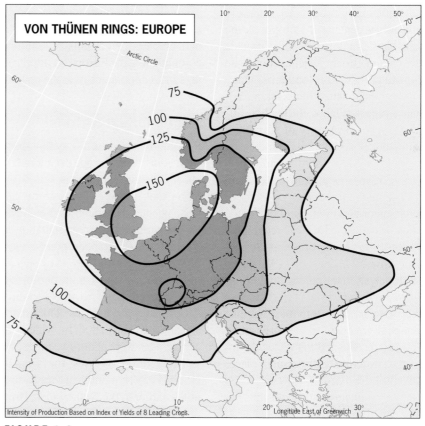

FIGURE 1-6

products. Soon the fruits of the Industrial Revolution were being exported, and the modern industrial spatial organization of Europe began to take shape. In Britain, manufacturing regions, densely populated and heavily urbanized, developed near coalfields in the English Midlands, at Newcastle to the northeast, in southern Wales, and along Scotland's Clyde River around Glasgow.

In mainland Europe, a belt of major coalfields extends from west to east, roughly along the southern margins of the North European Lowland, due eastward from southern England across northern France and Belgium, Germany (the Ruhr), western Bohemia in the Czech Republic, Silesia in southern Poland, and the Donets Basin (Donbas) in eastern Ukraine. Iron ore is found in a broadly similar belt, and the industrial map of Europe reflects the resulting concentrations of economic activity (Fig. 1-7). Another set of manufacturing zones emerged in and near the growing urban centers of Europe, as the same map demonstrates. London—already Europe's leading urban focus and Britain's richest domestic market—was typical of these developments. Many local industries were established here, taking advantage of the large supply of labor, the ready availability of capital, and the proximity of so great a number of potential buyers. Although the Industrial Revolution thrust other places into prominence, London did not lose its primacy: industries in and around the British capital multiplied.

The industrial transformation of Europe, like the agrarian revolution, became the focus of geographic research. One of the leaders in this area was the economic geographer Alfred Weber (1868–1958), who published a spatial analysis of the process titled *Concerning the Location of Industries* (1909). Unlike von Thünen, Weber focused on activities that take place at particular points rather than over large areas. His model, therefore, represented the factors of industrial location, the clustering or dispersal of places of intense manufacturing activity.

One of Weber's most interesting conclusions has to do with what he called *agglomerative* (concentrating) and *deglomerative* (dispersive) forces. It is often advantageous for certain industries to cluster together, sharing equipment, transport facilities, labor skills, and other assets of urban areas. This is what made London (as well as Paris and other cities that were not situated on rich deposits of industrial raw materials) attractive to many manufacturing plants that could benefit from agglomeration and from the large markets that these cities anchored. As Weber found, however, excessive agglomeration may lead to disadvantages such as competition for increasingly expensive space, congestion, overburdening of infrastructure, and environmental pollution. Manufacturers may then move away, and deglomerative forces will increase.

The Industrial Revolution spread eastward from Britain onto the European mainland throughout the middle and late nineteenth century (see inset, Fig. 1-7). Population skyrocketed, emigration mushroomed, and industrializing cities burst at the seams. European states already had acquired colonial empires before this revolution started; now colonialism gave Europe an unprecedented advantage in its dominance over the rest of the world.

Political Revolutions

Revolution in a third sphere—the political—had been going on in Europe even longer than the agrarian or industrial revolutions. Europe's *political revolution* took many different forms and affected diverse peoples and countries, but in general it headed toward parliamentary representation and democracy.

Historical geographers point to the Peace (Treaty) of Westphalia in 1648 as a key step in the evolution of Europe's state system, ending decades of war and recognizing territories, boundaries, and the sovereignty of countries. This treaty's stabilizing effect lasted until 1806, by which time revolutionary changes were again sweeping across Europe.

Most dramatic was the French Revolution (1789–1795), but political transformation had come much earlier to the Netherlands, Britain, and Scandinavian countries. Other parts of Europe remained under the control of authoritarian (dictatorial) regimes headed by monarchs or despots. Europe's patchy political revolution lasted into the twentieth century, and by then *nationalism* (national spirit, pride, patriotism) had become a powerful force in European politics.

When you look at the political map of Europe, the question that arises is how did so small a geographic realm come to be divided into so many political entities? Europe's map is a legacy of its feudal and royal periods, when powerful kings, barons, dukes, and other rulers, rich enough to fund armies and powerful enough to exact taxes and tribute from their domains, created bounded territories in which they reigned supreme. Royal marriages, alliances, and conquests actually simplified Europe's political map. In the early nineteenth century there still were 39 German states; Germany as we know it today did not emerge until the 1870s.

Europe's political revolution produced a form of political-territorial organization known as the **7** **nation-state**, a state embodied by its culturally distinctive population. But what is a nation-state **8** and what is not? The term **nation** has multiple meanings. In one sense it refers to a people with a single language, a common history, a similar ethnic background. In the sense of *nationality* it relates to legal membership in the state, that is, citizenship. Very few states today are so homogeneous culturally that the culture is conterminous with the state. Europe's prominent nation-states of a century ago—France, Spain, the United Kingdom, Italy—have become multicultural societies, their nations defined more by an intangible "national spirit" and emotional commitment than by cultural or ethnic homogeneity. Today, Poland, Hungary, and Sweden are among the few states that still satisfy the definition of the nation-state in Europe.

Mercantilism and colonialism empowered the states of Western Europe; the United Kingdom (Britain) was the superpower of its day. But all countries, even Europe's nation-states in their heyday, are subject to divisive stresses. Political geographers use **9** the term **centrifugal forces** to identify and measure the strength of such division, which may result

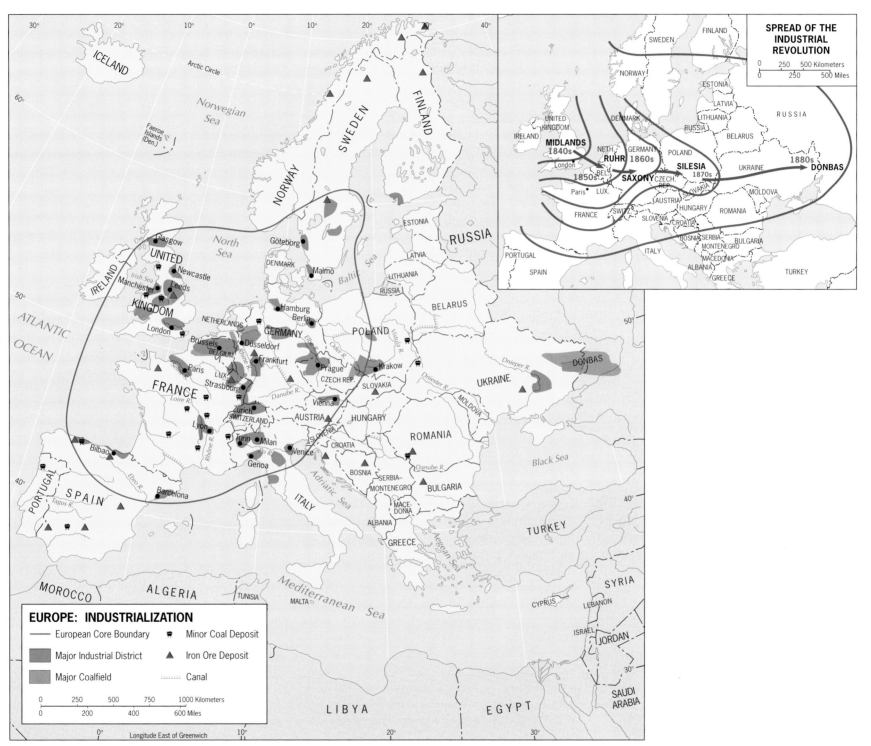

EUROPE: INDUSTRIALIZATION

— European Core Boundary
▮ Major Industrial District
▮ Major Coalfield
⚒ Minor Coal Deposit
▲ Iron Ore Deposit
⋮⋮⋮ Canal

0   250   500   750   1000 Kilometers
0   200   400   600 Miles

SPREAD OF THE INDUSTRIAL REVOLUTION

0   250   500 Kilometers
0   250   500 Miles

MIDLANDS 1840s
RUHR 1860s
1850s
SAXONY
SILESIA 1870s
1880s
DONBAS

FIGURE 1-7

from religious, racial, linguistic, political, or regional factors. In the United States, racial issues form a centrifugal force; during the Vietnam (Indochina) War (1964–1975) politics created strong and dangerous disunity.

10 Centrifugal forces are measured against **centripetal forces**, the binding, unifying glue of the state. General satisfaction with the system of government and administration, legal institutions, and other functions of the state (notably including its treatment of minorities) can ensure stability and continuity when centrifugal forces threaten. In the recent case of Yugoslavia, the centrifugal forces unleashed after the end of the Cold War exceeded the weak centripetal forces in that relatively young state, and it disintegrated.

Europe's political revolution continues. Today a growing group of European states is trying to create a realmwide union that might some day become a European superstate.

CONTEMPORARY EUROPE

Europe has been a regional laboratory of political revolution and evolution, and some of its nation-states were among the first of their kind to emerge on the world stage. Enriched and empowered by colonialism, European states competed and fought with each other, but Europe's nations survived and prospered. Strong cultural identities and historic durability gave European peoples a confidence that continues to mark the realm today, long after Europe's global empires collapsed.

The European realm exhibits only limited geographic homogeneity, which is a challenge that confronts those leading states that want to create a more unified Europe. As Figure 1-8 shows, most **11** Europeans speak **Indo-European languages**, but in fact Europe remains a veritable Tower of Babel. Not only are many of those Indo-European languages not mutually understandable, but peoples such as the Hungarians and the Finns have other linguistic sources. English has become Europe's *lingua franca*, but generally with declining effectiveness from west to east.

Christian religious traditions, another factor that might serve as a unifier, have instead been the source of endless conflict. Shared Christian values, for example, have done little to bring peace to Northern Ireland, where longstanding sectarian strife continues. Religion and politics remain closely connected, and some national and regional political parties still have religious names, such as Germany's Christian Democratic Union and Bavaria's Christian Social Union.

The name "European" is sometimes taken to refer not just to someone residing in Europe, but as a racial reference, describing a common ancestry. But here again, Europe's purported homogeneity is more apparent than real. In terms of physical characteristics, Europeans, from Swede to Spaniard and from Scot to Sicilian, are as varied as any of the world's other geographic realms.

Spatial Interaction

If not culture, what does unify Europe? The answer lies in this realm's outstanding opportunities for productive contact and profitable interaction. The ancient Romans knew it, but they centered their system on the imperial capital. Modern Europeans have seized the same opportunities to create a regionwide structure of *spatial interaction* that links regions, countries, and places in countless ways. The American geographer Edward Ullman conceptualized this process around three operating principles: (1) complementarity, (2) transferability, and (3) intervening opportunity.

12 **Complementarity** occurs when one area has a surplus of a commodity required by another area. The mere existence of a particular resource or product is no guarantee of trade: it must be needed elsewhere. One of Europe's countless examples (at various levels of scale) involves Italy, leader among Mediterranean countries in economic development but lacking adequate coal supplies. Italy imports coal from Western Europe, and in turn Italy exports to Western Europe its citrus fruits, olives, and grapes— which are in high demand on Western European markets. This is a case of *double complementarity*, and even the physical barrier of the Alps has not restricted this two-way trade by rail and road.

13 **Transferability** refers to the ease with which a commodity is transported between producer and consumer. Sheer distance, in terms of cost and time, may make it economically impractical to transfer a product. This is not a problem in modestly sized Europe, where distances are short and transport systems efficient.

14 **Intervening opportunity**, the third of Ullman's spatial interaction principles, holds that potential trade between two places, even if they are in a position of complementarity and do not have problems of transferability, will develop only in the absence of a closer, intervening source of supply. Using our current example, if a major coal reserve were discovered in southern Switzerland, Italy would avail itself of that (intervening) opportunity, reducing or eliminating its imports from Western Europe.

Europe's internal spatial interaction is facilitated by what in many respects is the world's most effective network of railroads, highways, waterways, and air routes. This network is continuously improving as tunnels, bridges, high-speed rail lines, and augmented airports are built. The continent's burgeoning cities and their environs, however, are increasingly troubled by severe congestion.

An Urbanized Realm

Overall, 73 percent of Europe's population resides in cities and towns, but this average is exceeded by far in the west and not attained in the east.

Large cities are the crucibles of their nations' cultures. In his 1939 study of the pivotal role of great cities in the development of national cultures, American geographer Mark Jefferson postulated the law **15** of the **primate city**, which stated that "a country's leading city is always disproportionately large and exceptionally expressive of national capacity and feeling." Though imprecise, this "law" can readily be demonstrated using European examples. Certainly

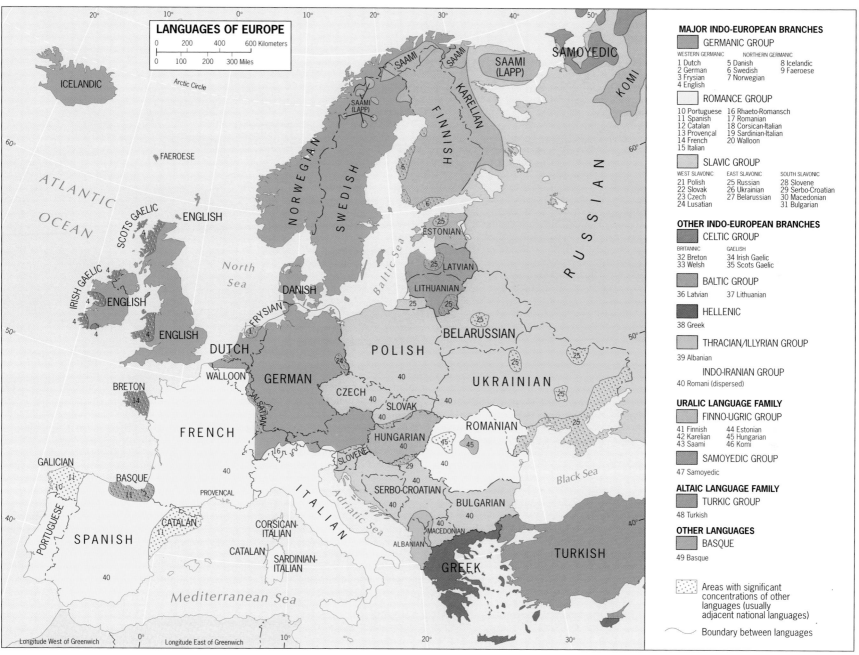

LANGUAGES OF EUROPE

0   200   400   600 Kilometers
0   100   200   300 Miles

MAJOR INDO-EUROPEAN BRANCHES

GERMANIC GROUP

WESTERN GERMANIC · NORTHERN GERMANIC
1 Dutch   5 Danish   8 Icelandic
2 German   6 Swedish   9 Faeroese
3 Frysian   7 Norwegian
4 English

ROMANCE GROUP

10 Portuguese   16 Rhaeto-Romansch
11 Spanish   17 Romanian
12 Catalan   18 Corsican-Italian
13 Provençal   19 Sardinian-Italian
14 French   20 Walloon
15 Italian

SLAVIC GROUP

WEST SLAVONIC   EAST SLAVONIC   SOUTH SLAVONIC
21 Polish   25 Russian   28 Slovene
22 Slovak   26 Ukrainian   29 Serbo-Croatian
23 Czech   27 Belarussian   30 Macedonian
24 Lusatian   31 Bulgarian

OTHER INDO-EUROPEAN BRANCHES

CELTIC GROUP

BRITANNIC   GAELISH
32 Breton   34 Irish Gaelic
33 Welsh   35 Scots Gaelic

BALTIC GROUP

36 Latvian   37 Lithuanian

HELLENIC

38 Greek

THRACIAN/ILLYRIAN GROUP

39 Albanian

INDO-IRANIAN GROUP

40 Romani (dispersed)

URALIC LANGUAGE FAMILY

FINNO-UGRIC GROUP

41 Finnish   44 Estonian
42 Karelian   45 Hungarian
43 Saami   46 Komi

SAMOYEDIC GROUP

47 Samoyedic

ALTAIC LANGUAGE FAMILY

TURKIC GROUP

48 Turkish

OTHER LANGUAGES

BASQUE

49 Basque

Areas with significant concentrations of other languages (usually adjacent national languages)

Boundary between languages

FIGURE 1-8

FROM THE FIELDNOTES

"Even Copenhagen was hot during the torrid summer of 2003. Having rented a bike to get around this sprawling city with its numerous waterways and bridges, I had to get used to the speeds on the paths . . . these Danes are fit! And they are also well served, as are European residents generally, by efficient public transport systems. In Europe, railroad stations still are important components of the urban landscape, and even the most remote corners of small countries are served by modern train and bus lines. I spent an hour watching thousands of commuters, schoolchildren, and other travelers leaving and boarding a steady stream of trains that arrived and departed on schedule at a dozen platforms—in the capital of a country with barely over 5 million inhabitants! Obviously this is one benefit derived from the high taxes Europeans pay."

Paris personifies France in countless ways, and nothing in England rivals London. In both of these primate cities, the culture and history of a nation and empire are indelibly etched in the urban landscape. Similarly, Vienna is a microcosm of Austria, Warsaw is the heart of Poland, Stockholm typifies Sweden, and Athens is Greece. Today, each of these (together with the other primate cities of Europe) sits atop a hierarchy of urban centers that has captured the lion's share of its national population growth since World War II.

Primate cities tend to be old, and in general the European cityscape looks quite different from the American. Seemingly haphazard inner-city street systems impede traffic; central cities may be picturesque, but they are also cramped. The urban layout of the London region (Fig. 1-9) reveals much about the internal spatial structure of the European **16** **metropolis** (the central city and its suburban ring). The metropolitan area remains focused on the large city at its center, especially the downtown *central business district (CBD)*, which is the oldest part of the urban agglomeration and contains the region's largest concentration of business, government, and shopping facilities as well as its wealthiest and most prestigious residences. Wide residential sectors radiate outward from the CBD across the rest of the central city, each one home to a particular income group. Beyond the central city lies a sizeable suburban ring, but residential densities are much higher here than in the United States because the European tradition is one of setting aside recreational spaces (in "greenbelts") and living in apartments rather than in de-

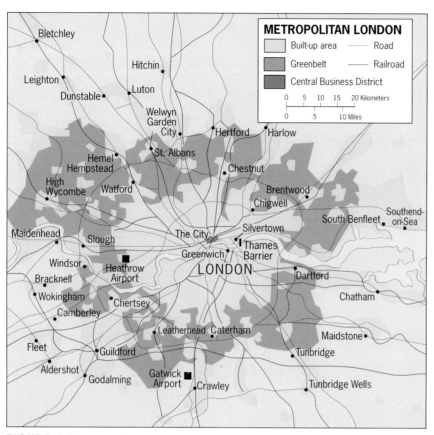

FIGURE 1-9

tached single-family houses. There is also a greater reliance on public transportation, which further concentrates the suburban development pattern. That has allowed many nonresidential activities to suburbanize as well, and today ultramodern outlying business centers increasingly compete with the CBD in many parts of urban Europe.

A Changing Population

When a population urbanizes, average family size declines, and so does the overall rate of natural increase. There was a time when Europe's population was (in the terminology of population geographers) exploding, sending millions to the New World and the colonies and still growing at home. But today Europe's indigenous population, unlike most of the rest of the world's, is actually shrinking. To keep a population from declining, the (statistically) average woman must bear 2.1 children. For Europe as a whole, that figure was 1.4 in 2003. Seven European countries, including Italy and Spain, recorded below 1.3—the lowest ever seen in any human population.

Such *negative population growth* poses serious challenges for any nation. When the population pyramid becomes top-heavy, the number of workers whose taxes pay for the social services of the aged goes down, leading to reduced pensions and dwindling funds for health care. Governments that impose tax increases endanger the business climate; their options are limited. Europe, and especially Western Europe, is experiencing a *population implosion* that will be a formidable challenge in decades to come.

Meanwhile, *immigration* is partially offsetting the losses European countries face. Millions of Turkish Kurds (mainly to Germany), Algerians (France), Moroccans (Spain), West Africans (Britain), and Indonesians (Netherlands) are changing the social fabric of what once were unicultural nation-states. One key dimension of this change is the spread of Islam in Europe (Fig. 1-10). Muslim populations in Eastern Europe (such as Albania's, Kosovo's, and Bosnia's) are local, Slavic communities converted during the period of Ottoman rule. The Muslim sectors of Western European countries, on the other hand, represent more recent immigrations.

The vast majority of these immigrants are intensely devout, politically aware, and culturally insular. They continue to arrive in a Europe where native populations are stagnant or declining, where religious institutions are weakening, where secularism is rapidly rising, where political positions often appear to be anti-Islamic, and where cultural norms are incompatible with Muslim traditions. Many young men, unskilled and uncompetitive in a European Union (EU) where unemployment is already high, get involved in petty crime or the drug trade, are harassed by the police, and turn to their faith for solace and reassurance. Unlike most other immigrant groups, Muslim communities also tend to resist assimilation, making Islam the essence of their identity. In Britain alone there are more than 1500 mosques, in themselves a transformation of local cultural landscapes.

In truth, European governments have not done enough to foster the very integration they see Muslims rejecting. The French assumed that their North African immigrants would aspire to assimilation as Muslim children went to public schools from their urban public-housing projects; instead they got into disputes over the dress codes of Muslim girls. The Germans for decades would not

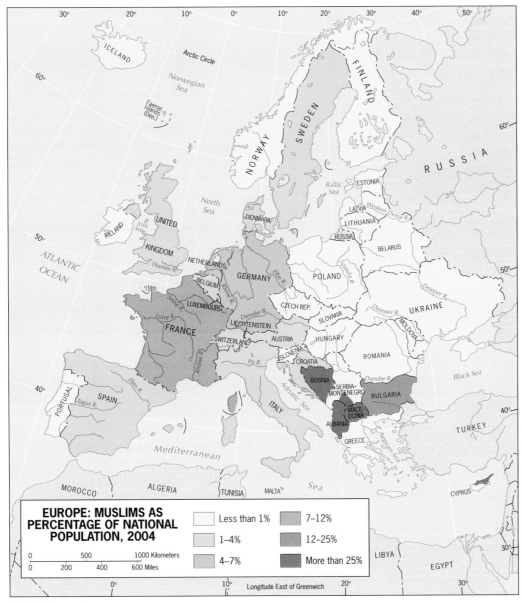

FIGURE 1-10

ages of Paris, Manchester, or Frankfurt are far different.

The social and political implications of Europe's demographic transformation are numerous and far-reaching. Long known for tolerance and openness, European societies are attempting to restrict immigration in various ways; political parties with anti-immigrant platforms are gaining ground. Multiculturalism poses a growing challenge in a changing Europe.

EUROPE'S MODERN TRANSFORMATION

At the end of World War II, much of Europe lay shattered, its cities and towns devastated, its infrastructure wrecked, its economies devastated. The Soviet Union had taken control over the bulk of Eastern Europe, and communist parties seemed poised to dominate the political life of major Western European countries.

In 1947, U.S. Secretary of State George C. Marshall proposed a European Recovery Program designed to counter all this dislocation and to create stable political conditions in which democracy would survive. Over the next four years, the United States provided about $13 billion in assistance to Europe (almost $100 billion in today's money). Because the Soviet Union refused U.S. aid and forced its Eastern European satellites to do the same, the Marshall Plan applied solely to 16 European countries, including defeated (West) Germany, and Turkey.

European Unification

The Marshall Plan did far more than stimulate European economies. It showed European leaders that their countries needed a joint economic-administrative structure not only to coordinate the financial assistance, but also to ease the flow of resources and products across Europe's mosaic of

award German citizenship to children of immigrant parents born on German soil. The British found their modest efforts at integration stymied by the Muslim practice of importing brides from Islamic countries, which had the effect of perpetu-

ating segregation. In many European cities, the housing projects that form the living space of Muslim residents (by now many of them born and retired there) are dreadful, barren, impoverished environments unseen by Europeans whose im-

Supranationalism in Europe

1944	Benelux Agreement signed.
1947	Marshall Plan created (effective 1948–1952).
1948	Organization for European Economic Co-operation (OEEC) established.
1949	Council of Europe created.
1951	European Coal and Steel Community (ECSC) Agreement signed (effective 1952).
1957	Treaty of Rome signed, establishing European Economic Community (EEC) (effective 1958), also known as the Common Market and "The Six." European Atomic Energy Community (EURATOM) Treaty signed (effective 1958).
1959	European Free Trade Association (EFTA) Treaty signed (effective 1960).
1961	United Kingdom, Ireland, Denmark, and Norway apply for EEC membership.
1963	France vetoes United Kingdom EEC membership; Ireland, Denmark, and Norway withdraw applications.
1965	EEC-ECSC-EURATOM Merger Treaty signed (effective 1967).
1967	European Community (EC) inaugurated.
1968	All customs duties removed for intra-EC trade; common external tariff established.
1973	United Kingdom, Denmark, and Ireland admitted as members of EC, creating "The Nine." Norway rejects membership in the EC by referendum.
1979	First general elections for a European Parliament held; new 410-member legislature meets in Strasbourg. European Monetary System established.
1981	Greece admitted as member of EC, creating "The Ten."
1985	Greenland, acting independently of Denmark, withdraws from EC.
1986	Spain and Portugal admitted as members of EC, creating "The Twelve." Single European Act ratified, targeting a functioning European Union in the 1990s.
1987	Turkey and Morocco make first application to join EC. Morocco is rejected; Turkey is told that discussions will continue.
1990	Charter of Paris signed by 34 members of the Conference on Security and Cooperation in Europe (CSCE). Former East Germany, as part of newly reunified Germany, incorporated into EC.
1991	Maastricht meeting charts European Union (EU) course for the 1990s.
1993	Single European Market goes into effect. Modified European Union Treaty ratified, transforming EC into EU.
1995	Austria, Finland, and Sweden admitted into EU, creating "The Fifteen."
1999	European Monetary Union (EMU) goes into effect. Helsinki summit discusses fast-track negotiations with six prospective members and applications from six others; prospects for Turkey considered in longer term.
2001	Denmark's voters reject EMU participation by 53 to 47 percent.
2002	The euro is introduced as historic national currencies disappear in 12 countries.
2003	First draft of a European Constitution is published to mixed reviews from member-states.
2004	Historic expansion of EU from 15 to 25 countries with the admission of Cyprus, the Czech Republic, Estonia, Hungary, Latvia, Lithuania, Malta, Poland, Slovakia, and Slovenia.

boundaries, to lower restrictive trade tariffs, and to seek ways to effect political cooperation.

For all these needs Europe's governments had some guidelines. While in exile in Britain, leaders of three small countries—Belgium, the Netherlands, and Luxembourg—had been discussing an association of this kind even before the end of the war. There, in 1944, they formulated and signed the Benelux Agreement, intended to achieve total economic integration. When the Marshall Plan was launched, the Benelux precedent helped speed the creation of the Organization for European Economic Cooperation (OEEC), which was established to coordinate the investment of America's aid (see box at left).

Soon the economic steps led to greater political cooperation as well. In 1949, the participating governments created the Council of Europe, the beginnings of what was to become a European Parliament meeting in Strasbourg, France. Europe was embarked on still another political revolution, the formation of a multinational union involving a growing number of European states. Geographers define **17 supranationalism** as the voluntary association in economic, political, or cultural spheres of three or more independent states willing to yield some measure of sovereignty for their mutual benefit. In later chapters we will encounter other supranational organizations, including the North American Free Trade Agreement (NAFTA), but none had reached the plateau achieved by the European Union (EU).

In Europe, the key initiatives arose from the Marshall Plan and lay in the economic arena; political integration came more haltingly. Under the Treaty of Rome, six countries joined to become the European Economic Community (EEC) in 1957, also called the "Common Market." In 1973 the United Kingdom, Ireland, and Denmark joined, and the renamed European Community (EC) had nine members. As Figure 1-11 shows, membership reached 15 countries in 1995, after the organization had been renamed yet one more time to become the European Union (EU).

The European Union is not just a paper organization for bankers and manufacturers. It has a major

FIGURE 1-11

impact on the daily lives of its member countries' citizens in countless ways. (You will see one of these ways when you arrive at an EU airport and find that EU-passport holders move through inspection at a fast pace, while non-EU citizens wait in long immigration lines.) Taxes tend to be high in Europe, and those collected in the richer member-states are used to subsidize growth and development in the less prosperous ones. This is one of the burdens of membership that is not universally popular in the EU, to say the least. But it has strengthened the economies of Portugal, Greece, and other national and regional economies to the betterment of the entire organization. Some countries also object to the terms and rules of the Common Agricultural Policy (CAP) which, according to critics, supports farmers far too much and, according to others, far too little. (France in particular obstructs efforts to move the CAP closer to consensus, subsidizing its agricultural industry relentlessly while arguing that this protects its rural cultural heritage as well as its farmers.)

New Money

Every since the first steps toward European unification were taken, EU planners dreamed of a time when the EU would have a single currency not only to symbolize its strengthening unity but also to establish a joint counterweight to the mighty American dollar. A European Monetary Union (EMU) became a powerful objective of EU planners, but many observers thought it unlikely that member-states would in the end be willing to give up their historic marks, francs, guilders, liras, and escudos. Yet it happened. In 2002, twelve of the (then) 15 EU countries withdrew their currencies and began using the euro, with only the United Kingdom (Britain), Denmark, and Sweden staying out (Fig. 1-11). Add non-member Norway to this group, and not a single Scandinavian country has converted to the euro.

Momentous Expansion

Expansion has always been an EU objective, and the subject has always aroused passionate debate. Will the incorporation of weaker economies undermine the strength of the whole? Despite such misgivings, negotiations to expand the EU have long been in progress, and in 2004 a momentous milestone was reached: ten new members were added, creating a greater European Union with 25 member-states. Geographically, these ten came in three groups: the three Baltic states (Estonia, Latvia, and Lithuania); five contiguous states in Eastern Europe extending from Poland and the Czech Republic through Slovakia, Hungary, and Slovenia; and two Mediterranean island-states, the ministate of Malta and the still-divided state of Cyprus (Fig. 1-11).

Numerous structural implications arise from this, affecting all EU countries. A common agricultural policy is now even more difficult to achieve, given the poor condition of farming in most of the new members. And some of the former EU's poorer states, which were on the receiving end of the subsidy program that aided their development, now will have to pay up to support the much poorer new eastern members. And disputes over representation at EU's Brussels headquarters arose quickly. Even before the new members' 2004 accession, Poland was demanding that the representative system favor medium-sized members (such as Poland and Spain) over larger ones such as Germany and France.

The geographic consequences of this momentous EU expansion for Europe as a whole are far-reaching. As Figure 1-11 shows, the EU's new eastern borders leave several groups of countries and peoples outside the Union: (1) the former Soviet republics of Ukraine, Belarus, and Moldova; (2) two major Eastern European states, Romania and Bulgaria; and (3) the states arising from the breakup of Yugoslavia along with impoverished Albania. Already, negotiations between the EU leadership and Romania and Bulgaria are in progress, and the entry of these countries is expected in 2007 or soon thereafter. In the former Yugoslavia, Croatia may eventually follow its neighbor Slovenia into the EU. But in the rest of Eastern Europe the process will undoubtedly slow down. Political instability, undemocratic regimes, weak economies, and other obstacles stand in the way.

We should take note of still another potential candidate for EU membership: Turkey. EU leaders would like to include a mainly Muslim country in what Islamic states sometimes call the "Christian Club," but Turkey needs progress on social standards, human rights, and economic policies before its accession can be contemplated. Still, the fact that the EU has now reached deeply into Eastern Europe, encompasses 25 members, has a common currency and a parliament, is developing a constitution, and is even considering expansion beyond the realm's borders constitutes a tremendous and historic achievement in this fragmented, fractious part of the world. The EU now has a combined population of more than 450 million constituting one of the world's richest markets; its member-states account for more than 40 percent of the world's exports. Although, as we will note later, Europe is currently experiencing budgetary troubles, it remains an economic power of global significance.

It is remarkable that all this has been accomplished in little more than half a century. Some of the EU's leaders want more than economic union: they envisage a United States of Europe, a political as well as an economic competitor for the United States. To others, such a "federalist" notion is an abomination not even to be mentioned (the British in general are especially wary of such an idea) and certainly not to be made part of the European constitution. Whatever happens, Europe is going through still another of its revolutionary changes, and when you study its evolving map you are looking at history in the making.

Centrifugal Forces

For all its dramatic progress toward unification, Europe remains a realm of geographic contradictions. Europeans are well aware of their history of conflict, division, and repeated self-destruction. Will supranationalism finally overcome the centrifugal forces that have so long and so frequently afflicted this part of the world?

Even as Europe's states have been working to join forces in the EU, many of those same states are confronting severe centrifugal stresses. The term **18** **devolution** has come into use to describe the powerful centrifugal forces whereby regions or peoples within a state, through negotiation or active rebellion, demand and gain political strength and sometimes autonomy at the expense of the center. Most states exhibit some level of internal regionalism, but the process of devolution is set into motion when a key centripetal binding force—the nationally accepted idea of what a country stands for—erodes to the point that a regional drive for autonomy, or for outright secession, is launched.

As Figure 1-12 shows, numerous European countries are affected by devolution. States large and small, young and old, EU members and non-EU members must deal with the problem. Even the long-stable United Kingdom is affected. England, the historic core area of the British Isles, dominates the UK in terms of population as well as political and economic power. The country's three other entities—Scotland, Wales, and Northern Ireland—were acquired over several centuries and attached to England (hence the "United" Kingdom). But neither time nor representative democratic government was enough to eliminate all latent regionalism in these three components of the UK. During the 1960s and 1970s the

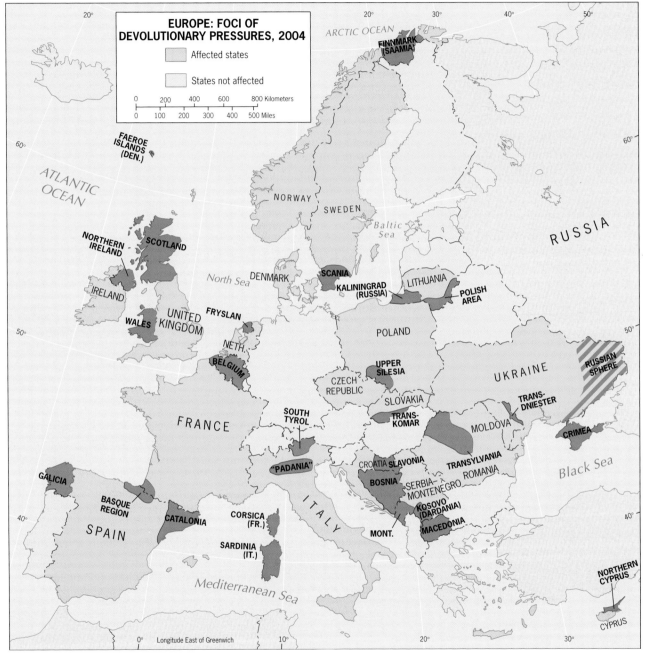

EUROPE: FOCI OF DEVOLUTIONARY PRESSURES, 2004

Affected states

States not affected

0   200   400   600   800 Kilometers
0   100   200   300   400   500 Miles

FIGURE 1-12

British government confronted a virtual civil war in Northern Ireland and rising tides of nationalism in Scotland and Wales. In 1997, the government in London gave the Scots and Welsh the opportunity to vote for greater autonomy in new regional parliaments that would have limited but significant powers over local affairs. The Scots voted overwhelmingly in favor, and the Welsh by a slim majority—and thus a major devolutionary step was taken in one of Europe's oldest, most durable, and most unified states.

Even while this devolutionary process was ongoing, the British government still joined the European Union on behalf not only of the English, but also of the Scots, Welsh, and Northern Irelanders seeking a new relationship with London. As the map shows, the UK is not alone in such contradiction. Spain faces severe devolutionary forces in its Basque area and lesser ones in Catalonia and Galicia; France contends with a secessionist movement on its island of Corsica; Belgium is driven by Flemish-Walloon separatism; Italy confronts devolutionary pressures in South Tyrol and Lombardy. In recent decades Eastern Europe has been a cauldron of devolution as Yugoslavia and Czechoslovakia collapsed, Moldova fragmented, and Ukraine suffered from the stresses of its historical (Russian-penetrated) geography.

Political devolution is not the only centrifugal force to buffet European states. As the European Union materialized, its freedoms (in the form of money flows, labor movements, and transferabilities) led to the emergence of powerful urban regions as hubs of economic power and influence, in some ways beyond the control of their national governments. Examples include the Rhône-Alpes region in France, centered on Lyon; Lombardy in Italy, focused on Milan; Catalonia in Spain, anchored by Barcelona; and Baden-Württemberg in Germany, headquartered **19** by Stuttgart. This group, known as the **Four Motors of Europe**, bypasses not only their national governments in dealing with each other but even extends their business channels to span the world.

The Japanese economist Kenichi Ohmae calls such **20** economic powerhouses **regional states**, entities that defy old borders and are shaped by the globalizing economy of which they have become a part.

Like devolutionary forces, these emerging regional states are changing the map of Europe. In some ways, they are beginning to supplant the old framework that arose from the nation-states of an earlier period. Elsewhere, economic development on both sides of international boundaries is having a similar effect, creating *Euroregions* that foster more localized, cross-border cooperation on a continent once rigidly partitioned.

All these developments underscore how far-reaching Europe's current transformation is. The Marshall Plan jump-started it; good economic times sustained it; and the end of the Cold War stimulated it. But Europe's gains remain clouded by violent discord in parts of the United Kingdom, Spain, France, and the Balkans, by devolutionary forces elsewhere, and by uncertainties over EU expansion and further political integration.

REGIONS OF EUROPE

Europe, though territorially small, is culturally and physically diverse. To comprehend its geography, we group its states into five regions: (1) Western Europe, (2) the British Isles, (3) Northern (Nordic) Europe, (4) Mediterranean Europe, and (5) Eastern Europe (Fig. 1-13). This regionalization employs the formal-region concept (see pp. 6-7), but we can approach it in other ways. Focusing on Europe's supranational and subnational developments in a functional-region context, we recognize a European core (see Fig. 1-13) and a highly varied periphery. For our purposes, however, the five-region delimitation works better.

◗ WESTERN EUROPE

Western Europe is the heart of the realm, the hub of its economic power, and the focus of its unifying drive. Germany and France dominate it, flanked by the three Benelux countries in the northwest and the three Alpine states in the east (Fig. 1-14). With a combined population of 185 million representing eight of the world's richest economies, Western Europe is a powerful force not only in Europe but also on the international stage.

Dominant Germany

Germany today is Europe's most populous country, its most powerful economy, and its strongest force for union. German leaders ceaselessly promote notions of greater European unity, and Germany has been the chief proponent of a Europe unified both politically and economically.

Twice during the twentieth century Germany plunged Europe and the world into war, until, finally, in 1945, the defeated and devastated German state was divided into two parts, West and East (see the

FROM THE FIELDNOTES

"Berlin, July 10, 2003. Took the train from Warnemunde, then a bus through the former East Berlin to the Friedrich-strasse (street) in the center, and started walking. The city reflects both its status as federal capital and the stagnant condition of the national economy. Even as giant construction projects fill newly cleared city blocks, parks show signs of neglect, weeds grow on tiled sidewalks, back streets need cleaning up. Near the Potsdamer Platz, however, where the Potsdamer Street meets the Leipziger Street and a repositioned section of the Berlin Wall forms a tourist attraction, Berlin is at its most modern and building continues. The Sony Center lies at the heart of this upscale district."

FIGURE 1-13

delimitation in red on Figure 1-15). Its eastern boundaries were also changed, leaving the industrial district of Silesia in newly-defined Poland, that of Saxony in communist-ruled and Soviet-controlled East Germany, and the Ruhr in West Germany (see Fig. 1-14). Aware that these were the industrial centers that had enabled Nazi Germany to seek world domination through war, the victorious allies laid out this new boundary framework to make sure this would not happen again.

In the aftermath of World War II, Soviet and Allied administration of East and West Germany differed. Soviet rule in East Germany was established on the Russian-communist model and, given the extreme hardships the USSR had suffered at German hands during the war, harshly punitive. The American-led authority in West Germany was less strict and aimed more at rehabilitation. When the Marshall Plan was instituted, West Germany was included, and its economy recovered rapidly. Meanwhile, West Germany was reorganized politically into a modern federal state on democratic foundations.

West Germany's economy thrived. Between 1949 and 1964 its GNP tripled while industrial output rose 60 percent. It absorbed millions of German-speaking refugees from Eastern Europe (and many escapees from communist East Germany as well). Since unemployment was virtually nonexistent, hundreds of thousands of Turkish and other foreign workers arrived to take jobs Germans could not fill or did not want.

Simultaneously, West Germany's political leaders participated enthusi-

Frankfurt (the leading financial hub as well) in the center, and Stuttgart in the south. West Germany exported huge quantities of iron, steel, motor vehicles, machinery, textiles, and farm products.

No economy grows without setbacks and slowdowns, however, and Germany experienced such problems in the 1970s and 1980s, when energy shortages, declining competitiveness on world markets, lagging modernization (notably in the aging Ruhr), and social dilemmas involving rising unemployment, an aging population, high taxation, and a backlash against foreign resident workers roiled West German society. And then, quite suddenly, the collapse of the communist Soviet Union opened the door to reunification with East Germany.

In 1990, West Germany had a population of about 62 million and East Germany 17 million. Communist misrule in the East had yielded outdated factories, crumbling infrastructures, polluted environments, drab cities, inefficient farming, and inadequate legal and other institutions. Reunification was more a rescue than a merger, and the cost to West Germany was enormous. When the West German government imposed sales-tax increases and an income-tax surcharge on its citizens, many Westerners doubted the wisdom of reunification. It was projected that it would take decades to reconstruct Virginia-sized East Germany: ten years later, exports from the former East still contributed only about 7 percent of the national total.

Once again, however, Germany proved its capacity to rebound. Helped by the initial weakness of the new currency, the euro, German exports

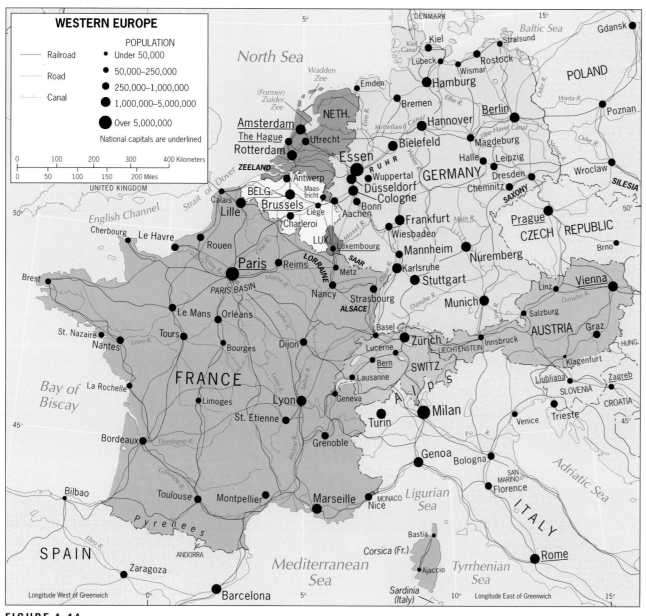

FIGURE 1-14

astically in the OEEC and in the negotiations that led to the six-member Common Market. Geography worked in West Germany's favor: it had common borders with all but one of the EEC member-states. Its transport infrastructure, rapidly rebuilt, was second to none in the realm. More than compensating for its loss of Saxony and Silesia were the expanding Ruhr (in the hinterland of the Dutch port of Rotterdam) and the newly emerging industrial complexes centered on Hamburg in the north,

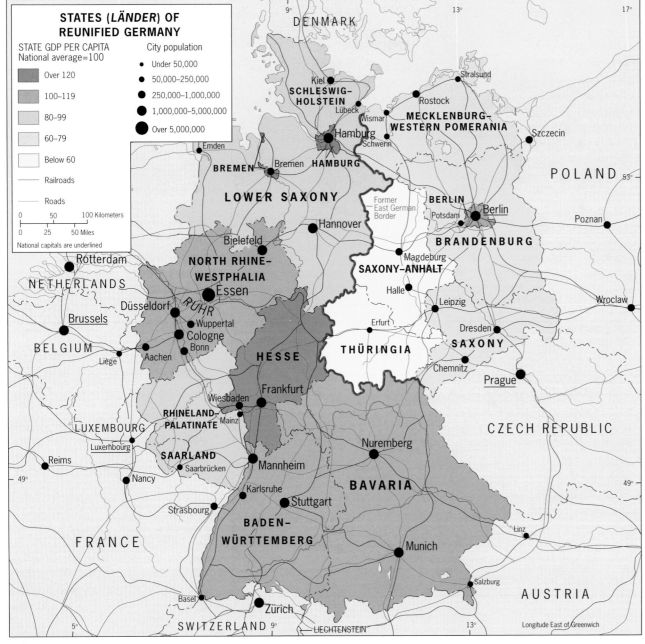

STATES (LÄNDER) OF REUNIFIED GERMANY

STATE GDP PER CAPITA
National average=100

- Over 120
- 100–119
- 80–99
- 60–79
- Below 60

City population
- Under 50,000
- 50,000–250,000
- 250,000–1,000,000
- 1,000,000–5,000,000
- Over 5,000,000

— Railroads
— Roads

0   50   100 Kilometers
0   25   50 Miles

National capitals are underlined

FIGURE 1-15

rebounded on world markets. More-over, conditions in the former East Germany began to improve, helped by the relocation of the national capital from Bonn to Berlin. Various statistical indicators that had taken a beating from the merger (lower life expectancies, reduced urbanization, smaller GNP) began to recover as heavy investment in the East took effect.

The Federal Republic

Before reunification, West Germany functioned as a federal state consisting of ten States or *Länder* (Figure 1-15). East Germany had been divided under communist rule into 15 districts including East Berlin. Upon reunification, East Germany was reorganized into six new States based on traditional provinces within its borders. Figure 1-15 makes a key point: regional disparity in terms of income per person remains a serious problem between the former East and West. Note that five of former East Germany's six States (Berlin being the sole exception) are in the two lowest income categories, while *none* of the ten former West German States is in these lowest categories. In the first decade of this century, Germany's economy was stagnant, raising unemployment and slowing former East Germany's recovery. Nevertheless, the gap continues to narrow, and with 82 million inhabitants including over 7 million foreigners and more than 4 million ethnic Germans born outside the country, Germany is again exerting its dominance over a mainland Europe in which it has no peer.

France

German dominance in the European Union is a constant concern in the other leading Western European country. The French and the Germans have been rivals in Europe for centuries. France is an old state, by most measures the oldest in Western Europe. Germany is a young country, created in 1871 after a loose association of German-speaking states had fought a successful war against . . . the French.

Territorially, France is much larger than Germany, and the map suggests that France has a superior relative location, with coastlines on the Mediterranean Sea, the Atlantic Ocean, and, at Calais, even a window on the North Sea. But France does not have any good natural harbors, and oceangoing ships cannot navigate its rivers and other waterways far inland. France has no equivalent to Rotterdam either internally or externally.

The map of Western Europe (Fig. 1-14) reveals a significant demographic contrast between France and Germany. France has one dominant city, Paris, at the heart of the Paris Basin, France's core area. No other city in France comes close to Paris in terms of population or centrality: Paris has 9.6 million residents, whereas its closest rival, Lyon, has

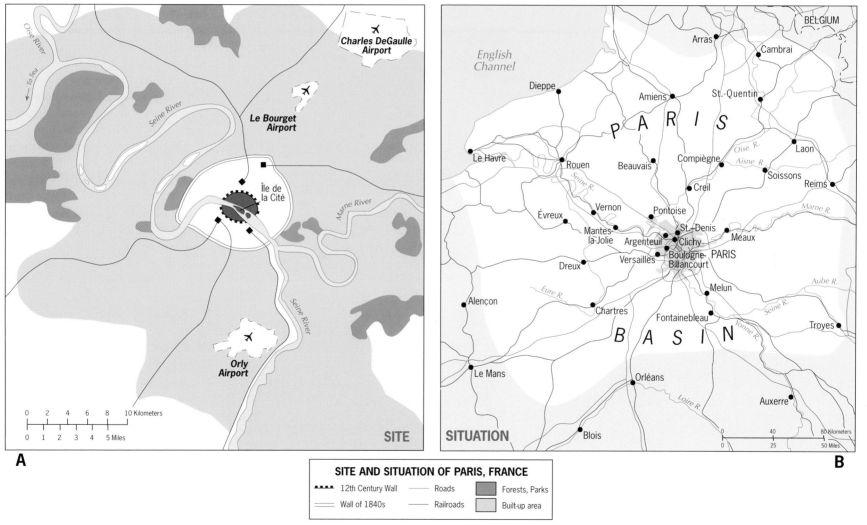

SITE AND SITUATION OF PARIS, FRANCE

▪▪▪▪ 12th Century Wall   — Roads   ▨ Forests, Parks

═══ Wall of 1840s   — Railroads   ▨ Built-up area

FIGURE 1-16

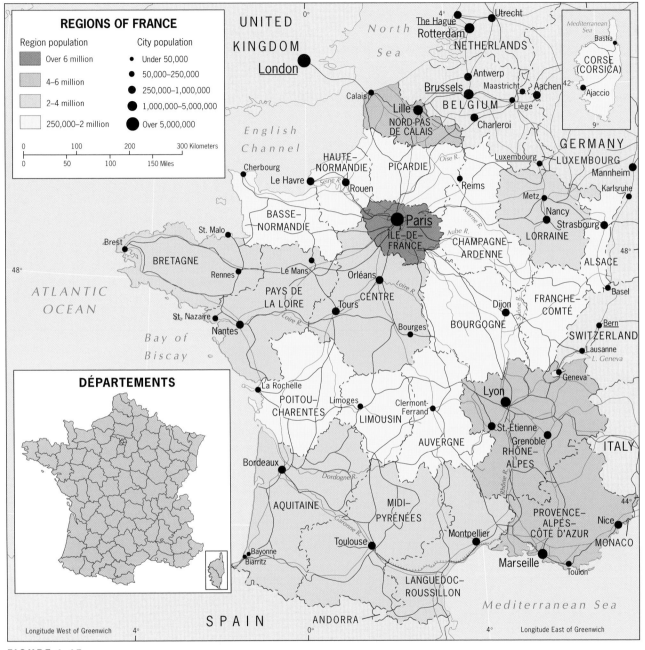

REGIONS OF FRANCE

Region population

	Over 6 million
	4–6 million
	2–4 million
	250,000–2 million

City population

- Under 50,000
- 50,000–250,000
- 250,000–1,000,000
- 1,000,000–5,000,000
- Over 5,000,000

0 100 200 300 Kilometers
0 50 100 150 Miles

DÉPARTEMENTS

FIGURE 1-17

only 1.4 million. Germany has no city to match Paris, but it does have a number of cities with populations between 1 and 5 million. And as Table G-1 shows, Germany is much more highly urbanized overall than France.

Why should Paris, without major raw materials nearby, have grown so large? Whenever geographers investigate the evolution of a city, they focus on two important locational **21** qualities: its **site** (the physical attributes of the place it occupies) and **22** its **situation** (its location relative to surrounding areas of productive capacity, other cities and towns, barriers to access and movement, and other aspects of the greater regional framework in which it lies).

The site of the original settlement at Paris lay on an island in the Seine River, a defensible place where the river was often crossed. This island, the *Île de la Cité*, was a Roman outpost 2000 years ago; for centuries its security ensured continuity. Eventually the island became overcrowded, and the city expanded along the banks of the river (Fig. 1-16A).

Soon the settlement's advantageous situation stimulated its growth and prosperity. Its fertile agricultural hinterland thrived, and, as an enlarging market, Paris's focality increased steadily. The Seine River is joined near Paris by several navigable tributaries (the Oise, Marne, and Yonne). When canals extended these waterways even farther, Paris was linked to the Loire Valley, the Rhône-Saône Basin, Lorraine (an industrial area in the northeast), and the northern border with Belgium. When Napoleon reorganized France and built a radial system of roads—followed later by railroads—that focused on Paris from all parts of the country, the city's primacy was assured (Fig. 1-16B). The only disadvantage in Paris's situation lies in its seaward access: oceangoing ships can sail up the Seine River only as far as Rouen.

Paris, in accordance with Weber's agglomeration principle, grew into one of Europe's greatest cities. French industrial development was less spectacular, but northern French agriculture remained Europe's most productive and varied, exploiting the country's wide range of soils and climates. Today France's economic geography is marked by new high-tech industries. It is a leading producer of high-speed trains, aircraft, fiber-optic communications systems, and space-related technologies. It also is the world leader in nuclear power, which currently supplies more than 75 percent of its electricity and thereby reduces its dependence on foreign oil.

When Napoleon reorganized France in the early 1800s, he broke up the country's large traditional subregions and established more than 80 small *départements* (additions and subdivisions later increased this number to 96). Each *département* had representation in Paris, but the power was concentrated in the capital, not in the individual *départements*. France became a highly centralized state and remained so for nearly two centuries (see inset map, Fig. 1-17). Only the island *département* of Corsica produced a rebel movement, whose violent opposition to French rule continued for decades and even touched the mainland. In 2003 the voters in Corsica rejected an offer of special status for their island, including limited autonomy. They wanted more, and trouble lies ahead.

Today, France is decentralizing. A new subnational framework of 22 historically significant provinces, groupings of *départements* called *regions* (Fig. 1-17), has been established to accommodate the devolutionary forces felt throughout Europe and, indeed, throughout the world. These regions, though still represented in the Paris government, have substantial autonomy in such areas as taxation, borrowing, and development spending. The cities that anchor them benefit because they are the seats of governing regional councils that can attract investment, not only within France but also from abroad.

Lyon, France's second city and headquarters of the region named Rhône-Alpes, has become a focus for growth industries and multinational firms. This region is evolving into a self-standing economic powerhouse that is becoming a driving force in the European economy; indeed, it is one of the Four Motors of Europe (see p. 49) with its own international business connections to countries as far away as China and Chile.

France has one of the world's most productive and diversified economies, based in one of humanity's richest cultures and vigorously promoted and protected (notably its heavily subsidized agricultural sector). And although France and Germany agree on many aspects of EU integration, they tend to differ on important issues. Old, historically centralized France is less eager than young, federal Germany to push political integration in supranational Europe.

Benelux

Three countries are crowded into the northwest corner of Western Europe: Belgium, the Netherlands, and tiny Luxembourg, collectively referred to by their first syllables (*Be-Ne-Lux*). The major differences that evolved between the agriculturally productive Netherlands and the industrially developed Belgium yielded a double complementarity that led to the 1944 Benelux customs union.

The Benelux countries are among the most densely populated on Earth. For centuries the Dutch have been expanding their living space—not by warring with their neighbors but by wresting land from the sea. The greatest project so far, the draining of almost the entire Zuider Zee (Southern Sea), began in 1932 and continues. In the province of Zeeland, islands are being connected by dikes and the water is being pumped out, adding more *polders* (reclaimed lands) to the total national territory.

The regional geography of the **Netherlands** is noted for the *Randstad*, a triangular urban core area

"Maastricht, May 8, 1995. No place in Europe displays the flag of the European Union as liberally as does this Dutch city, where the European Union Treaty of 1991 was signed. The EU flag, seen here at City Hall (symbolically also under construction!), shows twelve yellow stars, each one representing a signatory to the Treaty, in a circle against a blue background. When the Maastricht Treaty was drawn up, the EU consisted of twelve states, but three additional countries have been admitted to the organization this year (1995) and others may join later. The flag, however will remain as is, and will not, as the U.S. flag does, change to reflect changing times." (In 2004, the EU was poised to admit another ten member states.)

dominated by the constitutional capital, Amsterdam, Europe's largest port, Rotterdam, and the seat of government, The Hague. This **conurbation**, as geographers call large multimetropolitan complexes formed by the coalescence of two or more urban areas, forms a ring-shaped complex that encircles a still-rural center (*rand* means edge or margin).

Geographically, **Belgium** is marked by a cultural fault line that extends diagonally across the state, separating the Dutch-related Flemish in the northwest (55 percent of the population) from the Walloons in the southeast (33 percent). Devolution threatens here, but Belgium's great asset is its capital, Brussels, a headquarters of the European Union.

Luxembourg, small as it is, has the distinction of recording by far the highest per capita GNP in all of Europe (Table G-1). Financial, service, tourist, and information technology industries make this country, still run by a "Grand Duke," the most prosperous *ministate* in the realm.

The Alpine States

Switzerland, Austria, and the *microstate* of Liechtenstein on their border share an absence of coasts and the mountainous topography of the Alps—and little else (Fig. 1-14). Austria speaks one language; the Swiss speak German in their north, French in the west, Italian in the southeast, and even a bit of Rhaeto-Romansch in the remote central highlands (Fig. 1-8). Austria has a large primate city; multicultural Switzerland does not. Austria has a substantial range of domestic raw materials; Switzerland does not. Austria is twice the size of Switzerland and has a larger population, but far more trade crosses the Swiss Alps between Western and Mediterranean Europe than crosses Austria.

Switzerland, not Austria, is in most ways the leading state in Alpine Western Europe (Table G-1). 24 Mountainous terrain and **landlocked location** can constitute crucial barriers to economic development, tending to inhibit the dissemination of ideas and innovations, obstruct circulation, constrain farming, and divide cultures. That is why Switzerland is such an important lesson in human geography. Through the skillful maximization of their opportunities (including the transfer needs of their neighbors), the Swiss have transformed their seemingly restrictive environment into a prosperous state. They used the waters cascading from their mountains to generate hydroelectric power to develop highly specialized industries. Swiss farmers perfected ways to optimize the productivity of mountain pastures and valley soils. Swiss leaders converted their country's isolation into stability, security, and neutrality, making it a world banking giant, a global magnet for money. Zürich, in the German sector, is the financial center; Geneva, in the French sector, is one of the world's most international cities. The Swiss feel that they do not need to join the EU, and they have not done so.

Austria, which joined the EU in 1995, is a remnant of the Austro-Hungarian Empire and has a historical geography that is far more reminiscent of unstable Eastern Europe than that of Switzerland. Even Austria's physical geography seems to demand that the country look eastward: it is at its widest, lowest, and most productive in the east, where the Danube links it to Hungary, its old ally in the anti-Muslim wars of the past.

Vienna, by far the Alpine subregion's largest city, also lies on the country's eastern perimeter. One of the world's most expressive primate cities with magnificent architecture and monumental art, Vienna today is Western Europe's easternmost city, situated on the doorstep of fast-changing Eastern Europe. But Austria's cultural landscapes and economic and political standards are those of Western Europe, a nation-state of the European core, its relative location enormously enhanced by the EU's eastward expansion of 2004.

▶ THE BRITISH ISLES

Off the coast of mainland Western Europe lie two major islands, surrounded by a constellation of tiny ones, that constitute the British Isles, a discrete region of the European realm (Fig. 1-18). The larger of the two major islands, which also lies nearest to the mainland (a mere 21 miles, or 34 km, at the closest point), is the island called *Britain*; its smaller neighbor to the west is *Ireland*.

The names attached to these islands and the countries they encompass are the source of some confusion. They still are called the British Isles, even though British dominance over most of Ireland ended in 1921. The nation-state that occupies Britain and the northeastern corner of Ireland is officially called the United Kingdom of Great Britain and Northern Ireland—United Kingdom for short and UK by abbreviation. But this country often is referred to simply as Britain, and its people are known as the British. The nation-state of Ireland officially is the Republic of Ireland (*Eire* in Irish Gaelic), but it does not include the whole island of Ireland.

How convenient it would be if physical and political geography coincided! Unfortunately, the two do not. During the long British occupation of Ireland, which is overwhelmingly Catholic, many Protestants from northern Britain settled in northeastern Ireland. In 1921, when British domination ended, the Irish were set free—except in that corner in the north, where London kept control to protect the area's Protestant settlers. That is why the country to this day is officially known as the United Kingdom of Great Britain and Northern Ireland.

Northern Ireland (Fig. 1-18) was home not only to Protestants from Britain, but also to a substantial population of Irish Catholics who found themselves on the wrong side of the border when Ireland was liberated. Ever since, conflict has intermittently engulfed Northern Ireland and spilled over into Britain and even, in the form of terrorism, into Western Europe. It has been and remains one of Europe's costliest struggles.

Although all of Britain lies in the United Kingdom, political divisions exist here as well. England is the largest of these units, the center of power from which the rest of the region was originally brought under unified control. The English conquered Wales in the Middle Ages, and Scotland's link to England, cemented when a Scottish king ascended the English throne in 1603, was ratified by the Act of Union of

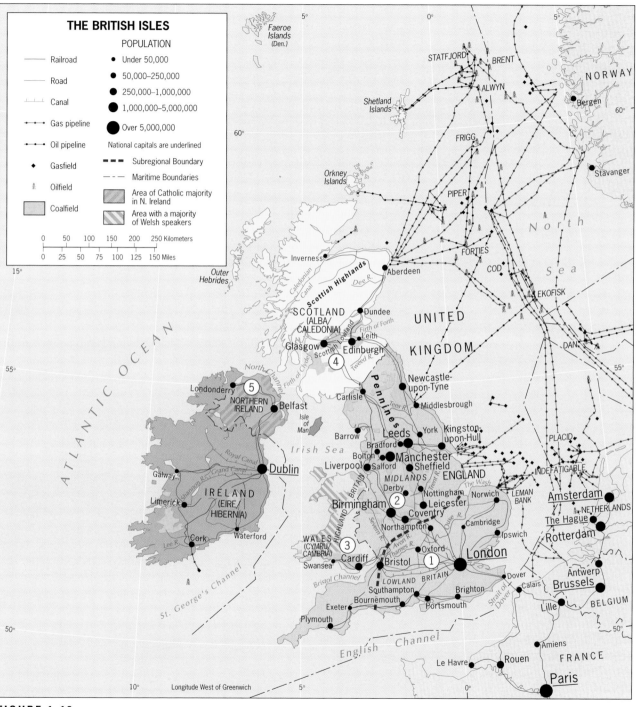

FIGURE 1-18

THE BRITISH ISLES

POPULATION

Railroad
Road
Canal
Gas pipeline
Oil pipeline
Gasfield
Oilfield
Coalfield

● Under 50,000
● 50,000–250,000
● 250,000–1,000,000
● 1,000,000–5,000,000
● Over 5,000,000

National capitals are underlined

--- Subregional Boundary
—·— Maritime Boundaries

Area of Catholic majority in N. Ireland
Area with a majority of Welsh speakers

0 50 100 150 200 250 Kilometers
0 25 50 75 100 125 150 Miles

1707. Thus England, Wales, Scotland, and Northern Ireland became the United Kingdom.

The British Isles form a distinct region of Europe for several reasons. Britain's insularity provided centuries of security from turbulent Europe, protecting the evolving British nation as it achieved a system of parliamentary government that had no peer in the Western world. Having united the Welsh, Scots, and Irish, the British set out to forge what would become the world's largest colonial empire. An era of mercantilism and domestic manufacturing (the latter based on water power from streams flowing off the Pennines, Britain's mountain backbone) foreshadowed the momentous Industrial Revolution, which transformed Britain— and much of the world. British cities became synonyms for specialized products as the smokestacks of factories rose like forests over the urban scene. London on the Thames River anchored an English core area that mushroomed into the headquarters of a global political, financial, and cultural empire. As recently as World War II, the narrow English Channel ensured the United Kingdom's impregnability against German invasion, giving the British time to organize their war machine. When the United Kingdom emerged from that conflict as a leading power among the victorious allies, it seemed that its superpower role in the postwar era was assured.

Two unanticipated developments changed that prospect: the worldwide collapse of colonial empires and the rapid resurgence of mainland Europe. Always ambivalent about the EC and EU, and with its first membership application vetoed by the French in 1963, Britain (admitted in 1973) has worked to restrain moves toward tighter integration. When most member-states adopted the new euro in favor of their national currencies, the British kept their pound sterling and delayed their participation in the EMU. As for a federalized Europe, to Britain this is out of the question. In this as in other respects, Britain's historic, insular standoffishness continues.

The United Kingdom

As we noted earlier, the British Isles as a region consists of two political entities: the United Kingdom and Ireland. The UK, with an area about the size of Oregon and a population of just over 60 million, is by European standards quite a large country. Based on a combination of physiographic, historical, cultural, economic, and political criteria, the United Kingdom can be divided into five subregions:

1. *Southern England.* Centered on the gigantic London metropolitan area, this is the UK's most affluent subregion, with one-third of the country's population. Financial, communications, engineering, and energy-related industries cluster in this economically and politically dominant area. London exemplifies the momentum of long-term agglomeration; today this subregion benefits anew not only from its superior links to the European mainland but also as one of the three leading "world cities" on the global scene.

2. *Northern England.* As the map suggests, this subregion should really be called Northern, Central, and Western England. Its center, appropriately called the Midlands, was the focus of the Industrial Revolution and the spectacular rise of Manchester, the source; Liverpool, the great port; Birmingham, unmatched industrial city; and many other manufacturing cities and towns. But here obsolescence has overtaken what was once ultramodern, and the North now

suffers from Rustbelt conditions and high unemployment among immigrants from South Asia, Africa, and the Caribbean.

3. *Wales.* This nearly rectangular, rugged territory was a refuge for ancient Celtic peoples, and in its western counties more than half the inhabitants still speak Welsh. Because of the high-quality coal reserves in its southern tier, Wales too was engulfed by the Industrial Revolution, and Cardiff, the capital, was once the world's leading coal exporter. But the fortunes of Wales also declined, and many Welsh emigrated. Among the 3 million who remained, however, the flame of Welsh nationalism survived, and in 1997 the voters approved the establishment of a Welsh Assembly to administer public services in Wales, a first devolutionary step.

4. *Scotland.* Nearly twice as large as the Netherlands and with a population about the size of Denmark's, Scotland is a major component of the United Kingdom. As Figure 1-18 suggests, most of Scotland's more than 5 million people live in the Scottish Lowlands anchored by Edinburgh, the capital, in the east and Glasgow in the west. Attracted there by the labor demands of the Industrial Revolution (coal and iron reserves lay in the area), the Scots developed a world-class shipbuilding industry. Decline and obsolescence were followed by high-tech development, notably in the hinterland of Glasgow, and Scottish participation in the exploitation of oil and gas reserves under the North Sea (Fig. 1-18), which transformed the eastern ports of Aberdeen and Leith (Edinburgh). But many Scots feel that they are disadvantaged within the UK and should play a major role in the EU. Therefore, when the British government put the option of a Scottish parliament before the voters in 1997, 74 percent approved. Many Scots still hope that total independence lies in their future.

5. *Northern Ireland.* Prospects of devolution in Scotland pale before the devastation caused by

political and sectarian conflict in Northern Ireland. With a population of 1.8 million occupying the northeastern one-sixth of the island of Ireland, this area represents the troubled legacy of British colonial rule. A declining majority, now about 54 percent of the people in Northern Ireland, trace their ancestry to Scotland or England and are Protestants; a growing minority, currently around 45 percent, are Roman Catholics, who share their Catholicism with virtually the entire population of the Irish Republic on the other side of the border. Although Figure 1-18 suggests that there are majority areas of Protestants and Catholics in Northern Ireland, no clear separation exists; mostly they live in clusters throughout the territory, including walled-off neighborhoods in the major cities of Belfast and Londonderry. Partition is no solution to a conflict that has raged for over three decades at a cost of thousands of lives; Catholics accuse London as well as the local Protestant-dominated administration of discrimination, whereas Protestants accuse Catholics of seeking union with the Republic of Ireland. Repeated efforts to achieve a settlement have ended in failure; the most recent mediation produced a Northern Ireland Assembly to which powers would be devolved from London. But in 2003 this political program failed, and the British government was forced to resume direct rule.

Republic of Ireland

What Northern Ireland is missing through its conflicts is shown by the Irish Republic itself: a growing, booming service-based economy, the fastest-growing in all of Europe for a time around the turn of this century, when Ireland was dubbed the "Celtic Tiger." Burgeoning cities and towns (and rising real-estate prices), mushrooming industrial parks, bustling traffic, and construction everywhere reflect this new era. For the first time ever, workers of Irish descent returned from foreign places to take jobs at home; non-Irish immigrants also arrived posing some new social problems for a closely knit, long-isolated society.

The Republic of Ireland fought itself free from British colonial rule just three generations ago. Its cool, moist climate had earlier led to the adoption of the American potato as the staple crop, but excessive rain and a blight in the late 1840s caused famine and cost over 1 million lives. Another two million Irish emigrated.

Independence in 1921 did not bring economic prosperity, and Ireland stagnated until the 1990s. Then, European telecommunications service industries saw labor and locational opportunities in Ireland, and soon Ireland was Europe's leading call center for the realm's rapidly expanding toll-free telephone market. Other service industries followed suit, benefiting from Ireland's well-educated but not highly paid labor pool. Changing economic conditions worldwide and in Europe slowed Ireland's economy after 2002, but its days as a regional backwater have ended.

NORTHERN (NORDIC) EUROPE

North of Europe's Western European core area lies a disconnected group of six countries that exemplify core-periphery contrasts. Northern Europe is a region of difficult environments: generally cold climates, poorly developed soils, limited mineral resources, long distances. Together, the six countries in this region—Sweden, Norway, Denmark, Finland, Estonia, and Iceland—contain just over 25 million inhabitants, which is a lower total population than that of Benelux and only one-seventh that of Western Europe. The overall land area, on the other hand, is almost the size of the entire European core. Here in peripheral *Norden*, as the people call their northerly domain, national core areas lie in the south: note the location of the capitals of Helsinki, Stockholm, and Oslo at approximately the same latitude (Fig. 1-19).

Northern Europe's peripheral situation is more than environmental. As viewed from Europe's core area, Norden is on the way to nowhere. No major shipping lanes lead from Western Europe past Norden to other productive areas of the world, limiting interaction of the sort that ties the British Isles to the mainland. Moreover, except for relatively small Denmark and Estonia, all of Norden lies separated from the European mainland by water. At all levels of spatial generalization, isolation is pervasive in Norden.

Northern Europe's remoteness, isolation, and environmental severity also have had positive effects for this region. The countries of the Scandinavian Peninsula lay removed from the wars of mainland Europe (although Norway was overrun by Nazi Germany during World War II). The three major languages—Danish, Swedish, and Norwegian—are mutually intelligible, which creates one of the criteria delimiting this region. Another regional criterion is the overwhelming adherence to the same Lutheran church in each of the Scandinavian countries (Norway, Sweden, and Denmark), Iceland, and Finland. Furthermore, democratic and representative governments emerged early, and individual rights and social welfare have long been carefully protected. Women participate more fully in government and politics here than in any other region of the world.

Sweden is the largest Nordic country in terms of both population and territory. Most Swedes live south of 60° North latitude (which passes through Uppsala), in what is climatically the most moderate part of the country (Fig. 1-19). Here lie the capital, core area, and, as Figure 1-7 shows, the main industrial districts; here, too, are the main agricultural areas that benefit from the lower relief, better soils, and milder climate.

Sweden long exported raw or semifinished materials to industrial countries, but today the Swedes are making finished products themselves, including automobiles, electronics, stainless steel, furniture, and glassware. Much of this production is based on local resources, including a major iron ore reserve at Kiruna in the far north (there is a steel mill at Luleå). Swedish manufacturing, in contrast to that of several Western European countries, is based on dozens of small and medium-sized towns specializing in particular products. Energy-poor Sweden was a pioneer in the development of nuclear power, but a national debate over the risks involved has reversed that course.

Norway does not need a nuclear power industry to supply its energy needs. It has found its economic opportunities on, in, and beneath the sea. Norway's fishing industry, now augmented by highly efficient fish farms, long has been a cornerstone of the economy, and its merchant marine spans the world. But since the 1970s, Norway's economic life has been transformed by the bounty of oil and natural gas discovered in its sector of the North Sea.

With its limited patches of cultivable soil, high relief, extensive forests, frigid north, and spectacularly fjorded coastline, Norway has nothing to compare to Sweden's agricultural or industrial development. Its cities, from the capital Oslo and the North Sea port of Bergen to the historic national focus of Trondheim as well as Arctic Hammerfest, lie on the coast and have difficult overland connections. The isolated northern province of Finnmark has even become the scene of an autonomy movement among the reindeer-herding indigenous Saami (Fig. 1-12). Norway has been described as a necklace, its beads linked by the thinnest of strands. But this has not constrained national development. Norway in 2003 had the second-lowest unemployment rate in Europe (after tiny Luxembourg). In terms of income per capita, Norway is the fourth-richest country in the world.

Norwegians have a strong national consciousness and a spirit of independence. In 1994, when Sweden and Finland voted to join the European Union, the Norwegians again said no. They did not want to trade their economic independence for the regulations of a larger, even possibly safer, Europe.

Denmark, territorially small by Scandinavian standards, has a population of 5.4 million, second largest in the Nordic region after Sweden. It consists of the Jutland Peninsula and several islands to the east at the gateway to the Baltic Sea; it is on one of these islands, Sjaelland, that the capital of Copenhagen is located. Copenhagen,

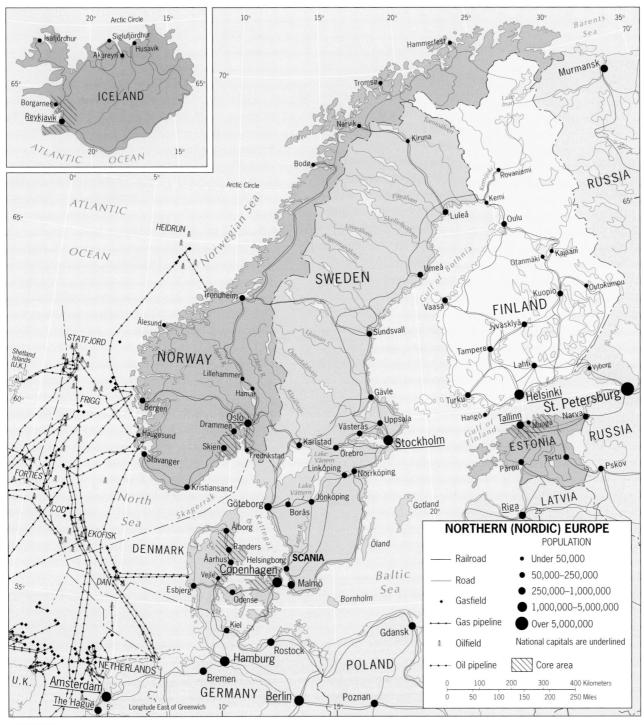

FIGURE 1-19

NORTHERN (NORDIC) EUROPE

POPULATION

—— Railroad	• Under 50,000
—— Road	● 50,000–250,000
♦ Gasfield	● 250,000–1,000,000
—— Gas pipeline	⬤ 1,000,000–5,000,000
⚒ Oilfield	⬤ Over 5,000,000
—— Oil pipeline	▨ Core area

National capitals are underlined

0   100   200   300   400 Kilometers
0   50   100   150   200   250 Miles

the "Singapore of the Baltic," has long been a port that collects, stores, and transships large quantities of goods. 25 This **break-of-bulk** function exists because many oceangoing vessels cannot enter the shallow Baltic 26 Sea, making the city an *entrepôt* where transfer facilities and activities prevail. The completion of the Øresund bridge-tunnel link to southern Sweden enhanced Copenhagen's situation in 2000 (Fig. 1-19).

Denmark remains a kingdom, and in centuries past Danish influence spread far beyond its present confines. Remnants of that period now challenge Denmark's governance. Greenland came under Danish rule after union with Norway (1380) and remained a Danish domain when that union ended (1814). In 1953, Greenland's status changed from colony to province, and in 1979 the 60,000 inhabitants were given home rule with an Inuit name: *Kalaallit Nunaat.* They promptly exercised their rights by withdrawing from the European Union, of which they had become a part when Denmark joined. Another restive dependency is the Faroe Islands, located between Scotland and Iceland. These 17 small islands and their 45,000 inhabitants were awarded self-government in 1948, complete with their own flag and currency, but even this was not enough to defuse demands for total independence. A referendum in mid-2001 confirmed that even Denmark is not immune from Europe's devolutionary forces (Fig. 1-12).

Finland, territorially almost as large as Germany, has only 5.2 million residents, most of them concentrated in the triangle formed by the capital, Helsinki, the textile-producing center, Tampere, and the shipbuilding center, Turku (Fig. 1-19). A land of evergreen forests and glacial lakes, Finland has an economy that has long been sustained by wood and wood product exports. But the Finns, being a skillful and productive people, have developed a diversified economy in which the manufacture of precision machinery and telecommunications equipment as well as the growing of staple crops are key.

As in Norway and Sweden, environmental challenges and relative location have created Nordic cultural landscapes in Finland, but the Finns are not a Scandinavian people; their linguistic and historic links are instead with the Estonians across the Gulf of Finland. As we will see in Chapter 2, ethnic groups speaking Finno-Ugric languages are widely dispersed across what is today western Russia.

Estonia, northernmost of the three "Baltic states," is part of Nordic Europe by virtue of its ethnic and linguistic ties to Finland. But during the period of Soviet control from 1940 to 1991, Estonia's demographic structure changed drastically: today about 30 percent of its 1.4 million inhabitants are Russians, most of whom came there as colonizers.

After a difficult period of adjustment, Estonia today is forging ahead of its Baltic neighbors (see Table G-1) and catching up with its Nordic counterparts. Busy traffic links Tallinn, the capital, with Helsinki, and a new free-trade zone at Muuga Harbor facilitates commerce with Russia. But more important for Estonia's future is its entry into the European Union in 2004.

Iceland, the volcanic, glacier-studded island in the frigid waters of the North Atlantic just south of the Arctic Circle, is the sixth country in this region. Inhabited by people with Scandinavian ancestries (population: 285,000), Iceland and its small neighboring archipelago, the Westermann Islands, are of special scientific interest because they lie on the Mid-Atlantic Ridge, where the Eurasian and North American tectonic plates of the Earth's crust are

diverging and new land can be seen forming (see Fig. G-4).

Iceland's population is almost totally urban, and the capital, Reykjavik, contains about half the country's inhabitants. The nation's economic geography is almost entirely oriented to the surrounding waters, whose seafood harvests give Iceland one of the world's highest standards of living—but at the risk of overfishing. Disputes over fishing grounds and fish quotas have intensified in recent decades; the Icelanders argue that, unlike the Norwegians or the British, they have little or no alternative economic opportunity.

MEDITERRANEAN EUROPE

South of Europe's core lie the six countries that comprise the Mediterranean region: Italy, Spain, Portugal, Greece, and the island countries of Cyprus and Malta (Fig. 1-20). Like Northern Europe, this is a region of peninsulas, and again it is discontinuous. It lies separated from the European core, and core-periphery contrasts in some areas are quite sharp. A degree of continuity, dating from Greco-Roman times, marks the region's languages and religion, lifeways, and cultural landscapes. As Figure 1-2 shows, natural environments in this region are dominated by a climatic regime that bears its very name—Mediterranean. Dry, hot summers are the norm, so that moisture is often in short supply during the growing season, and specially adapted plants mark local agriculture.

In terms of raw materials, Southern (Mediterranean) Europe is not nearly as well endowed as the European core, as reflected by the map of industrial complexes (Fig. 1-7). Only northern Italy and northern Spain (the former through massive imports of coal and iron ore) have become part of the core area. Furthermore, the region has been largely deforested, and the limited and highly seasonal water supply constrains Southern Europe's hydroelectric opportunities.

A key contrast between Northern and Southern Europe lies in their populations. Although territorially smaller than Northern Europe, Mediterranean Europe has nearly five times as many people (122 million in 2004). Population distribution continues to reflect the agricultural bases of the preindustrial era, with large concentrations in coastal lowlands and fertile river basins, although the growth of major industrial centers, notably in northern Italy and northern Spain, has superimposed a new mosaic. Still, urbanization in Southern Europe is far below that in Western or Northern Europe, or in the British Isles (Portugal's 48 percent is one of the lowest in the realm). Other data indicate that living standards in Mediterranean Europe also lag behind (Table G-1). All this is changing, of course, with the EU's expansion into Eastern Europe. As of 2004, Southern Europe no longer was the lowest-ranking EU region in terms of GNP, living standards, or urbanization.

Italy

Centrally located in the Mediterranean region, most populous of the Mediterranean states, best connected to the European core, and economically most advanced is Italy (58.1 million), a charter member of Europe's Common Market.

Administratively, Italy is organized into 20 regions, many with historic roots dating back centuries (Fig. 1-21). Several of these regions have become powerful economic entities centered on major cities, such as Lombardy (Milan) and Piedmont (Turin); others are historic hearths of Italian culture, including Tuscany (Florence) and Veneto (Venice). These regions in the northern half of Italy stand in strong social, economic, and political contrast to such southern regions as Calabria (the "toe" of the Italian "boot") and Italy's two major Mediterranean islands, Sicily and Sardinia. Not surprisingly, Italy is often described as two countries—a progressive north and a stagnant south, or *Mezzogiorno*.

North and south are bound by the ancient headquarters, Rome, which lies astride the narrow

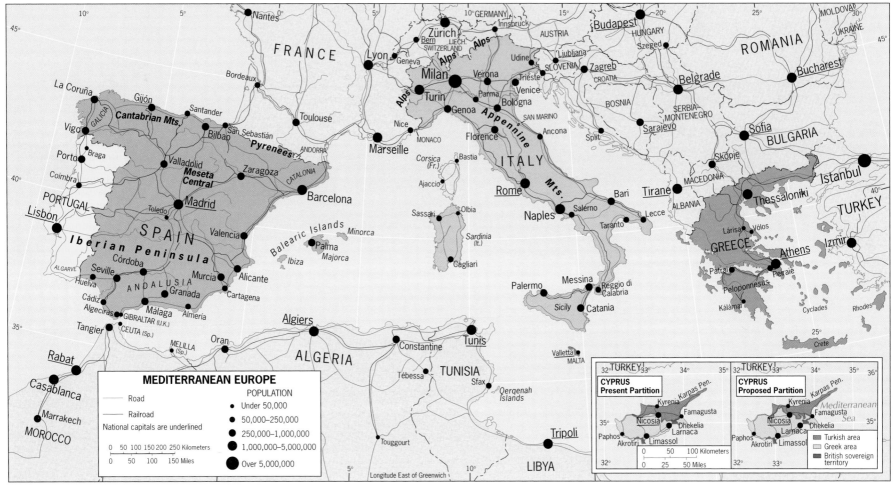

FIGURE 1-20

transition zone between Italy's contrasting halves. This zone is referred to as the *Ancona Line*, after the city on the Adriatic coast where the zone reaches the other side of the peninsula (Fig. 1-21). Whereas Rome remains Italy's capital and cultural focus, the functional core area of Italy has shifted northward into Lombardy in the basin of the Po River. Here lies Southern Europe's leading manufacturing complex, in which a large, skilled labor force and ample hydroelectric power from Alpine and Appennine slopes combine with a host of im-

ported raw materials to produce a wide range of machinery and precision equipment. The Milan–Turin–Genoa triangle exports appliances, instruments, automobiles, ships, and many specialized products. Meanwhile, the Po Basin, lying on the margins of the region's dominant Mediterranean climatic regime, enjoys a more even pattern of rainfall distribution throughout the year, making it a productive agricultural zone as well.

Metropolitan Milan embodies the new, modern Italy. Not only is Milan (at 4.3 million) Italy's

largest city and leading manufacturing center— making Lombardy one of Europe's Four Motors— but it also is the country's financial and service-industry headquarters. Today the Milan area, a cornerstone of the European core, has just 9 percent of Italy's population but accounts for one-third of the entire country's national income.

As in Germany, the lowest-income regions of Italy lie concentrated in one part of the country: the Mezzogiorno in the south. But Italy's north-south disparity continues to grow, which has led to tax-

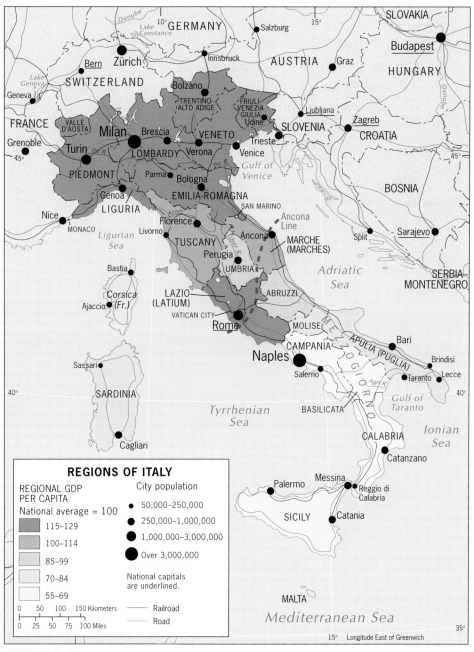

REGIONS OF ITALY

REGIONAL GDP
PER CAPITA

National average = 100

▓	115–129
▓	100–114
▒	85–99
░	70–84
□	55–69

City population

- • 50,000–250,000
- ● 250,000–1,000,000
- ⬤ 1,000,000–3,000,000
- ⬤ Over 3,000,000

National capitals
are underlined.

0   50   100   150 Kilometers
0   25   50   75   100 Miles

━━ Railroad
── Road

FIGURE 1-21

payers' revolts over the subsidies the state pays to the poorer southern regions (a separatist movement has even espoused an independent *Padania* in the north). In truth, the south receives the bulk of Italy's illegal immigrants, whether from Africa across the Mediterranean or from the former Yugoslavia and Albania across the Adriatic, and it is these workers, willing to work for low wages, who move north to take jobs in the factories of Milan and Turin. Italian southerners tend to stay where they were born. Sicily and the Mezzogiorno exemplify the problems of a periphery.

Iberia

At the western end of Southern Europe lies the Iberian Peninsula, separated from France and Western Europe by the rugged Pyrenees and from North Africa by the narrow Strait of Gibraltar (Fig. 1-20). Spain (population: 41.4 million) occupies most of this compact Mediterranean landmass. Portugal lies in its southwestern corner.

Imperial Romans, Muslim Moors, and Catholic kings left their imprints on Iberia, notably the boundary between Spain and Portugal dating from the twelfth century. The golden age of colonialism was followed by dictatorial rule and economic stagnation.

Today both Spain and Portugal are democracies, and their economies are doing well, helped by EU subsidies. As Figure 1-22 shows, **Spain** followed the leads of Germany and France and decentralized its administrative structure, creating 17 regions called Autonomous Communities (ACs). Each AC has its own parliament and administration that control planning, public works, cultural affairs, education, environmental matters, and even, to some extent, international commerce. Each AC can negotiate its own degree of autonomy with the central government in Madrid. This new system, however, has not been enough to defuse Spain's most problematic devolutionary issue, which involves the Basques in the AC

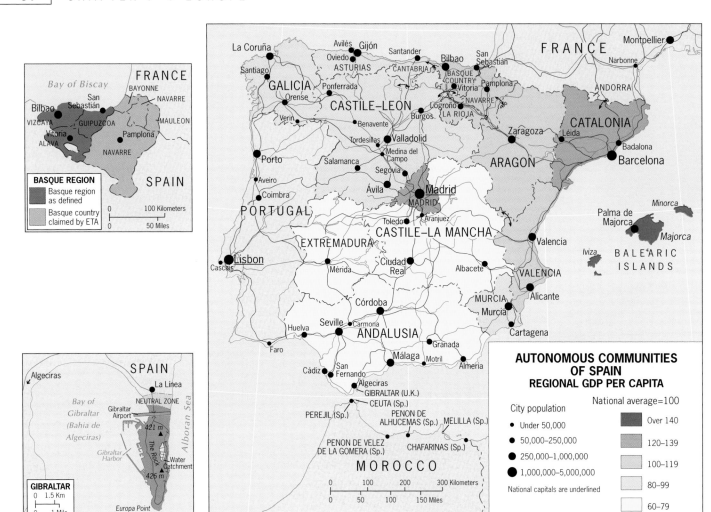

FIGURE 1-22

mapped as Basque Country (Fig. 1-22, especially the top inset map).

Particularly noteworthy is Catalonia, the triangular AC in Spain's northeastern corner, adjacent to France. Centered on prosperous, productive Barcelona, Catalonia is Spain's leading industrial area, an AC with a population of 6.5 million. It is imbued with a fierce nationalism, endowed with its own language and culture, and economically it is one of the Four Motors of Europe. Note that while most of Spain's industrial raw materials lie

in the northwest, its major industrial development has taken place in the northeast, where innovations and skills propel a high-technology-driven regional economy. In recent years, Catalonia—with 6 percent of Spain's territory and 16 percent of its population—has annually produced 25 percent of all Spanish exports and nearly 40 percent of its industrial exports. Such economic strength translates into political power, and in Spain the issue of Catalonian separatism is never far from the surface.

Here are two items that keep Catalans angry: first, the fact that Spain's first high-speed rail line, built with EU funds, runs south from Madrid to poor, agricultural Andalusia and not to rich, industrial Barcelona; and second, that the Madrid government is contemplating the southward diversion of river water from the Rio Ebro through dams and pipelines.

As Figure 1-22 indicates, Spain's capital and largest city, Madrid, lies near the geographic center of the state. It also lies along an economic-geographic divide. Catalonia and Madrid are Spain's most prosperous ACs; the contiguous group of five ACs between the Basque Country and Valencia rank next. Tourism (notably along the Mediterranean coast) and winegrowing (especially in La Rioja) contribute importantly here. Ranking below this cluster of ACs are the four such units of the northwest: Galicia, Castile-Leon, and industrialized Asturias and Cantabria. Incomes in these regions are below the national average because industrial obsolescence, dwindling raw material sources, and emigration have plagued their economies. Worst off, however, are the three large ACs to the south of Madrid: Extremadura, Castile-La Mancha, and especially Andalusia. Drought, inadequate land reform, scarce resources, and remoteness from Spain's fast-growing northeast are among the factors that inhibit development here. Even as EU members seek to reduce the differences among themselves, individual countries face the geographic consequences of focused growth that deepens internal divisions.

The state of **Portugal** (population: 10.4 million), a comparatively poor country that has benefited enormously from its admission to the EU, occupies the southwestern corner of the Iberian Peninsula. Since one rule of EU membership is that the richer members assist the poorer ones, Portugal shows the results in a massive renovation project in the capital, Lisbon, as well as in the modernization of surface transport routes.

Unlike Spain, which has major population clusters on its interior plateau as well as its coastal lowlands, the Portuguese are concentrated along and near the Atlantic coast. Lisbon and the second city, Porto, are coastal cities; the best farmlands lie in the moister western and northern zones of the country. But the farms are small and inefficient, and although Portugal remains dominantly rural, it must import as much as half of its foodstuffs. Exporting textiles, wines, corks, and fish, and running up an annual deficit, the indebted Portuguese economy remains a far cry from those of other European countries of similar dimensions.

Greece and Cyprus

The eastern segment of the region we define as Southern Europe is dominated by **Greece**, an outlier of both Mediterranean Europe and the European Union. Greece's land boundaries are with Turkey, Bulgaria, Macedonia, and Albania; as Figure 1-23 reveals, it also owns islands just offshore from mainland Turkey. Altogether, the Greek archipelago numbers some 2000 islands ranging in size from Crete (3218 square miles [8335 sq km]) to small specks of land in the Cyclades. In addition, Greeks represent the great majority on the now-divided island of Cyprus.

Ancient Greece was a cradle of Western civilization, and later it was absorbed by the expanding Roman Empire. For some 350 years beginning in the mid-fifteenth century, Greece was under the sway of the Ottoman Turks. Greece regained independence in 1827, but not until nearly a century later, through a series of Balkan wars, did it acquire its present boundaries. During World War II, Nazi Germany occupied and ravaged the country, and in the postwar period the Greeks have quarreled with the Turks, the Albanians, and the newly independent Macedonians. Today, Greece finds itself between the Muslim world of Southwest Asia and the Muslim communities of Eastern Europe; it is still the only noncontiguous mainland EU member even after the accession of ten more states in 2004.

Volatile as Greece's surroundings are, Greece itself is a country on the move, an EU success story to rival Ireland's. Political upheavals in the 1970s and economic stagnation in the 1980s are all but forgotten in the new century: now Greece, its economy booming, is described as the locomotive for the Balkans, a beacon for the EU in a crucial part of the world. Infrastructure improvements focused on the Athens urban area, where more than one-third of the population is concentrated, include new subways, a new beltway, and a new airport.

Greece will confront series challenges in the post–2004 period. With the accession of ten Eastern European members to the EU, Greece will lose much of its EU subsidy to these poorer countries. Greece's democratic institutions still need strengthening, and corruption remains a serious problem. Educational institutions need modernization. And while Greece is on better terms today with its fractious neighbors and invests in development projects in Macedonia and Bulgaria (for example, an oil pipeline from Thessaloniki to a Greek-owned refinery in Skopje, Macedonia), it will take skillful diplomacy to navigate the shoals of historic discord.

Modern Greece is a nation of 11 million centered on historic Athens, one of the realm's great cities. With its port of Piraeus, metropolitan Athens contains about 40 percent of the Greek population, making it one of Europe's most congested and polluted urban areas. Athens is the quintessential primate city; the monumental architecture of ancient Greece still dominates its cultural landscape. The Acropolis and other prominent landmarks attract a steady stream of visitors; tourism is one of Greece's leading sources of foreign revenues, and Athens is only the beginning of what the country has to offer.

Deforestation, soil erosion, and variable rainfall make farming difficult in much of Greece, but the country remains strongly agrarian. It is self-sufficient in staple foods, and farm products continue to figure strongly among exports. But other sectors of the economy, including manufacturing (textiles) and the service industries, are growing rapidly. The challenge for Greece will be to maintain its growth after the momentous EU expansion of 2004.

Cyprus lies in the far northeast corner of the Mediterranean Sea, much closer to Turkey than to Greece (Fig. 1-23), but is peopled dominantly by Greeks rather than Turks. In 1571, the Turks conquered Cyprus, then ruled by Venice, and controlled it until 1878 when the British took over. Most of the island's Turks arrived during the Ottoman period; the Greeks have been there longest.

When the British were ready to give Cyprus independence after World War II, the 80 percent Greek majority mostly preferred union with Greece. Ethnic conflict followed, but in 1960 the British granted Cyprus independence under a constitution

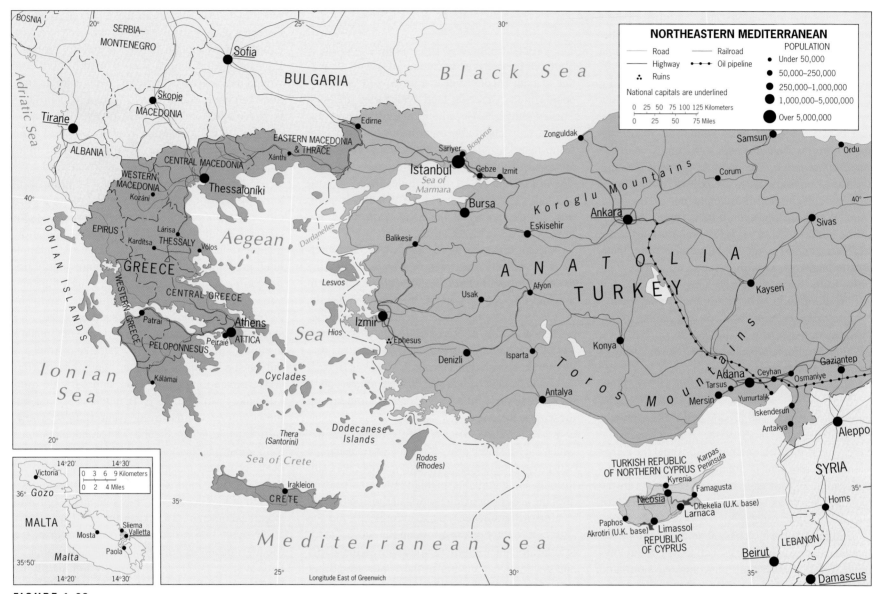

FIGURE 1-23

that prescribed majority rule but guaranteed minority rights.

This fragile order broke down in 1974, and civil war engulfed the island. Turkey sent in troops and massive dislocation followed, resulting in the partition of Cyprus into northern Turkish and southern Greek sectors (Fig. 1-23). In 1983, the 40 percent of Cyprus under Turkish control, with about 100,000 inhabitants (and some 30,000 Turkish soldiers), declared itself the independent Turkish Republic of Northern Cyprus. Only Turkey recognizes this ministate (which now contains a population of nearly 200,000); the international community recognizes the government on the Greek side as legitimate. With about 875,000 residents, a relatively prosperous economy based on agriculture and tourism, and strong links with Europe, the Greek-side government qualified the south for 2004 membership in the European Union.

The potential for serious conflict over Cyprus has not disappeared. In effect, the "Green Line" that separates the Turkish and Greek communities constitutes not just a regional border but a boundary between geographic realms.

The Mediterranean region contains one other ministate, **Malta**, located south of Sicily. Malta is a small archipelago of three inhabited and two uninhabited islands with a population under 400,000 (Fig. 1-23, inset). An ancient crossroads and culturally rich with Arab, Phoenician, Italian, and British infusions, Malta became a British dependency and served British shipping and its military. It suffered terribly during World War II bombings, but despite limited natural resources recovered strongly during the postwar period. Today Malta has a booming tourist industry and a relatively high standard of living, and is a new member of the European Union.

◗ EASTERN EUROPE

As Figure 1-13 shows, Eastern Europe is not only territorially the largest region in the European realm: it also contains more countries (17) than any other European region. Almost all of Eastern Europe lies outside the core, and the problems of the periphery affect many of its countries. From the North European Lowland in Poland to the rugged highlands of the south, this is a region of physiographic, cultural, and political fragmentation. Open plains, major rivers, strategic mountains, isolated valleys, and crucial corridors all have influenced Eastern Europe's tumultuous migrations, epic battles, foreign invasions, and imperial episodes. Illyrians, Slavs, Turks, Hungarians, and other peoples converged on this region from near and far. Ethnic and cultural differences have kept them in chronic conflict.

27 Geographers call this region a **shatter belt**, a zone of persistent splintering and fracturing. Geographic terminology uses several expressions to describe the breakup of established order, and these tend to have their roots in this part of the **28** world. One of them is **balkanization**. The southern half of Eastern Europe is referred to as the Balkans or Balkan Peninsula, after the name of a mountain range in Bulgaria. Balkanization denotes the recurrent division and fragmentation of this part of Eastern Europe, and it is now applied to any place where such processes take place. A more recent term is *ethnic cleansing*—the forcible ouster of entire populations from their homelands by a stronger power bent on taking their territories. The term may be new, but the process is as old as Eastern Europe itself.

Each episode in the historical geography of Eastern Europe has left its legacy in the cultural landscape. Twenty centuries ago the Roman Empire ruled much of it (Romania is a cartographic reminder of this period); during the past half-century, the Soviet Empire controlled almost all of it. In the intervening two millennia, Christian Orthodox church doctrines spread from the southeast, and Roman Catholicism advanced from the northwest. Turkish (Ottoman) Muslims invaded and created an empire that reached the environs of Vienna. By the time the Austro-Hungarian Empire ousted the Turks, millions of Eastern Europeans had been converted to Islam. Albania today remains a dominantly Muslim country. In the twentieth century, Eastern Europe became a battleground between superpowers, and the complicated map reflects the results through 1991 (Fig. 1-24).

The subsequent collapse of the Soviet Union freed several of Russia's neighbors, and with only one exception—Belarus—these countries turned their gaze from Moscow to the west, specifically to the European Union and its economic promise. This changed the map of Eastern Europe, repositioning the realm boundary eastward and adding five countries to the region (Fig. 1-25): Latvia, Lithuania, Belarus, Moldova, and Ukraine. Meanwhile, however, two of Eastern Europe's established states fell apart: Czechoslovakia peacefully, Yugoslavia violently.*

The Geographic Framework

Eastern Europe is changing so rapidly that long-established regional frameworks are buckling. Even the number of countries in this region may not be stable: the future of Kosovo is uncertain (this territory, under NATO administration, may become an independent state) and the possibility exists that Montenegro will separate from the Serbia-Montenegro of which it is currently a part.

One way to approach the complex geography of this region is to separate the countries that became part of the EU in 2004 from those that did not, because the exclusion of the latter had economic- and political-geographic causes. But this, too, is likely to change, because (as we noted earlier) Romania and Bulgaria are already in the waiting room and Croatia may be next. So, for the time being, we continue to group Eastern Europe's 17 countries (as of late 2004) into four subregions based on their relative location:

1. Countries Facing the Baltic Sea
2. The Landlocked Center
3. Countries Facing the Black Sea
4. Countries Facing the Adriatic Sea

---

*Between 1919 and 1991, the name *Yugoslavia* referred to the multicultural state shown in Figure 1-24. When civil ware destroyed the old order, parts of Yugoslavia became independent states, including Slovenia, Croatia, Bosnia, and Macedonia. The largest of these states, dominated by the Serbs, kept the name Yugoslavia. But in 2002 the then-Yugoslav government announced that it would agree to a name change, calling their country Serbia-Montenegro after its two main components. (The future of Kosovo, under United Nations administration but technically still a part of Serbia-Montenegro, remained unresolved in late 2004.)

Countries Facing the Baltic Sea

Poland dominates this subregion, which also includes Lithuania and Latvia and, by virtue of relative location, Belarus (although Belarus does not possess a Baltic Sea coast). Wedged between coastal Poland and Lithuania is the Russian *exclave* of Kaliningrad, which is not functionally a part of Eastern Europe (see Chapter 2).

Figure 1-24 displays Poland's most recent boundary shifts: in the aftermath of World War II, the whole country shifted westward, losing land to the (then) USSR in the east and gaining it at the expense of Germany to the west. Situated on the North European Lowland, Poland traditionally was an agrarian country, but during the Soviet-communist period Silesia (once part of Germany) became its industrial heartland, and Katowice, Wroclaw, and Krakow grew into major industrial cities amid some of the world's worst environmental pollution. The Soviets invested far less in agriculture, collectivizing farms without modernizing the technology and leaving post-Soviet farming in abysmal condition.

By many measures, Poland is the most important state that joined the EU in 2004: with nearly 40 million people, it has half the total population of the ten new entrants, territorially it is larger than all the others combined, and its economy is the largest by far.

Lithuania (3.5 million) and Latvia (2.3 million) face the Baltic Sea north of Poland. **Lithuania**, remnant of the Grand Duchy of Lithuania that once dominated Eastern Europe from the Baltic to the Black Sea, is centered on the interior capital of Vilnius. When Kaliningrad became a Russian **29** **exclave** as a World War II prize, Lithuania was left with only about 50 miles (80 km) of Baltic coastline and a small port (Klaipeda) that was not even connected to Vilnius by rail. But since Lithuania (like Latvia and Estonia) fell under Soviet domination, that seemed to matter little as its trade was with the Soviet Union, not the outside world.

Latvia, on the other hand, is centered on the Baltic port of Riga, but Latvia experienced far more Russian immigration than Lithuania did: today Latvians make up only about 54 percent of the population. Its economy was geared almost totally to Moscow, and after the half-century of Soviet domination over the Baltics (1940–1991) ended, Latvia was left with huge economic problems and few options. In recent years, though, Latvia's economy has reoriented and improved, and in 2003 Lithuania recorded the highest growth rate among European states, which enabled both countries to join the EU in 2004.

Belarus has no Baltic coast, but it borders all three of the coastal states in this subregion. About 80 percent of its nearly 10 million people are Belarussians ("White" Russians), a West Slavic people. Only some 13 percent are (East Slavic) Russians. Devastated during World War II, Belarus became one of Moscow's most loyal satellites, and the Soviets made Mensk (Minsk), the capital, into a large industrial center. But in the post-Soviet era, Belarus has lagged badly, and its government has retained powers reminiscent of the communist era. Unlike its neighbors, Belarus has no interest in joining the EU. On the contrary, its overtures have been toward Moscow: it seeks to rejoin Russia in some formal way.

FROM THE FIELDNOTES

"Turbulent history and prosperous past are etched on the cultural landscapes of Poland's port, Gdynia, and the old Hanseatic city of Gdansk nearby. Despite major wartime destruction, much of the old architecture survives, and we found restoration underway throughout these twin cities. Gdansk was the stage for the 1980s rebellion of the Solidarity labor union against Poland's communist regime; today city and country struggle through the difficult transition to a new economic and social order. Attracting foreign visitors is one way to raise revenues, and the historic old city on the Baltic is becoming a tourist draw."

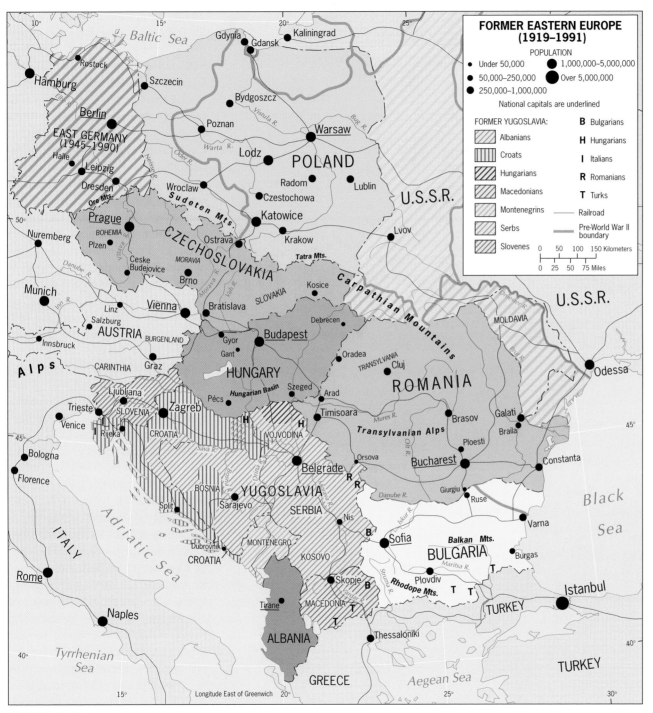

FIGURE 1-24

Map legend:

FORMER EASTERN EUROPE (1919–1991)

POPULATION
- Under 50,000
- 50,000–250,000
- 250,000–1,000,000
- 1,000,000–5,000,000
- Over 5,000,000

National capitals are underlined

FORMER YUGOSLAVIA:
- Albanians
- Croats
- Hungarians
- Macedonians
- Montenegrins
- Serbs
- Slovenes

- **B** Bulgarians
- **H** Hungarians
- **I** Italians
- **R** Romanians
- **T** Turks

— Railroad
Pre-World War II boundary

0 50 100 150 Kilometers
0 25 50 75 Miles

The Landlocked Center

Three states comprise this subregion: the Czech Republic, Slovakia, and Hungary. Until 1993, the first two formed the state of Czechoslovakia, but their "velvet divorce" broke it up. The Czechs got the better of the deal; Slovakia is this subregion's poorest country by far.

The **Czech Republic** (10.3 million) centers on Bohemia, the mountain-enclosed core area that contains the historic capital, Prague. This province always has been cosmopolitan and Western in its exposure, outlook, development, and linkages. Prague lies in the basin of the Elbe River, Bohemia's traditional outlet through northern Germany to the North Sea. It is a classic primate city, its cultural landscape faithful to Czech traditions; but it also is an industrial center. The encircling mountains contain many valleys with small towns that specialize, Swiss-style, in fabricating high-quality goods. In Eastern Europe the Czechs always were the leaders in technology and engineering; even during the communist period their products found markets in foreign countries near and far.

As Figure 1-25 shows, the Czech Republic incorporates an eastern province named Moravia, linked to Poland's Silesia through the Moravian Gate between the Sudeten and Carpathian Mountains. Here the Soviets brought their industrialization drive, making Ostrava and Brno important manufacturing cities but leaving behind obsolescence and environmental problems.

FIGURE 1-25

THE NEW EASTERN EUROPE
POPULATION
● Under 50,000
● 50,000–250,000
● 250,000–1,000,000
● 1,000,000–5,000,000
● Over 5,000,000

National capitals are underlined

—— Road
—— Railroad

Soviet-era realm boundary
Present realm boundary

0   100   200   300   400 Kilometers
0   50   100   150   200 Miles

Longitude East of Greenwich

The separation of the Czech and Slovak societies created a virtual nation-state in the Czech Republic, with only a small minority of several hundred thousand Roma, or gypsies, among the Czechs. The treatment of this minority became a human rights issue in the discussions leading to Czech membership in the EU, effective 2004.

As Figure 1-26 shows, **Slovakia** inherited Czechoslovakia's largest minority, the ethnic Hungarian community, which now constitutes about 11 percent of the country's 5.4 million inhabitants and is concentrated in the southern zone along the Danube River. The country's division resulted from ethnic and cultural factors, but the Slovaks also did not share the Czechs' enthusiasm for post-Soviet economic reforms. The Hungarian minority found itself at odds with the new government in Bratislava, but both communities saw the advantages of EU membership and found ways to improve their relationships. Slovakia, sure to be a major recipient of EU aid, was admitted in 2004.

Hungary also approaches the status of nation-state, but there are Hungarians not only in Slovakia but also in Yugoslavia (Serbia), Austria, and Romania (Fig. 1-24), all remnants of a time when Hungary ruled much of this region. Hungarian governments have repeatedly expressed support for these ethnic cohorts in neighboring countries, a practice referred to as **30** **irredentism**. The term derives from a nineteenth-century campaign by Italy to incorporate an Italian-speaking area of Austria, calling it

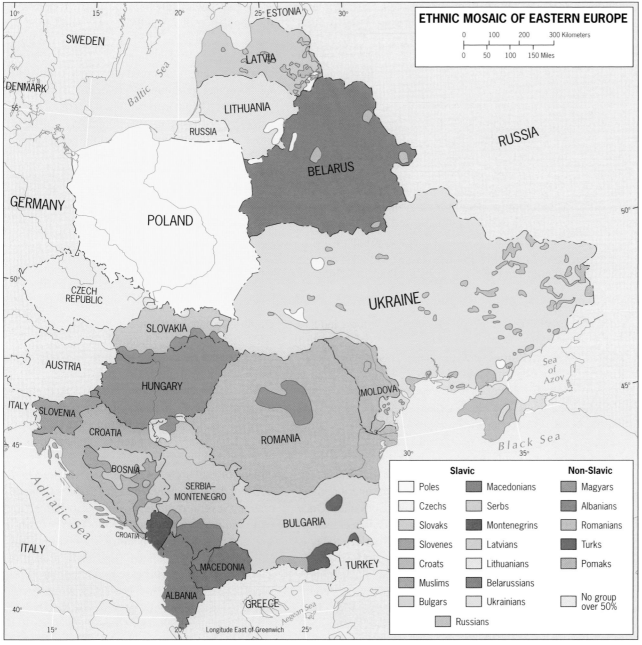

ETHNIC MOSAIC OF EASTERN EUROPE

Slavic		Non-Slavic
Poles	Macedonians	Magyars
Czechs	Serbs	Albanians
Slovaks	Montenegrins	Romanians
Slovenes	Latvians	Turks
Croats	Lithuanians	Pomaks
Muslims	Belarussians	
Bulgars	Ukrainians	No group over 50%
Russians		

FIGURE 1-26

Italia Irredenta (Unredeemed Italy). In the case of Hungary, its recent manifestation was the so-called "status law" giving persons of Hungarian ancestry living in neighboring countries certain rights (including work permits) in the "motherland." The governments of those neighboring countries did not like this "law" applying to its citizens, and neither did the EU leadership. It has been modified, but public opinion in Hungary still supports the notion.

The Hungarians (Magyars) moved into the middle Danube River Basin more than a thousand years ago from an Asian source; they have neither Slavic nor Germanic roots. They converted their fertile lowland into a thriving state while retaining their cultural and linguistic identity, eventually forging an imperial power in the region. The twin-cities capital astride the Danube, Buda and Pest (better known as Budapest), is a primate city nearly ten times the size of the next largest town in Hungary—reflecting the continuing rural character of the country.

With a population of 10 million and a considerable and varied resource base, Hungary has good economic prospects and qualified to join the European Union in 2004.

Countries Facing the Black Sea

Four countries form Eastern Europe's Black Sea quadrant: Ukraine, Moldova, Romania, and Bulgaria (Fig. 1-23). All except Moldova have coastlines on the Black Sea, but none of their core areas or capital cities lies on the coast. This reflects an inward orientation that characterizes the subregion as a whole. None qualified to join the EU in 2004.

Ukraine is Eastern Europe's most populous country (47.5 million); territorially, it is the largest state in the entire realm. Its capital, Kiev (Kyyiv), is a major historic, cultural, and political focus. Briefly independent before the communist takeover in Russia, Ukraine regained its sovereignty as a much-changed country in 1991. Once a land of farmers tilling its famously fertile soils, Ukraine emerged from the Soviet period with a huge industrial complex in its east—and with a large (over 20 percent) Russian minority. Ukraine's boundaries also changed during the Soviet era. In 1954, a Soviet dictator capriciously transferred the entire Crimea Peninsula, including its Russian inhabitants, to Ukraine as a reward for its productivity.

The Dnieper River forms a useful geographic reference to comprehend Ukraine's spatial division (Fig. 1-25). To its west lies agrarian, rural, mainly Roman Catholic Ukraine; in its great southern bend and eastward lies industrial, urban, Russified (and Russian Orthodox) Ukraine. Soviet planners made eastern Ukraine a communist Ruhr based on local coal and iron ore, making the Donets Basin (*Donbas* for short) a key industrial complex. Meanwhile, the Russian Soviet Republic supplied Ukraine with oil and gas.

Ukraine is a critically important country for Europe but is in trouble today. Political mismanagement and corruption, a faltering economy, falling social indicators, and rising crime are causing many Ukrainians to call for a reunion with Russia and a return to the "good old days." And yet this country has access to international shipping lanes, a large resource base, massive farm production, educated and skilled labor, and a large domestic market. A legacy of dependence is taking its toll, and there is no prospect of EU membership at present.

Moldova, Ukraine's small and impoverished neighbor, was a Romanian province seized by the Soviets in 1940 and made into a (deliberately land-locked) "Soviet Republic" of the USSR. Romanians remain in the majority among it 4.3 million people, but most of the Russians and Ukrainians (each about 13 percent) have moved to the small strip of land between the Dniester River and the Ukrainian border, proclaiming there a "Republic of Transdniestra" (also sometimes spelled "Transnistria"). Devolutionary forces form only one of Moldova's problems; its farm economy is declining steadily, and by many measures this is now Europe's poorest country.

During the decade following the collapse of the Soviet Empire, **Romania** was in dire political and economic straits. Political instability, corruption, flirtations with extremist Islamic groups, exhaustion of its oil and gas reserves, declining farm production, and quarrels with neighboring Hungary over the treatment of Romania's 1.6 million ethnic Hungarians darkened the country's prospects.

With 22 million people, a large territory, and five neighbors, Romania is an important component of Eastern Europe, and after the turn of the century conditions began to improve. The government started to sell off some of its unproductive enterprises, Romania joined the North Atlantic Treaty Organization (NATO), and public opinion strongly favored stronger ties with the United States. Still, EU membership is only in the preliminary discussion stage.

Romania's drab and decaying capital, Bucharest, and its surrounding core area lie in the interior linked by rail to the Black Sea port of Constanta. Once known as Eastern Europe's most civilized society, its capital the Paris of the region, Romania today is recovering from a half-century of misfortunes.

Across the Danube lies Romania's neighbor, **Bulgaria**, southernmost of this subregion's countries. The rugged Balkan Mountains form Bulgaria's physiographic backbone, separating the Danube and Maritsa basins. As the map shows, Bulgaria has five neighbors, several of which are in political turmoil.

The Bulgarian state appeared in 1878, when the Russian czar's armies drove the Turks out of this area. The Slavic Bulgars, who form 86 percent of the population of 7.7 million, were (unlike the Romanians) loyal allies of Moscow during the Soviet period. But they did not treat their Turkish minority, about 9 percent of the population, very kindly, closing mosques, prohibiting use of the Turkish language, and forcing Turkish families to adopt Slavic names. Conditions for the remaining Turks improved somewhat after the end of the Soviet period.

Bulgaria has a Black Sea coast and an outlet, the port of Varna, but the country does not generate much external trade; the capital, Sofia, lies near the Yugoslav border. Bulgaria needs economic reform, but like the rest of this subregion, its prospects are not bright.

Countries Facing the Adriatic Sea

As recently as 1990, only two Eastern European countries fronted the Adriatic Sea: Yugoslavia and Albania (Fig. 1-24). Albania survives, but the former Yugoslavia has splintered into five countries—and its disintegration may not be over.

We turn first to the former Yugoslavia ("Land of the South Slavs"), a country extending from Austria to Greece, thrown together in 1918 after World War I, containing 7 major and 17 smaller ethnic and cultural groups (Fig. 1-24). The Slovenes and Croats in the north were Roman Catholics; the Serbs in the south adhered to the Serbian Orthodox church. Several million Muslims also formed part of the cultural mosaic. Two alphabets were in use in separate regions of the country. At first the Royal House of Serbia dominated Yugoslavia; after 1945, a communist dictatorship personified by World War II hero Marshal Tito held Yugoslavia together. But when the communist system collapsed in Eastern Europe and the Soviet Union disintegrated, so did the Yugoslav state.

Communist social planners laid the groundwork for the disaster that befell Yugoslavia. They divided the country into six internal "republics" based on the Soviet model, each dominated (except Bosnia) by one major group. Since all these "republics" inevitably incorporated minorities, the state guaranteed their rights, albeit through autocratic means.

When the communist system collapsed, individual "republics" proclaimed their independence—that is, the majorities in these entities did so. What remained of the machine of state, still dominated by the Serbs, tried to halt this disintegration. That effort soon failed, and new countries named Slovenia, Croatia, Bosnia, Macedonia, and, by subtraction, Serbia appeared on the map. Minorities in these new states, however, objected and, in Croatia and Bosnia, rose against those who were advocating statehood. The result was a catastrophic conflict just when grandiose Euro-unification schemes were under way. Europe, which twice during the twentieth century had plunged the world into war and had vowed that genocide would "never again" cloud its horizons, failed this first test of its declarations. EU members, led by Germany, quickly accorded official recognition to the post-Yugoslavian states, even before the concerns of the frightened minorities could be considered. Then, European powers stood by as more than 250,000 people were killed, perhaps a million more were injured, ethnic cleansing and mass executions depopulated entire areas, refugees streamed into neighboring countries, and historic treasures were demolished.

As we will discover, the devolutionary process continues in the twenty-first century. Today, the still-evolving map emerging from the former Yugoslavia consists of five countries: Slovenia, Croatia, Bosnia, Macedonia, and the newly renamed Serbia-Montenegro (Fig. 1-25). Devolutionary pressures in the last of these five have wrested the Province of Kosovo from Serbian control and are now affecting its other province, Montenegro.

Slovenia was the first "republic" to secede and proclaim independence. Ethnically the most homogeneous but territorially small with only 2 million of the former Yugoslavia's 23 million citizens, Slovenia lay farthest from the Serbian power core and seized its opportunity. This Alpine-mountain state, with the most productive economy among the republics even before independence, today has this subregion's highest per-capita GNP by far. It qualified easily to join the EU in 2004.

Croatia's crescent-shaped territory, with prongs along the Hungarian border and along the Adriatic coast, only partially reflects the distribution of Croats in the former Yugoslavia (Fig. 1-26). About 85 percent of Croatia's 4.3 million people are Croats, but another 800,000 live in a broad zone within southern Bosnia. Croatia's Serb minority, about 12 percent when independence came, has dwindled under Croatian pressure to under 5 percent. Recent democratic reforms have improved Croatia's prospects.

Bosnia was the cauldron of calamity. Multicultural, effectively landlocked, and situated between the Croatian state to the west and the Serbian stronghold to the east, Bosnia fell victim to disastrous conflict among Serbs, Croats, and Bosniaks (now the official name for Bosnia's Muslims, who constituted about 50 percent of the population). In 1995, a U.S. diplomatic initiative resulted in a truce that partitioned Bosnia as shown in Figure 1-27, making the country a fragile federation.

FIGURE 1-27

Macedonia was the southernmost "republic" of the former Yugoslavia and became a state with 2 million inhabitants. Over 60 percent of these inhabitants are Macedonian Slavs, about 30 percent are Muslim Albanians, and the remainder are Turks, Serbs, and Roma (gypsies). Small, landlocked, poor, and powerless, Macedonia first faced the ire of Greece, which argued that the name *Macedonia* was Greek property, then confronted a massive flow of refugees entering from war-torn Kosovo, and most recently coped with an Albanian rebel movement in its northwestern corner, where its main minority is concentrated.

But it is in **Serbia-Montenegro** where the major problems of this subregion are now focused and where the unfinished devolutionary process continues. This geographic name now refers to what is left of the domain once ruled or dominated by the Serbs: *Serbia*, centered on the old Serb capital of Belgrade on the Danube River; *Vojvodina* in the north, part of which was once Hungarian territory and where a Hungarian minority numbers over 350,000; *Montenegro* on the Adriatic coast, once strongly supportive of the Serbs but recently trying to distance itself from Belgrade; and *Kosovo* in the south, one of the historic Muslim strongholds in the subregion, now under NATO administration following NATO's ouster of the repressive Serb authorities, and given the name *Dardania* ("Land of Pears") by its partisans.

The total population of this country-in-transition is nominally 10.7 million, but that of Serbia is just over 6 million, while Kosovo has about 2 million, Vojvodina some 1.7 million, and Montenegro 650,000. The two key politico-geographical problems involve the future of Kosovo and of Montenegro. When Serbian rule over Kosovo was ended by aerial bombardment and NATO's rule was instituted, there was no clear plan for the territory's future; many Muslims expect eventual independence, but the new government in Belgrade may demand something short of that. And in Montenegro a slender majority want independence from Belgrade as well—another development the Serbs may not be willing to accept. The devolutionary process in Yugoslavia has not yet run its course.

It is appropriate to save our discussion of **Albania** for last because even in this turbulent region Albania is unusual. It is the only dominantly Muslim state in Europe; some 70 percent of its 3.2 million people adhere to Islam. Albania also rivals Moldova as the poorest country in Europe, ranking lowest on several indices of well-being. It has by far the fastest rate of population growth in the realm. Most Albanians subsist on livestock herding and farming on the one-fifth of this mountainous, earthquake-prone country that can be used for agriculture. Thousands of Albanians have tried to reach Italy across the Adriatic Sea.

Kosovar and Macedonian Muslims look to Albania for support, but the government headquartered in Tirane cannot embark on irredentist campaigns. Albania has its own cultural divide between the poverty-stricken Gegs in the north and the somewhat-better-off Tosks in the south. National unity is its primary goal.

With 583 million inhabitants in 39 countries, including some of the world's highest-income economies, a politically stable and economically integrated Europe would be a superpower in the twenty-first century. But Europe's political geography is anything but stable, as devolutionary forces and cultural conflict continue to trouble the realm. Moreover, economic integration, despite the momentous EU expansion of 2004, still involves less than two-thirds of the realm's countries, a process that will become more difficult as the European Union confronts applications from marginally qualifying states in the east. Europe always has been a realm of revolutionary change, and it remains so today.

The *Concepts and Regions* Website, featuring *GeoDiscoveries*, offers many additional resources to enhance your understanding and experience of this chapter. Be sure to explore the following:

GeoDiscoveries Video Clips

The Capital of Europe

Qualifications for a Capital of Europe

Eastern Europe and the European Community

Berlin as a Capital of Europe

Weighing the Qualifications of Possible European capitals

GeoDiscoveries Animations

History of Supranationalism in Europe

Other Cities to Consider

The Four Motors of Europe

GeoDiscoveries Interactivities

Supranationalism

Choose the Capital of Europe

Photo Galleries

Flags of the European Union

Le Havre, France

Bourgogne, France

Gouda, Netherlands

London, UK

Oslo, Norway

Rome, Italy

Trevi Fountain, Italy

Malaga, Spain

Gdynia, Poland

Monaco

Ireland

Vienna, Austria

Athens, Greece

Seville, Spain

Rouen, France

Utrecht, Netherlands

Liverpool, UK

Dublin, Ireland

La Coruña, Spain

Valletta, Malta

More To Explore

Systematic Essay: Geomorphology

Issues in Geography: Supranationalism in Europe

Major Cities: Paris, London, Rome, Athens

Expanded Regional Coverage

Western Europe, The British Isles, Northern (Nordic) Europe, Mediterranean Europe, Eastern Europe

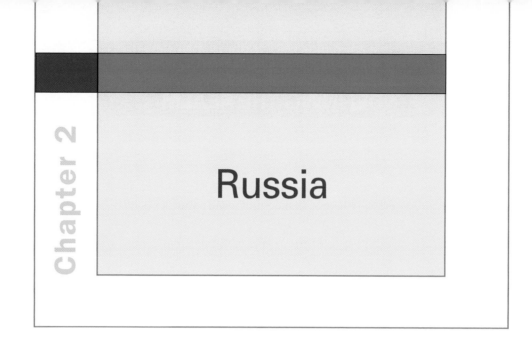

Chapter 2

Russia

CONCEPTS, IDEAS, AND TERMS

1. Climatology
2. Continentality
3. Weather
4. Tundra
5. Taiga
6. Permafrost
7. Forward capital
8. Colonialism
9. Imperialism
10. Russification
11. Federation
12. Collectivization
13. Command economy
14. Unitary state system
15. Distance decay
16. Population decline
17. Heartland theory
18. Core area

REGIONS

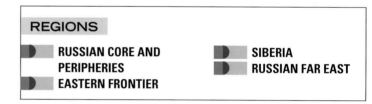

RUSSIAN CORE AND PERIPHERIES

EASTERN FRONTIER

SIBERIA

RUSSIAN FAR EAST

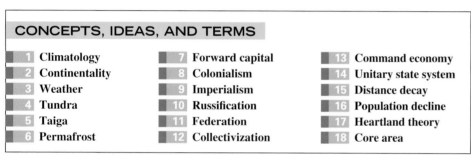

*I*S THE RUSSIAN state a geographic realm? A look at any world political map confirms the answer. Russia is the world's largest state territorially and has a substantial population; it has common borders with North Korea and Norway and extends from the Arctic to the shores of the Black and Caspian Seas. Three times as large as the contiguous United States and stretching across 11 time zones, Russia is the planet's giant.

But Russia also is a country in trouble. Its population is declining and will, if present demographic trends continue, decline from 144 million in 2004 to about 100 million in 2050. Tuberculosis, AIDS, alcohol poisoning, and chronic diseases are taking their toll on a people whose life expectancy is dropping relentlessly. And large numbers of people are poor, in sharp contrast to the comparatively few who have become rich in the post-Soviet era. Not only is Russia's social fabric fraying, its economy is excessively dependent on the export of one commodity: fuel in the form of oil and natural gas. Politically, Russia's democracy and federal system are struggling. Its armed forces are in disarray. All this while Russia still has huge arsenals of Soviet-era nuclear, chemical and biological weapons.

In the post-Cold War era, world attention has been focused elsewhere: on rising China, on unifying Europe, on terrorism, on global warming. But while Russia no longer occupies central stage, it remains a crucial component of the elusive New World Order, still a power to be reckoned with. We turn now to this troubled behemoth.

STUDY TOOLS

3D GLOBE
AREA AND DEMOGRAPHIC DATA

FLASHCARDS
MAP QUIZZES

AUDIO PRONUNCIATION GUIDE
CHAPTER QUIZ

ANNOTATED WEBLINKS
LONELY PLANET WEBLINKS

DEFINING THE REALM

Although the geographic realm under discussion is dominated by the giant Russian state, it also includes three small countries in the area between the Black Sea and the Caspian Sea: Georgia, Armenia, and Azerbaijan. For centuries Russia was a powerful empire whose influence reached far beyond its borders into Transcaucasia, Turkestan, Mongolia, and even China. Today, Russia retains strong links with all three of these Transcaucasian republics, placing the realm boundary along the borders with Turkey and Iran (Figs. G-2, G-11).

Russia is a land of vast distances, remote frontiers, bitter cold, and isolated outposts. Its core area lies west of the Urals, centered on Moscow. To the north, south, and east lie distant, problematic peripheries where economies falter and political systems fail. During the early twentieth century, Russia embarked on an immense social experiment with the goal of an egalitarian communist state that would be a model for the rest of the world. The experiment failed, but not before Russia and its Eurasian empire had been transformed at gigantic cultural, economic, and political cost. The geographic story of this chapter centers on Russia's postcommunist reorganization and its changing role in the world.

RUSSIAN ROOTS

The name *Russia* evokes cultural-geographic images of a stormy past: terrifying czars, conquering Cossacks, Byzantine bishops, rousing revolution-

---

MAJOR GEOGRAPHIC QUALITIES OF RUSSIA

1. Russia is the largest territorial state in the world. Its area is nearly twice as large as that of the next-ranking country (Canada).

2. Russia is the northernmost large and populous country in the world; much of it is cold and/or dry. Extensive rugged mountain zones separate Russia from warmer subtropical air, and the country lies open to Arctic air masses.

3. Russia was one of the world's major colonial powers. Under the czars, the Russians forged the world's largest contiguous empire; the Soviet rulers who succeeded the czars took over and expanded this empire.

4. For so large an area, Russia's population of under 145 million is comparatively small. The population remains heavily concentrated in the westernmost one-fifth of the country.

5. Development in Russia is concentrated west of the Ural Mountains; here lie the major cities, leading industrial regions, densest transport networks, and most productive farming areas. National integration and economic development east of the Urals extend mainly along a narrow corridor that stretches from the southern Urals region to the southern Far East around Vladivostok.

6. Russia is a multicultural state with a complex domestic political geography. Twenty-one internal Republics, originally based on ethnic clusters, function as politico-geographical entities.

7. Its large territorial size notwithstanding, Russia suffers from land encirclement within Eurasia; it has few good and suitably located ports.

8. Regions long part of the Russian and Soviet empires are realigning themselves in the postcommunist era. Eastern Europe and the heavily Muslim Southwest Asia realm are encroaching on Russia's imperial borders.

9. The failure of the Soviet communist system left Russia in economic disarray. Many of the long-term components described in this chapter (food-producing areas, railroad links, pipeline connections) broke down in the transition to the postcommunist order.

10. Russia long has been a source of raw materials but not a manufacturer of export products, except weaponry. Few Russian (or Soviet) automobiles, televisions, cameras, or other consumer goods reach world markets.

aries, clashing cultures. Russians repulsed the Tatar (Mongol) hordes, forged a powerful state, colonized a vast contiguous empire, defeated Napoleon, adopted communism, and, when the communist system failed, lost most of their imperial domain. Today, Russia seeks a place in the globalizing world.

Precommunist Russia may be described as a culture of extremes. It was a culture of strong nationalism, resistance to change, and despotic rule. Enormous wealth was concentrated in a small elite. Powerful rulers and bejeweled aristocrats perpetuated their privileges at the expense of millions of peasants and serfs who lived in dreadful poverty. The Industrial Revolution arrived late in Russia, and a middle class was slow to develop. Yet the Russian nation gained the loyalty of many of its citizens, and its writers and artists were among the world's greatest. Authors such as Tolstoy and Dostoyevsky chronicled the plight of the poor; composers celebrated the indomitable Russian people, and, as Tchaikovsky did in his *1812 Overture*, commemorated their victories over foreign foes.

Under the czars, Russia grew from nation into empire. The czars' insatiable demands for wealth, territory, and power sent Russian armies across the plains of Siberia, through the deserts of interior Asia, and into the mountains along Russia's rim. Russian pioneers ventured even farther, entering Alaska, traveling down the Pacific coast of North America, and planting the Russian flag near San Francisco in 1812. But as Russia's empire expanded, its internal weaknesses gnawed at the power of the czars. Peasants rebelled. Unpaid (and poorly fed) armies mutinied. When the czars tried to initiate reforms, the aristocracy objected. The empire at the beginning of the twentieth century was ripe for revolution, which began in 1905.

The last czar, Nicholas II, was overthrown in 1917, and civil war followed. The victorious communists led by V. I. Lenin soon swept away much of the Russia of the past. The Russian flag disappeared. The czar and his family were executed.

The old capital of Russia, St. Petersburg, was renamed Leningrad in honor of the revolutionary leader. Moscow, in the interior of the country, was chosen as the new capital for a country with a new name, the *Soviet Union*. Eventually, this Union consisted of 15 political entities, each a Soviet Socialist Republic. Russia was just one of these republics, and the name *Russia* disappeared from the international map.

But on the Soviet map, Russia was the giant, the dominant republic, the Slavic center. Not for nothing was the communist revolution known as the Russian Revolution. The other republics of the Soviet Union were for minorities the czars had colonized or for countries that fell under Soviet sway later, but none could begin to match Russia. The Soviet Empire was the legacy of czarist expansionism, and the new communist rulers were Russian first and foremost (with the exception of Josef Stalin, who was a Georgian). Russians moved by the millions to the non-Russian republics, where the Russification of the empire proceeded, just as the British and French and Dutch and Portuguese were also moving to their colonies in large numbers. The Soviet Union was a Russian colonial empire, and like all colonial empires it was doomed to failure.

Officially, the Union of Soviet Socialist Republics (USSR) endured from 1924 to 1991. The year 1924 was a fateful one in Russian and Soviet history, marking the death of Lenin, the ideological organizer, and the beginning of nearly three decades of rule by Stalin, the ruthless tyrant. During Stalin's time, many of the peoples under Russian control suffered unimaginably. In pursuit of communist reconstruction, Stalin and his henchmen starved millions of Ukrainian peasants to death, forcibly relocated entire ethnic groups (including a people known as the Chechens), exterminated "uncooperative" or "disloyal" peoples, and purged the Communist Party time and again. Death camps proliferated in Siberia; many of the country's most creative people were eliminated. The full extent of these horrors may never be known.

On December 25, 1991, the inevitable occurred: the Soviet Union ceased to exist, its economy a shambles, its political system shattered, the communist experiment a failure. The Soviet hammer-and-sickle flag flying atop the Kremlin was lowered for the last time and replaced by the white, red, and blue Russian tricolor. The republics were set free, and world maps had to be redrawn. Now Russia and its former colonies had to adjust to the realities of a changing world order. It proved to be a difficult, sometimes desperate journey.

RUSSIA'S PHYSICAL ENVIRONMENTS

Russia's physiography is dominated by vast plains and plateaus rimmed by rugged mountains (Fig. 2-1). Only the Ural Mountains break an expanse of low relief that extends from the Polish border to eastern Siberia. In the western part of this huge plain, where the northern pine forests meet the midlatitude grasslands, the Slavs established their domain.

The historical geography of Russia is the story of Slavic expansion from its populous western heartland across interior Eurasia to the east, and into the mountains and deserts of the south. This eastward march was hampered not only by vast distances but also by harsh natural conditions. As the northernmost populous country on Earth, Russia has virtually no natural barriers against the onslaught of Arctic air. Moscow lies farther north than Edmonton, Canada, and St. Petersburg lies at latitude 60° North—the latitude of the southern tip of Greenland. Winters are long, dark, and bitterly cold in most of Russia; summers are short and growing seasons limited. Many a Siberian frontier outpost was doomed by cold, snow, and hunger.

It is therefore useful to view Russia's past, **1** present, and future in the context of its **climatology**. This field of geography investigates not only the distribution of climatic conditions over the Earth's surface but also the processes that generate

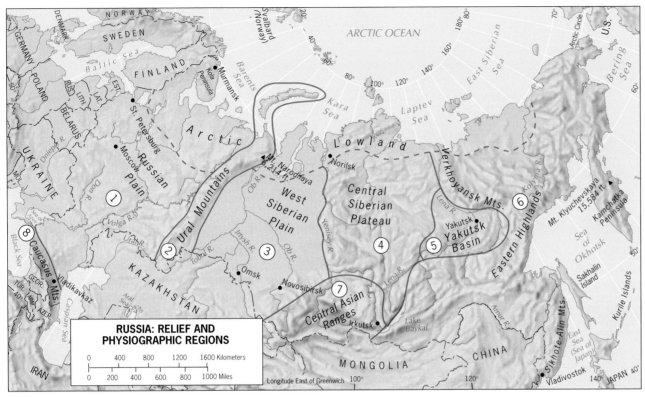

FIGURE 2-1

this spatial arrangement. The Earth's atmosphere traps heat received as radiation from the Sun, but this *greenhouse effect* varies over planetary space—and over time. As we noted in the introductory chapter, much of what is today Russia was in the grip of a glaciation until the onset of the warmer Holocene. But even today, with a natural warming cycle in progress augmented by human activity, Russia still suffers from severe cold and associated drought. If the enhanced global warming cycle continues, Russia (in contrast to low-lying countries faced by rising sea levels) may benefit. But that will take generations.

Currently precipitation totals, even in western Russia, range from modest to minimal because the warm, moist air carried across Europe from the North Atlantic Ocean loses much of its warmth and

moisture by the time it reaches Russia. Figures G-7, G-8, and 2-2 reveal the consequences. Russia's 2 climatic **continentality** (inland climatic environment remote from moderating and moistening maritime influences) is expressed by its prevailing *Dfb* and *Dfc* conditions. Compare the Russian map to that of North America (Fig. G-8), and you note that, except for a small corner off the Black Sea, Russia's climatic conditions resemble those of the U.S. Upper Midwest and Canada. Along its entire north, Russia has a zone of *E* climates, the most frigid on the planet. In these Arctic latitudes originate the polar air masses that dominate its environments.

By studying Russia's climates we can begin to understand what the map of population distribution (Fig. G-9) shows. The overwhelming majority of the country's more than 143 million people are

concentrated in the west and southwest, where climatic conditions were least difficult at a time when farming was the mainstay for most of the people. The map is a legacy that will mark Russia's living space for generations to come.

Climate and weather (there is a distinction: *climate* is a long-term 3 average, whereas **weather** refers to existing atmospheric conditions at a given place and time) have always challenged Russia's farmers. Conditions are most favorable in the west, but even there temperature extremes, variable and undependable rainfall, and short growing seasons make farming difficult. During the Soviet period, fertile and productive Ukraine supplied much of Russia's food needs, but even then Russia often had to import grain. Soviet rulers wanted to reduce their country's dependence on imported food, and their communist planners built major irrigation projects to increase crop yields in the colonized republics of Central Asia. As we will see in Chapter 7, some of these attempts to overcome nature's limitations spelled disaster for the local people.

Physiographic Regions

To assess the physiography of this vast country, refer again to Figure 2-1. Note how mountains and deserts encircle Russia: the Caucasus in the southwest ⑧; the Central Asian Ranges in the center ⑦; the Eastern Highlands facing the Pacific from the Bering Sea to the East Sea (Sea of Japan) ⑥. The Kamchatka Peninsula has a string of active volcanoes in one of the world's most earthquake-prone zones (Fig. G-5). Warm subtropical air thus has little opportunity to penetrate Russia, while cold Arctic air sweeps southward without impedi-

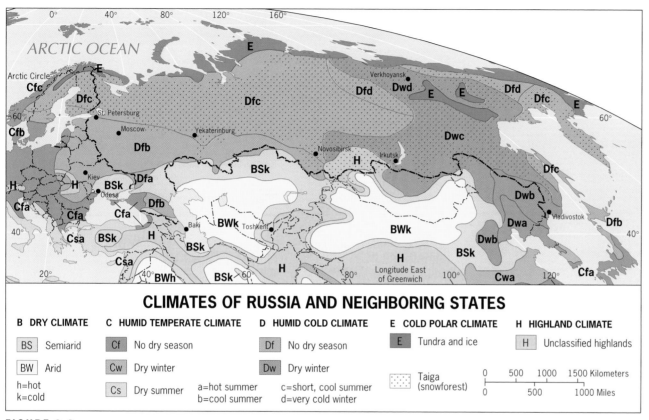

CLIMATES OF RUSSIA AND NEIGHBORING STATES

B DRY CLIMATE	C HUMID TEMPERATE CLIMATE	D HUMID COLD CLIMATE	E COLD POLAR CLIMATE	H HIGHLAND CLIMATE
BS Semiarid	Cf No dry season	Df No dry season	E Tundra and ice	H Unclassified highlands
BW Arid	Cw Dry winter	Dw Dry winter		
h=hot k=cold	Cs Dry summer	a=hot summer b=cool summer	c=short, cool summer d=very cold winter	Taiga (snowforest)

0   500   1000   1500 Kilometers
0   500   1000 Miles

FIGURE 2-2

ment. Russia's Arctic north is a gently sloping lowland broken only by the Urals and Eastern Highlands.

Russia's vast and complex physical stage can be divided into eight physiographic regions, each of which, at a larger scale, can be subdivided into smaller units. In the Siberian region, one criterion for such subdivision would be the vegetation. The Russian language has given us two terms to **4** describe this vegetation: **tundra**, the treeless plain along the Arctic shore where mosses, lichens, **5** and some grasses survive, and **taiga**, the mostly coniferous forests that begin to the south of where the tundra ends, and extend over vast reaches of Siberia, which means the "sleeping land" in Russian.

The Russian Plain ① is the eastward continuation of the North European Lowland, and here the Russian state formed its first core area. Travel north from Moscow at its heart, and the countryside soon is covered by needleleaf forests like those of Canada; to the south lie the grain fields of southern Russia and, beyond, those of Ukraine. Note the Kola Peninsula and Barents Sea in the far north: warm water from the North Atlantic comes around northern Norway and keeps the port of Murmansk ice free most of the year. The Russian Plain is bounded on the east by the Ural Mountains ②; though not a high range, it is topographically prominent because it separates two extensive plains. The range of the Urals is more than 2000 miles (3200 km) long and reaches from the shores of the Kara Sea to the border with Kazakhstan. It is not a barrier to east-west transportation, and its southern end is densely populated. Here the Urals yield minerals and fossil fuels.

East of the Urals lies Siberia. The West Siberian Plain ③ has been described as the world's largest unbroken lowland; this is the vast basin of the Ob and Irtysh rivers. Over the last 1000 miles (1600 km) of its course to the Arctic Ocean, the Ob falls less than 300 feet (90 m). In Figure 2-1, note the dashed line that extends from the Finnish border to the East Siberian Sea and offsets the Arctic Lowland. North of the line, water in the ground is **6** permanently frozen; this **permafrost** creates another obstacle to permanent settlement. Looking again at the West Siberian Plain, we see that the north is permafrost-ridden and the central zone is marshy. The south, however, has such major cities as Omsk and Novosibirsk, within the corridor of the Trans-Siberian Railroad.

East of the West Siberian Plain the country begins to rise, first into the central Siberian Plateau ④, another sparsely settled, remote, permafrost-affected region. Here winters are long and cold, and summers are short; the area remains barely touched by human activity. Beyond the Yakutsk Basin ⑤, the terrain becomes mountainous and the relief high. The Eastern Highlands ⑥ are a jumbled mass of ranges and ridges, precipitous valleys, and volcanic mountains. Lake Baykal lies in a trough that is over 5000 feet (1500 m) deep—the deepest rift lake in the world (see Chapter 6). On the Kamchatka Peninsula, volcanic Mount Klyuchevskaya reaches nearly 15,600 feet (4750 m).

The northern part of region ⑥ is Russia's most inhospitable zone, but southward along the Pacific coast the climate is less severe. Nonetheless, this is a true frontier region. The forests provide opportunities for lumbering, a fur trade exists, and there are gold and diamond deposits.

Mountains also mark the southern margins of much of Russia. The Central Asian Ranges ⑦, from the Kazakh border in the west to Lake Baykal in the east, contain many glaciers whose annual meltwaters send alluvium-laden streams to enrich farmlands at lower elevations. The Caucasus ⑧, in the land corridor between the Black and Caspian seas, form an extension of Europe's Alpine Mountains and exhibit a similarly high *relief* (range of elevations). Here Russia's southern border is sharply defined by *topography* (surface configuration).

As our physiographic map suggests, the more habitable terrain in Russia becomes latitudinally narrower from west to east. Beyond the southern Urals, the zone of settlement in Russia becomes discontinuous (in Soviet times it extended into northern Kazakhstan, where a large Russian population still lives). Isolated towns did develop in Russia's vast eastern reaches, even in Siberia, but the tenuous ribbon of settlement does not widen again until it reaches the country's Far Eastern Pacific Rim.

EVOLUTION OF THE RUSSIAN STATE

Four centuries ago, when European kingdoms were sending fleets to distant shores to search for riches and capture colonies, there was little indication that the largest of all empires would one day center on a city in the forests halfway between Sweden and the Black Sea. The plains of Eurasia south of the taiga had seen waves of migrants sweep from east to west: Scythians, Sarmatians, Goths, Huns, and others came, fought, settled, and were absorbed or driven off. Eventually, the Slavs, heirs to these Eurasian infusions, emerged as the dominant culture in what is today Ukraine, south of Russia and north of the Black Sea.

From this base on fertile, productive soils, the Slavs expanded their domain, making Kiev (Kyyiv)

their southern headquarters and establishing Novgorod on Lake Ilmen in the north. Each was the center of a state known as a *Rus*, and both formed key links on a trade route between the German-speaking Hanseatic ports of the Baltic Sea and the trading centers of the Mediterranean. During the eleventh and twelfth centuries, the Kievan Rus and the Novgorod Rus united to form a large and comparatively prosperous state astride forest and steppe (short-grass prairie).

The Mongol Invasion

Prosperity attracts attention, and far to the east, north of China, the Mongol Empire had been building under Genghis Khan. In the thirteenth century, the Mongol (Tatar) "hordes" rode their horses into the Kievan Rus, and the state fell. By mid-century, Slavs were fleeing into the forests, where the Mongol horsemen had less success than on the open plains of Ukraine. Here the Slavs reorganized, setting up several new Russes. Still threatened by the Tatars, the ruling princes paid tribute to the Mongols in exchange for peace.

Moscow, deep in the forest on the Moscow River, was one of these Russes. Its site was remote and defensible. Over time, Moscow established trade and political links with Novgorod and became the focus of a growing area of Slavic settlement. When the Mongols, worried about Moscow's expanding influence, attacked near the end of the fourteenth century and were repulsed, Moscow emerged as the unchallenged center of the Slavic Russes. Soon its ruler, now called a Grand Duke, took control of Novgorod, and the stage was set for further growth.

Grand Duchy of Muscovy

Moscow's geographic advantages again influenced events. Its centrality, defensibility, links to Slavic settlements including Novgorod, and open frontiers to the north and west, where no enemies threatened, gave Moscow the potential to expand at the cost of its old enemies, the Mongols. Many of these invaders had settled in the basin of the

Volga River and elsewhere, where the city of Kazan had become a major center of Islam, the Tatars' dominant religion. During the rule of Ivan the Terrible (1547–1584), the Grand Duchy of Muscovy became a major military power and imperial state. Its rulers called themselves czars and claimed to be the heirs of the Byzantine emperors. Now began the expansion of the Russian domain (Fig. 2-3), marked by the defeat of the Tatars at Kazan and the destruction of hundreds of mosques.

The Cossacks

This eastward expansion of Russia was spearheaded by a relatively small group of seminomadic peoples who came to be known as Cossacks and whose original home was in present-day Ukraine. Opportunists and pioneers, they sought the riches of the eastern frontier, chiefly fur-bearing animals, as early as the sixteenth century. By the mid-seventeenth century they reached the Pacific Ocean, defeating Tatars in their path and consolidating their gains by constructing *ostrogs* (strategic fortified waystations) along river courses. Before the eastward expansion halted in 1812 (Fig. 2-3), the Russians had moved across the Bering Strait to Alaska and down the western coast of North America into what is now northern California.

Czar Peter the Great

When Peter the Great became czar (he ruled from 1682 to 1725), Moscow already lay at the center of a great empire—great, at least, in terms of the territories it controlled. The Islamic threat had been ended with the defeat of the Tatars. The influence of the Russian Orthodox Church was represented by its distinctive religious architecture and powerful bishops.

Peter consolidated Russia's gains and hoped to make a modern, European-style state out of his loosely knit country. He built St. Petersburg as a ▣ **forward capital** on the doorstep of Swedish-held Finland, fortified it with major military installations, and made it Russia's leading port.

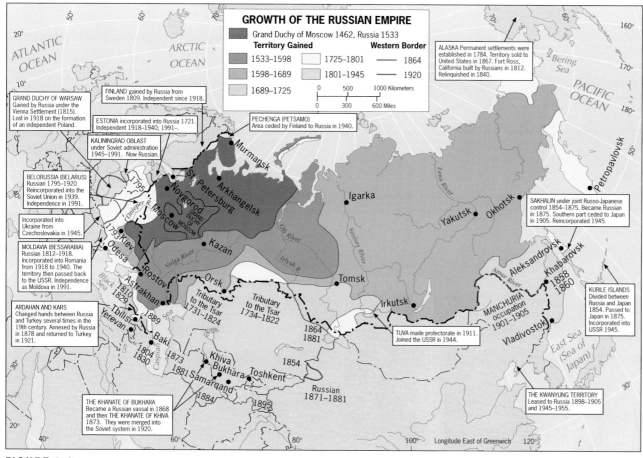

GROWTH OF THE RUSSIAN EMPIRE

Grand Duchy of Moscow 1462, Russia 1533

Territory Gained   **Western Border**

1533–1598   1725–1801   —— 1864

1598–1689   1801–1945   —— 1920

1689–1725

0   500   1000 Kilometers

0   300   600 Miles

FINLAND gained by Russia from Sweden 1809. Independent since 1918.

GRAND DUCHY OF WARSAW Gained by Russia under the Vienna Settlement (1815). Lost in 1918 on the formation of an independent Poland.

ESTONIA incorporated into Russia 1721. Independent 1918–1940; 1991–.

PECHENGA (PETSAMO) Area ceded by Finland to Russia in 1940.

ALASKA Permanent settlements were established in 1784. Territory sold to United States in 1867. Fort Ross, California built by Russians in 1812. Relinquished in 1840.

KALININGRAD OBLAST under Soviet administration 1945–1991. Now Russian.

BELORUSSIA (BELARUS) Russian 1795–1920. Reincorporated into the Soviet Union in 1939. Independence in 1991.

Incorporated into Ukraine from Czechoslovakia in 1945.

MOLDAVIA (BESSARABIA) Russian 1812–1918. Incorporated into Romania from 1918 to 1940. The territory then passed back to the USSR. Independence as Moldova in 1991.

SAKHALIN under joint Russo-Japanese control 1854–1875. Became Russian in 1875. Southern part ceded to Japan in 1905. Reincorporated 1945.

ARDAHAN AND KARS Changed hands between Russia and Turkey several times in the 19th century. Annexed by Russia in 1878 and returned to Turkey in 1921.

KURILE ISLANDS Divided between Russia and Japan 1854. Passed to Japan in 1875. Incorporated into USSR 1945.

TUVA made protectorate in 1911. Joined the USSR in 1944.

THE KWANYUNG TERRITORY Leased to Russia 1898–1905 and 1945–1955.

THE KHANATE OF BUKHARA Became a Russian vassal in 1868 and then THE KHANATE OF KHIVA 1873. They were merged into the Soviet system in 1920.

Tributary to the Tsar 1731–1824

Tributary to the Tsar 1734–1822

Russian 1871–1881

FIGURE 2-3

Czar Peter the Great, an extraordinary leader, was in many ways the founder of modern Russia. In his desire to remake Russia—to pull it from the forests of the interior to the waters of the west, to open it to outside influences, and to relocate its population—he left no stone unturned. Prominent merchant families were forced to move from other cities to St. Petersburg. Ships and wagons entering the city had to bring building stones as an entry toll. The czar himself, aware that to become a major power Russia had to be strong at sea as well as on land, went to the Netherlands to work as a laborer in the famed Dutch shipyards to learn the most efficient method for building

ships. Meanwhile, the czar's forces continued to conquer people and territory: Estonia was incorporated in 1721, widening Russia's window to the west, and major expansion soon occurred south of the city of Tomsk (Fig. 2-3).

Czarina Catherine the Great

Under Czarina Catherine the Great, who ruled from 1760 to 1796, Russia's empire in the Black Sea area grew at the expense of the Ottoman Turks. The Crimea Peninsula, the port city of Odesa (Odessa), and the whole northern coastal zone of the Black Sea fell under Russian control. Also during this

period, the Russians made a fateful move: they penetrated the area between the Black and Caspian seas, the mountainous Caucasus with dozens of ethnic and cultural groups, many of which were Islamized. The cities of Tbilisi (now in Georgia), Baki (Baku) in Azerbaijan, and Yerevan (Armenia) were captured. Eventually, the Russian push toward an Indian Ocean outlet was halted by the British, who held sway in Persia (modern Iran), and also by the Turks.

Meanwhile, Russian colonists entered Alaska, founding their first American settlement at Kodiak Island in 1784. As they moved southward, numerous forts were built to protect their tenuous holdings against indigenous peoples. Eventually, they reached nearly as far south as San Francisco Bay where, in the fateful year 1812, they erected Fort Ross.

Catherine the Great had made Russia a colonial power, but the Russians eventually gave up on their North American outposts. The sea-otter pelts that had attracted the early pioneers were running out, European and white American hunters were cutting into the profits, and Native American resistance was growing. When U.S. Secretary of State William Seward offered to purchase Russia's Alaskan holdings in 1867, the Russian government quickly agreed—for $7.2 million. Thus Alaska and its Russian-held panhandle became U.S. territory and, ultimately, the forty-ninth State.

A Russian Empire

Although Russia withdrew from North America, Russian expansionism during the nineteenth century continued in Eurasia. While extending their empire southward, the Russians also took on the Poles, old enemies to the west, and succeeded in taking most of what is today the Polish state, including the capital of Warsaw. To the northwest, Russia took over Finland from the Swedes in 1809. During most of the nineteenth century, however, the Russians were preoccupied with Central Asia—the region between the Caspian Sea and western China—where Toshkent (Tashkent) and Samarqand (Samarkand) came under St. Petersburg's control (Fig. 2-3). The Russians here were still bothered by raids of nomadic horsemen, and they sought to establish their authority over the Central Asian steppe country as far as the edges of the high mountains that lay to the south. Thus Russia gained many Muslim subjects, for this was Is-

lamic Asia they were penetrating. Under czarist rule, these people retained some autonomy. Much farther to the east, a combination of Japanese expansionism and a decline of Chinese influence led Russia to annex from China several provinces east of the Amur River. Soon thereafter, in 1860, the port of Vladivostok on the Pacific was founded.

Now began the events that were to lead to the first involuntary halt to the Russian drive for territory. As Figure 2-3 shows, the most direct route from western Russia to the port of Vladivostok lay across northeastern China, the territory then still called Manchuria. The Russians had begun construction of the Trans-Siberian Railroad in

1892, and they wanted China to permit the track to cross Manchuria. But the Chinese resisted. Then, taking advantage of the Boxer Rebellion in China in 1900 (see Chapter 9), Russian forces occupied Manchuria so that railroad construction might proceed.

This move, however, threatened Japanese interests in this area, and the Japanese confronted the Russians in the Russo-Japanese War of 1904–1905. Not only was Russia defeated and forced out of Manchuria: Japan even took possession of the southern part of Sakhalin Island, which they named Karafuto and retained until the end of World War II.

FROM THE FIELDNOTES

"Not only the city of St. Petersburg itself, but also its surrounding suburbs display the architectural and artistic splendor of czarist Russia. The czars built opulent palaces in these outlying districts (then some distance from the built-up center), among which the Catherine Palace, begun in 1717 and completed in 1723 followed by several expansions, was especially majestic. During my first visit in 1994 the palace, parts of which had been deliberately destroyed by the Germans during World War II, was still being restored; large black and white photographs in the hallways showed what the Nazis had done and chronicled the progress of the repairs during communist and postcommunist years. A return visit in 2000 revealed the wealth of sculptural decoration on the magnificent exterior (left) and the interior detail of a set of rooms called the 'golden suite' of which the Amber Room (right) exemplifies eighteenth-century Russian Baroque at its height."

THE COLONIAL LEGACY

Thus Russia, like Britain, France, and other European **8** powers, expanded through **colonialism**. Yet whereas the other European powers expanded overseas, Russian influence traveled overland into Central Asia, Siberia, China, and the Pacific coastlands of the Far East. What emerged was not the greatest empire but the largest territorially contiguous empire in the world. At the time of the Japanese war, the Russian czar controlled more than 8.5 million square miles (22 million sq km), just a tiny fraction less than the area of the Soviet Union after the 1917 Revolution. Thus the communist empire, to a large extent, was the legacy of St. Petersburg and European Russia, not the product of Moscow and the socialist revolution.

The czars embarked on their imperial conquests in part because of Russia's relative location: Russia always lacked warm-water ports. Had the Revolution not intervened, their southward push might have reached the Persian Gulf or even the Mediterranean Sea. Czar Peter the Great envisaged a Russia open to trading with the entire world; he developed St. Petersburg on the Baltic Sea into Russia's leading port. But in truth, Russia's historical geography is one of remoteness from the mainstreams of change and progress, as well as one of self-imposed isolation.

An Imperial, Multinational State

Centuries of Russian expansionism did not confine itself to empty land or unclaimed frontiers. The Russian state became an imperial power that annexed and incorporated many nationalities and cultures. This was done by employing force of arms, by overthrowing uncooperative rulers, by annexing territory, and by stoking the fires of ethnic conflict. By the time the ruthless Russian regime began to face revolution among its own people, czarist Russia was **9** a hearth of **imperialism**, and its empire contained peoples representing more than 100 nationalities. The winners in the ensuing revolutionary struggle—the communists who forged the Soviet Union—did not liberate these subjugated peoples.

Rather, they changed the empire's framework, binding the peoples colonized by the czars into a new system that would in theory give them autonomy and identity. In practice, it doomed those peoples to bondage and, in some cases, extinction.

When the Soviet system failed and the Soviet Socialist Republics became independent states, Russia was left without the empire that had taken centuries to build and consolidate—and that contained crucial agricultural and mineral resources. No longer did Moscow control the farms of Ukraine and the oil and natural gas reserves of Central Asia. But look again at Figure 2-3 and you will see that, even without its European and Central Asian colonies, Russia remains an empire. Russia lost the "republics" on its periphery, but Moscow still rules over a domain that extends from the borders of Finland to North Korea. Inside that domain Russians are in the overwhelming majority, but many subjugated nationalities, from Tatars to Yakuts, still inhabit ancestral homelands. Accommodating these many indigenous peoples is one of the challenges facing the Russian Federation today.

On the basis of physiographic, ethnic, historic, and cultural criteria, therefore, Russia constitutes a geographic realm, although transition zones rather than sharp boundaries mark its limits in some areas. Encircled by mountains and deserts, ruled by climatic continentality, unified by a dominant culture, and unmatched on Earth in terms of dimensions, Russia stands apart—from Europe to the west, China to the east, Central Asia to the south. But Russia, as we will see, is a society in transition. The realm's boundaries are unstable, still changing. To the west, Belarus at the opening of the twenty-first century was redirecting its interests from Europe toward Moscow. In the Caucasus, Armenia and Georgia effectively were in the Russian realm's orbit. In Central Asia, millions of Russians still live across the border in northern Kazakhstan.

In the 1990s, Russia began to reorganize in the aftermath of the collapse of the Soviet Union. This reorganization cannot be understood without reference to the seven decades of Soviet communist rule that went before. We turn next to this crucial topic.

THE SOVIET LEGACY

The era of communism may have ended in the Soviet Empire, but its effects on Russia's political and economic geography will long remain. Seventy years of centralized planning and implementation cannot be erased overnight; regional reorganization toward a market economy cannot be accomplished in a day.

While the world of capitalism celebrates the failure of the communist system in the former Soviet realm, it should remember why communism found such fertile ground in the Russia of the 1910s and 1920s. In those days Russia was infamous for the wretched serfdom of its peasants, the cruel exploitation of its workers, the excesses of its nobility, and the ostentatious palaces and riches of the czars. Ripples from the Western European Industrial Revolution introduced a new age of misery for those laboring in factories. There were workers' strikes and ugly retributions, but when the czars finally tried to better the lot of the poor, it was too little too late. There was no democracy, and the people had no way to express or channel their grievances. Europe's democratic revolution passed Russia by, and its economic revolution touched the czars' domain only slightly. Most Russians, and tens of millions of non-Russians under the czars' control, faced exploitation, corruption, starvation, and harsh subjugation. When the people began to rebel in 1905, there was no hint of what lay in store; even after the full-scale Revolution of 1917, Russia's political future hung in the balance.

The Russian Revolution was no unified uprising. There were factions and cliques; the Bolsheviks ("Majority") took their ideological lead from Lenin, while the Mensheviks ("Minority") saw a different, more liberal future for their country. The so-called Red army factions fought against the Whites, while both battled the forces of the czar. The country stopped functioning; the people suffered terrible deprivations in the countryside as well as the cities. Most Russians (and other nationalities within the empire as well) were ready for radical change.

That change came when the Revolution succeeded and the Bolsheviks bested the Mensheviks, most of whom were exiled. In 1918, the capital was moved from Petrograd (as St. Petersburg had been renamed in 1914, to remove its German appellation) to Moscow. This was a symbolic move, the opposite of the forward-capital principle: Moscow lay deep in the Russian interior, not even on a major navigable waterway (let alone a coast), amid the same forests that much earlier had protected the Russians from their enemies. The new Soviet Union would look inward, and the communist system would achieve with Soviet resources and labor the goals that had for so long eluded the country. The chief political and economic architect of this effort was the revolutionary leader who prevailed in the power struggle: V. I. Lenin (born Vladimir Ilyich Ulyanov).

The Political Framework

Russia's great expansion had brought many nationalities under czarist control; now the revolutionary government sought to organize this heterogeneous ethnic mosaic into a smoothly functioning state. The czars had conquered, but they had done little to bring Russian culture to the peoples they ruled. The Georgians, Armenians, Tatars, and residents of the Muslim states of Central Asia were among dozens of individual cultural, linguistic, and religious groups that had not been "Russified." In 1917, however, the Russians themselves constituted only about one-half of the population of the entire realm. Thus it was impossible to establish a Russian state instantly over this vast political region, and these diverse national groups had to be accommodated.

The question of the nationalities became a major issue in the young Soviet state after 1917. Lenin, who brought the philosophy of Karl Marx to Russia, talked from the beginning about the "right of self-determination for the nationalities." The first response by many of Russia's subject peoples was to proclaim independent republics, as they did in Ukraine, Georgia, Armenia, Azerbaijan, and even in Central Asia. But Lenin had no intention of permitting the Russian state to break up. In 1923, when

his blueprint for the new Soviet Union went into effect, the last of these briefly independent units was fully absorbed into the sphere of the Moscow regime. Ukraine, for example, declared itself independent in 1917 and managed to sustain this initiative until 1919. But in that year the Bolsheviks set up a provisional government in Kiev, the Ukrainian capital, thereby ensuring the incorporation of the country into Lenin's Soviet framework.

The Bolsheviks' political framework for the Soviet Union was based on the ethnic identities of its many incorporated peoples. Given the size and cultural complexity of the empire, it was impossible to allocate territory of equal political standing to all the nationalities; the communists controlled the destinies of well over 100 peoples, both large nations and small isolated groups. It was decided to divide the vast realm into Soviet Socialist Republics (SSRs), each of which was delimited to correspond broadly to one of the major nationalities. At the time, Russians constituted about half of the developing Soviet Union's population, and, as Figure 2-4 shows, they also were (and still are) the most widely dispersed ethnic group in the realm. The Russian Republic, therefore, was by far the largest designated SSR, comprising just under 77 percent of total Soviet territory.

Within the SSRs, smaller minorities were assigned political units of lesser rank. These were called Autonomous Soviet Socialist Republics (ASSRs), which in effect were republics within republics; other areas were designated Autonomous Regions or other nationality-based units. It was a complicated, cumbersome, often poorly designed framework, but in 1924 it was launched officially under the banner of the Union of Soviet Socialist Republics (USSR).

Eventually, the Soviet Union came to consist of 15 SSRs (shown in Fig. 2-5), including not only the original republics of 1924 but also such later acquisitions as Moldova (formerly Moldavia in Romania), Estonia, Latvia, and Lithuania. The internal political layout often was changed, sometimes at the whim of the communist empire's dictators. But no communist apartheid-like system of segregation could accommodate the shifting multinational mosaic of the Soviet realm. The republics quarreled

among themselves over boundaries and territory. Demographic changes, migrations, war, and economic factors soon made much of the layout of the 1920s obsolete. Moreover, the communist planners made it Soviet policy to relocate entire peoples from their homelands in order to better fit the grand design, and to reward or punish—sometimes capriciously. The overall effect, however, was to move minority peoples eastward and to replace them with **10** Russians. This **Russification** of the Soviet Empire produced substantial ethnic Russian minorities in all the non-Russian republics.

11 The Soviet planners called their system a **federation**. We focus in more detail on this geographic concept in Chapter 11, but we note here that federalism involves the sharing of power between a country's central government and its political subdivisions (provinces, States, or, in the Soviet case, "Socialist Republics"). Study the map of the former Soviet Union (Fig. 2-5), and an interesting geographic corollary emerges: every one of the 15 Soviet Republics had a boundary with a non-Soviet neighbor. Not one was spatially locked within the others. This seemed to give geographic substance to the notion that any Republic was free to leave the USSR if it so desired. Reality, of course, was different. Moscow's control over the republics made the Soviet Union a federation in theory only.

The centerpiece of the tightly controlled Soviet "federation" was the Russian Republic. With half the vast state's population, the capital city, the realm's core area, and over three-quarters of the Soviet Union's territory, Russia was the empire's nucleus. In other republics, "Soviet" often was simply equated with "Russian"—it was the reality with which the lesser republics lived. Russians came to the other republics to teach (Russian was taught in the colonial schools), to organize (and often dominate) the local Communist Party, and to implement Moscow's economic decisions. This was colonialism, but somehow the communist disguise—how could socialists, as the communists called themselves, be colonialists?—and the contiguous spatial nature of the empire made it appear to the rest of the world as something else. Indeed, on the world stage

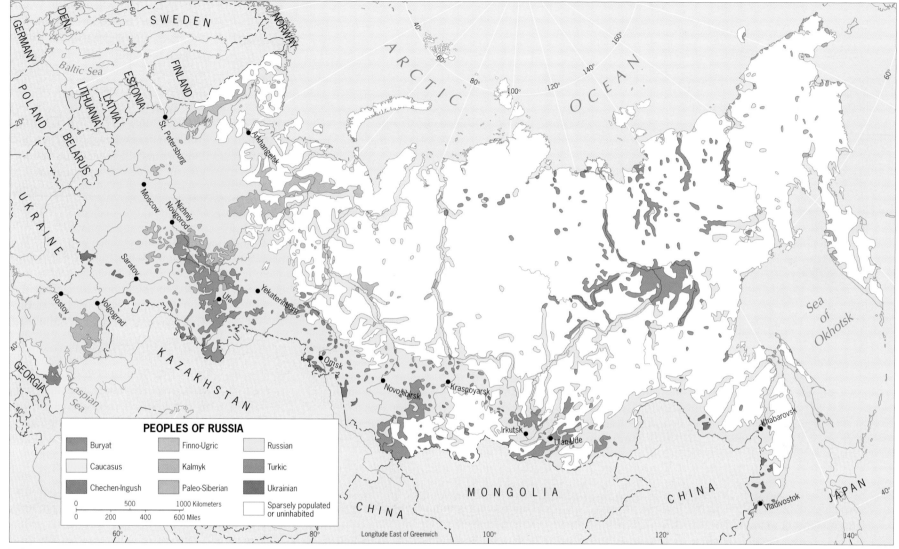

FIGURE 2-4

the Soviet Union became a champion of oppressed peoples, a force in the decolonization process. It was an astonishing contradiction that would, in time, be fully exposed.

The Soviet Economic Framework

The geopolitical changes that resulted from the founding of the Soviet Union were accompanied by a gigantic economic experiment: the conversion of the empire from a czarist autocracy with a capitalist veneer to communism. From the early 1920s onward, the country's economy would be centrally planned—the communist leadership in Moscow would make all decisions regarding economic planning and development. Soviet planners had two principal objectives: (1) to accelerate industrialization **12** and (2) to **collectivize** agriculture. For the first

time ever on such a scale, and for the first time in accordance with Marxist-Leninist principles, an entire country was organized to work toward national goals prescribed by a central government.

The Soviet planners believed that agriculture could be made more productive by organizing it into huge state-run enterprises. The holdings of large landowners were expropriated, private farms were taken away from the farmers, and the land was

FIGURE 2-5

consolidated into collective farms. Initially, all such land was meant to be part of a *sovkhoz*, literally a grain-and-meat factory in which agricultural efficiency, through maximum mechanization and minimum labor requirements, would be at its peak. But many farmers opposed the Soviets and tried to sabotage the program in various ways, hoping to retain their land.

The farmers and peasants who obstructed the communists' grand design suffered a dreadful fate. In the 1930s, for instance, Stalin confiscated Ukraine's agricultural output and then ordered part of the border between the Russian and Ukrainian republics sealed—thereby leading to a famine that killed millions of farmers and their families. In the

Soviet Union under communist totalitarianism, the ends justified the means, and untold hardship came to millions who had already suffered under the czars. In his book *Lenin's Tomb*, David Remnick estimates that between 30 and 60 million people lost their lives from imposed starvation, political purges, Siberian exile, and forced relocation. It was an incalculable human tragedy, but the secretive character of Soviet officialdom made it possible to hide it from the world.

The Soviet planners hoped that collectivized and mechanized farming would free hundreds of thousands of workers to labor in factories. Industrialization was the prime objective of the regime, and here the results were superior. Productivity

rose rapidly, and when World War II engulfed the empire in 1941, the Soviet manufacturing sector was able to produce the equipment and weapons needed to repel the German invaders.

Yet even in this context, the Soviet grand design entailed liabilities for the future. The USSR **13** practiced a **command economy**, in which state planners assigned the production of particular manufactures to particular places, often disregarding the rules of economic geography. For example, the manufacture of railroad cars might be assigned (as indeed it was) to a factory in Latvia. No other factory anywhere else would be permitted to produce this equipment—even if supplies of raw materials would make it cheaper to build them near, say, Volgograd 1200 miles away. Despite an expanded and improved transport network (Fig. 2-6), such practices made manufacturing in the USSR extremely expensive, and the absence of competition made managers complacent and workers less productive than they could be.

Of course, the Soviet planners never imagined that their experiment would fail and that a market-driven economy would replace their command economy. When that happened, the transition was predictably difficult; indeed, it is far from over, and it is putting severe strains on the now more democratic state.

RUSSIA'S CHANGING POLITICAL GEOGRAPHY

When the USSR dissolved in 1991, Russia's former empire devolved into 14 independent countries, and Russia itself was a changed nation. Russians now made up about 83 percent of the population of under 150 million, a far higher pro-

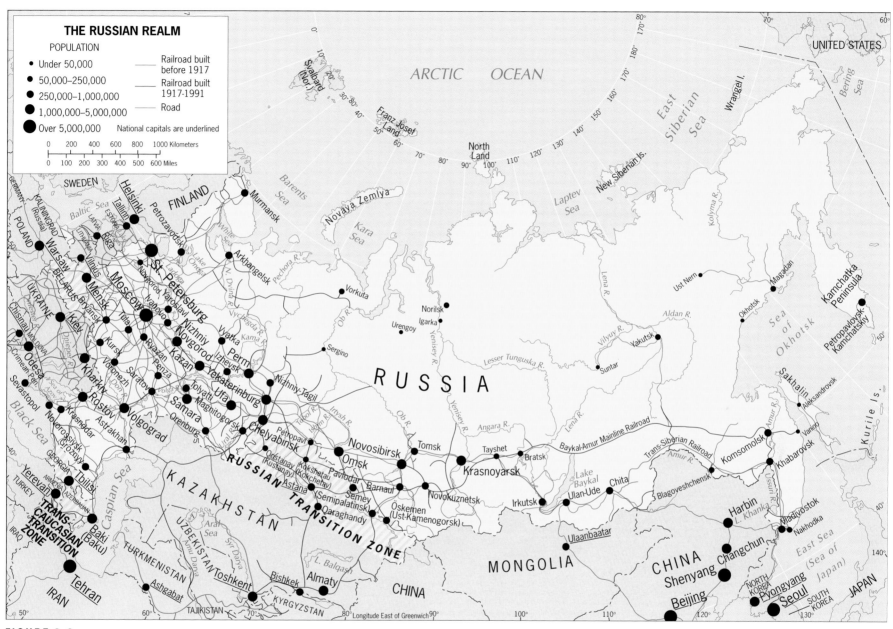

FIGURE 2-6

portion than in the days of the Soviet Union. But numerous minority peoples remained under Moscow's new flag, and millions of Russians found themselves under new governments in the former Republics.

Soviet planners had created a complicated administrative structure for their "Russian Soviet Federative Socialist Republic," and Russia's postcommunist leaders had to use this framework to make their country function. In 1992, most of

Russia's internal "republics," autonomous regions, oblasts, and krays (all of them components of the administrative hierarchy) signed a document known as the Russian Federation Treaty, committing them to cooperate in the new federal

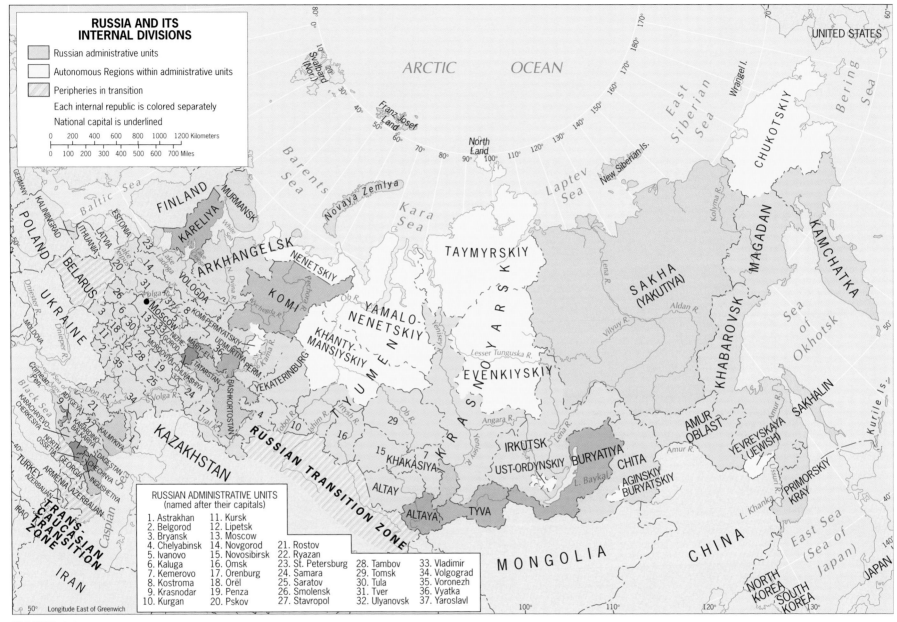

RUSSIA AND ITS INTERNAL DIVISIONS

Russian administrative units

Autonomous Regions within administrative units

Peripheries in transition

Each internal republic is colored separately

National capital is underlined

RUSSIAN ADMINISTRATIVE UNITS
(named after their capitals)

1. Astrakhan	11. Kursk		
2. Belgorod	12. Lipetsk		
3. Bryansk	13. Moscow		
4. Chelyabinsk	14. Novgorod		
5. Ivanovo	15. Novosibirsk		
6. Kaluga	16. Omsk		
7. Kemerovo	17. Orenburg		
8. Kostroma	18. Orël		
9. Krasnodar	19. Penza		
10. Kurgan	20. Pskov	21. Rostov	
		22. Ryazan	
		23. St. Petersburg	28. Tambov	33. Vladimir
		24. Samara	29. Tomsk	34. Volgograd
		25. Saratov	30. Tula	35. Voronezh
		26. Smolensk	31. Tver	36. Vyatka
		27. Stavropol	32. Ulyanovsk	37. Yaroslavl

FIGURE 2-7

system. At first a few units refused to sign, including Tatarstan, scene of Ivan the Terrible's brutal conquest more than four centuries ago, and a republic in the Caucasus periphery, then known as Chechenya-Ingushetia, where Muslim rebels waged a campaign for independence. As the map shows, Chechnya-Ingushetiya split into two separate republics whose names at the time were spelled Chechenya and Ingushetia (Fig. 2-7). Eventually, only Chechnya refused to sign the Russian Federation Treaty, and subsequent Russian military intervention led to a prolonged and violent conflict, with disastrous consequences for Chechnya's people and infrastructure (the capital, Groznyy, was completely destroyed). The Chechnya war continues today and is a disaster for Russia's government as well.

The Federal Framework of Russia

The spatial framework of the still-evolving Russian Federation is as complex as that of the Russian Federative Socialist Republic of communist times. As the twenty-first century opened, the Federation consisted of 89 entities: 2 Autonomous Federal Cities, 21 Republics, 11 Autonomous Regions (Okrugs), 49 Provinces (Oblasts), and 6 Territories (Krays). Moscow and St. Petersburg are the two Autonomous Federal Cities. The 21 Republics, recognized to accommodate substantial ethnic minorities in the population, lie in several clusters (Fig. 2-7).

Russia's post-1991 governments have faced complicated administrative problems. A multinational, multicultural state that had been accustomed to authoritarian rule and government control over virtually everything—from factory production to everyday life—now had to be governed in a new way. Democratization of the political system, transition to a market economy, sale of state-owned industries (privatization), and liberation of the press and other media must all take place in an orderly way, or the country would, literally, fall apart.

Russia's leaders knew that their options were limited. They could continue to hold as much power as possible at the center, making decisions in Moscow that would apply to all the Republics, Regions, and other subdivisions of the state. Such **14** a **unitary state system**, with its centralized government and administration, marked authoritarian kingdoms of the past and serves totalitarian dictatorships of the present. Or they could share power with the Republics and Regions, allowing elected regional leaders to come to Moscow to represent the interests of their people. This is the federal system Russia chose as the only way to accommodate the country's economic and cultural diversity.

In a *federal system*, the national government usually is responsible for matters such as defense, foreign policy, and foreign trade. The Regions (or provinces, States, or other subdivisions) retain authority over affairs ranging from education to transportation. A federal system does not create unity out of diversity, but it does allow diverse components of the state to coexist, their common interests represented by the national government and their regional interests by their local administrations. Some countries owe their survival as coherent states to their federal frameworks. India and Australia are cases in point.

But to maintain a generally acceptable balance of power between the center and the Regions (or States) is difficult. Disputes over "States' rights" continue to roil the American political scene more than two centuries after the Constitution was adopted. Russia, which also owes its continuity to a grand federal experiment, is barely more than ten years old. Power shifts continue, and the framework of the Russian Federation will undoubtedly change over time.

An additional problem arises from Russia's sheer size. Geographers refer to the principle of **15** **distance decay** to explain how increasing distances between places tend to reduce interactions among them. Because Russia is the world's largest country, distance is a significant factor in the relationships between the capital and outlying areas. Furthermore, Moscow lies in the far west of the giant country, half a world away from the shores of the Pacific. Not surprisingly, one of the most obstreperous Regions has been remote Primorskiy, the Region of Vladivostok.

Still another problem is reflected in Figure 2-7: the enormous size variation among the Republics (and to a lesser extent, the Regions). Whereas the (territorially) smallest Republics are concentrated in the Russian core area, the largest lie far to the east, where Sakha is nearly a thousand times as large as Ingushetiya. On the other hand, the populations of the huge eastern Republics are tiny compared to those of the smaller ones in the core. Such diversity spells administrative difficulty.

But perhaps the most serious current problem for Russia is the growing social disharmony between Moscow and the subnational political entities. Almost wherever one goes in Russia today, Moscow is disliked and often berated by angry locals. The capital is seen as the privileged playground for those who have benefited most from the post-Soviet transition—bureaucrats and hangers-on whose economic policies have driven down standards of living, whose greed and corruption have hurt the economy, whose actions in Chechnya have been a disaster, who have allowed foreigners (prominently the United States) to erode Russian power and prestige, who fail to pay the wages of workers toiling in industries still owned by the state, and who do not represent the Russian people. Complaints about the capital are not uncommon in free countries, but in Russia the mistrust between capital and subordinate areas has become so serious that it constitutes a key challenge for the government.

In 2000, the Putin administration moved to diminish the influence of the Regions by creating a new spatial framework that combines the 89 Regions, Republics, and other entities into seven new administrative units—not to enhance their influence in Moscow, but to increase Moscow's authority over them (Fig. 2-8). Each of these new federal administrative districts has its capital city, which will become the conduit for Moscow's "guidance," as the official plan puts it. This framework will counter some powerful vested interests in the Republics and Regions, however, and it will reverse a trend toward ever-greater local self-determination. It is far from certain that the system will work.

CHANGING SOCIAL GEOGRAPHIES

The political-geographical machinations discussed above give no indication of what is happening in the daily lives of most of Russia's citizens. But

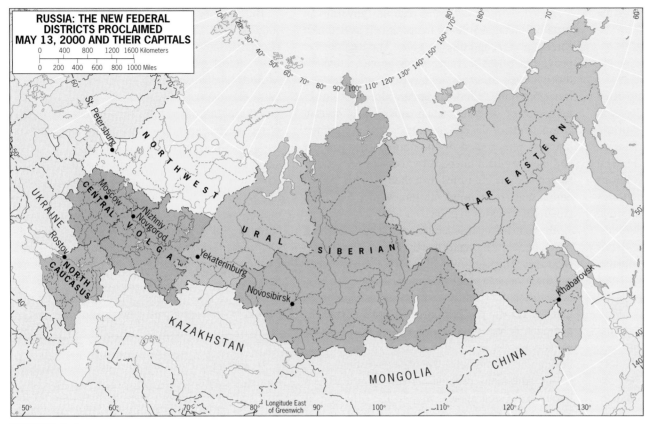

FIGURE 2-8

(Map caption on image)

RUSSIA: THE NEW FEDERAL
DISTRICTS PROCLAIMED
MAY 13, 2000 AND THEIR CAPITALS

lion, extrapolated to 10 million by 2020. Deaths attributed to AIDS were less than 1000 by mid-2003, but UN projections warned of 100,000 deaths annually by 2010 and 500,000 by 2020.

Coupled with this distress is the decline of national institutions. The armed forces are in disrepair. Major universities lack modern facilities. Leading industries are failing; the energy industry is in the grip of powerful magnates who flout the rules. Legal systems, banking operations, and property regulations are not sufficiently developed to ensure orderly procedures. Corruption and cronyism are rife, and some observers describe organized crime as Russia's most successful enterprise.

Yet Russia remains a formidable nuclear power with a vast arsenal of weapons. Russia's quest for social and political recovery and stability is a matter of global interest.

when you see statistics on the country's population, its general health, financial problems, life expectancy, and related conditions, the extent of the country's post-Soviet dislocation becomes clear. When the Soviet Union broke apart, the Russian Republic had a population of almost 150 million, still growing albeit very slowly. By 2004 it was down to slightly under 144 million. **16** If this **population decline** continues, there will be barely 100 million Russian citizens by 2050. Life expectancy for males, over 64 ten years ago, is now 59, and for females it has declined from over 74 to under 72. Among men, alcoholism, suicide, and other manifestations of social disorder drive down life expectancy. Deaths

from alcohol poisoning rose from about 16,000 in 1991 to over 41,000 in 2001. In the general population, drug abuse, heavy smoking, and poor diets are to blame.

The incidence of diseases is staggering for a country that not long ago was a world power. Tuberculosis may now affect more than 200,000 people (WHO and Russian sources vary widely), and deaths in just one recent year were nearly 30,000 compared to less than 1000 in the United States with double Russia's population. Rates of heart disease are rising. HIV cases can only be estimated because Russian sources are unreliable. In 2003 Russia officially reported 240,000 people infected, but UN estimates hovered around 1 mil-

RUSSIA'S PROSPECTS

When Peter the Great began to reorient his country toward Europe and the outside world, he envisaged a Russia with warm-water ports, a nation no longer encircled by Swedes, Lithuanians, Poles, and Turks, and a force in European affairs. Core-periphery relationships always have been crucial to Russia, and they remain so today.

Russia's relative location has long been the subject of study and conjecture by geographers. Exactly a century ago, the British geographer Sir Halford Mackinder (1861–1947) argued in a still-discussed article entitled "The Geographical Pivot of History" that western Russia and Eastern Europe enjoyed a combination of natural protection and resource wealth that would someday propel its occupants to world power. The protected core area, he reasoned, overshadowed the exposed periphery. Eventually, this pivotal interior region of Eurasia, which he later called the *heartland*, would become a stage for world domination.

17 Mackinder's **heartland theory** was published when Russia was a weak, economically backward society ripe for revolution, but Mackinder stuck to his guns. When Russia, in control of Eastern Europe and with a vast colonial empire, emerged from World War II as a superpower, Mackinder's conjectures seemed prophetic.

But not all geographers agreed. Probably the first scholar to use the term *rimland* (today *Pacific Rim* is part of everyday language) was Nicholas Spykman, who in 1944 countered Mackinder by calculating that Eurasia's periphery, not its core, held the key to global power. Spykman foresaw the rise of rimland states as superpowers and viewed Japan's emergence to wealth and power as just the beginning of that process. In that perspective, Russia's twentieth-century superpower status was just a temporary phenomenon.

Russia, Europe, and the World

The expansion of the European Union in 2004 has reached Russia's borders. Former components of the Soviet Empire now are members of NATO. How will Russia's relationship with Europe evolve, and what are the prospects for Russia in the wider world?

To Europeans, Russia has always been the enigmatic colossus to the east, alternately bungling and threatening but never in tandem with the West. When the Berlin Wall came down and the Soviet banner folded, there were hopeful predictions of a "European Russia" finally joining its erstwhile ideological adversaries in a regional partnership that would extend from the Atlantic to the Pacific. Such forecasts were based on real as well as idealized evidence: Russians' ethnic ties with Eastern Europe;

the revival of Christian churches after decades of communist atheism; the sprouting of Russian democracy; the budding of a market economy; the long-term dependence of Europe on Russian energy supplies.

Reality, however, was rather different. Russian democracy remained tinged by an authoritarian streak that scared investors. The media continued to face government interference. The Russian Orthodox Church suppressed efforts by other churches to become reestablished. Russian leaders objected vigorously, then acquiesced as NATO expanded toward its borders. Russia was unhelpful during efforts to mitigate the collapse of Yugoslavia. And Russia's treatment of minorities and migrants seemed to violate EU standards. All this occurred against the background of a military nuclear arsenal still capable of destroying the world in an afternoon.

Nevertheless, Russia has more in common with Europe than it has with any other realm of the world, and the "Europeanization of Russia," whatever form it takes, is likely to be a hallmark of the century just started. The process was given an inadvertent boost by al-Qaeda's 9/11 attack on New York and Washington D.C., when Russian, European, and American leaders realized that they faced a common enemy far more capable than they had estimated. One result was a softening of Western criticism of Russian efforts to suppress Islamic dissidents in Chechnya and elsewhere in the Federation. Over the longer term, however, Russia's well-being, and its ability to reverse its social deterioration discussed above, will depend in large measure on its integration into Europe and its adoption of European norms.

Russia's enormous territory endows it with numerous neighbors, and its imperial history leaves it with many border problems. In Central Asia, millions of Russians still reside in the former Soviet "republics" including neighboring Kazakhstan (see Chapter 7). In East Asia, China and Russia have settled a number of the boundary disputes shown in Figure 2-9, but for the future the larger question is likely to involve China's "lost" territo-

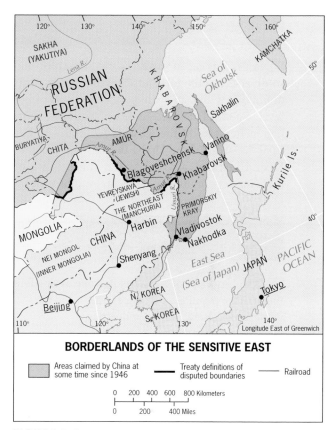

BORDERLANDS OF THE SENSITIVE EAST

Areas claimed by China at some time since 1946 — Treaty definitions of disputed boundaries — Railroad

0   200 400 600   800 Kilometers

0   200   400 Miles

FIGURE 2-9

ries east of the Amur and Ussuri rivers, taken by Russia more than a century ago.

Other significant external problems Russia confronts include: (1) the ownership of oil and gas reserves in the Caspian Basin, where maritime boundaries in the Caspian Sea have not yet been satisfactorily delimited; (2) settlement of the Kurile Islands issue with Japan, a legacy of World War II still not settled (see Chapter 9); (3) relations with its Transcaucasian neighbor Georgia in context of that area's Islamic militancy (see page 97); and (4) efforts by the authoritarian regime of Belarus to reunite in some formal way with the Russian state. But perhaps Russia's greatest challenge of all, in the international arena, is to sustain its residual position as a credible force in world affairs.

REGIONS OF THE REALM

So vast is Russia, so varied its physiography, and so diverse its cultural landscape that regionalization requires a small-scale perspective and a high level of generalization. Figure 2-10 outlines a four-region framework: the Russian Core and its Peripheries west of the Urals, the Eastern Frontier, Siberia, and the Far East. As we will see, each of these massive regions contains major subregions.

RUSSIAN CORE AND PERIPHERIES

18 The heartland of a state is its **core area**. Here much of the population is concentrated, and here lie its leading cities, major industries, densest transport networks, most intensively cultivated lands, and other key components of the country. Core areas of long standing strongly reflect the imprints of culture and history. The Russian core area, broadly defined, extends from the western border of the Russian realm to the Ural Mountains in the east (Fig. 2-10). This is the Russia of Moscow and St. Petersburg, of the Volga River and its industrial cities.

Central Industrial Region

At the heart of the Russian Core lies the Central Industrial Region (Fig. 2-11). The precise definition of this subregion varies, for all regional definitions are subject to debate. Some geographers prefer to call this the Moscow Region, thereby emphasizing that for over 250 miles (400 km) in all directions from the capital, everything is oriented toward this historic focus of the state. As Figure

2-6 shows, Moscow has maintained its decisive centrality: roads and railroads converge in all directions from Ukraine in the south; from Mensk (Belarus) and the rest of Eastern Europe in the west; from St. Petersburg and the Baltic coast in the northwest; from Nizhniy Novgorod (formerly Gorkiy) and the Urals in the east; from the cities and waterways of the Volga Basin in the southeast (a canal links Moscow to the Volga, Russia's most important navigable river); and even to the subarctic northern periphery that faces the Barents Sea, where the strategic naval port of Murmansk and lumber-exporting Arkhangelsk lie.

Moscow (population: 8.1 million) is the focus of an area that includes some 50 million inhabitants (more than one-third of the country's total population), many of them concentrated in such major cities as Nizhniy Novgorod, the automobile-producing "Soviet Detroit;" Yaroslavl, the tire-producing center; Ivanovo, the heart of the textile industry; and Tula, the mining and metallurgical center where lignite (brown coal) deposits are worked.

St. Petersburg (the former Leningrad) remains Russia's second city, with a population of 4.5 million. Under the czars, St. Petersburg was the focus of Russian political and cultural life, and Moscow was a distant second city. Today, however, St. Petersburg has none of Moscow's locational advantages, at least not with respect to the domestic market. It lies well outside the Central Industrial Region

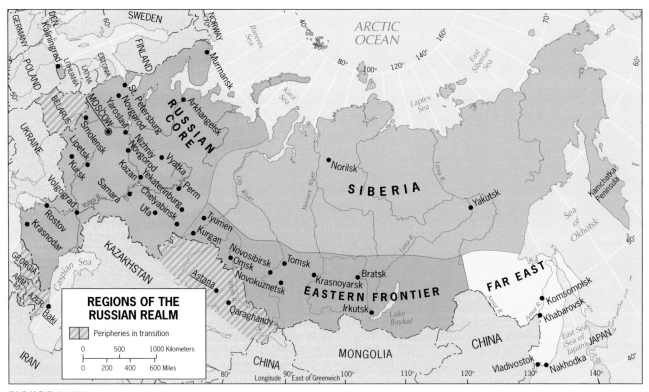

FIGURE 2-10

FROM THE FIELDNOTES

"June 25, 1964. We left Moscow this morning after our first week in the Soviet Union, on the road to Kiev. Much has been seen and learned after we arrived in Leningrad from Helsinki (also by bus) last Sunday. Some interesting contradictions: in a country that launched sputnik nearly a decade ago, nothing mechanical seems to work, from elevators to vehicles. The arts are vibrant, from the ballet in the Kirov Theatre to the Moscow Symphony. But much of what is historic here seems to be in disrepair, and in every city and town you pass, rows and rows of drab, gray apartment buildings stand amid uncleared rubble, without green spaces, playgrounds, trees. My colleagues here talk about the good fortune of the residents of these buildings: as part of the planning of the 'socialist city,' their apartments are always close to their place of work, so that they can walk or ride the bus a short distance rather than having to commute for hours. Our bus broke down as we entered the old part of Tula, and soon we learned that it would be several hours before we could expect to reboard. We had stopped right next to a row of what must have been attractive shops and apartments half a century ago; everything decorative seems to have been removed, and nothing has been done to paint, replace rotting wood, or fix roofs. I walked up the street a mile or so, and there was the 'new' Tula—the Soviet imprint on the architecture of the place. This scene could have been photographed on the outskirts of Leningrad, Moscow, or any other urban center we had seen. What will life be like in the Soviet city of the distant future?"

near the northwestern corner of the country, 400 miles (650 km) from Moscow. Neither is it better off than Moscow in terms of resources: fuels, metals, and foodstuffs must all be brought in, mostly from far away. The former Soviet emphasis on self-sufficiency even reduced St. Petersburg's asset of being on the Baltic coast because some raw materials could have been imported much more cheaply across the Baltic Sea from foreign sources than from domestic sites in distant Central Asia (only bauxite deposits lie nearby, at Tikhvin).

Yet St. Petersburg was at the vanguard of the Industrial Revolution in Russia, and its specialization and skills have remained important. Today, the city and its immediate environs contribute about 10 percent of the country's manufacturing, much of it through fabricating high-quality machinery.

Povolzhye: The Volga Region

A second region lying within the Russian Core is the *Povolzhye*, the Russian name for an area that extends along the middle and lower valley of the Volga River. It would be appropriate to call this the Volga Region, for that greatest of Russia's rivers is its lifeline and most of the cities that lie in the Povolzhye are on its banks (Fig. 2-11). In the 1950s, a canal was completed to link the lower Volga with the lower Don River (and thereby the Black Sea).

The Volga River was an important historic route in old Russia, but for a long time neighboring regions overshadowed it. The Moscow area and Ukraine were far ahead in industry and agriculture. The Industrial Revolution that came late in the nineteenth century to the Moscow Region did not have much effect in the Povolzhye. Its major function remained the transit of foodstuffs and raw materials to and from other regions.

This transport function is still important, but the Povolzhye has changed. First, World War II brought furious development because the Volga River, located east of Ukraine, was far from the German armies that invaded from the west. Second, in the postwar era the Volga-Urals Region for some time was the largest known source of petroleum and natural gas in the entire Soviet Union. From near Volgograd (formerly Stalingrad) in the southwest to Perm on the Urals' flank in the northeast lies a belt of major oilfields (Fig. 2-12).

Third, the transport system has been greatly expanded. The Volga-Don Canal directly connects the Volga waterway to the Black Sea; the Moscow Canal extends the northern navigability of this river system into the heart of the Central Industrial Region; and the Mariinsk canals provide a link to the Baltic Sea. Today, the Volga Region's population exceeds 25 million, and the cities of Samara (formerly Kuybyshev), Volgograd, Kazan, and Saratov all have populations between 1.0 and 1.3 million. Manufacturing has also expanded into the middle Volga Basin, emphasizing more specialized engineering industries. The huge Fiat-built auto assembly plant in Tolyatti, for example, is one of the world's largest of its kind.

The Urals Region

The Ural Mountains form the eastern limit of the Russian Core. They are not particularly high; in the north they consist of a single range, but southward they broaden into a hilly zone. Nowhere are they an obstacle to east-west transportation. An enormous storehouse of metallic mineral resources located in and near the Urals has made this area a natural place for industrial development. Today, the Urals Region, well connected to the Volga and Central Industrial Region, extends from Serov in the north to Orsk in the south (Fig. 2-11).

The Central Industrial, Volga, and Urals regions form the anchors of the Russian core area. For decades they have been spatially expanding to-

ward one another, their interactions ever more intensive. These regions of the Russian Core stand in sharp contrast to the comparatively less developed, forested, Arctic north and the remote upland to the south between the Black and Caspian seas. Thus even within this Russian coreland, frontiers still await growth and development.

The Internal Southern Periphery

We turn now to an area that has generated some of the most difficult politico-geographical problems the new Russia has faced since the collapse of the Soviet Union. Between the Black Sea to the west and the Caspian Sea to the east, and with the Caucasus Mountains to the south, Moscow seeks to stabilize and assert its authority over a tier of minority "republics," holdovers from the Soviet era (Fig. 2-13).

A combination of geographic factors causes the problems Moscow confronts in this peripheral zone. Here, the Russians (and later the Soviets) met and stalled the advance of Islam. Here the cultural mosaic is as intricate as anywhere in the Federation: Dagestan, the southernmost republic fronting the Caspian Sea, has 2 million inhabitants comprising some 30 ethnic groups speaking about 80 languages. And here Russia and its Caspian holdings penetrate the reservoirs of oil and natural gas that this huge, energy-rich basin contains. Groznyy, the capital of Chechnya which adjoins Dagestan to the west, was a crucial oil-industry center, pipeline junction, and service hub during the Soviet era (Fig. 2-13, inset map).

But Chechnya also contains a sizeable Muslim population, and fiercely independent Muslim Chechens used Caucasus mountain hideouts to resist Russian colonization during the nineteenth century. Accused of collaboration with the Nazis during World War II, on Stalin's orders the Soviets exiled the entire Chechen population to Central Asia, with much loss of life. Rehabilitated and allowed to return by Khrushchev in the 1950s, the

Chechens seized their opportunity in 1991 after the collapse of the Soviet Union and fought the Russian army to a stalemate. But Moscow never granted Chechnya independence (one quarter of the population of 1.2 million was Russian), and an uneasy standoff continued. Meanwhile, Chechen hit-and-run attacks on targets in neighboring republics continued, and in 1999 three apartment buildings in Moscow were bombed, resulting in 230 deaths. Russian authorities blamed Chechnyan terrorists for these bombings and ordered a full-scale attack on Chechnya's Muslim holdouts. Groznyy, already severely damaged in the earlier conflict, was totally devastated and the rebels were driven into the mountains, but the Russian armed forces, still taking substantial losses, were unable to establish unchallenged control of all of the Republic.

And the Chechens were able to continue carrying their war of terror to Russia's core. Late in 2002, a team of 41 armed Chechens invaded a Moscow theater in mid-performance, taking all 700 audience members hostage. After a three-day standoff, a botched rescue effort left nearly 130 theater-goers and all the terrorists dead.

Russia's costly problems in Chechnya illustrate the fractious nature of its internal southern periphery, from Buddhist Kalmykiya to also-Muslim Ingushetiya. Here the still-evolving federal framework will be put to the test for many years to come.

The External Southern Periphery

Beyond Russia's tier of internal republics lies still another periphery, legally outside the Russian sphere but closely tied to it nevertheless. This is *Transcaucasia*, historically a battleground for Christians and Muslims, Armenians and Turks, Russians and Persians. Today it is a subregion containing three former Soviet Socialist Republics: landlocked Armenia, coastal Georgia on the Black Sea, and Azerbaijan on the land-encircled Caspian Sea. Although these three countries are indepen-

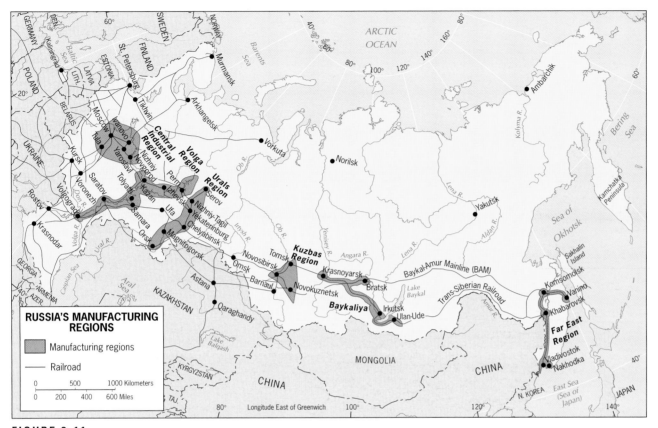

FIGURE 2-11

acknowledging the cultural (Christian) distinctiveness of this cluster of Armenians, nevertheless gave (Muslim) Azerbaijan jurisdiction over it.

That was a recipe for trouble: in the ensuing conflict, Armenian troops entered Azerbaijan and gained control over the exclave, even ousting Azerbaijanis from the zone between the main body of Armenia and Nagorno-Karabakh (Fig. 2-13). The international community, however, has not recognized Armenia's occupation, and officially the territory remains a part of Azerbaijan. After the turn of the twenty-first century, the matter still remained unresolved.

Georgia

Of the three former Soviet Republics in Transcaucasia, only Georgia has a Black Sea coast and thus an outlet to the wider world. Smaller than South Carolina, Georgia is a country of high mountains and fertile valleys. Its social and political geographies are complicated. The population of 4.4 million is more than 70 percent Georgian but also includes Armenians (8%), Russians (6%), Ossetians (3%), and Abkhazians (2%). The Georgian Orthodox Church dominates the religious community, but about 10 percent of the people are Muslims, most of them concentrated in Ajaria in the southwest.

Unlike Armenia and Azerbaijan, Georgia has no exclaves, but its political geography is problematic nonetheless (Fig. 2-13). Within Georgia's borders lie three minority-based autonomous entities: the Abkhazian and Ajarian Autonomous Republics, and the South Ossetian Autonomous Region.

Sakartvelos, as the Georgians call their country, has a long and turbulent history. Tbilisi, the capital for 15 centuries, lay at the core of an empire around the turn of the thirteenth century, but the Mongol

dent states today, they are so closely bound up with Russia that they are, functionally, part of the Russian realm (Fig. 2-13).

Armenia

As Figure 2-1 shows, landlocked Armenia (population: 3.8 million) occupies some of the most rugged and mountainous terrain in the earthquake-prone Transcaucasus. The Armenians are an embattled people who adopted Christianity 17 centuries ago and for more than a millennium sought to secure their ancient homeland here on the margins of the Muslim world. During World War I, the Ottoman Turks massacred much of the Christian Armenian minority and drove the survivors from

eastern Anatolia and what is now Iraq into the Transcaucasus. At the end of that war in 1918, an independent Armenia arose, but its autonomy lasted only two years. In 1920, Armenia was taken over by the Soviets; in 1936, it became one of the 15 constituent republics of the Soviet Union. The collapse of the Soviet Empire gave Armenia what it had lost three generations earlier: independence.

Or so it seemed. Soon afterward, the Armenians found themselves at war with neighboring Azerbaijan over the fate of some 150,000 Armenians living in Nagorno-Karabakh, a pocket of territory surrounded by Azerbaijan. Such a separated territory is called an *exclave*, and this one had been created by Soviet sociopolitical planners who, while

FIGURE 2-12

Caucasus. Today, Georgia remains a dysfunctional state.

Azerbaijan

Azerbaijan is the name of an independent state *and* of a province in neighboring Iran. The Azeris (short for Azerbaijanis) on both sides of the border have the same ancestry: they are a Turkish people divided between the (then) Russian and Persian empires by a treaty signed in 1828. By that time, the Azeris had become Shi'ite Muslims, and when the Soviet communists laid out their grand design for the USSR, they awarded the Azeris their own republic. On the Persian side, the Azeris were assimilated into the Persian Empire, and their domain became a province. Today, the former Soviet Socialist Republic is the independent state of Azerbaijan (population: 8.3 million), and the 10 million Azeris to the south live in the Iranian province.

During the brief transition to independence and at the height of their war with the Armenians, the dominantly Muslim Azeris tended to look southward, toward Iran. But geographic realities dictate a more practical orientation. Azerbaijan possesses huge reserves of oil and natural gas; under the Soviets it was one of Moscow's chief regional sources of fuels. The center of the oil industry is Baki (Baku), the capital on the shore of the Caspian Sea—but the Caspian Sea is a lake. To export its oil, Azerbaijan needs pipelines, but those of Soviet vintage link Baki to Russia's Black Sea terminal of Novorossiysk. During the mid-1990s, when Azerbaijan's leaders announced plans to build new pipelines via Georgia or Iran, Russia objected and even meddled in Azeri politics to stymie such plans.

invasion ended that era. Next, the Christian Georgians found themselves in the path of wars between Islamic Turks and Persians. Turning northward for protection, the Georgians were annexed by the Russians, who were looking for warm-water ports, in 1800. Like other peoples overpowered by the czars, the Georgians took advantage of the Russian Revolution to reassert their independence; but the Soviets reincorporated Georgia in 1921 and proclaimed a Georgian Soviet Socialist Republic in 1936. Josef Stalin, the communist dictator who succeeded Lenin, was a Georgian.

Georgia is renowned for its scenic beauty, warm and favorable climates, agriculture (especially tea), timber, manganese, and other products. Georgian wines, tobacco, and citrus fruits are much in demand. The diversified economy could support a viable state.

Unfortunately, Georgia's political geography is loaded with centrifugal forces. Georgia declared its independence in 1991, but in the Autonomous Region of South Ossetia, a movement for union with (Russian) North Ossetia created havoc. Abkhazia proclaimed independence. The price of all this instability was Russian intervention, and just when things calmed down and Russian military bases were closed, Moscow charged that Georgia was not doing enough to deter Muslim extremists operating in the

**SOUTHERN RUSSIA:
INTERIOR AND EXTERIOR PERIPHERIES**

POPULATION

- Under 50,000
- 50,000–250,000
- 250,000–1,000,000
- 1,000,000–5,000,000
- Over 5,000,000

National capitals are underlined

Railroad
Road
Canal
Oil pipeline
Proposed oil pipeline

0   50   100   150   200   250   300   350 Kilometers
0   50   100   150   200 Miles

FIGURE 2-13

A look at the map (Fig. 2-13) shows Azerbaijan's options. A pipeline could be routed from Baki through Georgia to the Black Sea coast near Batumi, avoiding Russian territory altogether. Another route could run through Georgia and across Turkey to the Mediterranean terminal at Ceyhan. Still another option would run the pipeline across Iran to a terminal on the Persian Gulf. Moscow objects to such alternatives, but Azerbaijan now has powerful international allies. American, French, British, and Japanese oil companies are developing Azerbaijan's offshore Caspian reserves, and U.S. interests prefer the Turkish route.

All these alternate routes are in the future, however; for the present, Azerbaijan's oil must flow across Russian territory, and Azerbaijan must cooperate with Moscow. Given its Muslim roots and its location, the time may come when Azerbaijan turns toward the Islamic world. For the moment, however, this country is inextricably bound up with its neighbors in Transcaucasia and the north.

THE EASTERN FRONTIER

From the eastern flanks of the Ural Mountains to the headwaters of the Amur River, and from the latitude of Tyumen to the northern zone of neighboring Kazakhstan, lies Russia's vast Eastern Frontier Region, product of a gigantic experiment in the eastward extension of the Russian Core (Fig. 2-10). As the maps of cities and surface communications suggest, this eastern frontier is more densely peopled and more fully developed in the west than in the east; at the longitude of Lake Baykal, settlement has become linear, marked by ribbons and clusters along the east-west railroads. Two subregions dominate the geography: the Kuznetsk Basin in the west and the Lake Baykal area in the east.

The Kuznetsk Basin (Kuzbas)

Some 900 miles (1450 km) east of the Urals lies another of Russia's primary regions of heavy manufac-

turing resulting from the communist period's national planning: the Kuznetsk Basin, or *Kuzbas* (Fig. 2-11). In the 1930s, it was opened up as a supplier of raw materials (especially coal) to the Urals, but that function became less important as local industrialization accelerated. The original plan was to move coal from the Kuzbas west to the Urals and allow the returning trains to carry iron ore east to the coalfields. However, good-quality iron ore deposits were subsequently discovered near the Kuznetsk Basin itself. As the new resource-based Kuzbas industries grew, so did its urban centers. The leading city, located just outside the region, is Novosibirsk, which stands at the intersection of the Trans-Siberian Railroad and the Ob River as the symbol of Russian enterprise in the vast eastern interior. To the northeast lies Tomsk, one of the oldest Russian towns in all of Siberia, founded in the seventeenth century and now caught up in the modern development of the Kuzbas Region. Southeast of Novosibirsk lies Novokuznetsk, a city that produces steel for the region's machine and metal-working plants and aluminum products from Urals bauxite.

The Lake Baykal Area (Baykaliya)

East of the Kuzbas, development becomes more insular, and distance becomes a stronger adversary. North of the Tyva Republic and eastward around Lake Baykal, larger and smaller settlements cluster along the two railroads to the Pacific coast (Fig. 2-11). West of the lake, these rail corridors lie in the headwater zone of the Yenisey River and its tributaries. A number of dams and hydroelectric projects serve the valley of the Angara River, particularly the city of Bratsk. Mining, lumbering, and some farming sustain life here, but isolation dominates it. The city of Irkutsk, near the southern end of Lake Baykal, is the principal service center for a vast Siberian region to the north and for a lengthy east-west stretch of southeastern Russia.

Beyond Lake Baykal, the Eastern Frontier really lives up to its name: this is southern Russia's most

rugged, remote, forbidding country. Settlements are rare, many being mere camps. The Buryat Republic (Fig. 2-7) is part of this zone; the territory bordering it to the east was taken from China by the czars and may become an issue in the future. Where the Russian-Chinese boundary turns southward, along the Amur River, the region called the Eastern Frontier ends and Russia's Far East begins.

SIBERIA

Before we assess the potential of Russia's Pacific Rim, we should remember that the ribbons of settlement just discussed hug the southern perimeter of this giant country, avoiding the vast Siberian region to the north (Fig. 2-10). Siberia extends from the Ural Mountains to the Kamchatka Peninsula—a vast, bleak, frigid, forbidding land. Larger than the conterminous United States but inhabited by only an estimated 15 million people, Siberia quintessentially symbolizes the Russian environmental plight: vast distances, cold temperatures worsened by strong Arctic winds, difficult terrain, poor soils, and limited options for survival.

But Siberia also has resources. From the days of the first Russian explorers and Cossack adventurers, Siberia's riches have beckoned. Gold, diamonds, and other precious minerals were found. Later, metallic ores including iron and bauxite were discovered. Still more recently, the Siberian interior proved to contain massive quantities of oil and natural gas (Fig. 2-12) and began to contribute significantly to Russia's energy supply.

As the physiographic map (Fig. 2-1) shows, major rivers—the Ob, Yenisey, and Lena—flow gently northward across Siberia and the Arctic Lowland into the Arctic Ocean. Hydroelectric power development in the basins of these rivers has generated electricity used to extract and refine local ores, and in the lumber mills that have been set up to exploit the vast Siberian forests.

The human geography of Siberia is fragmented, and much of the region is virtually uninhabited (Fig.

2-4). Ribbons of Russian settlement have developed; the Yenisey River, for instance, can be traced on this map of Soviet peoples (a series of small settlements north of Krasnoyarsk), and the upper Lena Valley is similarly fringed by ethnic Russian settlement. Yet hundreds of miles of empty territory separate these ribbons and other islands of habitation.

The political geography of eastern Siberia is marked by the growing identity of Sakha (the Yakut Republic). As additional resources are discovered here (including oil and natural gas), this Republic, centered on the capital, Yakutsk, will become more important.

Siberia, Russia's freezer, is stocked with goods that may become mainstays of future national development. Already, precious metals and mineral fuels are bolstering the Russian economy. In time, we may expect Siberian resources to play a growing role in the economic development of the Eastern Frontier and the Russian Far East as well. One step in that process was already taken during Soviet times: the completion of the BAM (Baykal-Amur Mainline) Railroad in the 1980s. This route, lying north of and parallel to the old Trans-Siberian Railroad, extends 2200 miles (3540 km) eastward from Tayshet (near the important center of Krasnoyarsk) directly to the Far East city of Komsomolsk (Fig. 2-6). In the post-Soviet era, the BAM Railroad has been beset by equipment breakdowns and workers' strikes. Nonetheless, it is a key element of the infrastructure that will serve the Eastern Frontier's economic growth in the twenty-first century.

THE RUSSIAN FAR EAST

Imagine this: a country with 5000 miles (8000 km) of Pacific coastline, two major ports, large interior cities nearby, huge reserves of resources ranging from minerals to fuels to timber, directly across from one of the world's largest economies—all this at a time when the Asian Pacific Rim was the world's fastest-growing economic region. Would not that country have burgeoning cities, busy harbors, growing industries, and expanding trade?

In the Russian Far East (Fig. 2-10), the answer is—no. Activity in the port of Vladivostok is a shadow of what it was during the Soviet era, when it was the communists' key naval base. The nearby container terminal at Nakhodka suffers from breakdowns and inefficiencies. The railroad to western Russia carries just a fraction of the trade it did during the 1970s and 1980s. Cross-border trade with China is minimal. Trade with Japan is inconsequential. The region's cities are grimy, drab, moribund. Utilities are shut off for hours at a time because of fuel shortages and system breakdowns. Outdated factories are shut, their workers dismissed. Political relations with Moscow are poor. There is potential here, but little of it has been realized.

As a region, the Russian Far East consists of two parts: the mainland area extending from Vladivostok to the Eastland Highlands and the large

FROM THE FIELDNOTES

"Standing in an elevated doorway in the center of Vladivostok, you can see some of the vestiges of the Soviet period: the omnipresent, once-dominant, GUM department store behind the blue seal on the left, and the communist hammer-and-sickle on top of the defunct hotel on the right. But in the much more colorful garb of the people, and the private cars in the street, you see reflections of the new era. Once a city closed to foreigners, Vladivostok now throbs with visitors—and has become a major point of entry for contraband goods."

island of Sakhalin (Figs. 2-1; 2-14). This is cold country: icebreakers have to keep the ports of Vladivostok and Nakhodka open throughout the winter. Winters here are long and bitterly cold; summers are brief and cool. Although the population is small (about 7 million), food must be imported because not much can be grown. Most of the region is rugged, forested, and remote. Vladivostok, Khabarovsk, and Komsomolsk are the only cities of any size. Nakhodka and the newer railroad terminal at Vanino are smaller towns; the entire population of Sakhalin is about 700,000 (on an island the size of Caribbean Hispaniola [16.6 million]). Offshore lie productive fishing grounds, and Russian fleets from Vladivostok and points north catch salmon, herring, cod, and mackerel to be frozen or canned and shipped to local and distant markets.

The Soviet regime rewarded people willing to move to this region with housing and subsidies. The communists, like the czars before them, realized the importance of this frontier (Vladivostok means "We Own the East"), and they used every possible incentive to develop it and link it more strongly to Russia's distant western core. Freight rates on the Trans-Siberian Railroad, for example, were about 10 percent of their real costs; the trains were always loaded in both directions. Vladivostok was a military base and a city closed to foreigners, and Moscow invested heavily in its infrastructure. Komsomolsk in the north and Khabarovsk near the region's center were endowed with state-owned industries using local resources: iron ore from Komsomolsk, oil from Sakhalin, timber from the ubiquitous forests. The steel, chemical, and furniture industries sent their products westward by train, and they received food and other consumer goods from the Russian heartland.

For several reasons, the post-Soviet transition has been especially difficult here in the Far East. The new economic order has canceled the region's communist-era advantages: the Trans-Siberian Railroad now must charge the real cost of transporting products from the Eastern Frontier to the Russian Core. State-subsidized industries must compete on market principles, their subsidies having ended. The decline of Russia's armed forces has hit Vladivostok hard. The fleet lies rusting in port; service industries have lost their military markets; the shipbuilding industry has no government contracts. Coal miners in the Bureya River Valley (a tributary of the Amur) go unpaid for months and go on strike; coal-fired power

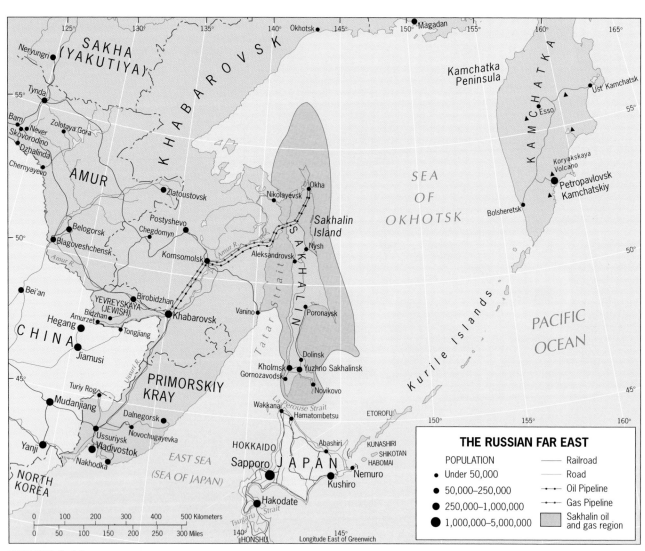

FIGURE 2-14

plants do not receive fuel shipments, and cities and towns go dark.

Locals put much of the blame for their region's failure on Moscow, and with reason. As Figure 2-7 shows, the Far East contains only five administrative regions: Primorskiy Kray, Khabarovsk Kray, Amur Oblast, Sakhalin Oblast, and Yevreyskaya, originally the Jewish Autonomous Region. This does not add up to much political clout, which is the way Moscow appears to want it. Compare Figures 2-8 and 2-14 and you will note that the city of Khabarovsk, not the Primorskiy Kray capital of Vladivostok, has been made the headquarters of the Far Eastern Region as defined by Moscow.

For all its stagnation, Russia's Far East will figure prominently in Russian (and probably world) affairs. Here, Russia meets China on land and Japan at sea (an unresolved issue between Russia and Japan involves several Kurile islands). Here lie vast resources ranging from Sakhalin's fuels to Siberia's lumber, Sakha's gold to Khabarovsk's metals. Here, Russia has a foothold on the Pacific Rim, a window on the ocean on whose shores the world is being transformed.

DISCOVERY TOOLS   www.wiley.com/college/deblij

The *Concepts and Regions* Website, featuring *GeoDiscoveries*, offers many additional resources to enhance your understanding and experience of this chapter. Be sure to explore the following:

GeoDiscoveries Video Clips

Legacies of Central Planning

East of Eden: Vladivostok

GeoDiscoveries Interactivities

Putin Russia on the Map

Placing Places in Russia

Cities on the Right Side of the Tracks

Many Rivers Run Through It

Manufacturing Regions in Russia

Made in Russia

Photo Galleries

St. Petersburg

Mount Elbrus

Amur River

Moscow

Vladivostok

Catherine's Palace, St. Petersburg

Tula

More To Explore

Systematic Essay: Climatology

Issues in Geography: The Soviet Union, 1924–1991; Russians in North America

Major Cities: Moscow, St. Petersburg

Expanded Regional Coverage

Russian Core and Peripheries; Eastern Frontier; Siberia; Russian Far East

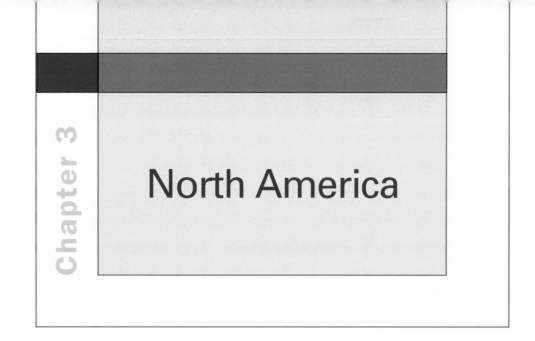

Chapter 3

North America

CONCEPTS, IDEAS, AND TERMS

1 Cultural pluralism
2 Physiographic province
3 Rain shadow effect
4 Migration
5 Push/pull factors
6 Sunbelt
7 American Manufacturing Belt

8 Ghetto
9 Outer city
10 Suburban downtown
11 Urban realms
12 Mosaic culture
13 Productive activities
14 Fossil fuels

15 Economies of scale
16 Postindustrialism
17 Technopole
18 Ecumene
19 Regional state
20 Pacific Rim

REGIONS

NORTH AMERICAN CORE
MARITIME NORTHEAST
FRENCH CANADA
CONTINENTAL INTERIOR
SOUTH

SOUTHWEST
WESTERN FRONTIER
NORTHERN FRONTIER
PACIFIC HINGE

NORTH AMERICA AS a geographic name means different things to different people. To physical geographers, North America refers to the landmass that extends from Alaska to Panama. As Figure G-4 shows, the North American continent is the dominant feature of the North American tectonic plate, which reaches almost—but not quite—to Panama. To human geographers, North America (sometimes inappropriately called "Anglo" America) signifies a geographic realm consisting of two large countries with much in common, Canada and the United States. We therefore recognize three realms in the Americas: (1) *North America*, propelled by the United States; (2) *Middle America*, anchored by Mexico; and (3) *South America*, dominated by Brazil.

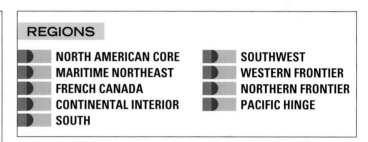

DEFINING THE REALM

Defined in the context of human geography, North America is constituted by two of the world's most highly advanced countries by virtually every measure of human development. Blessed by an almost endless range of natural resources and bonded by trade as well as culture, Canada and the United States are locked in a mutually productive embrace that is reflected by the statistics. In an average year of the recent past, 85 percent of Canadian exports went to

MAJOR GEOGRAPHIC QUALITIES OF NORTH AMERICA

1. North America encompasses two of the world's biggest states territorially (Canada is the second largest in size; the United States is third).

2. Both Canada and the United States are federal states, but their systems differ. Canada's is adapted from the British parliamentary system and is divided into ten provinces and three territories. The United States separates its executive and legislative branches of government, and it consists of 50 States, the Commonwealth of Puerto Rico, and a number of island territories under U.S. jurisdiction in the Caribbean Sea and the Pacific Ocean.

3. Both Canada and the United States are plural societies. Although ethnicity is increasingly important, Canada's pluralism is most strongly expressed in regional bilingualism. In the United States, major divisions occur along racial lines.

4. Many of Quebec's French-speaking citizens support a movement that seeks independence for the province. The movement's high-water mark was reached in the 1995 referendum in which (minority) non-French-speakers were the difference in the narrow defeat of separation. The prospects for a break-up of the Canadian state have diminished significantly since 2000.

5. North America's population, not large by international standards, is highly urbanized and the world's most mobile. Largely propelled by a continuing wave of immigration, the realm's population total is expected to grow by more than 40 percent over the next half-century.

6. By world standards, this is a rich realm where high incomes and high rates of consumption prevail. North America possesses a highly diversified resource base, but nonrenewable fuel and mineral deposits are consumed prodigiously.

7. North America is home to one of the world's great manufacturing complexes. The realm's industrialization generated its unparalleled urban growth, but a new postindustrial society and economy are rapidly maturing in both countries.

8. The two countries heavily depend on each other for supplies of critical raw materials (e.g., Canada is the leading source of U.S. energy imports) and have long been each other's chief trading partners. Today, the North American Free Trade Agreement (NAFTA), which also includes Mexico, is linking all three economies ever more tightly as barriers to international trade and investment flow are steadily dismantled.

9. North Americans are the world's most mobile people. Although plagued by recurrent congestion problems, the realm's networks of highways, commercial air routes, and cutting-edge telecommunications are the most efficient on Earth.

the United States and 74 percent of Canada's imports came from its southern neighbor. For the United States, Canada is its leading export market and its number one source of imports.

Both countries also rank among the world's most highly urbanized, with at least 75 percent of their populations concentrated in cities that have evolved into metropolitan agglomerations. The old core cities are now encircled by rings of suburban development, which draw an ever-greater proportion of the metropolitan population and activity base into their newly urbanized environments. Nothing symbolizes the North American city as strongly as the skyscrapered panoramas of New York, Chicago, and Toronto—or the vast, beltway-connected suburban expanses of Los Angeles, Washington, D.C., and Houston.

North Americans also are the most mobile people in the world. Commuters stream into and out of suburban business centers and central-city downtowns by the millions each working day; most of them drive cars, whose numbers have multiplied more than six times faster than the human population since 1970. Moreover, each year nearly one out of every six individuals changes his or her residence.

Although Canada and the United States share many historical, cultural, and economic qualities, they also differ in significant ways, as can be seen on the map. The United States, somewhat smaller territorially than Canada, occupies the heart of the North American continent and, as a result, encompasses a greater environmental range. The U.S. population is dispersed across most of the country, forming major concentrations along both the (north-south-trending) Atlantic and Pacific coasts. The overwhelming majority of Canadians, however, live in an interrupted east-west corridor that lies across southern Canada, mainly within 200 miles (320 km) of the U.S. border. The United States also encompasses North America's northwestern extension, Alaska. (Offshore Hawai'i, however, belongs in the Pacific Realm).

Demographic contrasts between the United States and Canada are substantial. The United States is the world's third most populous country with 290.9 million in 2004; Canada ranks 35th with 31.6 million. Comparatively small as its population may be, however, Canada has varied cultures and traditions ranging from multilingualism (French has equal status with English) to ethnic diversity that, as we will note later, have strong spatial expression. Thus Canada shares with the United States a characteristic feature of this realm: the prevalence of **1** **cultural pluralism**. In the United States, this is exhibited by a range of ethnic backgrounds from European and African to Asian and Native American. And while the civil rights movement and other initiatives have eliminated *de jure* (legal) segregation in the United States, ethnic clustering and *de facto* (real-world) self-segregation continue to mark the social landscape. In part, this is due to a troubling aspect of this wealthiest of all realms: poverty and deprivation still afflict a substantial minority as income gaps between the successful rich and the struggling poor continue to widen.

NORTH AMERICA'S PHYSICAL GEOGRAPHY

Before we examine the human geography of the United States and Canada more closely, we need to consider the setting in which they are rooted. The North American continent extends from the Arctic Ocean to Panama (Fig. 3-1), but we will confine ourselves here to the territory north of Mexico—a geographic realm that still stretches from the near-tropical latitudes of southern Florida and Texas to subpolar Alaska and Canada's far-flung northern periphery. The remainder of the North American continent comprises a separate realm, *Middle America*, which is covered in Chapter 4.

Physiography

North America's physiography is characterized by its clear, well-defined division into physically **2** homogeneous regions called **physiographic provinces**. Each region is marked by considerable uniformity in relief, climate, vegetation, soils, and other environmental conditions, resulting in a scenic sameness that comes readily to mind. For example, we identify such regions when we refer to the Rocky Mountains or the Great Plains, and the Appalachian Highlands. However, not all the physiographic provinces of North America are so easily delineated.

Figure 3-2 maps the complete layout of the continent's physiography and includes a cross-sectional terrain profile along the 40th parallel. The most obvious aspect of this map of North America's physiographic provinces is the north-south alignment of the continent's great mountain backbone, the Rocky Mountains, whose rugged topography dominates the western segment of the continent from Alaska to New Mexico. The major feature of eastern North America is another, much lower chain of mountain ranges called the Appalachian Highlands, which also trend approximately north-south and extend from Canada's Atlantic Provinces to Alabama. The orientation of the Rockies and Appalachians is important because, unlike Europe's Alps, they do not form a topographic barrier to polar or tropical air masses flowing southward or northward, respectively, across the continent's interior.

Between the Rocky Mountains and the Appalachians lie North America's vast interior plains, which extend from the Mackenzie Delta on the Arctic Ocean to the Gulf of Mexico. We can subdivide these into several provinces: (1) the great Canadian Shield, which is the geologic core area containing North America's oldest rocks; (2) the Interior Lowlands, covered largely by glacial debris laid down by ice, meltwater, and wind during the Pleistocene glaciation; and (3) the Great Plains, the extensive sedimentary surface that rises slowly westward toward the Rocky Mountains. Along the southern margin, these interior plainlands merge into the Gulf-Atlantic Coastal Plain, which stretches from southern Texas along the seaward margin of the Appalachian Highlands and the neighboring Piedmont until it ends at Long Island just to the east of New York City.

On the western side of the Rocky Mountains lies the zone of Intermontane Basins and Plateaus. Within the conterminous United States, this physiographic province includes: (1) the Colorado Plateau in the south, with its thick sediments and spectacular Grand Canyon; (2) the lava-covered Columbia Plateau in the north, which forms the watershed of the Columbia River; and (3) the central Basin-and-Range country (Great Basin) of Nevada and Utah, which contains several extinct lakes from the glacial period as well as the surviving Great Salt Lake. This province is called *intermontane*, because of its position between the Rocky Mountains to the east and the Pacific coast mountain system to the west.

From the Alaskan Peninsula to Southern California, the west coast of North America is dominated by an almost unbroken corridor of high mountain ranges that originated from the contact between the North American and Pacific Plates (Fig. G-4). The major components of this coastal mountain belt include California's Sierra Nevada, the Cascades of Oregon and Washington, and the long chain of highland massifs that line the British Columbia and southern Alaska coasts. Three broad valleys—which contain dense populations—are the only noteworthy interruptions: California's Central (San Joaquin-Sacramento) Valley; the Cowlitz-Puget Sound lowland of Washington State, which extends southward into western Oregon's Willamette Valley; and the

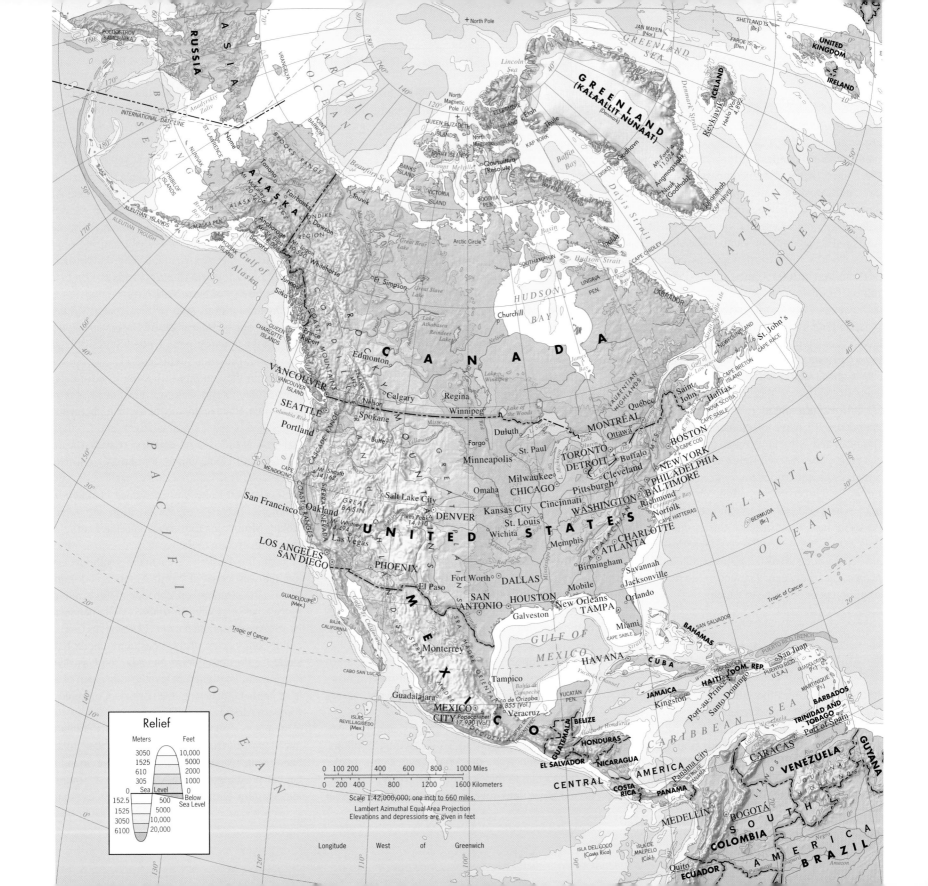

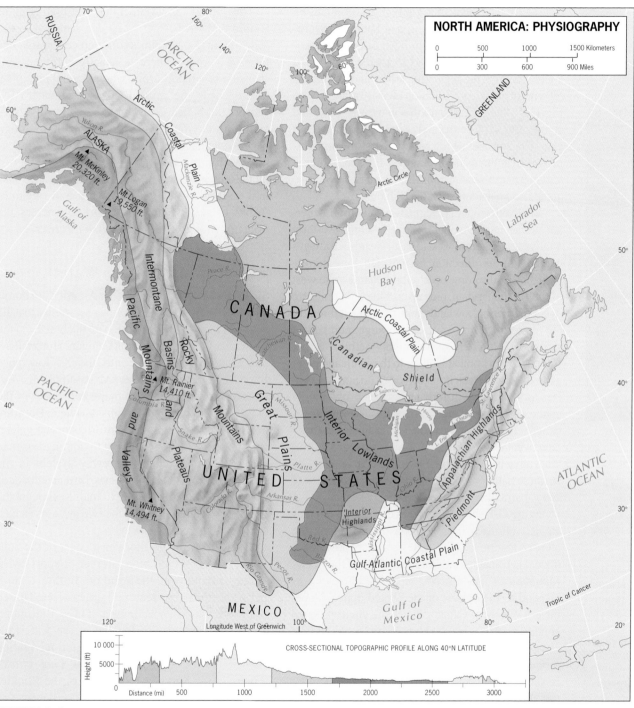

FIGURE 3-2

FIGURE 3-1

lower Fraser Valley, which slices through southern British Columbia's coast range.

Climate

The world climate map (Fig. G-8) clearly depicts the various climatic regimes and regions of North America. In general, temperature varies latitudinally—the farther north one goes, the cooler it gets. Regional land-and-water-heating differentials, however, distort this broad pattern. Because land surfaces heat and cool far more rapidly than water bodies, yearly temperature ranges are much larger where *continentality* (interior remoteness from the sea) is greatest.

Precipitation generally tends to decline toward the west (except for the Pacific coastal strip itself) as a result **3** of the **rain shadow effect**. This occurs because Pacific air masses, driven by prevailing winds, carry their moisture onshore but soon collide with the Sierra Nevada-Cascades wall, forcing them to rise—and cool—in order to crest these mountain ranges. Such cooling is accompanied by major condensation and precipitation, so that by the time these air masses descend and warm along the eastern slopes to begin their journey across the continent's interior, they have already deposited much of their moisture. Thus the mountains produce a downwind "shadow" of dryness, which is reinforced whenever eastward-moving air must surmount other ranges farther inland, especially the massive Rockies. Indeed, this semiarid (and in places truly arid) environment extends so deeply into the

"Drive westward from Denver on winding back roads and avoid the interstate highway, and you are rewarded with magnificent vistas of the high-relief landscape of the Southern Rocky Mountains. Spectacular scenery like this, sunny climes, overcrowding in California, job losses in the Midwest, and booming high-tech and service industries continue to attract numerous migrants from west as well as east, and the Western Frontier (see pp. 132–133) is no longer a remote mining and livestock-raising region with poor internal circulation and limited external connections. Today, this is a tourist haven, a telecommunications and financial-services complex, and a region of fast-growing, modern cities. But nature's grandeur, from the Continental Divide seen here to the Grand Canyon, dominates its image."

central United States that a broad division can be made between Arid (western) and Humid (eastern) America, which face each other along a fuzzy boundary that is best viewed as a wide transitional zone. Although the separating criterion of 20 inches (50 cm) of annual precipitation is easily mapped (see Fig. G-7), that generally north-south *isohyet* (the line connecting all places receiving exactly 20 inches per year) can and does swing widely across the drought-prone Great Plains from year to year because highly variable warm-season rains from the Gulf of Mexico come and go in unpredictable fashion.

On the other hand, precipitation in Humid America is far more regular. The prevailing westerly winds (blowing from west to east—winds are always named for the direction *from* which they come), which normally come up dry for the large zone west of the 100th meridian, pick up considerable moisture over the Interior Lowlands, and distribute it throughout eastern North America. A large number of storms develop here on the highly active weather front between tropical Gulf air to the south and polar air to the north. Even if major storms do not materialize, local weather disturbances created by sharply contrasting temperature differences are always a danger. There are more tornadoes (nature's most violent weather) in the central United States each year than anywhere else on Earth. And in winter the northern half of this region receives large amounts of snow, particularly around the Great Lakes.

Figure G-8 shows the absence of humid temperate (*C*) climates from Canada (except along the narrow Pacific coastal zone) and the prevalence of cold in Canadian environments. East of the Rocky Mountains, Canada's most *moderate* climates correspond to the *coldest* of the United States. Nonetheless, southern Canada shares the environmental conditions that mark the Upper Midwest and Great Lakes areas of the United States, so that agricultural productivity in the Prairie Provinces and in Ontario is substantial. Canada is a leading food exporter (chiefly wheat), as is the United States, despite its comparatively short growing season.

The broad environmental partitioning into Humid and Arid America is also reflected in the distribution of the realm's soils and vegetation. For farming purposes there is usually sufficient soil moisture to support crops where annual precipitation exceeds the critical 20 inches; where the yearly total is less, soils may still be fertile (especially in the Great Plains), but irrigation is often necessary to achieve their full agricultural potential. As for vegetation, the Humid/Arid America dichotomy is again a valid generalization: the natural vegetation of areas receiving more than 20 inches of water annually is *forest*, whereas the drier climates give rise to a *grassland* cover.

Hydrography (Surface Water)

Surface water patterns in North America are dominated by the two major drainage systems that lie between the Rockies and the Appalachians: (1) the five Great Lakes (Superior, Michigan, Huron, Erie, and Ontario) that drain into the St. Lawrence River, and (2) the mighty Mississippi-Missouri river network, fed by such major tributaries as the Ohio, Tennessee, and Arkansas rivers. Both are products of the last episode of Pleistocene glaciation, and together they amount to nothing less than the best natural inland waterway system in the world. Human intervention has further enhanced this network of navigability, mainly though the building of canals that link the two systems as well as the St. Lawrence Seaway.

Elsewhere, the northern east coast of the continent is well served by a number of short rivers leading inland from the Atlantic. In fact, a number of northeastern seaboard cities of the United States—such as Washington, D.C., Baltimore, and Philadelphia—are located at the waterfalls that marked the limit to tidewater navigation (hence their designation as *Fall Line cities*). Rivers in the Southeast and west of the Rockies at first offered little practical value because of their orientation and the difficulty of navigating them. In the Far West, however, the Colorado and Columbia rivers have become important as suppliers of drinking and irrigation water as well as hydroelectric power.

INDIGENOUS NORTH AMERICA

When the first Europeans set foot on North American soil, the continent was occupied by millions of people whose ancestors had reached the Americas from Asia, via Alaska and probably also across the Pacific, more than 13,000 years before (and possibly as long as 30,000 years ago). In search of Asia, the Europeans misnamed them "Indians," but the historic affinities of these earliest Americans were with the peoples of eastern and northeastern Asia, not India. In North America these *Native Americans* or *First Nations*—as they are now called in the United States and Canada, respectively—had organized themselves into hundreds of nations with a rich mosaic of languages and a great diversity of cultures (Fig. 3-3). Farmers grew crops the Europeans had never seen; other nations depended chiefly on fishing, herding, hunting, or some combination of these. Elaborate houses, efficient watercraft, effective weaponry, decorative clothing, and wide-ranging art forms distinguished the aboriginal nations. Certain nations had formulated sophisticated health and medical practices; ceremonial life was complex and highly developed; and political institutions were mature and elaborate.

The eastern nations were the first to bear the brunt of the European invasion. By the end of the eighteenth century, ruthless, land-hungry settlers had driven most of the Native American peoples living along the Atlantic and Gulf coasts from their homes and lands, beginning a westward push that was to devastate indigenous society. The U.S. Congress in 1789 proclaimed that "Indian . . . land and property shall never be taken from them without

FROM THE FIELDNOTES

"A short climb to an overlook near the Oregon-California border provides an instructive perspective over a geologically active segment of the Pacific coastline. This is quite a contrast to the smooth-sloping shores of most of eastern North America: deep water, unobstructed (wind) fetch, and pounding waves drive back this shore along steep cliffs. The owners of those apartments must have dramatic ocean views, but future generations will face the reality of nature's onslaught."

their consent," but in fact this is just what happened. One of the sorriest episodes in American history involved the removal of the eastern Cherokee, Chickasaw, Choctaw, Creek, and Seminole from their homelands in forced marches a thousand miles westward to Oklahoma. One-fourth of the entire Cherokee population died along the way from exposure, starvation, and disease, and the others fared little better. Again, Congress approved treaties that would at least protect the native peoples of the Plains (Fig. 3-3) and those farther to the west, but after the mid-nineteenth century the white settlers ignored those guarantees as well. A half-century of war left what remained of North America's nations

with about 4 percent of U.S. territory in the form of mostly impoverished reservations.

In what is today Canada, too, the comparatively small First Nations population was overwhelmed by the numbers and power of European settlers, and decimated by the diseases they introduced. Efforts at restitution and recognition of First Nations rights, however, have gone farther in Canada than in the United States.

As we noted earlier, two large countries, with geographic similarities as well as differences, constitute the North American realm. Canada and the United States share physiographic provinces as well

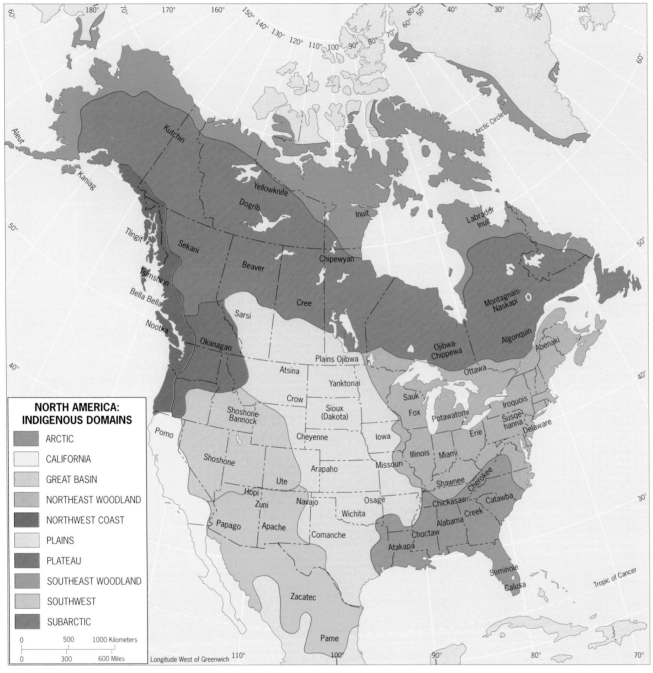

NORTH AMERICA: INDIGENOUS DOMAINS

- ARCTIC
- CALIFORNIA
- GREAT BASIN
- NORTHEAST WOODLAND
- NORTHWEST COAST
- PLAINS
- PLATEAU
- SOUTHEAST WOODLAND
- SOUTHWEST
- SUBARCTIC

FIGURE 3-3

as border-straddling economic and cultural subregions (for example, large-scale grain cultivation in the Great Plains). It would be possible to base the regionalization of this realm on contrasts between the two states; but as we will see in the second part of this chapter, quite a different set of regions emerges from our geographic analysis. To fully appreciate this regional framework, we must examine in some detail the changing human geography of the United States and Canada individually.

THE UNITED STATES

As we note throughout this book, the administrative components of states are assuming ever-greater importance in the scheme of things. In Europe, all the major states have reorganized their political-economic systems, forged regions from smaller units, and devolved power to the provinces. As a result, it is increasingly important to know not just Spain but also Catalonia, not just France but also Rhône-Alpes, not just Italy but also Lombardy. So it is in the United States. We should have a clear mental map of the functional layout of this federation, which is why it may be helpful to take a few moments to review Figure 3-4.

Population in Time and Space

The current population distribution of the United States is shown in Figure 3-5, a map that is the latest "still"

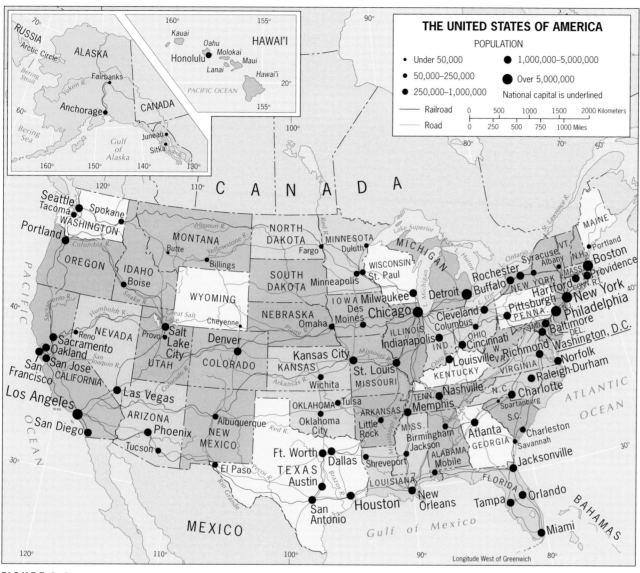

FIGURE 3-4

sorted themselves out to maximize their proximity to existing economic opportunities, and they have shown little resistance to relocating as the nation's evolving economic geography has successively favored different sets of places over time. These movements continue to redistribute millions of Americans, but before we trace the evolution of the population map we need to acquaint ourselves with the process of migration.

The Role of Migration

4 **Migration** refers to a change in residential location intended to be permanent, and it has long played a key role in the human geography of this realm. After thousands of years of native settlement, European explorers reached North American shores beginning with Columbus in 1492 (and probably centuries before that). The first permanent colonies were established in the early 1600s, and from them evolved the modern United States of America. Between 1835 and 1935, perhaps as many as 75 million Europeans departed for distant shores—most of them bound for the Americas (Fig. 3-6). Some sought religious freedom; others escaped poverty and famine; still others simply hoped for a better life.

Studies of the *migration decision* indicate that migration flows vary in size with: (1) the perceived degree of difference between one's home, or source, and the destination; (2) the effectiveness of flows of information about the destination which migrants sent to those who stayed behind waiting to decide; and (3) the distance between the source and the destination. More than a century ago, the British social scientist

in a motion picture that has been unreeling for nearly four centuries since the founding of the first permanent European settlements on the northeastern coast. Slowly at first and then with accelerating speed after 1800, Americans (and Canadians) took charge of their remarkable continent and pushed the settlement frontier westward to the Pacific.

To understand the current population map, we need to review the major forces that have shaped, and continue to shape, the distribution of Americans and their activities. Since its earliest days, the United States has attracted a steady influx of immigrants who were rapidly assimilated into the societal mainstream. Within the country, people have

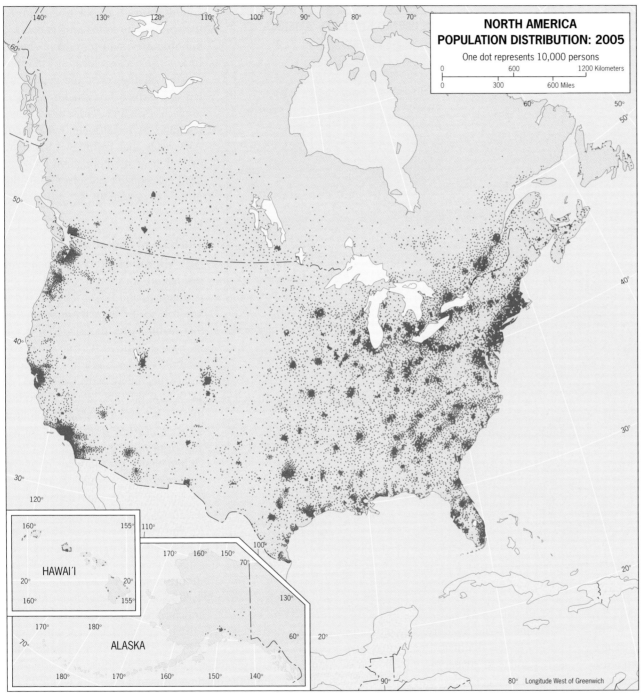

**NORTH AMERICA
POPULATION DISTRIBUTION: 2005**

One dot represents 10,000 persons

| 0 | 600 | 1200 Kilometers |
| 0 | 300 | 600 Miles |

HAWAI′I

ALASKA

Longitude West of Greenwich

FIGURE 3-5

Ernst Georg Ravenstein studied the migration process, and many of his conclusions remain valid today. For example, every migration stream from source to destination produces a counter-stream of returning migrants who cannot adjust. Studies of migration also conclude that several factors are at work in the migration process. **5** **Push factors** motivate people to move away; **pull factors** attract them to new destinations. To those early Europeans the United States was a new frontier, a place where one might acquire a piece of land, and the opportunities were reported to be unlimited. That perception has never changed, and international immigration continues to significantly shape the demographic complexion of the United States in the twenty-first century (with Middle America and eastern Asia now replacing Europe as the dominant source).

Within the United States today, the leading migration flow that continues to transform the population map is the persistent drift of people and livelihoods toward the West and **6** South—the so-called **Sunbelt**. In addition, a pair of lesser but long-standing migratory flows still play a major role: (1) the growth of metropolitan areas, first triggered by the late-nineteenth-century Industrial Revolution, which since the 1960s has been largely rechanneled from the central cities to the suburban ring; and (2) the movement of African-Americans from the rural South to the urban North, which since the 1970s has become a stronger return flow.

Let us now look more closely at the historical geography of these changing

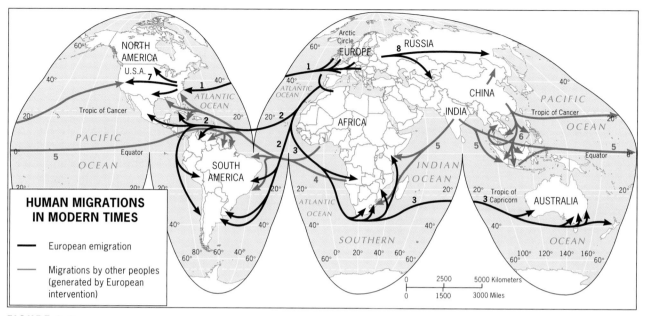

FIGURE 3-6

central United States (as the arrows in Fig. 3-7 show). The northern half of this vast interior space soon became well unified as its infrastructure steadily improved following the introduction of the railroad in the 1830s. The American South, however, did not wish to integrate itself economically with the North, preferring to export tobacco and cotton from its plantations to overseas markets. Its insistence on preserving slavery to support this system soon led the South into secession, disastrous Civil War (1861–1865), and a dismal aftermath that took a full century to overcome.

The second half of the nineteenth century saw the frontier cross the western United States, and by 1869 agriculturally booming California was linked to the rest of the nation by transcontinental railroad (these same steel tracks also opened up the bypassed, semiarid Great Plains). When the American frontier closed in the 1890s, today's rural settlement pattern was firmly in place, anchored to a set of enduring national agricultural regions (discussed later in this chapter). By then, however, the exodus of rural Americans toward the burgeoning cities had begun in response to the Industrial Revolution that had taken hold after 1870.

Post–1900 Industrial Urbanization

The Industrial Revolution occurred almost a century later in the United States than in Europe, but when it did cross the Atlantic in the 1870s, it took hold and advanced so robustly that only 50 years later America was surpassing Europe as the world's mightiest industrial power. The impact of industrial urbanization occurred simultaneously at two levels of generalization. At the national or *macroscale*, a system of new cities swiftly emerged, specializing in the collection, processing, and distribution of

population patterns, considering first the initial rural influence and then the decisive impacts of industrial urbanization.

Pre-Twentieth-Century Population Patterns

The current distribution of the U.S. population is rooted in the colonial era of the seventeenth and eighteenth centuries that was dominated by England and France. The French sought mainly to organize a lucrative fur-trading network, while the English established settlements along what is now the northeastern U.S. seaboard. These British colonies quickly became differentiated in their local economies, a diversity that later shaped American cultural geography. The northern colony of New England specialized in commerce; the southern Chesapeake Bay colony emphasized the plantation farming of tobacco; and the Middle Atlantic area lying in between was home to a number of smaller, independent-farmer colonies.

These neighboring colonies soon thrived and yearned to expand, but the British government responded by closing the inland frontier and tightening economic controls. By 1783 this had led to colonial unification, British defeat in the Revolutionary War, and independence for the newly formed United States of America. The western frontier of the fledgling nation now swung open, and the zone north of the Ohio River was promptly settled following the discovery that the soils of the Interior Lowlands were more favorable for farming than those of the Atlantic seaboard. This triggered the rapid growth of trans-Appalachian agriculture and coastal-interior trading ties as U.S. spatial organization assumed national-scale proportions.

By the time the westward-moving frontier swept across the Mississippi Valley in the 1820s, the three former seaboard colonies (**A**, **B**, and **C** in Fig. 3-7) had become separate *culture hearths*—primary source areas and innovation centers from which migrants carried cultural traditions into the

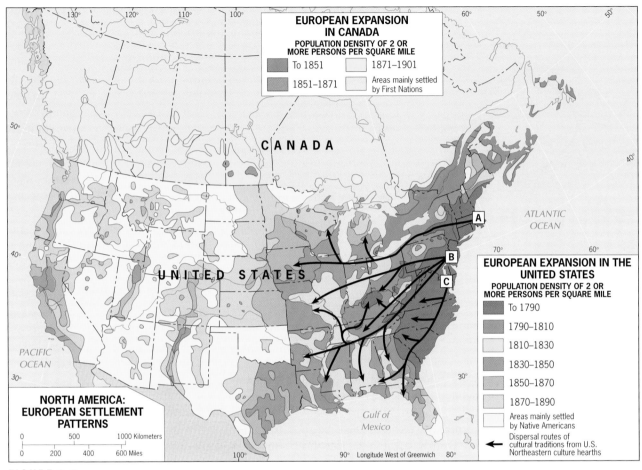

EUROPEAN EXPANSION IN CANADA
POPULATION DENSITY OF 2 OR MORE PERSONS PER SQUARE MILE

To 1851
1851–1871
1871–1901
Areas mainly settled by First Nations

EUROPEAN EXPANSION IN THE UNITED STATES
POPULATION DENSITY OF 2 OR MORE PERSONS PER SQUARE MILE

To 1790
1790–1810
1810–1830
1830–1850
1850–1870
1870–1890
Areas mainly settled by Native Americans
Dispersal routes of cultural traditions from U.S. Northeastern culture hearths

NORTH AMERICA: EUROPEAN SETTLEMENT PATTERNS

0   500   1000 Kilometers
0   200   400   600 Miles

FIGURE 3-7

raw materials and manufactured goods, linked together by an efficient web of railroad lines. Within that urban network, at the local or *microscale*, individual cities prospered in their new roles as manufacturing centers, generating an internal structure that still forms the framework of most large central cities. We now examine the urban trend at both of these scales.

Evolution of the U.S. Urban System. The rise of the national urban system was based on the traditional external role of cities: providing goods and

services for their hinterlands in exchange for raw materials. Because people, commercial activities, investment capital, and transport facilities were already agglomerated in existing cities, new industrialization favored such locations. Their growing incomes permitted industrially intensifying cities to develop their infrastructures, public services, and housing—and thereby convert each round of industrial expansion into a new stage of urban growth.

The evolution of the national urban system has been studied by John Borchert, who identified five

epochs of metropolitan evolution based on transportation technology and industrial energy: (1) *The Sail-Wagon Epoch* (1790–1830), marked by primitive overland and waterway circulation; the leading cities were the northeastern ports oriented to the European overseas trade; (2) *The Iron Horse Epoch* (1830–1870), dominated by the spread of the steam-powered railroad; a nationwide transport system appeared and the national urban system took shape, with New York City emerging as the primate city; (3) *The Steel-Rail Epoch* (1870–1920), spanning the Industrial Revolution; new forces shaping growth were the increasing scale of manufacturing and the introduction of steel rails that enabled trains to travel faster and haul heavier cargoes; (4) *The Auto-Air-Amenity Epoch* (1920–1970), encompassing the later stage of U.S. industrial urbanization; key elements were the automobile and the airplane, the expansion of white-collar services jobs, and the growing locational pull of *amenities* (pleasant environments) that favored suburbs and certain Sunbelt locales; (5) *The Satellite-Electronic-Jet Propulsion Epoch* (1970–   ), shaped by advancements in information management, computer technologies, global communications, and intercontinental travel; favors globally oriented metropolises, particularly those functioning as international gateways.

Industrialization and the accompanying growth of the urban system reconfigured the realm's economic landscape. The most notable regional transformation was the emergence of the North American **7** Core, or **American Manufacturing Belt**, which contained the lion's share of industrial activity in both the United States and Canada. As Figure 3-8

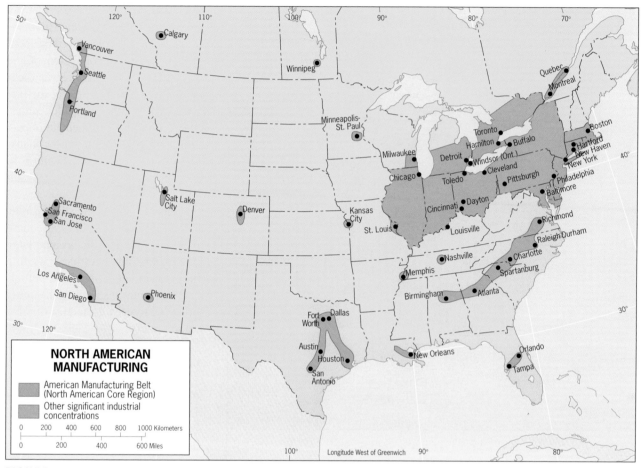

FIGURE 3-8

Miami), *Texas* (Houston-Dallas/Fort Worth-San Antonio), *California* (San Diego-Los Angeles-San Francisco), and the *Pacific Northwest* (Portland-Seattle-Vancouver). Note that the last spills across the border into Canada, which has also spawned its own nationally predominant conurbation—*Main Street* (Windsor-Toronto-Montreal-Quebec City).

The Changing Structure of the U.S. Metropolis. The internal structure of the metropolis reflected the same mixture of forces that shaped the national urban system. Rails—in this case, lighter street-rail lines—again shaped spatial organization as horse-drawn trolleys were succeeded by electric streetcars in the late nineteenth century. The mass introduction of the automobile after World War I changed all that, and America increasingly turned from building compact cities to widely dispersed metropolises. By 1970, the new intraurban expressway network had equalized location costs throughout the metropolis, setting the stage for the suburban ring to transform itself from a residential preserve into a complete outer city. Newly urbanized suburbs now began to capture major economic activities, and their rise came at the expense of the central city which saw its status diminish to that of coequal.

This growth process was conceptualized into a four-stage model by John Adams, who identified the four eras of intraurban structural evolution that are diagrammed in Figure 3-10. Stage I, prior to 1888, was the *Walking-Horsecar Era*, which produced a compact pedestrian city in which everything had to be within a 30-minute walk, a layout only slightly augmented when horse-drawn trolleys

shows, the geographic form of the Core Region—which includes southern Ontario—was a near rectangle whose four corners were Boston, Milwaukee, St. Louis, and Baltimore. Within this region, manufacturing is heavily concentrated into a dozen districts centered on the cities mapped in Figure 3-8.

At the subregional scale, as transportation breakthroughs permitted progressive urban decentralization and *megalopolitan growth*, the expanding peripheries of major cities soon coalesced to form a number of conurbations. The most important of these by far is the *Atlantic Seaboard Megalopolis* (Fig. 3-9), the 600-mile (1000-km) urbanized northeastern coastal strip extending from southern Maine to Virginia that contains metropolitan Boston, New York, Philadelphia, Baltimore, and Washington. This was the economic heartland of the Core; the seat of U.S. government, business, and culture; and the trans-Atlantic trading interface between much of North America and Europe. Six other primary conurbations have also emerged: *Lower Great Lakes* (Chicago-Detroit-Cleveland-Pittsburgh), *Piedmont* (Atlanta-Charlotte-Raleigh/Durham), *Florida* (Jacksonville-Tampa-Orlando-

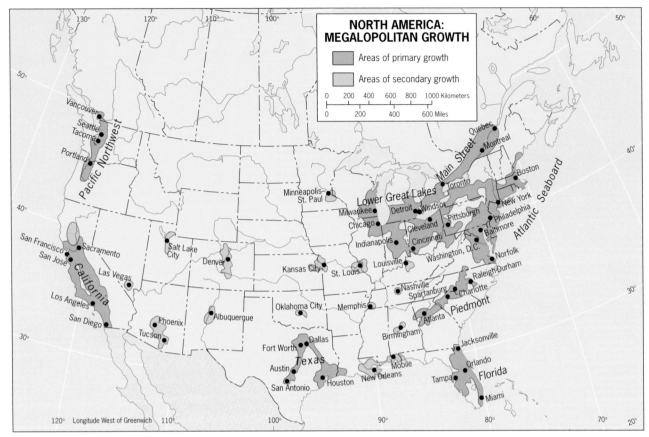

**NORTH AMERICA:
MEGALOPOLITAN GROWTH**

Areas of primary growth

Areas of secondary growth

FIGURE 3-9

hoods. When the United States sharply curtailed immigration in the 1920s, industrial managers discovered the large African-American population of the rural South and began to recruit these workers by the thousands for the factories of Manufacturing Belt cities. This influx had an immediate impact on the social geography of the industrial city: whites were unwilling to share their living space, and the result was the segregation of these newest migrants into geographically separate, all-black areas. By the 1950s, these mostly inner-city areas became large 8 expanding **ghettos**, speeding the departure of many white central-city communities and reinforcing the trend toward a racially divided urban society.

As ties to the central city loosened, the suburban ring was transformed 9 into a full-fledged **outer city**. Its independence was accelerated by the rise of major new suburban nuclei, multipurpose activity nodes that grew up around large shopping centers with prestigious images that attracted industrial parks, office campuses and highrises, entertainment facilities, and even major league sports stadiums and arenas. 10 These burgeoning new **suburban downtowns**, in fact, are nothing less than an automobile-age version of the CBD. The newest spatial elements of the contemporary urban complex are assembled in the model displayed in Figure 3-11. The outer city today anchors a *multicentered* metropolis consisting of the traditional CBD and a constellation of coequal suburban downtowns, with each activity center serving a discrete and self-sufficient surrounding area. James 11 Vance defined these tributary areas as **urban realms**, recognizing in his studies that each such realm maintains a separate, distinct economic, so-

began to operate after 1850. The 1888 invention of the electric traction motor launched Stage II, the Electric Streetcar Era (1888–1920); higher speeds enabled the 30-minute travel radius and the urbanized area to expand considerably along new outlying trolley corridors; in the older core city, the central business district (CBD), industrial, and residential land uses differentiated into their modern form. Stage III, the *Recreational Automobile Era* (1920–1945), was marked by the initial impact of cars and highways that steadily improved the accessibility of the outer metropolitan ring, thereby launching a wave of mass suburbanization that further extended the urban frontier. During this era,

the still-dominant central city experienced its economic peak and the partitioning of its residential space into neighborhoods sharply defined by income, ethnicity, and race. Stage IV, the *Freeway Era* (under way since 1945), saw the full impact of automobiles, with the metropolis turning inside-out as expressways pushed suburban development more than 30 miles (50 km) from the CBD.

The social geography of the evolving industrial metropolis was marked by the development of a residential mosaic of ever-more-specialized groups. The electric streetcar, which introduced affordable transit for all, allowed the immigrant-dominated city population to sort itself into ethnically uniform neighbor-

"Monitoring the urbanization of U.S. suburbs for the past three decades has brought us to Tyson's Corner, Virginia on many a field trip and data-gathering foray. It is now hard to recall from this scene that only 40 years ago this place was merely a near-rural crossroads. But as nearby Washington, D.C. steadily decentralized, *Tyson's* capitalized on its unparalleled regional accessibility (its Capital Beltway location at the intersection with the radial Dulles Airport Toll Road) to attract a seemingly endless parade of high-level retail facilities, office complexes, and a plethora of supporting commercial services. We would rank it today among the three largest and most influential suburban downtowns (or 'edge cities') in North America."

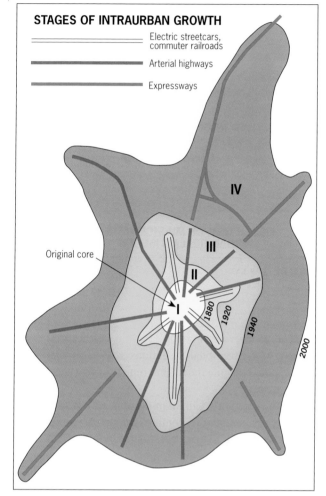

FIGURE 3-10

cial, and political significance and strength. Figure 3-12 applies the urban realms model to Los Angeles.

The position of the central city within the new multinodal metropolis of realms is eroding. The CBD increasingly serves the less affluent residents of the innermost realm and those working downtown. As manufacturing employment declined, many large cities adapted by shifting to service industries, embodied by downtown commercial revitalization. Residential reinvestment has also occurred in many downtown-area neighborhoods, but beyond the CBD the vast inner city remains the problem-ridden domain of low- and moderate-income people, with most forced to reside in ghettos.

Cultural Geography

In the United States over the past two centuries, the contributions of a wide spectrum of immigrant groups have shaped—and continue to shape—a rich and varied cultural complex. Great numbers of these newcomers were willing to set aside their original cultural baggage in favor of assimilation into the emerging culture of their adopted homeland, which itself was a hybrid nurtured by constant infusions of new influences.

Language and Religion

Although linguistic variations play a far more important role in Canada, more than one-eighth of the U.S. population speaks a primary language other than English. Differences in English usage are also

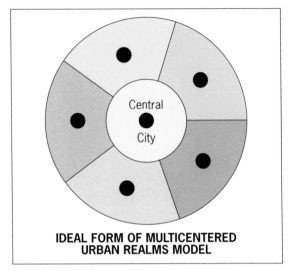

**IDEAL FORM OF MULTICENTERED
URBAN REALMS MODEL**

FIGURE 3-11

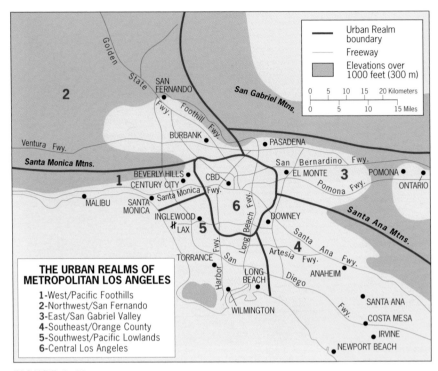

**THE URBAN REALMS OF
METROPOLITAN LOS ANGELES**
1-West/Pacific Foothills
2-Northwest/San Fernando
3-East/San Gabriel Valley
4-Southeast/Orange County
5-Southwest/Pacific Lowlands
6-Central Los Angeles

FIGURE 3-12

evident at the subnational level in the United States, where regional variations (*dialects*) can still be noted despite the recent trend toward a truly national society.

North America's Christian-dominated kaleidoscope of religious faiths contains important spatial variations. Many major Protestant denominations are clustered in particular regions, with Southern Baptists localized in the southeastern quadrant of the United States, Lutherans in the Upper Midwest, and Mormons focused on Utah. Roman Catholics are most visibly concentrated in Manufacturing Belt metropolises and the Mexican borderland zone. Jews are most heavily clustered in the suburbs of Megalopolis, Southern California, South Florida, and the Midwest.

Ethnic Patterns

Ethnicity (national ancestry) has always played a key role in American cultural geography. Today, whites of European background no longer dominate the increasingly diverse U.S. ethnic tapestry, with ethnics of color and non-European origin comprising a steadily expanding proportion

that will surpass 50 percent by 2050. In the late 1990s, Hispanic-Americans surpassed African-Americans to become the nation's largest minority, and today they account for more than 13 percent of the U.S. total. The spatial distribution of the four largest ethnic minorities is mapped in Figure 3-13.

Immigration has long influenced the ethnic complexion of the United States, and at the end of the 1990s nearly 700,000 legal immigrants annually entered the country. The source areas, however, have changed dramatically over the past half-century. During the 1950s, just over 50 percent came from Europe, 25 percent from Middle and South America, and 15 percent from Canada; today, approximately 50 percent come from Middle and South America, 30 percent from Asia, and only about 15 percent from Europe and Canada.

The Emerging Mosaic Culture

American cultural geography continues to evolve. What is now taking place is a new fragmentation **12** into the emerging nationwide **mosaic culture**, an increasingly heterogeneous complex of separate, uniform "tiles" that cater to more specialized groups than ever before. No longer based solely on such broad divisions as income, race, and ethnicity, today's residential communities of interest are also forming along the dimensions of age, occupational status, and especially lifestyle.

The Changing Geography of Economic Activity

The economic geography of the United States today is the product of all of the foregoing, with people and activities overcoming the tyranny of distance to organize a continentwide spatial econ-

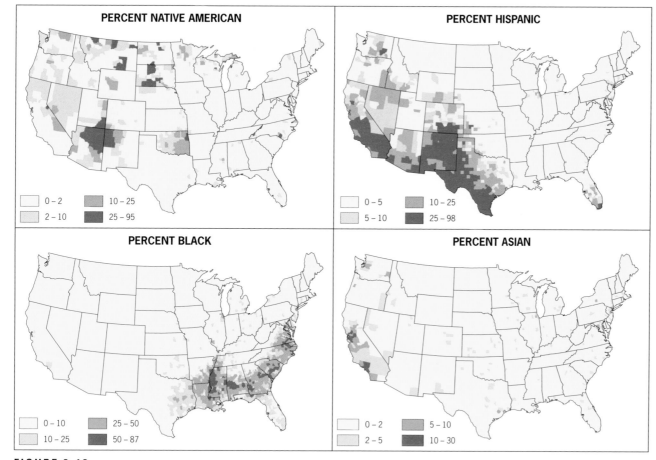

PERCENT NATIVE AMERICAN

0 – 2
2 – 10
10 – 25
25 – 95

PERCENT HISPANIC

0 – 5
5 – 10
10 – 25
25 – 98

PERCENT BLACK

0 – 10
10 – 25
25 – 50
50 – 87

PERCENT ASIAN

0 – 2
2 – 5
5 – 10
10 – 30

FIGURE 3-13

omy that took full advantage of agricultural, industrial, and urban development opportunities. Despite these past achievements, today American economic geography is again in the throes of restructuring as the transition is completed from industrial to postindustrial society.

Major Components of the Spatial Economy

Economic geography is heavily concerned with the **13** locational analysis of **productive activities**. Four major sets may be identified:

- **Primary activity**: the extractive sector of the economy in which workers and the environment come into direct contact, especially in *mining* and *agriculture*.

- **Secondary activity**: the *manufacturing* sector, in which raw materials are transformed into finished industrial products.

- **Tertiary activity**: the *services* sector, including a wide range of activities from retailing to finance to education to routine office-based jobs.

- **Quaternary activity**: today's dominant sector, involving the collection, processing, and manipulation of *information*; a subset, sometimes referred to as **quinary activity**, is managerial decision-making in large organizations.

Historically, each of these activities has successively dominated the U.S. labor force for a period over the past 200 years, with the quaternary sector now dominant. The approximate current breakdown by major sector of employment is agriculture, 2 percent; manufacturing, 15 percent; services, 18 percent; and quaternary, 65 percent (with about 10 percent in the quinary sector). We now review these major productive components of the spatial economy in the following coverage of resource use, agriculture, manufacturing, and the postindustrial revolution.

Resource Use

The United States (and Canada) was blessed with abundant deposits of mineral and energy resources. North America's mineral resources are localized in three zones: the Canadian Shield north of the Great Lakes, the Appalachian Highlands, and scattered areas across the mountain ranges of the West. The Shield's most noteworthy minerals are iron ore, nickel, uranium, and copper. Besides vast deposits of coal, the Appalachian region also contains iron ore in central Alabama. The western mountain zone contains significant deposits of coal, copper, lead, zinc, and gold.

The realm's most vital energy resources are its petroleum (oil), natural gas, and coal deposits **14** (mapped in Fig. 3-14); these are the **fossil fuels**, so named because they were formed by the geologic compression of tiny organisms that lived hundreds of millions of years ago. The leading *oil*-production areas of the United States are located along and offshore from the Texas-Louisiana Gulf Coast; in the Midcontinent district, extending through western Texas-Oklahoma-eastern Kansas;

FIGURE 3-14

and along Alaska's North Slope facing the Arctic Ocean. (Canada's leading oilfields lie in a crescent curving southeastward from northern Alberta to southern Manitoba.) The distribution of *natural gas* supplies resembles the geography of oilfields because both fuels are usually found in the floors of ancient shallow seas. The realm's *coal* production zones, which rank among the greatest on Earth, are found in Appalachia, the northern U.S. Great Plains/southern Alberta, and southern Illinois/western Kentucky.

Agriculture

Despite the post–1900 emphasis on developing the nonprimary sectors of the spatial economy, agriculture remains an important element in America's human geography. Vast expanses of the U.S. landscape are clothed with fields of grain or support great herds of livestock that are sustained by pastures and fodder crops. In recent decades, high-technology mechanization has revolutionized farming and has been accompanied by a sharp reduction in the agricultural workforce (today only about 1.5 percent of the U.S. population still resides on farms).

The regionalization of U.S. agricultural production is shown in Figure 3-15, its spatial organization developed largely within the framework of the *von Thünen model* (see pp. 36–37). As in Europe (Fig. 1-6), the early-nineteenth-century, original-scale model of town and hinterland expanded outward (driven by constantly improving transportation technology) from a locally "isolated state" to encompass the entire continent by 1900.

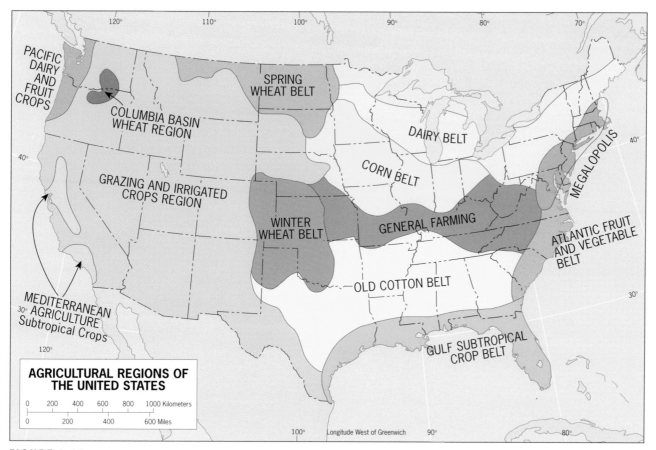

FIGURE 3-15

The "supercity" anchoring this macro-Thünian regional system was the northeastern Megalopolis, already the dominant food market and transport focus of the entire country. Although the circular rings of the model are not apparent in Figure 3-15, many spatial regularities can be observed in this real-world application. Most significant is the sequence of farming regions as distance from the national market increases, especially westward from Megalopolis toward central California (the main directional thrust of the interior penetration of the United States). The Atlantic Fruit and Vegetable Belt, Dairy Belt, Corn Belt, Wheat Belts, and Grazing Region are indeed consistent with the

model's logical structure, each zone successively farther inland astride the main transcontinental routeway.

Manufacturing

The geography of North America's industrial production has long been dominated by the Manufacturing Belt (Fig. 3-8). The emergence of this region was propelled by (1) superior access to the Megalopolis national market that formed its eastern edge, and (2) proximity to industrial resources, particularly iron ore and coal for the pivotal steel industry that arose in its western half. We noted earlier that

manufacturers had a strong locational affinity for cities, and the internal structure of the Belt became organized around a dozen urban-industrial districts interconnected by a dense transportation network. As these industrial centers expanded, they swiftly **15** achieved **economies of scale**, savings accruing from large-scale production in which the cost of manufacturing a single item was further reduced as factories mechanized assembly lines, specialized their workforces, and purchased raw materials in massive quantities.

This production pattern served the nation well throughout the remainder of the industrial age. Today, however, much of the Manufacturing Belt is aging and the distribution of American industry is changing. As transportation costs equalize among U.S. regions, as energy costs now favor the south-central oil- and gas-producing States, as high-technology manufacturing advances steadily reduce the need for lesser skilled labor, and as locational decision-making intensifies its attachment to noneconomic factors, industrial management has increasingly demonstrated its willingness to relocate to regions it perceives as more desirable in the South and West (as well as certain foreign locations where conditions may be even more favorable).

Nonetheless, parts of the Manufacturing Belt have been resisting this trend. This is particularly true of the industrial Midwest, whose prospects have improved since the "Rustbelt" days of the 1970s. Shedding that label in recent years, Midwestern manufacturers are reinventing their operations by rooting out inefficiencies, investing in cutting-edge factories and technologies, and maintaining their competitiveness in the global marketplace.

The Postindustrial Revolution

16 The signs of **postindustrialism** are visible throughout the United States today, and they are popularly grouped under such labels as "the computer age" or "the new economy." High-technology, white-collar, office-based activities are the leading growth industries of the postindustrial economy. Most are relatively footloose and are therefore responsive to such noneconomic locational forces as geographic prestige, local amenities, and proximity to recreational opportunities. Northern California's *Silicon Valley*—the world's leading center for computer research and development and the headquarters of the U.S. microprocessor industry—epitomizes the blend of locational qualities that attract a critical mass of high-tech companies to a given locality. These include: (1) a world-class research university (Stanford); (2) a large pool of highly skilled labor; (3) proximity to a cosmopolitan urban center (San Francisco); (4) abundant venture capital; and (5) a locally-based network of global business linkages.

The development of Silicon Valley is so significant to the new postindustrial era that it has **17** been conceptualized as the first **technopole**. Technopoles are planned techno-industrial complexes that innovate, promote, and manufacture the hardware and software products of the new informational economy. On the landscape of the outer suburban city, where almost all of these complexes are located, the signature of a technopole is a low-density cluster of ultramodern buildings laid out as a campus. From Silicon Valley, technopoles have spread in all directions, and many will be noted in the regional section of this chapter. Technopoles are also becoming a global phenomenon, and several are mentioned in other chapters.

CANADA

Like the United States, Canada is a federal state, but it is organized differently. Canada is divided into ten provinces and three territories (Fig. 3-16).

The provinces—where almost all Canadians live—range in territorial size from tiny, Delaware-sized Prince Edward Island to sprawling Quebec, more than twice the area of Texas. Beginning in the east, the four Atlantic Provinces are Nova Scotia, New Brunswick, Prince Edward Island, and Newfoundland and Labrador. To their west lie Quebec and Ontario, Canada's two biggest provinces. Most of western Canada is covered by the three Prairie Provinces—Manitoba, Saskatchewan, and Alberta. In the far west, facing the Pacific, lies British Columbia.

The three territories—Yukon, the Northwest Territories, and Nunavut—together occupy a massive area half the size of Australia but are inhabited by only about 100,000 people. Nunavut is the newest addition to Canada's political map and deserves special mention. Created in 1999, this new territory is the outcome of a major aboriginal land claim agreement between the Inuit people (formerly called Eskimos) and the federal government, and encompasses all of Canada's eastern Arctic.

In population size, Ontario (12.0 million) and Quebec (7.5 million) are again the leaders; British Columbia ranks third with 4.3 million; next come the three Prairie Provinces with a combined total of 5.3 million; the Atlantic Provinces are the smallest, together containing 2.4 million. Canada's total population of 31.6 million is only slightly larger than one-tenth the size of the U.S. population.

Population in Time and Space

The map showing the distribution of Canada's population (Fig. 3-5, p. 114) reveals that only about one-eighth of this enormous country can be classified **18** as its **ecumene**—the inhabitable zone of permanent settlement. As Figure 3-16 indicates, the Canadian ecumene is dominated by a discontinuous strip of population clusters that lines the U.S. border. We can identify four such clusters on the map, the largest by far being Main Street (home to more than six out of every ten Canadians). As noted earlier (Fig. 3-9), *Main Street* is the conurbation that stretches across southernmost Quebec and Ontario, from Quebec City on the lower St. Lawrence River southwest through Montreal and Toronto to Windsor on the Detroit River. The three lesser clusters are: (1) the Saint John-Halifax crescent in central New Brunswick and Nova Scotia; (2) the prairies of southern Alberta, Saskatchewan, and Manitoba; and (3) the southwestern corner of British Columbia, focused on Canada's third-largest metropolis, Vancouver.

Pre-Twentieth-Century Canada

As Figure 3-7 shows, compared to European expansion in the United States, penetration of the Canadian interior lagged well behind. In terms of political geography, Canada did not unify before the last third of the nineteenth century—and then mainly because of fears the United States was about to expand in a northerly direction.

The evolution of modern Canada is deeply rooted in a longstanding bicultural division. Its origin lies in the fact that it was the French, not the British, who were the first European colonizers of present-day Canada, with *New France* growing to encompass the St. Lawrence Basin, the Great Lakes region, and the Mississippi Valley. A series of wars between the English and French subsequently ended in France's defeat and the cession of New France to Britain in 1763. By the time London took control of its new possession, the French had made considerable progress in their North American domain. The British, anxious to avoid a war of suppression and preoccupied with problems in their other American colonies, gave former French Quebec—the territory extending from the Great Lakes to the mouth of the St. Lawrence—the right to retain its legal and land-tenure systems as well as freedom of religion.

After the American War for Independence, London was left with a region it called British North America but whose cultural imprint still was decidedly French. The Revolutionary War drove many thousands of English refugees northward, and soon

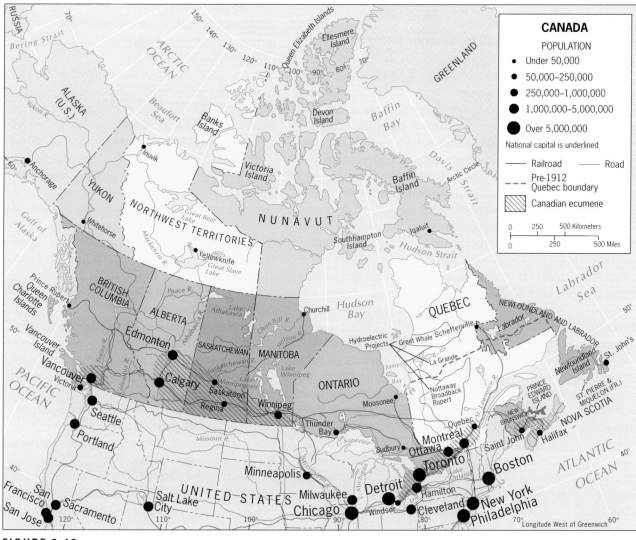

FIGURE 3-16

unchanged, and the French language was protected in Parliament and in the courts.

Canada Since 1900

By 1900, the Canadian federation was making major strides toward regional development and the spatial integration of a continentwide economy. The transcontinental Canadian Pacific Railway had been completed to Vancouver, along the way spawning the settlement of the fertile Prairie Provinces whose wheat-raising economy expanded steadily as immigrants arrived from the east and abroad. Industrialization also began to stir, and by 1920, Canadian manufacturing had surpassed agriculture as the leading source of national income. As noted earlier (Fig. 3-8), the dominant zone of industrial activity is the Toronto-Hamilton-Windsor corridor of southern Ontario, the crucible of Canada's Industrial Revolution that took hold during World War I (1914–1918).

As in the United States, industrial intensification was accompanied by the rise of a national urban system. Along the lines of Borchert's epochs of American metropolitan evolution, Maurice Yeates constructed a similar multistage model of development that divides the telescoped Canadian experience into three eras. The initial *Frontier-Staples Era* (prior to 1935) encompasses the century-long transition from a frontier-mercantile economy to one oriented to staples (production of raw materials and agricultural goods for export), with increasing manufacturing activity in the budding industrial heartland. By 1930, Montreal and Toronto had emerged as the two leading cities atop the national urban hierarchy (thus Canada has no single primate city).

difficulties arose between them and the French. In 1791, heeding appeals by these new settlers, the British Parliament divided Quebec into two provinces: Upper Canada, the region upstream from Montreal centered on the north shore of Lake Ontario, and Lower Canada, the valley of the St. Lawrence. Upper and Lower Canada became, respectively, the provinces of (English-speaking) Ontario and (French-speaking or *Francophone*)

Quebec (Fig. 3-16). This earliest cultural division did not work well and in 1867 finally led to the British North America Act, which established the Canadian federation (consisting initially of Upper and Lower Canada, New Brunswick, and Nova Scotia, later to be joined by the other provinces and territories). Under this Act, Ontario and Quebec were again separated, but this time Quebec was given important guarantees: the French civil code was left

Next came the *Era of Industrial Capitalism* (1935–1975), during which Canada achieved U.S.-style prosperity. A major stimulus was the investment of U.S. corporations in Canadian branch-plant construction, especially in the automobile industry in Ontario near the automakers' Detroit-area headquarters. In western Canada, the growth of oil and natural gas production fueled Alberta's urban development. The post–World War II period also saw the ascent of Main Street, which on less than 2 percent of Canada's land quickly came to contain more than 60 percent of its people, contributed two-thirds of its national income, and claimed nearly 75 percent of its manufacturing jobs.

The third stage, ongoing since 1975, is the *Era of Global Capitalism*, signifying the rise of additional foreign investment from the Asian Pacific Rim and Europe. This, of course, is also the era of transformation into a postindustrial economy and society, and in the process Canada is experiencing many of the same upheavals as the United States. Most of this development is occurring in the form of new suburbanization, a departure from the past because the pre-1990 Canadian metropolis had experienced far less automobile-generated deconcentration than the United States. But today Canada's large cities are turning inside-out, and the new intraurban geography is symbolized by the suburban downtowns that anchor the ultra-modern business complexes lining the Highway 401 freeway north of Toronto and Alberta's West Edmonton megamall, the world's biggest shopping center.

Cultural/Political Geography

The historic cleavage between Canada's French- and English-speakers has resurfaced in the past three decades to dominate the country's cultural and political geography. By the time the Canadian federation observed its centennial in 1967, it had become evident that Quebecers regarded themselves as second-class citizens; they believed that bilingualism meant that French-speakers had to learn English but not vice versa; and they perceived that Quebec was not getting its fair share of the country's wealth. Since the 1960s, the intensity of ethnic feelings in Quebec has risen in surges despite the federal government's efforts to satisfy the province's demands. During the 1970s, while a separatist political party came to power in Quebec, a new federal constitution was drawn up in Ottawa. In 1980, Quebec's voters solidly rejected independence when given that choice in a referendum. But the new constitution did *not* satisfy the Quebecers, and throughout the 1980s and early 1990s the Ottawa government struggled unsuccessfully to devise a plan, acceptable to all the provinces, that would keep Quebec in the Canadian federation.

By 1995, with Canada's interest in constitutional reform exhausted, a second referendum on Quebec's sovereignty could no longer be put off. With the reenergized separatist party again leading the way, the Francophone-dominated electorate very nearly approved independence. Subsequently, despite calls by many separatists for a follow-up vote that might turn narrow defeat into victory, plans for a third referendum have been shelved because opinion polls in Quebec have shown a steady erosion in public support for secession. Several reasons lie behind that changing attitude: (1) the implementation of provincial laws that firmly established the use of French and the primacy of Quebecois culture; (2) the increased bilinguality of Quebec's Anglophones (English speakers), which has calmed Francophone fears concerning assimilation into Canada's English-speaking mainstream; and (3) the arrival of a new wave of immigrants from Asia, Africa, Eastern Europe, and the Caribbean, who have accelerated the weakening of language barriers by settling in both French-speaking and English-speaking neighborhoods.

After what many Canadians still call their "near-death experience," the federal government seized the initiative in late 1995 by asking Canada's Supreme Court to review the constitutionality of Quebec's separation. The court ruled that if a clear majority of Quebec's electorate voted to secede, the Ottawa government and the other provinces would be obliged to negotiate the terms of separation.

Among the issues that an exiting Quebec would have to negotiate, most significant from a geographic standpoint is that the French linguistic region does *not* coincide with the province's territorial boundaries. Thus, as the distribution of the "no" vote on separation in the 1995 referendum strongly suggests (Fig. 3-17; areas colored pink), there are numerous non-French communities located within the Francophone region. Dozens of English-speaking municipalities have, in fact, spearheaded a partitionist movement to stay in Canada, arguing they have the same right to secede from Quebec that Quebec has to secede from Canada. Even more important is an identical movement among the First Nations of Quebec's northern frontier, the Cree, whose historic domain covers more than half of the province. As Figure 3-16 shows, the territory of the Cree is no unproductive wilderness: it contains vital facilities of the James Bay Hydroelectric Project, a massive scheme of dams and artificial lakes that generates electrical power for a huge market within and outside the province—an enterprise that would be an economic cornerstone in any plan for an independent Quebec.

Finally, the events of the past quarter-century have increasingly impacted Canada's political landscape, and not only in Quebec. Regionalism has also intensified in the west, whose leaders oppose federal concessions to Quebec. The most recent federal elections clearly reveal the emerging fault lines that surround Quebec and set the western provinces off from the rest of the country. Even the remaining Ontario-led center and the eastern bloc of Atlantic Provinces voted divergently, and today the politico-geographical hypothesis of "Four Canadas" may be on its way toward becoming reality. Thus, with its national unity under pressure, Canada today confronts the coalescing forces of *devolution* that threaten to transform the new fault lines into permanent fractures.

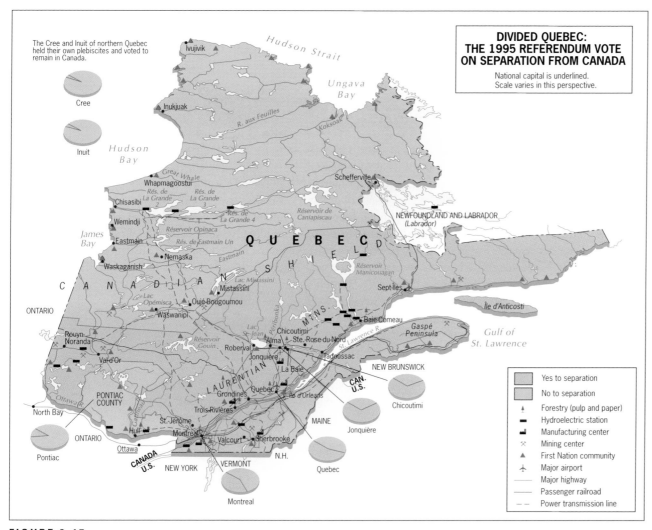

The Cree and Inuit of northern Quebec held their own plebiscites and voted to remain in Canada.

Cree

Inuit

DIVIDED QUEBEC: THE 1995 REFERENDUM VOTE ON SEPARATION FROM CANADA

National capital is underlined.
Scale varies in this perspective.

Yes to separation

No to separation

▲ Forestry (pulp and paper)
▬ Hydroelectric station
▬ Manufacturing center
✕ Mining center
▲ First Nation community
✈ Major airport
— Major highway
— Passenger railroad
- - Power transmission line

Pontiac

Chicoutimi

Jonquière

Quebec

Montreal

FIGURE 3-17

Economic Geography

As in the United States, the growth of Canada's spatial economy has been supported by a diversified *resource base*. We noted earlier that the Canadian Shield is endowed with major mineral deposits and that oil and natural gas are extracted in sizeable quantities in Alberta. Canada has long been a leading *agricultural* producer and exporter, especially of wheat and other grains from its breadbasket in the Prairie Provinces. Postindustrialization has caused substantial employment decline in the *manufacturing sector*, with Southern Ontario's industrial heartland being the most adversely affected. On the other hand, Canada's robust *tertiary and quaternary sectors* (which today employ more than 70 percent of the total workforce) are creating a host of new economic opportunities.

Canada's economic future is also going to be strongly affected by the continuing development of its trading relationships. These include the 1989 United States-Canada Free Trade Agreement (today more than four-fifths of Canada's exports go to the United States, from which it also derives about three-quarters of its imports), and the 1994 North American Free Trade Agreement (NAFTA) which added Mexico to the trading partnership.

Because these free-trade agreements increasingly impact the Canadian spatial economy, they are likely to weaken domestic east-west linkages and strengthen international north-south ties. Since many local cross-border linkages built on geographical and historical commonalities are already well developed, they can be expected to intensify in the future: the Atlantic Provinces with neighboring New England; Quebec with New York State; Ontario with Michigan; the Prairie Provinces with the Upper Midwest; and British Columbia with the (U.S.) Pacific Northwest. Such functional reorientations, of course, constitute yet another set of powerful devolutionary forces confronting the Ottawa government because most of these potential economic fault lines coincide with those that politically demarcate the "Four Canadas."

The rising importance of this framework of transnational regions straddling the U.S.-Canadian border recalls Kenichi Ohmae's regional state concept **19** introduced in Chapter 1. A **regional state** is a "natural economic zone" that defies old borders, and is shaped by the global economy of which it is a part; its leaders deal directly with foreign partners and negotiate the best terms they can with the national governments under which they operate. Writing about Canada, Ohmae identified a Pacific

Northwest (the Seattle-Vancouver axis) and a Great Lakes regional state (the intertwined Ontario-Michigan industrial complex), and warned that the manner in which Ottawa's leaders dealt with these new economic entities was critical to the survival of the Canadian state. Indeed, these growing international interactions are increasingly evident in the regional configuration of North America, to which we now turn.

REGIONS OF THE REALM

The ongoing transformation of North America's human geography is fully reflected in its internal regional organization. We now examine that spatial arrangement within a framework of nine regions (Fig. 3-18).

THE NORTH AMERICAN CORE

The Core Region (Fig. 3-18)—synonymous with the American Manufacturing Belt—was introduced earlier. This region was the workshop for the linked spatial economies of the United States and Canada during the century (1870–1970) dominated by industry. In the postindustrial era, however, that linchpin regional role is diminishing as strengthening challengers siphon key functions away from the Core. But make no mistake: this is still the geographic heart of North America. Here, in each country, one finds over one-third of the population, the capital, and the largest city—as well as the leading business complexes, cultural centers, and busiest transportation facilities.

Although manufacturing remains highly important within the transformed American economy, productivity and obsolescence problems in the Core Region have erased many of its competitive advantages over emerging industrial areas in the southern and western United States. In parts of the Core, employment has dropped sharply as factories have closed or relocated, speeding the economic decline of surrounding communities. But other areas have fought back, particularly the Midwest portion of the Manufacturing Belt, which reinvented itself by pursuing the high-tech upgrading of its aged industrial base and becoming more competitive in the international marketplace. The recent renaissance of its key automobile industry is a prime example, propelled by Toyota's investments in the Interstate-64 corridor between St. Louis and northern West Virginia, which became the axis of a new automaking complex centered on Lexington, Kentucky. Detroit's Big Three (GM, Ford, and Daimler-Chrysler) were forced to keep pace by modernizing fabrication technologies, which dispersed many operations away from Michigan to locations elsewhere in the Midwest and as far away as Mexico.

Postindustrial development has also spawned new growth centers in the Core, especially in the northeastern Megalopolis. The Boston area, richly endowed with research facilities, continues to attract innovative high-tech businesses to the Route 128 freeway corridor that girdles the central city; and Greater New York remains the national leader in finance, advertising, and corporate decision-making activity. The Core Region metropolis that has gained the most from postindustrialization is Washington, D.C. As quaternary activities have blossomed, and as the U.S. federal government extended its ties to the private sector, Washington and its surrounding outer city of Maryland and Virginia suburbs (interconnected by the 66-mile [105-km] Beltway encircling the capital city) has amassed an enormous complex of office, research, high-tech, and consulting firms. The most important development has occurred along the expressway corridor between Tyson's Corner on the Beltway and booming Dulles International Airport.

Since 1995, many telecommunications and Internet companies have been attracted to this part of northern Virginia, a leading fiber-optic cable hub thanks to its local cluster of federal intelligence agencies. This major, still-developing technopole is already being called the "technological capital" of the United States, because more technical workers are now employed here than in Silicon Valley.

THE MARITIME NORTHEAST

The Maritime Northeast consists of upper New England (Vermont, New Hampshire, Maine) and the neighboring Atlantic Provinces of easternmost Canada (Fig. 3-18). New England, one of the realm's culture hearths, has retained a strong regional identity for almost 400 years. Even with the urbanized southern half (Massachusetts, Connecticut, Rhode Island) lying in the Core, the six New England States still share many characteristics. The Maritime Northeast region also extends northeastward to encompass most of Canada's four Atlantic Provinces of New Brunswick, Nova Scotia, Prince Edward Island, and Newfoundland and Labrador.

A long association based on economic and cultural similarities has tied northern New England to Atlantic Canada. Both have a strong maritime orientation, are rural in character, possess difficult environments with limited land resources, and were historically bypassed in favor of more fertile inland areas. Economic growth in upper New England has always lagged behind the rest of the

REGIONS OF THE NORTH AMERICAN REALM

| 0 | 500 | 1000 | 1500 Kilometers |
| 0 | 300 | 600 | 900 Miles |

FIGURE 3-18

realm, with development centered on fishing (once) rich offshore waters, forestry, and farming in the few fertile lowlands available. Recreation and tourism have boosted the regional economy in recent times, with scenic coasts and mountains attracting millions from the neighboring Core Region.

Since 1980, New England's roller-coaster economy has experienced prosperity, hard times, and, most recently, cautious recovery. To achieve greater stability, New England is developing a more diversified economic base built around telecommunications, financial services, health care, and biotechnology. Nonetheless, the economic revival has yet to take hold in northern New England. Most of the benefits of the recovery are found in the zone closest to metropolitan Boston, which includes a wide swath of fringe areas spilling over into southern Maine and New Hampshire.

The Atlantic Provinces have also experienced hard economic times in recent years. Most adversely affected was the groundfish industry as offshore stocks of flounder, haddock, and cod became severely depleted through overfishing. New opportunities, never easy to come by here, are most promising in remote Newfoundland Island, where major offshore oil deposits (see Fig. 3-14) were discovered in the 1990s. The construction and opening of seabed drilling platforms and coastal support facilities has been swift, and the hope is that oil can transform the economy of Newfoundland and Labrador along the lines of Norway's over the past quarter-century.

FROM THE FIELDNOTES

"It is January, 2004 on a frigid day in one of the coldest winters people here can remember, but the work continues at the fish pier of Chatham, Massachusetts. Trucks load the fresh catch for dispatch to markets from Boston to New York, but all is not well with the fishing industry of New England. Overfishing, declining stocks, and limitations on the legal catch have idled boats and laid off thousands of workers. When I walked down to the pier, I saw several vehicles that had a sticker saying 'Chatham: a drinking town with a fishing problem,' but at least this particular town has an alternative: tourism, a seasonal yet high-yield industry. That's not the case in some of the less scenic fishing ports elsewhere along the region's coast, where fishing is the only game in town and where the decline of the industry is hitting hard."

for each other's languages as the strict equality of a new, government-mandated system of bilingualism.

THE CONTINENTAL INTERIOR

The Continental Interior extends across the center of both the conterminous United States and the southern tier of Canada (Fig. 3-18). With few exceptions—most notably in the region's northeastern corner where the barren but mineral-rich Canadian Shield meets the Great Lakes—agriculture is the predominant feature of the landscape. Because the eastern half of the Continental Interior lies in Humid America and closer to the national food market on the northeastern seaboard (Fig. 3-15), mixed crop-and-livestock farming wins out over less competitive wheat raising, which is relegated to the fertile but semiarid environment of the central and western Great Plains on the dry side of the 100th meridian. The latter area also contains Canada's agricultural heartland north of the 49th-parallel border. Although these fertile Prairie Provinces are less subject to drought than the U.S. high plains to the south, they are situated at a more northerly latitude and thus have a shorter growing season.

The distribution of North American corn and wheat farming is mapped in Figure 3-19, and the edges of the most productive zones of these two leading Continental Interior crops coincide with several of the region's boundaries. As we saw earlier in Figure 3-15, the Corn Belt is most heavily focused on Iowa and northern Illinois, while the Spring and Winter Wheat Belts are respectively centered on North Dakota and Kansas. This separation of the Wheat Belts is also the basis for subdividing the Continental Interior region, as the dashed line in Figure 3-18 indicates. The North subregion encompasses the Upper Midwest, the northern Great Plains, and the cultivable southern portions of Canada's three Prairie Provinces; the Spring Wheat Belt is its leading agricultural com-

FRENCH CANADA

Francophone Canada constitutes the effectively settled, southern portion of Quebec, which straddles the central and lower St. Lawrence Valley from where that river crosses the Ontario-Quebec border to its mouth in the Gulf of St. Lawrence. Also included is a sizeable concentration of French speakers, known as the Acadians, who reside just beyond Quebec's provincial boundary in neighboring New Brunswick (Fig. 3-18). The Old World charm of Quebec's cities is matched by an equally unique rural settlement landscape introduced by the French: narrow rectangular farms, known as *long lots*, are laid out in sequence perpendicular to the St. Lawrence, other rivers, and the roads that parallel them, thereby allowing each farm access to an adjacent routeway.

The economy of French Canada, however, is no longer rural and exhibits urbanization rates similar to those of the rest of the country. Industrialization is widespread, supported by cheap hydroelectric energy generated at huge dams in northern Quebec, but relatively little of the region's manufacturing could be classified as high-tech. Tertiary and postindustrial commercial activities are concentrated around Montreal, and tourism and recreation are also important to the regional economy. But the health of these sectors is tied to the resolution of Quebec's political status within Canada.

French Canada also includes Acadia, Canada's largest cluster of Francophones outside Quebec. Here in northernmost New Brunswick, the approximately 250,000 French-speakers constitute one-third of the province's population. The Acadians, however, reject the notion of independence and actively promote all efforts to keep Quebec within the Canadian federation. Today, a strengthening relationship (known as "cohabitation") has been forged between the Acadians and New Brunswick's Anglophone community, based as much on mutual respect

ponent and is so named because crops are planted in the spring and harvested in early autumn. The South subregion contains a major part of the Corn Belt in its eastern half and all of the Winter Wheat Belt in its western half. The former should probably be renamed the *Corn-Soybean Belt* because it also produces huge soybean crops that are often grown in rotation with corn. The latter is centered in the southern Great Plains, and its name is derived from the practice of planting wheat in the fall and harvesting it by late spring; the advantage of this growth cycle is that the crop is harvested before the hot, dry summer takes hold in this semiarid environment.

Throughout the Continental Interior, economic activity is oriented toward farming. Its leading metropolises—Kansas City, Minneapolis-St. Paul, Winnipeg, Omaha, and even Denver—are major processing and marketing points for pork and beef packing, flour milling, and soybean, sunflower, and canola oil production (all of which are increasingly exported). Although the family farm still dominates the rural landscape, the aggressive incursion of large-scale corporate farming now

threatens that way of life. As individual agricultural opportunities decline, the accelerating exodus of younger people leaves behind a poorer and ever more elderly population widely dispersed across a constellation of stressed rural communities struggling to survive the depopulation trend.

Despite the key role that agriculture plays in the regional economy, the nonfarming sectors of the Continental Interior continue to develop and diversify in and around major metropolitan centers. Minnesota has been the most successful in promoting such growth, capitalizing on the reputations of that State's highly educated workforce and reputation as a product innovator. Other new pursuits include telecommunications, telemarketing, and catalog sales.

THE SOUTH

The American South occupies the realm's southeastern corner, extending from the lush Bluegrass Basin of northern Kentucky to the swampy bayous of Louisiana's Gulf Coast, and from the knobby

hills of West Virginia to the flat sandy islands off southernmost Florida (Fig. 3-18). Of the realm's nine regions, none has undergone more overall change during the past half-century. For more than 100 years after the ruinous Civil War, the South languished in economic stagnation, but the 1970s finally witnessed a reassessment of the nation's perception of the region that launched a still-ongoing wave of growth and change.

Propelled by the forces that created the Sunbelt phenomenon, people and activities streamed into the urban South. Cities such as Atlanta, Charlotte, Miami, and Tampa became booming metropolises practically overnight, and conurbations swiftly formed in such places as southeastern Florida, the Carolina Piedmont, and the Texas-Louisiana Gulf Coast. Yet for all the growth that has taken place, the South remains a region beset by many economic problems because the geography of its development has been quite uneven. Although several metropolitan and a few favored farming areas have benefited, many others containing sizeable populations have not, and the juxtaposition of progress and backwardness is frequently encountered in the Southern landscape. Not surprisingly, the gap between rich and poor is wider here than in any other U.S. region. This economic disparity also carries racial dimensions: even though institutionalized racial segregation has been dismantled, compared to whites, blacks still exhibit higher poverty rates and rank behind in median family income.

Within this checkerboard spatial framework of development, the South's new affluence is largely concentrated in its central-city CBDs and burgeoning suburban rings. At the regional scale, another geographic pattern of inequality emerges: except for centrally located Atlanta and the Interstate-85 "Boombelt" corridor extending from there northeastward into central North Carolina, a significant proportion of the South's recent economic growth has taken place on its periphery in the Washington D.C. suburbs of northern Virginia, on the western Gulf Coast, and along Florida's central and southern-coast corridors.

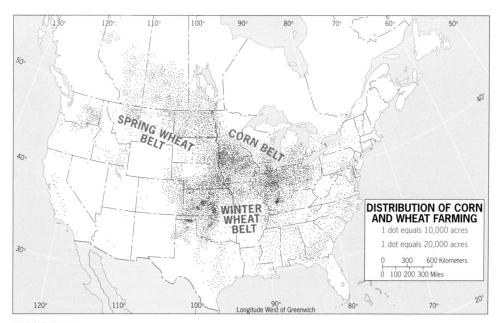

FIGURE 3-19

Three decades of rapid change have filled the tiles of the Southern population mosaic with an ever more diverse array of social and lifestyle groups, a clear signal that this once outcast region is completing its convergence with the rest of the country. Indeed, this 30-year transition period has ended with the South not only fully absorbed into the *national* economy and society, but also becoming a player on the *international* stage. During the 1990s, European and Japanese manufacturers were active investors in the region, but industrial jobs are proving hard to hang onto as factories now relocate outside the United States to take advantage of cheaper foreign labor. In terms of relative location, the most promising postindustrial opportunities in this era of NAFTA lie to the south in Middle and South America. South Florida has been in the vanguard here, and Miami appropriately bills itself as "the gateway to Latin America." Lately, other cities of the South have indicated an interest in undercutting this monopoly, most notably Atlanta and New Orleans.

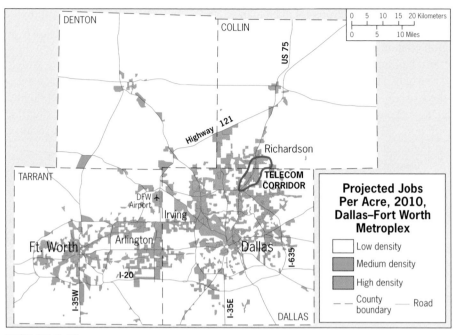

FIGURE 3-20

THE SOUTHWEST

Until recently, most geographers did not identify a distinct southwestern region, but this booming area—constituted by the States of Texas, New Mexico, and Arizona—has now earned its place on the regional map of North America (Fig. 3-18). The still-emerging Southwest is also unique in the United States because it is a bicultural regional complex where in-migrating Anglo-Americans join a rapidly expanding Mexican-American population, which traces its roots to the Spanish colonial era. In fact, with its large Native American population added in, the Southwest can be regarded as a *tricultural* region.

Recent development in the Southwest is based on: (1) huge amounts of electricity to power air conditioners through the extremely hot summers; (2) sufficient water to enable a large population to reside in this dry environment; and (3) the automobile, which supports the region's dispersed, low-density settlement fabric. The first and third of these have been

rather easily attained because the eastern half of the Southwest is well endowed with oil and natural gas (Fig. 3-14). But the future of water supplies is far more problematical; with few remaining sources to draw upon, the region cannot maintain its current pace of growth much longer.

The economic growth of the Southwest since 1990 has been led by Texas, whose eastern triangle (cornered by Dallas-Fort Worth, Houston, and San Antonio) has become a highly productive postindustrial complex, specializing in business-information and health-care services and high-tech manufacturing. Austin, lying near the heart of the triangle, has evolved into a technopole second only to Silicon Valley as a leader in computer research and fabrication. To the north, at the triangle's apex, the Dallas-Fort Worth Metroplex is keeping pace with its own expanding cluster of high-tech industries focused on the *Telecom Corridor* technopole just north of Dallas (the world's largest concentration of telecommunications firms). As Figure 3-20 shows, the geography of employment is increasingly oriented to booming suburban freeway corridors.

The western flank of the Southwest—whose development is mostly concentrated in southern Arizona—is unique. Equaled only by parts of southern Israel's Negev, this is the most technologically transformed desert environment on Earth. But while the Negev is devoted to agriculture, Arizona's far more intensive development is marked by vast urbanized landscapes focused on two of the largest western U.S. cities, Phoenix and Tucson.

THE WESTERN FRONTIER

To the north of the western half of the Southwest lies the realm's newest region: the burgeoning Western Frontier. In its east-west extent, this mountain-studded plateau region stretches between (and includes) the Rocky Mountains and the Sierra Nevada-Cascades chain; its longer north-south axis reaches from northern Arizona's Grand Canyon country to the edge of the subarctic snowforest west of the Canadian Rockies (Fig. 3-18). The region's heartland encompasses Utah, Nevada, Idaho, and western Colorado, and it is here that the old inter-mountain West is most actively reinventing itself. Until recently, such an opportunity would have been unthinkable because of this region's remoteness, dryness, and sparse population. But the Western Frontier's geographic prospects are changing significantly as the postindustrial national economy matures, advances in communications and transportation erode interregional differences in the costs of doing business, and amenities become the leading locational preference of employers. Proximity to the increasingly expensive, overcrowded, and problem-plagued Pacific coast also plays a role: well over 1 million disenchanted Californians have migrated into the Western Frontier since 1990.

The label *Western Frontier* further emphasizes that the transformation under way is being driven by the influx of people and activities from outside the region. Indeed, the populations of the four States of its heartland all ranked among the five fastest-growing of the 1990s. In the economic-geographic sphere, the region's identity is being reforged as traditional resource-based industries (mining, timber, livestock grazing) decline, while in-migrating and new home-grown companies create thousands of high-tech manufacturing and specialized services jobs.

The spatial pattern of the region's development is focused on its mushrooming urban areas. This creates an unusual frontier: instead of a single line of advancing settlement, a dispersed set of population clusters is simultaneously expanding around large- and middle-sized cities, high-amenity zones, and interstate freeway corridors. The hottest growth area is the Las Vegas Valley of southernmost Nevada, whose boom was triggered not only by the recreation industry (Las Vegas attracts some 30 million tourists annually) but is also the product of new economic development dominated by the influx of firms that specialize in computers, finance, and health care. Among the other urban areas experiencing this accelerated growth are the Western Frontier's two largest metropolises, Denver and Salt Lake City. Metropolitan Denver has expanded rapidly along the Front Range of the Rocky Mountains, building a diverse, globally oriented economy based on telecom-munications, computer software, and financial services. On a smaller scale, Salt Lake City now centers a thriving conurbation that includes the budding technopole of *Software Valley*, one of the realm's largest concentrations of information technology companies.

Even with most of the Western Frontier's development confined to its urban areas, the effects of this rapid growth are rippling deeply into the region's vast rural zones. Some of the problems are caused by the spillover effects of urbanization on adjacent areas. But many others result from the collision of lifestyles as recreation-seeking city dwellers invade ever more deeply into a countryside whose long-time, nonaffluent residents increasingly struggle to preserve a way of life based on such declining activities as mining and extensive agriculture.

THE NORTHERN FRONTIER

The northern half of the realm, lying poleward of roughly 52°N latitude, constitutes the Northern Frontier region (Fig. 3-18). This is by far North America's biggest regional subdivision, covering almost 90 percent of Canada and all of the U.S. State of Alaska. Indeed, if this huge territory were a separate country, it would rank as the world's fifth largest in size.

The latitudinal extent of this region leaves no doubt as to why it is designated *northern*, and most of its people and activities are understandably concentrated in its southern half. Unquestionably, this region is also a classic *frontier* that represents a thrust into undeveloped country, one marked by the long-term advance of a line across Canada's northern periphery that absorbs the areas it crosses into the national spatial economy. Boom and bust cycles over the past century have controlled the rate of movement of this line, and the still sparsely populated Northern Frontier is mostly a region of recent European settlement based on the extraction of newly discovered resources.

As noted earlier, the Canadian Shield, which covers the eastern half of the Northern Frontier (Fig. 3-2), is one of the world's richest storehouses of mineral resources. Most developing areas are focused on mining complexes along the Shield's southern margins that extract such metallic ores as nickel, copper, and zinc. The Northern Frontier is also endowed with an abundance of other resources (fossil fuels, hydroelectric power opportunities, timber, fisheries), which are dominated by substantial oil and gas reserves lying east of the Canadian Rockies between the Yukon and the U.S. border (Fig. 3-14).

The productive activities of the Northern Frontier are linked together within a far-flung network of mines, oil- and gas fields, pulp mills, and hydropower stations that have spawned hundreds of settlements and thousands of miles of transport and communications facilities. Although dominant, this commercial sector is part of a wider, dual regional economy that also includes a native sector. This native economy, which adheres to the traditional values of the indigenous peoples (First Nations), has long been tied to the commercial economy as a source of wages. It has also been heavily land-based, and in recent years First Nations throughout the region have become more assertive and filed legal actions to recover lands that Europeans took in the past without negotiating treaties.

The Northern Frontier also contains Alaska, whose geographic opportunities and challenges differ somewhat from those of the Canadian North. Besides being the largest U.S. State, Alaska earns a sizeable income from oil production on its north-eastern Arctic slope (Fig. 3-14). Fossil fuels are likely to continue driving Alaska's development even though supplies are dwindling in the key Prudhoe Bay oilfield. Besides the existence of significant additional oil reserves on the North Slope—most notably in the controversial Arctic National Wildlife Refuge—enormous supplies of natural gas are available for development.

THE PACIFIC HINGE

The Pacific coastlands of the conterminous United States and southwesternmost Canada, which comprise the realm's ninth region (Fig. 3-18), have been a powerful lure to migrants since the Oregon Trail was pioneered more than 150 years ago. Unlike the remainder of western North America south of 50°N latitude, the strip of land between the Sierra Nevada-Cascade mountain wall and the sea receives adequate moisture. It also possesses a far more hospitable environment, with generally delightful weather south of San Francisco, highly productive farmlands in California's Central Valley, and such scenic glories as the spectacular waters surrounding San Francisco and Vancouver. Most of the major development here took place during the post–World War II era, accommodating enormous population and economic growth along this outermost edge of the conterminous United States.

As the twenty-first century opens, it is clear that, in terms of its economic geography, the west coast no longer represents an end but rather a beginning— a gateway to an abundance of growing opportunities that in recent times have blossomed on many of the distant shores encircling the Pacific Basin. That is why we use the term *hinge*, for now this region increasingly forms an interface between North America and the booming, still-emerging **Pacific Rim**.

The regional term *Pacific Rim* has come into use to describe this dramatic development, which over the past quarter-century has seen the rise of a string of economic miracles that has redrawn the map of the Pacific periphery. Although led by its foci on the Pacific's western margins—Japan, coastal China, South Korea, Taiwan, Thailand, Malaysia, and Singapore—this far-flung, still-discontinuous region also includes the United States and Canada, and even such Southern Hemisphere locales as Australia and South America's Chile. The Pacific Rim is a superb example of a *functional region*: economic activity in the form of capital flows, raw-material movements, and trade linkages are generating urbanization, industrialization, and labor migration. (We will continue to discuss this important regional development and its spatial impacts in later chapters, wherever the Pacific Rim intersects Pacific-bordering geographic realms.)

Fifty-plus years of unrelenting growth have taken their toll, and whereas the three west coast States and British Columbia are now savoring the prospects of their Pacific Rim location, they must also confront the less pleasant consequences of regional maturity. This is especially true in California, where the massive development of America's most populated and multiethnic State (36 million in 2004, 14 percent larger than all of Canada) has been overwhelmingly concentrated in the teeming conurbation extending south from San Francisco through San Jose, the San Joaquin Valley, the Los Angeles Basin, and the southwestern coast into San Diego at the Mexican border. Environmental hazards bedevil this entire corridor, including droughts, mudslides, wildfires, and particularly earthquakes—with the ominous San Andreas Fault practically the axis of megalopolitan coalescence. To all these hazards humans have added their own abuses of California's habitat, from overuse of water supplies to chronic urban air pollution to the overcrowding of suburban landscapes (that has produced widespread traffic congestion, inflated real-estate prices, and even shortages of electrical power).

These challenges notwithstanding, Californians over the past half-century have built one of the realm's most productive economic machines. Indeed, if the Golden State were an independent country, its economy would today rank as the fifth largest on Earth. Southern California constitutes one of the State's two leading subregions. Here, its cornerstone Los Angeles-area economy practically reinvented itself during the 1990s after the sudden end to the Cold War devastated key local aerospace and other defense industries. Showing remarkable resilience, Southern California diversified its economic base to swiftly develop such new growth sectors as the entertainment industry, (Pacific Rim) foreign trade, and the nation's largest single metropolitan manufacturing complex (with products ranging from high-tech medical equipment to the clothing churned out by the burgeoning local garment industry). Northern California, the State's other major subregion focused on the San Francisco Bay Area, experienced a smoother path

FROM THE FIELDNOTES

"Minutes after takeoff from the Seattle-Tacoma International Airport on the way to Tokyo we got a superb view of central Seattle from the ferry terminals on Elliott Bay to Interstate 5 linking California to Canada and beyond. Like Sydney, Australia, Seattle has suburbs across the water reached most easily by boat; this is one of America's most scenic and vibrant cities. Seattle has transformed itself repeatedly through economic cycles reflected by the urban landscape: the dominance of precision-equipment manufacture, long dominated by Boeing Aircraft, is ending in favor of a new era of computer technology (Microsoft has its headquarters in a suburb) and related industries of the 'new economy.'" (There is something, though, that dates this photograph . . . what is it?)

during the 1990s, driven by the steady expansion of the world's leading technopole, Silicon Valley.

Continuing up the coast, the northern portion of the region is centered on the Pacific Northwest, which extends northward from Oregon's Willamette Valley through the Cowlitz-Puget Sound lowland of western Washington State into the adjoining coastal zone of British Columbia beyond the Canadian border. This subregion of the Pacific Hinge is also deeply involved in the pioneering and production of computer technology, especially in the technopole centered on suburban Seattle's Redmond. Originally developed by timber and fishing, the Pacific Northwest found its impetus for industrialization in the Columbia River dams built in the 1930s and 1950s, which generated cheap hydroelectricity that attracted aluminum and aircraft manu-

facturers. North of the Canadian border, British Columbia's development clusters around its economic core, Vancouver. Vancouver is the most Asianized metropolis in North America: it not only lies closer to East Asia on the air and sea routes than the cities to its south, but it also got a head start in forging trade and investment linkages to the western Pacific Rim thanks to its large and growing Asian community (ethnic Chinese residents now constitute more than 20 percent of the metropolitan population).

With North America ever more tightly enmeshed in the global economy, nowhere are the realm's international linkages more apparent today than in the Pacific Hinge. From Vancouver to San Diego, its geographic advantages as a gateway to the Pacific Rim are providing opportunities undreamed of barely a decade ago. Asianizing British

Columbia now exports more than 40 percent of its goods to the countries of the western Pacific Rim. The rest of the Vancouver-Seattle-Portland corridor is following suit, led by its high-technology industries. In California, the foreign-trade sector of the economy has tripled since 1990, and today more than 15 percent of its jobs are export-dependent. Both Northern and Southern California are expected to thrive as these trends continue, with metropolitan Los Angeles becoming the financial, manufacturing, and trading capital of the eastern Pacific Rim as well as its transport hub. Thus it is abundantly clear today that the mid-latitude west coast no longer serves as North America's back door. Rather, it forms the hinges of a new front door to the Pacific arena thrown wide open to foster international interactions of every kind.

DISCOVERY TOOLS   www.wiley.com/college/deblij

The *Concepts and Regions* Website, featuring GeoDiscoveries, offers many additional resources to enhance your understanding and experience of this chapter. Be sure to explore the following:

GeoDiscoveries Video Clips

Leaving Downtown

What's That Topping on My Pizza?

Suburbs: The Next Generation

Farming: Solely Economics or a Way of Life?

GeoDiscoveries Interactivities

The Look of the North American City

Where is the Greatest Amount of Sprawl in Your State?

Putting Sprawl Under the Microscope

Photo Galleries

North Star Steel Company Minimill

Port of New Orleans

The Las Vegas Valley

Fenway Park, Boston, Massachusetts

Miami, Florida

Glacier National Park, Alaska

Virtual Field Trip

The World's Breadbasket: U.S. Wheat Production

More To Explore

Systematic Essay: Urban Geography

Issues in Geography: The Migration Process; North American Free Trade Agreement; Geography of the Mighty Soybean; The Pacific Rim Connection

Major Cities: Toronto, Chicago, New York, Montreal, Los Angeles

Expanded Regional Coverage

North American Core; Maritime Northeast; French Canada; Continental Interior; South; Southwest; Western Frontier; Northern Frontier; Pacific Hinge

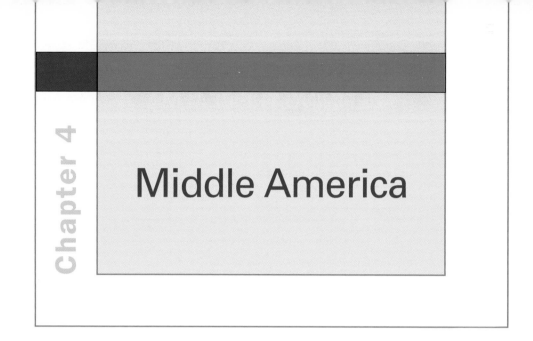

Chapter 4

Middle America

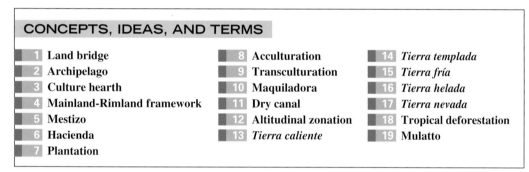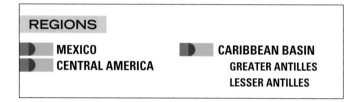

IDDLE AMERICA IS a realm of vivid contrasts, turbulent history, continuing political turmoil, and an uncertain future. The realm encompasses all the lands and islands between the United States to the north and South America to the south. Its regions are mapped in Figure 4-1 and include: (1) the substantial landmass of *Mexico*;

(2) the narrowing strip of land to its southeast that constitutes *Central America*; and (3) the many large and small islands—respectively known as the *Greater Antilles* and *Lesser Antilles*—of the Caribbean Sea to the east.

Middle America is a realm of soaring volcanoes and forested plains, of mountainous islands and flat coral cays. Moist tropical winds sweep in from the east, watering windward (wind-facing) coasts while leaving leeward (wind-protected) areas dry. Soils vary from fertile volcanic to desert barren. Spectacular scenery abounds, and tourism is one of the realm's leading industries.

DEFINING THE REALM

Is Middle America a discrete geographic realm? Some geographers combine Middle and South America into a realm they call "Latin" America, citing the dominant Spanish (and in Brazil, Portuguese) heritage and the prevalence of Roman Catholicism. But these criteria apply more strongly to South America. In Middle America, large populations exhibit African and Asian as well as European ancestries. And nowhere in South America has native, Amerindian culture contributed to modern civilization as strongly as it has in Mexico. The Caribbean Basin is a patchwork of independent states, territories in political transition, and residual colonial dependencies. The Dominican Republic speaks Spanish, adjacent Haiti uses French; Dutch is spoken in Aruba, while English is spoken in Jamaica. Middle America thus gives vivid definition to concepts of cultural-geographical pluralism.

Middle America is occasionally called *Central America*, but that name actually refers to a region within the Middle American realm (Fig. 4-1). Central America comprises the republics that occupy the strip of mainland between Mexico and Panama: Guatemala, Belize, Honduras, El Salvador, Nicaragua, and Costa Rica. Although Panama itself is regarded here as belonging to Central America, it should be noted that many Central Americans still do not consider Panama to be part of their region because for most of its history that country was a part of South America's Colombia. Again, as we define it, *Middle America* includes all the mainland and island countries and territories that lie between the United States and the continent of South America.

PHYSIOGRAPHY

As Figure 4-2 shows, Middle America is a realm is of high relief, fragmented territory, and crustal instability studded with active volcanoes. Figure G-4 reminds us why: the Caribbean, North American, South American, and Cocos tectonic plates converge here, creating spectacular and dangerous landscapes subject to earthquakes and landslides. Add to this the realm's location in the path of Atlantic hurricanes, and it amounts to some of the highest-risk topography on Earth.

The funnel-shaped mainland, a 3800-mile (6000-km) connection between North and South America, is wide enough in the north to contain two major mountain chains and a vast interior plateau, but narrows to a slim 40-mile (65-km) ribbon of land in Panama. Here this strip of land, or *isthmus*, bends eastward so that Panama's orientation is east-west. Thus mainland Middle America is what physical geographers call a **land bridge**, an isthmian link between continents.

If you examine a globe, you can see other present and former land bridges: Egypt's Sinai Penin-

---

MAJOR GEOGRAPHIC QUALITIES OF MIDDLE AMERICA

1. Middle America is a fragmented realm that consists of all the mainland countries from Mexico to Panama and all the islands of the Caribbean Basin to the east.

2. Middle America's mainland constitutes a crucial barrier between Atlantic and Pacific waters. In physiographic terms, this is a land bridge that connects the continental landmasses of North and South America.

3. Middle America is a realm of intense cultural and political fragmentation. The political geography defies unification efforts, but countries and regions are beginning to work together to solve mutual problems.

4. Middle America's cultural geography is complex. African influences dominate the Caribbean, whereas Spanish and Amerindian traditions survive on the mainland.

5. The realm contains the Americas' least-developed territories. New economic opportunities may help alleviate Middle America's endemic poverty.

6. In terms of area, population, and economic potential, Mexico dominates the realm.

7. Mexico is reforming its economy and has experienced major industrial growth. Its hopes for continuing this development are tied to overcoming its remaining economic problems and to expanding trade with the United States and Canada under the North American Free Trade Agreement (NAFTA).

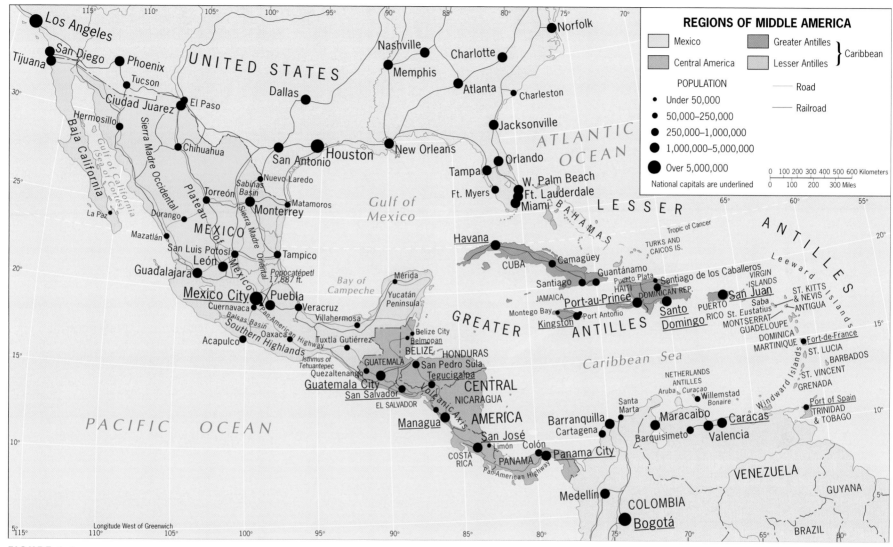

FIGURE 4-1

sula between Asia and Africa, the (now-broken) Bering land bridge between northeasternmost Asia and Alaska, and the shallow waters between New Guinea and Australia. Such land bridges, though temporary surface features in geologic time, have played crucial roles in the dispersal of animals and humans across the planet. But even though mainland Middle America forms a land bridge, its internal fragmentation has always inhibited movement. Mountain ranges, swampy coastlands, and dense rainforests make contact and interaction difficult.

Island Chains

As shown in Figure 4-1, the approximately 7000 islands of the Caribbean Sea stretch in a lengthy arc from Cuba and the Bahamas east and south to Trinidad, with numerous outliers outside (such as Barbados) and inside (e.g., the Cayman Islands) the main chain. The four large islands—Cuba, Hispaniola (containing Haiti and the Dominican Republic), Puerto Rico, and Jamaica—are called the *Greater Antilles*. All the remaining smaller islands are called the *Lesser Antilles*.

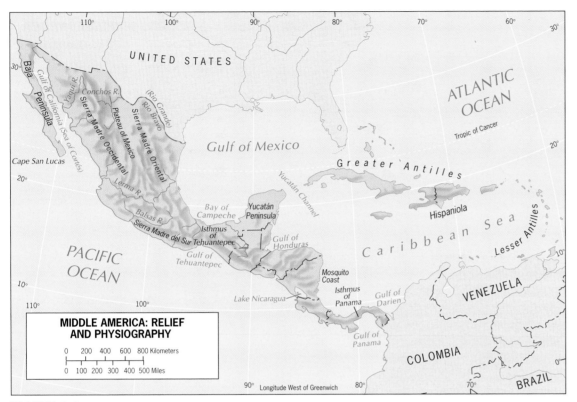

FIGURE 4-2

2 The entire Antillean **archipelago** (island chain) consists of the crests and tops of mountain chains that rise from the floor of the Caribbean, the result of collisions between the Caribbean Plate and its neighbors (Fig. G-4). Some of these crests are relatively stable, but elsewhere they contain active volcanoes, and almost everywhere in this realm earthquakes are an ever-present danger—in the islands as well as on the mainland.

LEGACY OF MESOAMERICA

Mainland Middle America was the scene of the emergence of a major ancient civilization. Here lay **3** one of the world's true **culture hearths** (see Fig. 7-3), a source area from which new ideas radiated and whose population could expand and make significant material and intellectual progress. Agricultural specialization, urbanization, and transport networks developed, and writing, science, art, and other spheres of achievement saw major advances. Anthropologists refer to the Middle American culture hearth as *Mesoamerica*, which extended southeast from the vicinity of present-day Mexico City to central Nicaragua. Its development is especially remarkable because it occurred in very different geographic environments, each presenting obstacles that had to be overcome in order to unify and integrate large areas. First, in the low-lying tropical plains of what is now northern Guatemala, Belize, and Mexico's Yucatán Peninsula, and perhaps simultaneously in Guatemala's highlands to the south, the Maya civilization arose more than 3000 years ago. Later, far to the northwest on the high plateau in central Mexico, the Aztecs founded a major civilization centered on the largest city ever to exist in pre-Columbian times.

The Lowland Maya

The Maya civilization is the only one in the world that ever arose in the lowland tropics. Its great cities, with their stone pyramids and massive temples, still yield archeological information today. Maya culture reached its zenith from the third to the tenth centuries AD.

The Maya civilization, anchored by a series of city-states, unified an area larger than any of the modern Middle American countries except Mexico. Its population probably totaled between 2 and 3 million; certain Maya languages are still used in the area to this day. The Maya city-states were marked by dynastic rule that functioned alongside a powerful religious hierarchy, and the great cities that now lie in ruins were primarily ceremonial centers. We also know that Maya culture produced skilled artists and scientists, and that these people achieved a great deal in agriculture and trade. They grew cotton, created a rudimentary textile industry, and exported cotton cloth by seagoing canoes to other parts of Middle America in return for valuable raw materials. They domesticated the turkey and grew cacao, developed writing systems and studied astronomy.

The Highland Aztecs

In what is today the intermontane highland zone of Mexico, significant cultural developments were also taking place. Here, just north of present-day Mexico City, lay Teotihuacán, the first true urban center in the Western Hemisphere, which prospered for nearly seven centuries after its founding around the beginning of the Christian era. Later the Toltecs migrated into this area from the north, conquered and absorbed the local Amerindian peoples, and formed a powerful state. When, about

300 years later, the Toltec state itself was penetrated by new elements from the north, it was already in decay; but its technology was readily adopted and developed by the conquering Aztecs.

The Aztec state, the pinnacle of organization and power in pre-Columbian Middle America, is thought to have originated in the early fourteenth century with the founding of a settlement on an island in a lake that lay in the *Valley of Mexico* (the area surrounding what is now Mexico City). This urban complex, a functioning city as well as a ceremonial center, named Tenochtitlán, was soon to become the greatest city in the Americas and the capital of a large powerful state. The Aztecs soon gained control over the entire Valley of Mexico, a pivotal 30-by-40-mile (50-by-65-km) mountain-encircled basin that is still the heart of the modern state of Mexico. Both elevation and interior location moderate its climate; for a tropical area, it is quite dry and very cool. The basin's lakes formed a valuable means of internal communication, and the Aztecs built canals to connect several of them. This fostered a busy canoe traffic, bringing agricultural produce to the cities, and tribute was paid by their many subjects to the headquarters of the ruling nobility.

With its heartland consolidated, the new Aztec state soon began to conquer territories to the east and south. The Aztecs' expansion of their empire was driven by their desire to subjugate peoples and towns in order to extract taxes and tribute. As Aztec influence spread throughout Middle America, the state grew ever richer, its population mushroomed, and its cities thrived and expanded.

The Aztecs produced a wide range of impressive accomplishments, although they were better borrowers and refiners than they were innovators. They developed irrigation systems, and they built elaborate walls to terrace slopes where soil erosion threatened. Indeed, the greatest contributions of Mesoamerica's Amerindians surely came from the agricultural sphere. Corn (maize), the sweet potato, various kinds of beans, the tomato, squash, cacao beans (the raw material of chocolate), and tobacco are just a few of the crops that grew in Mesoamerica when the Europeans first made contact.

COLLISION OF CULTURES

We in the Western world all too often believe that history began when the Europeans arrived in some area of the world and that the Europeans brought such superior power to the other continents that whatever existed there previously had little significance. Middle America confirms this misperception: the great, feared Aztec state apparently fell before a relatively small band of Spanish invaders in an incredibly short period of time (1519–1521). But let us not lose sight of a few realities. At first, the Aztecs believed the Spaniards were "White Gods" whose arrival had been predicted by Aztec prophecy. Hernán Cortés, for all his 508 soldiers, did not singlehandedly overthrow this powerful empire: he ignited a rebellion by Amerindian peoples who had fallen under Aztec domination and had seen their relatives carried off for human sacrifice to Aztec gods. Led by Cortés with his horses and guns, these peoples rose against their Aztec oppressors and joined the band of Spaniards headed toward Tenochtitlán (where thousands of them would die in combat against Aztec defenders).

Effects of the Conquest

Spain's defeat of Middle America's dominant indigenous state opened the door to Spanish penetration and supremacy. Throughout the realm, the confrontation between Hispanic and native cultures spelled disaster for the Amerindians: a catastrophic decline in population (perhaps as high as 90 percent), rapid deforestation, pressure on vegetation from grazing animals, substitution of Spanish wheat for maize (corn) on cropland, and the concentration of Amerindians into newly built towns.

The Spaniards were ruthless colonizers but not more so than other European powers that subjugated other cultures. True, the Spaniards first enslaved the Amerindians and were determined to destroy the strength of indigenous society. But biology accomplished what ruthlessness could not have achieved in so short a time: diseases introduced by the Spaniards and slaves imported from Africa decimated millions of Amerindians.

Middle America's cultural landscape—its great cities, its terraced fields, its dispersed aboriginal villages—was thus drastically modified. Unlike the Amerindians, who had used stone as their main building material, the Spaniards employed great quantities of wood and used charcoal for heating, cooking, and smelting metal. The onslaught on the forests was immediate, and rings of deforestation swiftly expanded around the colonizers' towns. The Spaniards also introduced large numbers of cattle and sheep, and people and livestock now had to compete for available food (requiring the opening of vast areas of marginal land that further disrupted the region's food-production balance). Moreover, the Spaniards introduced their own crops (notably wheat) and farming equipment, and soon large fields of wheat began to encroach upon the small plots of corn that the natives cultivated.

The Spaniards' most far-reaching cultural changes derived from their traditions as town dwellers. To facilitate domination, the Amerindians were moved off their land into nucleated villages and towns that the Spaniards established and laid out. In these settlements, the Spaniards could exercise the kind of rule and administration to which they were accustomed (Fig. 4-3). The internal focus of each Spanish town was the central *plaza* or market square, around which both the local church and government buildings were located. The surrounding street pattern was deliberately laid out in *gridiron* form, so that any insurrections by the resettled Amerindians could be contained by having a small military force seal off the affected blocks and then root out the troublemakers. Each town was located near what was thought to be good agricultural land (which was often not so good), so that the Amerindians could go out each day and work in the fields. Packed tightly into these towns

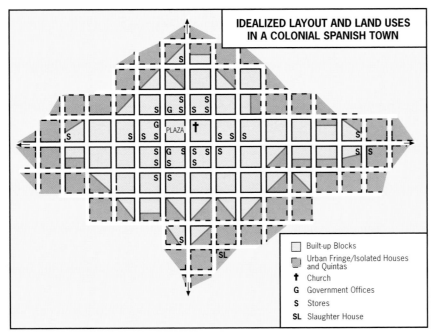

IDEALIZED LAYOUT AND LAND USES IN A COLONIAL SPANISH TOWN

Built-up Blocks
Urban Fringe/Isolated Houses and Quintas
† Church
G Government Offices
S Stores
SL Slaughter House

FIGURE 4-3

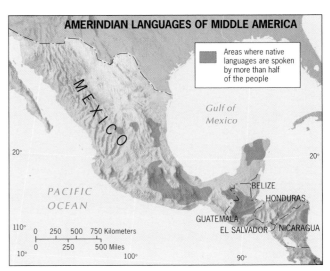

AMERINDIAN LANGUAGES OF MIDDLE AMERICA

Areas where native languages are spoken by more than half of the people

FIGURE 4-4

and villages, they came face to face with Spanish culture. Here they learned the Europeans' Roman Catholic religion and Spanish language, and they paid their taxes and tribute to a new master. Nonetheless, the nucleated indigenous village survived under colonial (and later postcolonial) administration and is still a key feature of remote Amerindian areas in southeastern Mexico and inner Guatemala, where to this day native languages prevail over Spanish (Fig. 4-4).

Once the indigenous population was conquered and resettled, the Spaniards were able to pursue another primary goal in their New World territory: the exploitation of its wealth for their own benefit. Lucrative trade, commercial agriculture, livestock ranching, and especially mining were avenues to affluence. And wherever the Spaniards ruled—in towns, farms, mines, or indigenous villages—the Roman Catholic Church was the supreme cultural force transforming Amerindian society. With Jesuits and soldiers working together to advance the

frontiers of New Spain, in the wake of conquest it was the church that controlled, pacified, organized, and acculturated the Amerindian peoples.

MAINLAND AND RIMLAND

In Middle America outside Mexico, only Panama, with its twin attractions of interoceanic transit and gold deposits, became an early focus of Spanish activity. From there, following the Pacific side of the isthmus, Spanish influence radiated northwestward through Central America and into Mexico. The major arena of international competition in Middle America, however, lay not on the Pacific side but on the islands and coasts of the Caribbean Sea. Here the British gained a foothold on the mainland, controlling a narrow coastal strip that extended southeast from Yucatán to what is now Costa Rica. As the colonial-era map (Fig. 4-5) shows, in the Caribbean the Spaniards faced the

British, French, and Dutch, all interested in the lucrative sugar trade, all searching for instant wealth, and all seeking to expand their empires.

Later, after centuries of European colonial rivalry in the Caribbean Basin, the United States entered the picture and made its influence felt in the coastal areas of the mainland, not through colonial conquest but through the introduction of widespread, large-scale, banana plantation agriculture. The effects of these plantations were as far-reaching as the impact of colonialism on the Caribbean islands. Because the diseases the Europeans had introduced were most rampant in these hot, humid lowlands (as well as the Caribbean islands to the east), the Amerindian population that survived was too small to provide a sufficient workforce. This labor shortage was remedied through the generation of the trans-Atlantic slave trade from Africa that transformed the Caribbean Basin's demography (see map, p. 203).

These contrasts between the Middle American highlands and the coastal areas and Caribbean islands were conceptualized by John Augelli into ▪ the **Mainland-Rimland framework** (Fig. 4-6). Augelli recognized (1) a Euro-Amerindian **Mainland**, which consisted of continental Middle Amer-

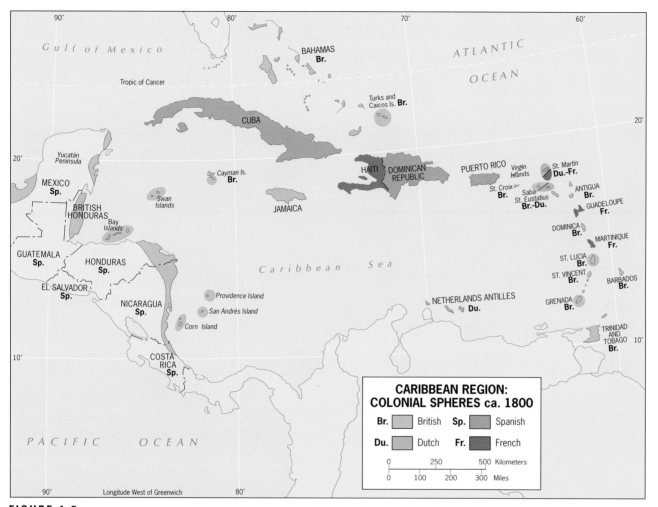

FIGURE 4-5

high accessibility, of seaward exposure, and of maximum cultural contact and mixture. The Mainland, being farther removed from these contacts, was an area of greater isolation. The Rimland was the region of the great *plantation*, and its commercial economy was therefore susceptible to fluctuating world markets and tied to overseas investment capital. The Mainland was the region of the *hacienda*, which was more self-sufficient and less dependent on external markets.

The Hacienda

This contrast between plantation and hacienda land tenure in itself constitutes strong evidence for the Rimland-Mainland division. The hacienda was a Spanish institution, but the modern plantation, Augelli argued, was the concept of Europeans of more **6** northerly origin. In the **hacienda**, Spanish landowners possessed a domain whose productivity they might never push to its limits: the very possession of such a vast estate brought with it social prestige and a comfortable lifestyle. Native workers lived on the land—which may once have been *their* land—and had plots where they could grow their own subsistence crops. All this is written as though it is mostly in the past, but the legacy of the hacienda, with its inefficient use of land and labor, is still visible throughout mainland Middle America.

The Plantation

7 The **plantation** was conceived as something entirely different from the hacienda. Robert West and John Augelli list five characteristics of Middle

ica from Mexico to Panama, with the exception of the Caribbean coastal belt from mid-Yucatán southeastward; and (2) a Euro-African **Rimland**, which included this coastal zone as well as the islands of the Caribbean. The terms *Euro-Amerindian* and *Euro-African* underscore the cultural heritage of each region. On the Mainland, European (Spanish) and Amerindian influences are paramount and also **5** include **mestizo** sectors where the two ancestries mixed. In the Rimland, the heritage is European and African.

As Figure 4-6 shows, the Mainland is subdivided into several areas based on the strength of the Amerindian legacy. The Rimland is also subdivided, with the most obvious division the one between the mainland-coastal plantation zone and the islands. Note, too, that the islands themselves can be classified according to their cultural heritage (Figs. 4-5, 4-6).

Supplementing these contrasts are regional differences in outlook and orientation. The Rimland was an area of sugar and banana plantations, of

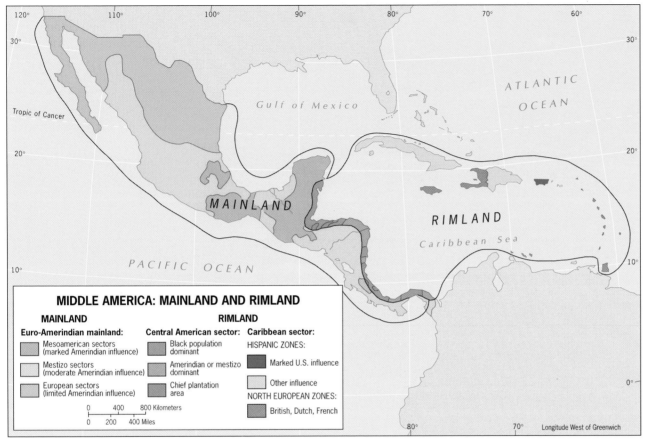

FIGURE 4-6

American plantations that illustrate the differences between hacienda and plantation: (1) plantations are located in the humid tropical coastal lowlands; (2) plantations produce for export almost exclusively—usually a single crop; (3) capital and skills are often imported so that foreign ownership and an outflow of profits occur; (4) labor is seasonal—needed in large numbers mainly during the harvest period—and such labor has been imported because of the scarcity of Amerindian workers; and (5) with its "factory-in-the-field" operation, the plantation is more efficient in its use of land and labor than the hacienda. The objective was not self-sufficiency but profit, and wealth rather than social prestige is a dominant motive for the plantation's establishment and operation.

POLITICAL DIFFERENTIATION

Today continental Middle America is fragmented into eight countries, all but one of which have Hispanic origins. The largest of them all is Mexico—the giant of Middle America—whose 756,000 square miles (1,960,000 sq km) constitute more than 70 percent of the realm's entire land area (the Caribbean region included), and whose 106 million people outnumber those of all the other coun-tries and islands of Middle America combined.

The cultural variety in Caribbean Middle America is much greater. Here Hispanic-influenced Cuba dom-inates: its area is almost as large as that of all the other islands put to-gether, and its population of 11.4 mil-lion is well ahead of the next-ranking country—the Dominican Republic (9.2 million), also of Spanish her-itage. As we noted, however, the Caribbean is hardly an arena of ex-clusive Hispanic cultural heritage: for example, Cuba's southern neigh-bor, Jamaica (population 2.7 million, mostly black), has a legacy of British involvement, while to the east in Haiti (7.4 million, overwhelmingly black) the strongest imprints have been Af-rican and French. The Lesser Antilles also exhibit great cultural diversity. There are the (once Danish) U.S. Virgin Islands; French Guadeloupe and Martinique; a group of British-influenced islands, including Barba-dos, St. Lucia, and Trinidad and To-bago; and the Dutch St. Maarten (shared with the French) and A-B-C islands of the Netherlands Antilles—Aruba, Bonaire, and Curaçao—off the northwest-ern Venezuelan coast.

Independence movements stirred Middle Amer-ica at an early stage. On the mainland, revolts against Spanish authority (beginning in 1810) achieved independence for Mexico by 1821, and for the Central American republics by the end of the 1820s. The colonial map of the Caribbean was a patchwork of holdings by British, Dutch, French, Spanish, and Danish interests (Fig. 4-5). The United States, concerned over European designs in the realm, proclaimed the Monroe Doctrine in 1823 to deter any European power from reasserting its authority in newly independent republics or from further expanding its existing domains. By the end of the nineteenth century, the United States itself had become a major force in Middle America. The Spanish-American War of 1898 made Cuba independent and put Puerto Rico under the U.S. flag; soon afterward, the Americans were in Panama constructing the Panama Canal. Meanwhile, with U.S. corporations driving a boom based on huge banana plantations, the Central American republics had become colonies of the United States in all but name.

Independence came to the Caribbean Basin in fits and starts. Afro-Caribbean Jamaica as well as Trinidad and Tobago, where the British had brought a large South Asian population, attained full sovereignty from the United Kingdom in 1962; other British islands (among them Barbados, St. Vincent, and Dominica) became independent later. France, however, retains Martinique and Guadeloupe as overseas *départements* of the French Republic, and the Dutch islands are at various stages of autonomy.

REGIONS OF THE REALM

Middle America consists of four geographic regions: (1) Mexico, the giant of the realm in every respect; (2) Central America, the string of seven small republics occupying the land bridge to South America; (3) the four large islands that constitute the Greater Antilles of the Caribbean; and (4) the numerous islands of the Caribbean's Lesser Antilles (Fig. 4-1).

MEXICO

By virtue of its physical size, large population, cultural identity, resource base, economy, and relative location, the state of Mexico by itself constitutes a geographic region in this multifaceted realm. The U.S.-Mexican boundary on the map crosses the continent from the Pacific to the Gulf, but Mexican cultural influences penetrate deeply into the southwestern States and American impacts reach far into Mexico. To Mexicans the border is a reminder of territory lost to the United States in historic conflicts; to Americans it is a symbol of economic contrasts and illegal immigration. Along the Mexican side of this border the effects of NAFTA (the North American Free Trade Agreement among Canada, the United States, and Mexico) are transforming Mexico's economic geography.

Physiography

The physiography of Mexico is reminiscent of that of the western United States, although its environments are more tropical. Figure 4-7 shows several prominent features: the elongated Baja (Lower) California Peninsula in the northwest, the far eastern Yucatán Peninsula, and the Isthmus of Tehuantepec in the southeast where the Mexican landmass tapers to its narrowest extent. Here in the southeast, Mexico most resembles Central America physiographically; a mountain backbone forms the isthmus, curves southeast into Guatemala, and extends northwest toward Mexico City. Shortly before reaching the capital, this mountain range divides into two chains, the Sierra Madre Occidental in the west and the Sierra Madre Oriental in the east (Figs. 4-2, 4-7). These diverging ranges frame the funnel-shaped Mexican heartland, the center of which consists of the rugged, extensive Plateau of Mexico (the Valley of Mexico lies near its southeastern end). As Figure G-8 reveals, Mexico's climates are marked by dryness, particularly in the broad, mountain-flanked north. Most of the better-watered areas lie in the southern half of the country where the major population concentrations have developed.

Population Patterns

Mexico's population grew rapidly during the last three decades of the twentieth century, doubling in just 28 years; but demographers have recently noted a sharp drop in fertility, and they are predicting that Mexico's population (currently 106 million) will stop growing altogether by about 2050. This will have enormous implications for the country's economy, and it will reduce the cross-border migration that is of so much concern at present.

The distribution of population across Mexico's 31 internal States is shown in Figure 4-8. The largest concentration, containing more than half the Mexican people, extends across the densely populated "waist" of the country from Veracruz State on the eastern Gulf coast to Jalisco State on the Pacific. The center of this corridor is dominated by the most populous State, Mexico (**3** on the map), at whose heart lies the Federal District of Mexico City (**9**). In the dry and rugged terrain to the north of this central corridor lie Mexico's least-populated States. Southern Mexico also exhibits a sparsely peopled periphery in the hot and humid lowlands of the Yucatán Peninsula, but here most of the highlands of the continental spine contain sizeable populations.

Another major feature of Mexico's population map is urbanization, driven by the *pull* of the cities (with their perceived opportunities for upward mobility) in tandem with the *push* of the economically stagnant countryside. Today, 74 percent of the Mexicans reside in towns and cities, a surprisingly high proportion for a developing country. Undoubtedly, these numbers are affected by the explosive recent growth of Mexico City, which now totals 28 million (making it the largest urban concentration on Earth) and is home to 26 percent of the national population. Among the other leading cities are Guadalajara, Puebla, and León in the central population corridor, and Monterrey, Ciudad Juarez, and

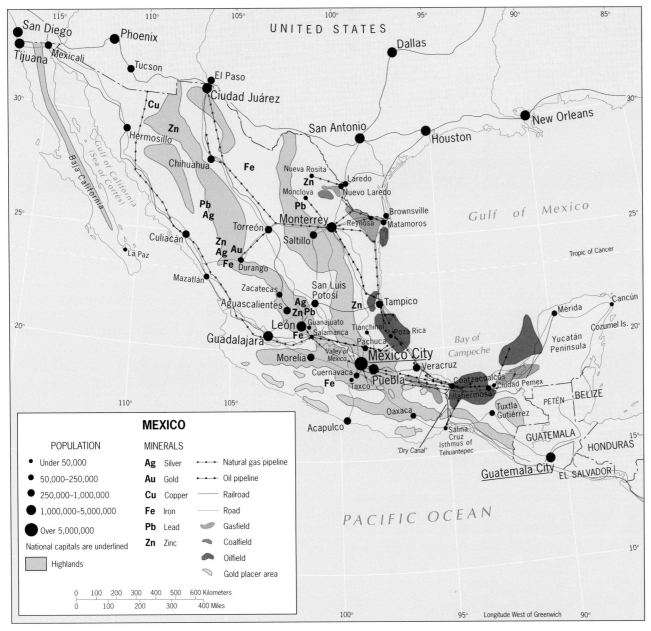

FIGURE 4-7

cent are Europeans. Certainly the Mexican Amerindian has been Europeanized, but the Amerindianization of modern Mexican society is so powerful that it would be inappropriate here to speak of one-way, **8** European-dominated **acculturation**. Instead, what took place in **9** Mexico is **transculturation**—the two-way exchange of culture traits between societies in close contact. In the southeastern periphery (Fig. 4-4), several hundred thousand Mexicans still speak only an Amerindian language, and millions more still use these languages in everyday conversation even though they also speak Mexican Spanish. The latter has been strongly shaped by Amerindian influences, as have Mexican modes of dress, foods and cuisine, sculpture and painting, architectural styles, and folkways. This fusion of heritages, which makes Mexico unique, is the product of an upheaval that began to reshape the country nearly a century ago.

Revolution and Its Aftermath

Modern Mexico was forged in a revolution that began in 1910 and set into motion events that are still unfolding today. At its heart, this revolution was about the redistribution of land, an issue that had not been resolved after Mexico freed itself from Spanish colonial control in the early nineteenth century. As late as 1900, more than 8000 haciendas blanketed virtually all of Mexico's good farmland, and about 95 percent of all rural families owned no land whatsoever and toiled as *peones* (landless, constantly indebted serfs) on the haciendas. The triumphant revolution produced a new constitu-

Tijuana in the northern U.S. border zone (Fig. 4-8). Urbanization rates at the other end of Mexico, however, are at their lowest in those remote uplands where Amerindian society has been least touched by modernization.

Nationally, the Amerindian imprint on Mexican culture remains quite strong. Today, 60 percent of all Mexicans are mestizos, 20 percent are predominantly Amerindian, and about 10 percent are full-blooded Amerindians; only 9 per-

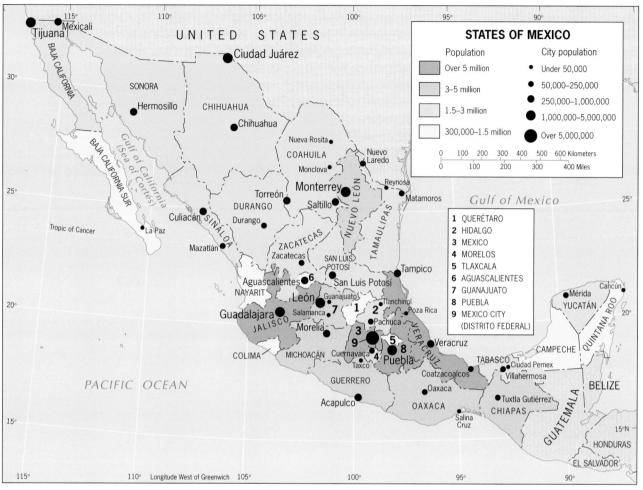

FIGURE 4-8

Chiapas, the poorest of the 31 States, lies wedged against the Guatemalan border and has more in common with Central America than with Mexico. The complexion of its rapidly growing population is heavily Amerindian and dominated by families of peasant farmers who eke out a precarious existence by cultivating tiny plots of land in the rainforested hills. For centuries, the better valley soils have been incorporated into the estates of the large landholders, a system that endures here virtually unaffected by the land redistribution that reshaped so much of rural Mexico. In response to this lack of development, during the 1980s the Mexican government had pledged to introduce new services and programs in Chiapas, but its main effort was an ineffective coffee-raising scheme and a feeble attempt to privatize Amerindian lands. This only intensified the longstanding bitterness of the Chiapans, and a radical group of Mayan peasant farmers began to organize to resume the historic struggle of Amerindian *peones* to gain land and fair treatment.

On January 1, 1994, this organization, now calling itself the Zapatista National Liberation Army (ZNLA), ignited a guerrilla war with coordinated attacks on several Chiapan towns. By taking the name of Zapata (a legendary leader of the 1910 revolution), and by timing its insurgency to coincide with the birth of NAFTA, the ZNLA achieved maximum impact and publicity. The Mexican army's response was heavy-handed, and suddenly Mexico confronted a major domestic challenge.

As the map shows, Chiapas lies on Mexico's periphery, far from the burgeoning core area and even farther from NAFTA developments. The Zapatistas demanded greater political autonomy (on the European devolutionary model) and control

tion in 1917 that launched a program of expropriation and parceling out of the haciendas to rural communities.

Since 1917, more than half the cultivated land of Mexico has been redistributed, mostly to peasant communities consisting of 20 families or more. On such farmlands (known as *ejidos*) the government holds title to the land, and use rights are parceled out to villages and then individuals for cultivation. Most of these *ejidos* lie in central and southern Mexico, where Amerindian agricultural traditions survived, but the newly created landholdings were too finely fragmented to provide a

real development opportunity. Despite an understandable temporary decline in agricultural productivity during the transition, the miracle is that land reform was carried off without major dislocation. Although considerable poverty persisted in the countryside, it was recognized that Mexico, alone among Middle America's countries with large Amerindian populations, had made significant strides toward solving the land question. Just how much farther Mexico has to go, however, became evident during the 1990s when the issue resurfaced at the heart of a rebellion that broke out in southeasternmost Chiapas State.

over local affairs. And they brought to the fore an uncomfortable reality of Mexican culture: the low and disadvantaged status of Mexico's 10 percent ethnic Amerindians, who choose to preserve their pre-Hispanic cultural traditions, and whose ancestral homes are widely distributed throughout the country (Fig. 4-9).

Regions of Mexico

Physiographic, demographic, economic, historical, and cultural criteria combine to reveal a re-

gionally diverse Mexico extending from the lengthy ridge of Baja California to the tropical lowlands of the Yucatán Peninsula, and from the economic frenzy of the NAFTA North to the Amerindian traditionalism of the Chiapas southeast (Fig. 4-10). In the Core Area, anchored by Mexico City, and the West, centered on Guadalajara, lies the transition zone from the more Hispanic-mestizo north to the more Amerindian-infused mestizo south. East of the Core Area lies the Gulf Coast, once dominated by major irrigation projects and huge livestock-raising schemes but now the main-

land center of Mexico's petroleum industry. The dry, scrub-vegetated Balsas Lowland separates the Core Area from the rugged, Pacific-fronting Southern Highlands, where Acapulco's luxury hotels stand in stark contrast to the Amerindian villages and *ejidos* of the interior, scene of major land reform in the wake of the 1910 revolution.

The dry, vast north stands in sharp contrast to these southern regions: only in the Northwest was there significant sedentary Amerindian settlement when the Spanish arrived. Huge haciendas, major irrigation projects, and some large cities separated by great distances mark this area, now galanized by the impact of NAFTA. The NAFTA North region is still formative and discontinuous, but is changing northern Mexico significantly.

The Changing Geography of Economic Activity

During the last two decades of the twentieth century and in the first years of the twenty-first, Mexico's economic geography has changed, and in some respects progressed—though not without setbacks. During the early 1990s, the implementation of NAFTA led to an economic boom as Mexico became part of a free-trade zone and market comprising over 400 million people. This boom transformed urban landscapes along the 1936-mile (3115-km) border between Mexico and the United States, but it could not, of course, close the economic gap between the two sides.

Today, Mexico is in progressive transition in many spheres: its democratic institutions are strengthening; its economy is more robust than it was at the time of NAFTA's inception; its once-rapid population growth is declining; its social fabric (notably rela-

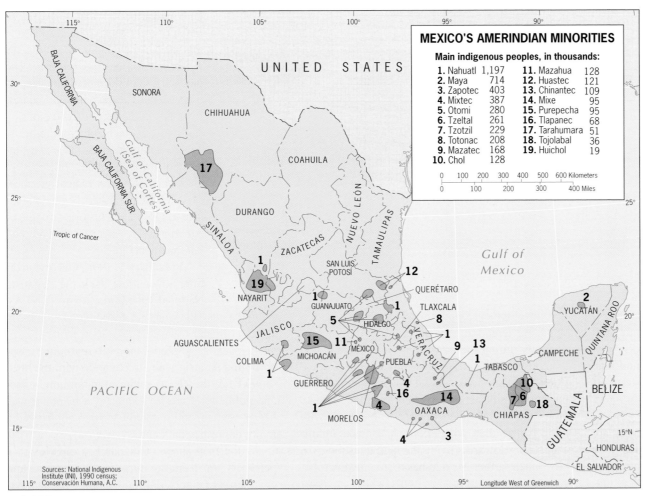

MEXICO'S AMERINDIAN MINORITIES

Main indigenous peoples, in thousands:

1. Nahuatl	1,197		11. Mazahua	128
2. Maya	714		12. Huastec	121
3. Zapotec	403		13. Chinantec	109
4. Mixtec	387		14. Mixe	95
5. Otomi	280		15. Purepecha	95
6. Tzeltal	261		16. Tlapanec	68
7. Tzotzil	229		17. Tarahumara	51
8. Totonac	208		18. Tojolabal	36
9. Mazatec	168		19. Huichol	19
10. Chol	128			

Sources: National Indigenous Institute (INI), 1990 census; Conservación Humana, A.C.

FIGURE 4-9

with fruit and vegetable cultivation attracting foreign investors.

But many Mexican small farmers are having a tough time of it, for two reasons. First, cheap corn from the United States, grown by American farmers heavily subsidized by the government, floods Mexico's markets, so that local farmers cannot make a profit. Second, the U.S. tried to create barriers against the import of low-priced Mexican produce such as avocados and tomatoes—violating the very rules NAFTA is supposed to stand for.

Energy Resources

Possessing major reserves of oil and natural gas is no guarantee of national prosperity, as many countries from Venezuela to Nigeria have learned. Mexico's economy was transformed by the discovery of major energy resources along the Gulf Coast from Matamoros in the north through Tampico, Poza Rica, and Villahermosa on the Bay of Campeche to the western shore of the Yucatán Peninsula (Fig. 4-7). Domestic refineries and a network of oil and gas pipelines supply all the energy Mexico needs, and exports bring substantial foreign revenues. But such major assets require good management, and this has not always happened. To finance domestic development projects, Mexico borrowed heavily against future oil revenues when international prices were high, only to find itself unable to make payments when prices fell. Poor planning, corruption, and other failures in the state-controlled industry undermined one of the country's most valuable assets.

Industrialization

Manufacturing got an early start in Mexico: the country's first iron and steel plant began production

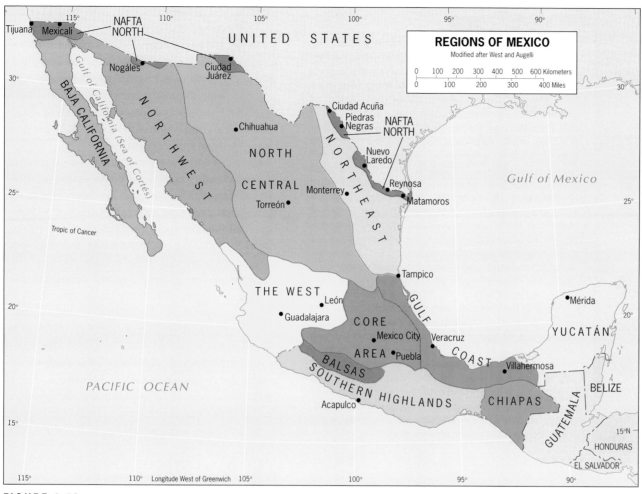

FIGURE 4-10

tions with Amerindian minorities) is improving. The familiar problems of a country in transition, such as unchecked urbanization, inadequate infrastructure, corruption, and violent crime, will continue to afflict Mexico for decades to come. But, as the following discussion confirms, Mexico is a far stronger economy and society today than it was just one generation ago.

Agriculture

Although traditional subsistence agriculture and the output of the inefficient *ejidos* have not changed a great deal in the poorer areas of rural Mexico, larger-scale commercial agriculture has diversified during the past three decades and made major gains with respect to both domestic and export markets. The country's arid northern tier has led the way as major irrigation projects have been built on streams flowing down from the interior highlands. Along the booming northwest coast of the mainland, which lies within a day's drive of Southern California, mechanized large-scale cotton production now supplies an increasingly profitable export trade. Here, too, wheat and winter vegetables are grown,

"As soon as I arrived in the city of Mérida I got the sense that many people here in Yucatán—the Yucatecos, as they call themselves—feel culturally distinct from, even aloof from, the Mexico of which they are a part. After hearing complaints about the behavior of visitors from Mexico City and about the failure of NAFTA to do much for the local economy, I spent several days in Mérida's Maya hinterland, learning more about the area's indigenous history. About an hour from the city I stopped at this hacienda, which invited visitors and promised lunch, and got an earful about local opinion. 'We brought the advanced hacienda economy to this area,' I was told. 'This hacienda had land as far as the eye can see from the top of the roof. We raised cattle and planted henequen and gave the Indians land to plant their food crops. Did you see those magnificent villas in the city? That's what built them. But then this so-called Agrarian Revolution happened and the land was taken away and put into *ejidos*, and the economy collapsed. Well, the Indians have their land back, but they don't use it like we did. So now we're down to serving lunch and running weddings, and farming what we have left to keep things going. And we've got a government in Mexico City that doesn't seem to care what happens over here.' I walked around the establishment's numerous buildings before leaving, and it appeared to be prospering, whatever its current economic mainstays. But Yucatecos certainly have an attitude."

strengthened, however slightly, the maquiladoras' competitiveness declined. While factories producing bulky and heavy items like cars and refrigerators were still better off right across the border, others producing smaller and lighter goods such as electronic equipment and cameras moved to China, Thailand, Vietnam, and other countries where wages are as little as one-fourth of the (U.S.) $2.00 per hour many Mexican workers earn.

How can Mexico counter this trend? Education of a more highly skilled workforce is one solution; the training of more Mexican managers is another. Bureaucratic red tape and corruption need addressing. And the concentration of maquiladoras along the U.S. border needs to be countered through a program of factory dispersal to other parts of Mexico. Further, the country's infrastructure must be upgraded to international standards, particularly its telecommunications, transportation, and electric-power networks. One megaproject that has Mexican **11** planners occupied is the so-called **dry canal** across the 150-mile (250-km) Tehuantepec isthmus to move containerized goods between the Pacific Ocean port of Salina Cruz and the Mexican Gulf port of Coatzacoalcos, a potential rival for the aging Panama Canal (Fig. 4-7).

Uneven Regional Development

Despite such efforts, an economic divide is deepening between the southern and northern halves of Mexico. South of the capital, technological and social development lags far behind the rest of the country as traditional productive activities (low-output farming, mining, logging) continue to dominate. North of Mexico City, growth corridors are stirring that increasingly exhibit the landscape of the global economy in the form of new manufacturing facilities, technology training academies, and even the beginnings of a "Silicon Valley of Mexico" on the edge of Guadalajara.

Burgeoning Monterrey—150 miles (250 km) inside Mexico, yet close enough to the Texas border to have benefited from all the recent develop-

in the northern city of Monterrey in 1903. More recently, however, Mexican manufacturing became synonymous with NAFTA when hundreds **10** of factories called **maquiladoras** opened in northernmost Mexico, just on the Mexican side of the U.S. border. Maquiladoras are factories that assemble imported, duty-free components and raw materials into finished products, most of which are then re-exported back to the United States market. Foreign (mainly U.S. and Canadian) factory owners profit from Mexico's lower wages, while Mexico benefits through job creation, foreign investment, and the transfer of new technologies into the country. On the negative side, factory workers get few fringe benefits, work long hours and long work

weeks, and can only afford to live in the most basic shacks and slum dwellings that surround the towns where the factories are located.

Still, the number of maquiladoras grew rapidly, from about 1800 plants when NAFTA was initiated to some 4000 with over 1.2 million employees just after the turn of this century, accounting for nearly one-third of Mexico's industrial jobs and 45 percent of its total exports (Fig. 4-11). After 2000, however, a disturbing trend appeared: hundreds of American and other foreign corporations that had moved their factories to Mexico because of the cheaper labor there found even greater advantage in relocating once more: to East and Southeast Asia. As Mexican wages rose, however slowly, and as Mexico's peso

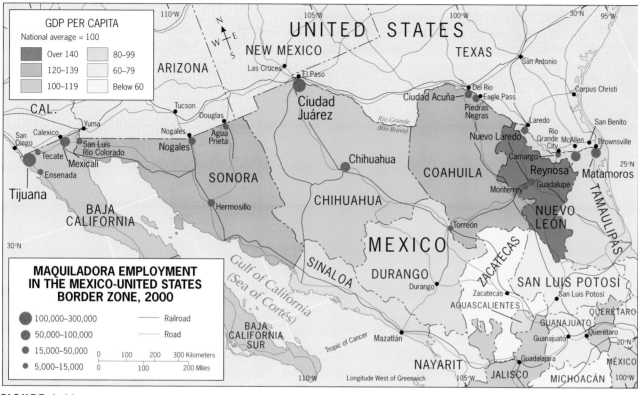

GDP PER CAPITA
National average = 100

Over 140	80–99
120–139	60–79
100–119	Below 60

MAQUILADORA EMPLOYMENT
IN THE MEXICO-UNITED STATES
BORDER ZONE, 2000

- 100,000–300,000
- 50,000–100,000
- 15,000–50,000
- 5,000–15,000

Railroad
Road

FIGURE 4-11

ment trends—is frequently singled out as a model of what a future Mexican growth center should be. Among this city's assets are a highly educated and well-paid labor force, a stable international business community, and a thriving high-technology complex of ultramodern industrial facilities that has attracted blue-chip multinational companies. In addition, a new expressway link to the Rio Grande is helping to forge an international growth corridor between Monterrey and Dallas-Fort Worth, which is becoming a primary axis of cross-border trade with Texas.

NAFTA and Continuing Challenges

The launching of the North American Free Trade zone in 1994 was supposed to mark an economic turning point for Mexico. Instead, NAFTA's early years were plagued by a series of crises that included the Chiapas rebellion, monetary devaluation, economic recession, crime waves associated with uncontrolled drug trafficking, and political scandals and transitions. Still, according to a report in *The Economist*, in 1990 United States imports and exports from Canada and Mexico combined accounted for about one-quarter of all external trade; today, this has risen to one-third (remember that total trade also grew significantly over this period, making NAFTA's success in this regard quite remarkable).

While economists laud the achievements of NAFTA, however, others, including social and economic geographers, see a different picture. Mexico's overall economy may have benefited in the early days of NAFTA, especially when expanding trade with North America helped it overcome the financial crisis of 1994, but since then growth in Mexico has been very slow, averaging about 1 percent per year (compare this to China's more than 7 percent and South Korea's more than 4 percent). Despite predictions that NAFTA would alleviate poverty in Mexico, consider this: real wages have been dropping there every year over the past ten, and income disparities between the U.S. and Mexico have risen by more than 10 percent during the same period.

Meanwhile, unemployment in the once-booming maquiladora cities is growing as jobs are being lost when corporations move their factories to East Asia. Visit the Mexican "twins" on the opposite side of American cities on the Rio Bravo (Grande), and you will be reminded of conditions in the harsh, concrete-jungle manufacturing complexes on China's Pacific Rim—except that workers in Mexico, unable to pay the rent for decent apartments, huddle in even more squalid shacks in vast slums that encircle the factory clusters. Unemployment, crime (especially against women), insecurity, and poverty cloud the NAFTA picture in Mexico's maquiladora cities, which have grown far beyond capacity and cannot cope with the resulting needs.

Notions that the string of maquiladora0 cities with their job opportunities would stem the tide of illegal migration from Mexico across the U.S. border faded quickly. In 2004 U.S. President Bush proposed a program that would begin to substitute a form of legalization for the failed interdiction effort (as estimated 10 million Mexican citizens are in the U.S. today). What would really help stem this tide, however, would be rapid economic growth of the Asian Pacific Rim variety in Mexico itself. Some positive steps have indeed been taken—democracy and open government have progressed, for example—but Mexico, as we noted earlier, needs other advances ranging from fairer and more comprehensive tax collection (which would be a byproduct of a national attack on corruption) to better education. In the first decade of this new century Mexico remains at a crossroads, the NAFTA promise fading and the alternatives beckoning, but as yet beyond reach.

THE CENTRAL AMERICAN REPUBLICS

Crowded onto the narrow segment of the Middle American land bridge between Mexico and the South American continent are the seven countries of Central America (Fig. 4-12). Territorially, they are all quite small; their population sizes range from Guatemala's 12.8 million down to Belize's

295,000. Physiographically, the land bridge here consists of a highland belt flanked by coastal lowlands on both the Caribbean and Pacific sides (Fig. 4-2). These highlands are studded with volcanoes, and local areas of fertile volcanic soils are scattered throughout them. From earliest times, the region's inhabitants have been concentrated in this upland zone, where tropical temperatures are moderated by elevation and rainfall is sufficient to support a variety of crops.

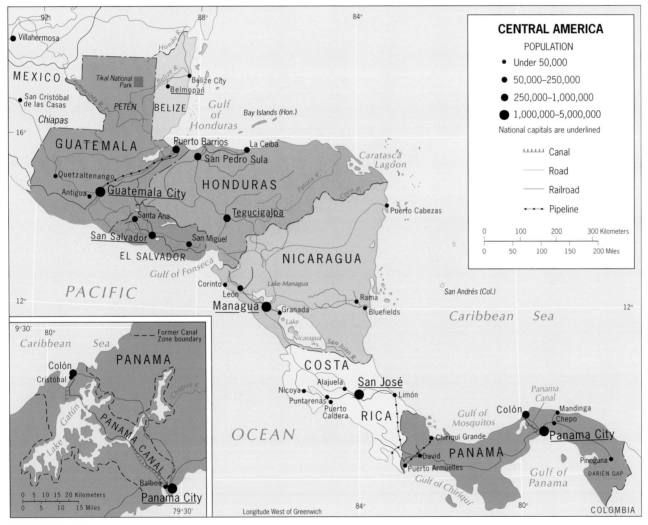

FIGURE 4-12

Altitudinal Zonation of Environments

Continental Middle America and the western margin of South America are areas of high relief and strong environmental contrasts. Even though settlers have always favored temperate intermontane basins and valleys, people also cluster in hot tropical lowlands as well as high plateaus just below the snow line in South America's Andes Mountains. In each of these zones, distinct local climates, soils, vegetation, crops, domestic animals, and modes of life prevail. Such **12** **altitudinal zones** (diagrammed in Fig. 4-13) are known by specific names as if they were regions with distinguishing properties—as, in reality, they are.

The lowest of these vertical zones, from sea level to 2500 feet (about **13** 750 m), is known as the *tierra caliente*, the "hot land" of the coastal plains and low-lying interior basins where tropical agriculture predominates. Above this zone lie the tropical highlands containing Middle and South America's largest population **14** clusters, the *tierra templada* of temperate land reaching up to about 6000 feet (1800 m). Temperatures

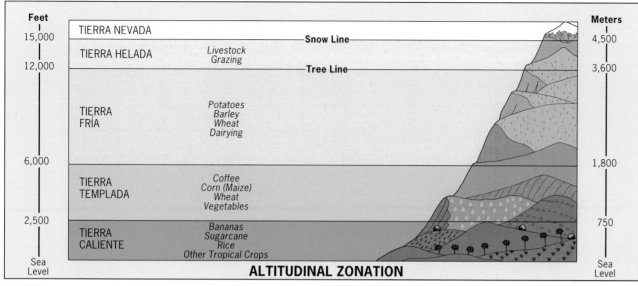

FIGURE 4-13

here are cooler; prominent among the commercial crops is coffee, while corn (maize) and wheat are the staple grains. Still higher, from about 6000 feet **15** to nearly 12,000 feet (3600 m), is the ***tierra fría***, the cold country of the higher Andes where hardy crops such as potatoes and barley are mainstays. Above the tree line, which marks the upper **16** limit of the *tierra fría*, lies the ***tierra helada***; this fourth altitudinal zone, extending from about 12,000 to 15,000 feet (3600 to 4500 m), is so cold and barren that it can support only the grazing of sheep and other hardy livestock. The highest zone **17** of all is the ***tierra nevada***, a zone of permanent snow and ice associated with the loftiest Andean peaks. As we will see, the varied cultural geography of Middle and western South America is closely related to these diverse environments.

Population Patterns

Even at its comparatively small scale, Figure G-9 indicates that Central America's population tends to concentrate in the uplands of the *tierra templada* and that population densities are generally greater to-

ward the Pacific than toward the Caribbean side. The most significant exception is El Salvador, whose political boundaries confine its people mostly to its tropical *tierra caliente*, tempered here by the somewhat cooler Pacific offshore. On the opposite side of the isthmus, Belize has the typically sparse population of the hot, wet Caribbean coastlands and their infertile soils. Panama is the only other exception to the rule, but for different reasons. Economic development has focused on the Panama Canal and the coasts, although new settlement is moving onto the mountain slopes of the Pacific side.

Central America, we noted earlier, actually begins within Mexico, in Chiapas and in Yucatán, and the region's republics face many of the same problems as less-developed parts of Mexico. Population pressure is one of these problems: a population explosion began in the mid-twentieth century, increasing the region's human inhabitants from 9 million to nearly 40 million by the start of the twenty-first. Unlike Mexico, Central America's population growth is not yet slowing down except in Costa Rica and Panama, and population geographers talk of a demographic catastrophe in the making.

Emergence from a Turbulent Era

Devastating inequities, repressive governments, external interference, and the frequent unleashing of armed forces have destabilized Central America for much of its modern history. The roots of these upheavals are old and deep, and today the region continues its struggle to emerge from a period of turmoil that lasted through the 1980s into the mid-1990s.

Central America is not a large region, but because of its physiography it contains many isolated, comparatively inaccessible locales. Conflicts between Amerindian population clusters and mestizo groups are endemic to the region, and contrasts between the privileged and the poor are especially harsh. Dictatorial rule by local elites followed authoritarian rule by Spanish colonizers, and the latest episode of violent confrontation was simply another manifestation of this persistent polarization.

Today there are signs of progress but, as Figure G-10 and Table G-1 underscore, this region as a whole remains one of the world's poorest, a legacy of disadvantage that typifies the global periphery. Central American countries produce little the world needs (except Panama, whose Canal income gives it the region's highest GNP). Its farmers face subsidized competition on world markets. There is little or no money to maintain, let alone improve, infrastructure. Intraregional trade is minimal; the Pan American Highway is in disrepair and does not even reach Panama's eastern border. The turbulence of decades past may be in abeyance, but its causes remain.

The Seven Republics

Guatemala, the westernmost of Central America's republics, has more land neighbors than any other.

Straight-line boundaries across the tropical forest mark much of the border with Mexico, creating the box-like region of Petén between Chiapas State on the west and Belize on the east; also to the east lie Honduras and El Salvador (Fig. 4-12). This heart of the ancient Maya Empire, which remains strongly infused by Amerindian culture and tradition, has just a small window on the Caribbean but a longer Pacific coastline. Guatemala was still part of Mexico when the Mexicans threw off the Spanish yoke, and although independent from Spain after 1821, it did not become a separate republic until 1838. Mestizos, not the Amerindian majority, secured the country's independence.

Most populous of the seven republics with 12.8 million inhabitants (mestizos are in the majority with 56 percent, Amerindians 44 percent), Guatemala has seen much conflict. Repressive regimes made deals with U.S. and other foreign economic interests that stimulated development,

but at a high social cost. Over the past half-century, military regimes have dominated political life. The deepening split between the wretchedly poor Amerindians and the better-off mestizos, who here call themselves *ladinos*, generated a civil war that started in 1960 and has since claimed more than 200,000 lives as well as 50,000 "disappearances." An overwhelming number of the victims have been of Mayan descent; the mestizos control the government, army, and land-tenure system.

The tragedy of Guatemala is that its economic geography has considerable potential but has long been shackled by the unending internal conflicts that have kept the income of 80 percent of the population below the poverty line. The country's mineral wealth includes nickel in the highlands and oil in the lower-lying north. Agriculturally, soils are fertile and moisture is ample over highland areas large enough to produce a wide range of crops including excellent coffee.

Belize, strictly speaking, is not a Central American republic in the same tradition as the other six. Until 1981, this country, a wedge of land between northern Guatemala, Mexico's Yucatán Peninsula, and the Caribbean, was a dependency of the United Kingdom known as British Honduras. Slightly larger than Massachusetts and with a minuscule population of only 295,000 (many of African descent), Belize has been more reminiscent of a Caribbean island than of a continental Middle American state. Today, all that is changing as the demographic complexion of Belize is being reshaped. Thousands of residents of African descent have recently emigrated (many went to the United States) and were replaced by tens of thousands of Spanish-speaking immigrants. Most of the latter are refugees from strife in nearby Guatemala, El Salvador, and Honduras, and their proportion of the Belizean population has risen from 33 to nearly 50 percent since 1980. Within the next few years the

FROM THE FIELDNOTES

"We spent Monday and Tuesday upriver at Lamanai, a huge, still mostly overgrown Maya site deep in the forest of Belize. On Wednesday we drove from Belize City to Altun Ha, which represents a very different picture. Settled around 200 BC, Altun Ha flourished as a Classic Period center between AD 300 and 900, when it was a thriving trade and redistribution center for the Caribbean merchant canoe traffic and served as an entrepôt for the interior land trails, some of them leading all the way to Teotihuacán. Altun Ha has an area of about 2.5 square miles (6+ sq km), with the main structures, one of which is shown here, arranged around two plazas at its core. I climbed to the top of this one to get a perspective, and sat down to have my sandwich lunch, imagining what this place must have looked like as a bustling trade and ceremonial center when the Roman Empire still thrived, but a more urgent matter intruded. A five-inch tarantula emerged from a wide crack in the sun-baked platform, and I noticed it only when it was about two feet away, apparently attracted by the crumbs and a small piece of salami. A somewhat hurried departure put an end to my historical-geographical ruminations."

newcomers will be in the majority, Spanish will become the *lingua franca*, and Belize's cultural geography will exhibit an expansion of the Mainland at the expense of the Rimland.

The Belizean transformation extends to the economic sphere as well. No longer just an exporter of sugar and bananas, Belize is producing new commercial crops, and its seafood-processing and clothing industries have become major revenue earners. Also important is tourism, which annually lures more than 150,000 vacationers to the country's Mayan ruins, resorts, and newly legalized casinos; a growing speciality is ecotourism, based on the natural attractions of the country's near-pristine environment. Belize is also known as a center for *offshore banking*—a financial haven for foreign companies and individuals who want to avoid paying taxes in their home countries.

Honduras is a country on hold as it struggles to rebuild its battered infrastructure and economy. In 1998 Hurricane Mitch struck Honduras, and the consequences were catastrophic as massive floods and mudslides were unleashed across the country, killing 9200 people, demolishing more than 150,000 homes, destroying 21,000 miles of roadway and 335 bridges, and rendering 2 million homeless. Also devastated was the critical agricultural sector that employed two-thirds of Honduras's labor force, accounted for nearly a third of its gross domestic product, and earned more than 70 percent of its foreign revenues.

With 7.1 million inhabitants, about 90 percent mestizo, bedeviled Honduras still has years to go even to restore what was already the third poorest economy in the Americas (after Haiti and Nicaragua). Agriculture, livestock, forestry, and limited mining formed the mainstays of the pre-1998 economy, with the familiar Central American products—bananas, coffee, shellfish, apparel—earning most of the external income.

Honduras, in direct contrast to Guatemala, has a lengthy Caribbean coastline and a small window on the Pacific (Fig. 4-12). The country also occupies a critical place in the political geography of Central America, flanked as it is by Nicaragua, El Salvador, and Guatemala—all continuing to grapple with the aftermath of years of internal conflict and, most recently, natural disaster. The road back to economic viability is an arduous one, but once traversed will still leave four out of five Hondurans deeply mired in poverty and the country with little overall improvement in its development prospects.

El Salvador is Central America's smallest country territorially, smaller even than Belize, but with a population about 25 times as large (6.9 million) it is the most densely peopled. Again, like Belize, it is one of only two continental republics that lack coastlines on both the Caribbean and Pacific sides (Fig. 4-12). El Salvador adjoins the Pacific in a narrow coastal plain backed by a chain of volcanic mountains, behind which lies the country's heartland. Unlike neighboring Guatemala, El Salvador has a quite homogeneous population (94 percent mestizo and just 5 percent Amerindian). Yet ethnic homogeneity has not translated into social or economic equality or even opportunity. Whereas other Central American countries were called banana republics, El Salvador was a coffee republic, and the coffee was produced on the huge landholdings of a few landowners and on the backs of a subjugated peasant labor force. The military supported this system and repeatedly suppressed violent and desperate peasant uprisings.

From 1980 to 1992, El Salvador was torn by a devastating civil war that was worsened by outside arms supplies from the United States (supporting the government) and Nicaragua (aiding the Marxist rebel forces). But ever since the negotiated end to that war, efforts have been under way to prevent a recurrence because El Salvador is having difficulty overcoming its legacy of searing inequality. The civil war did have one positive result: affluent citizens who left the country and did well in the United States and elsewhere send substantial funds back home, which now provide the largest single source of foreign revenues. This has helped stimulate such industries as apparel and footwear manufacturing, as well as food processing. But a major stumbling block to revitalization of the agricultural sector has again been land reform, and El Salvador's future still hangs in the balance.

Nicaragua is best approached by reexamining the map (Fig 4-12), which underscores the country's pivotal position in the heart of Central America. The Pacific coast follows a southeasterly direction, but the Caribbean coast is oriented north-south so that Nicaragua forms a triangle of land with its lakeside capital, Managua, located in a valley on the mountainous, earthquake-prone, Pacific side (the country's core area has always been located here). The Caribbean side, where the uplands yield to a coastal plain of rainforest, savanna, and swampland, has for centuries been home to Amerindian peoples such as the Miskito, who have been remote from the focus of national life.

Until the end of the 1970s, Nicaragua was the typical Central American republic, ruled by a dictatorial government and exploited by a wealthy land-owning minority, its export agriculture dominated by huge plantations owned by foreign corporations. It was a situation ripe for insurgency, and in 1979, leftist rebels overthrew the government; but the new regime quickly produced its own excesses, resulting in civil war through most of the 1980s. That conflict ended in 1990, and more democratic governments have since been voted into office.

Nicaragua's economy has been a leading casualty of this turmoil, and for the past two decades it has ranked as continental Middle America's poorest. Hurricane Mitch struck here too, devastating the country's farms and driving tens of thousands into the impoverished towns just at a time when the agricultural sector was recovering and land reform promised a better life for some 200,000 peasant families.

Nicaragua's options are limited: none of the billions of aid dollars headed for Bosnia, Iraq, and Subsaharan Africa will be matched here. For years there has been talk of a *dry canal* transit role for this country, but other potential overland routes are more promising. Meanwhile, Nicaragua's population explosion continues (2004 total: 5.7 million), dooming hopes for a rise in standards of living.

Costa Rica underscores what was said about Middle America's endless variety and diversity because it differs significantly from its neighbors and from the norms of Central America as well.

Bordered by two volatile countries (Nicaragua to the north and Panama to the east), Costa Rica is a nation with an old democratic tradition and, in this cauldron, no standing army for the past half-century! Although the country's Hispanic imprint is similar to that found elsewhere on the Mainland, its early independence, its good fortune to lie re-

mote from regional strife, and its leisurely pace of settlement allowed Costa Rica the luxury of concentrating on its economic development. Perhaps most important, internal political stability has prevailed over much of the past 175 years.

Like its neighbors, Costa Rica is divided into environmental zones that parallel the coasts. The most densely settled is the central highland zone, lying in the cooler *tierra templada*, whose heartland is the Valle Central (Central Valley), a fertile basin that contains the country's main coffee-growing area and the leading population cluster focused on San José—the most cosmopolitan urban center between Mexico City and the primate cities of northern South America. To the east of the highlands are the hot and rainy Caribbean lowlands, a sparsely populated segment of Rimland where many plantations have been abandoned and replaced by subsistence farming. Between 1930 and 1960, the U.S.-based United Fruit Company shifted most of the country's banana plantations from this crop-disease-ridden coastal plain to Costa Rica's third zone—the plains and gentle slopes of the Pacific coastlands. This move gave the Pacific zone a major boost in economic growth, and it is now an area of diversifying and expanding commercial agriculture.

The long-term development of Costa Rica's economy has given it the region's highest standard of living, literacy rate, and life expectancy. Agriculture continues to dominate (with bananas, coffee, seafood, and tropical fruits the leading exports), and tourism is expanding steadily. Costa Rica is widely known for its superb scenery and for its efforts to protect what is left of its diverse tropical flora and **18** fauna. **Tropical deforestation** is the regionwide price of human population growth and forest exploitation, and even here as much as 70 percent of the original forest has vanished—but enough remains to attract more than a million visitors annually.

Still, the country's veneer of development cannot mask serious problems. In terms of social structure, about one-quarter of the population is trapped in an unending cycle of poverty, and the huge gap be-

FROM THE FIELDNOTES

"The Panama Canal remains an engineering marvel 90 years after it opened in August, 1914. The parallel lock chambers each are 1000 feet long and 110 feet wide, permitting vessels as large as the Queen Elizabeth II to cross the isthmus. Ships are raised by a series of locks to Gatún Lake, 85 feet above sea level. We watched as tugs helped guide the QEII into the Gatún Locks, a series of three locks leading to Gatún Lake, on the Atlantic side. A container ship behind the QEII is sailing up the dredged channel leading from the Limón Bay entrance. The lock gates are 65 feet wide and seven feet thick, and range in height from 47 to 82 feet. The motors that move them are recessed in the walls of the lock chambers. Once inside the locks, the ships are pulled by powerful locomotives called *mules* that ride on rails that ascend and descend the system. It was still early morning, and a major fire, probably a forest fire, was burning near the city of Colón, where land clearing was in progress. This was the beginning of one of the most fascinating days ever."

tween the poor and the affluent is constantly widening. With volatile neighbors and an economy that remains insufficiently diversified against risk, Costa Rica is only one step ahead of its regional partners.

Panama owes its existence to the idea of a canal connecting the Atlantic and Pacific oceans to avoid the lengthy circumnavigation of South America. In the 1880s, when Panama was still an extension of neighboring Colombia, a French company tried and failed to build such a waterway here. By the turn of the twentieth century, U.S. interest in a Panama canal rose sharply, and the United States in 1903 proposed a treaty that would permit a renewed effort at construction across Colombia's Panamanian isthmus. When the Colombian Senate refused to go along, Panamanians rebelled and the United States supported this uprising by preventing Colombian forces from intervening. The Panamanians, at the behest of the United States, declared their independence from Colombia, and the new republic immediately granted the United States rights to the Canal Zone, averaging about 10 miles (16 km) in width and just over 50 miles (80 km) in length.

Soon canal construction commenced, and this time the project succeeded as American technology and medical advances triumphed over a formidable set of obstacles. The Panama Canal (see inset map, Fig. 4-12) was opened in 1914, a symbol of U.S. power and influence in Middle America. The Canal Zone was held by the United States under a treaty that granted it "all the rights, powers, and authority" in the area "as if it were the sovereign of the territory." Such language might suggest that the United States held rights over the Canal Zone in perpetuity, but the treaty nowhere stated specifically that Panama permanently yielded its own sovereignty in that transit corridor. In the 1970s, as the canal was transferring more than 14,000 ships per year (that number is now only slightly lower, but the cargo tonnage is up significantly) and generating hundreds of millions of dollars in tolls, Panama sought to terminate U.S. control in the Canal Zone. Delicate negotiations began. In 1977,

an agreement was reached on a staged withdrawal by the United States from the territory, first from the Canal Zone and then from the Panama Canal itself (a process completed on December 31, 1999).

Panama today reflects some of the usual geographic features of the Central American republics. Its population of 3 million is more than two-thirds mestizo and also contains substantial Amerindian, white, and black minorities. Spanish is the official language, but English is also widely used. Ribbonlike and oriented east-west, Panama's topography is mountainous and hilly. Eastern Panama, especially Darien Province adjoining Colombia, is densely forested, and here is the only remaining gap in the intercontinental Pan American Highway. Most of the rural population lives in the uplands west of the canal; there, Panama produces bananas, shrimps and other seafood, sugarcane, coffee, and rice. Much of the urban population is concentrated in the vicinity of the waterway, anchored by the cities at each end of the canal.

Near the northern end of the Panama Canal lies the city of Colón, site of the Colón Free Zone, a huge trading entrepôt designed to transfer and distribute goods bound for South America. It is augmented by the Manzanillo International Terminal, an ultramodern port facility capable of transshipping more than 1000 containers a day. By 2002 China had become the third-largest user of the Canal (after the United States and Japan) and accounted for more than 20 percent of the cargo entering the Colón Free Zone.

The Panama Canal has political as well as economic implications for Panama. For many years, Panama has recognized Taiwan as an independent entity and has sponsored resolutions to get Taiwan readmitted to the United Nations (where it was ousted in favor of China in 1971). In return, Taiwan has spent hundreds of millions of dollars in Panama in the form of investment and direct aid. But now, China's growing presence in the Canal and the Colón Free Zone are causing a conundrum. Panama has awarded a contract to a Hong Kong-based (and thus Chinese) firm to operate and modernize the

ports at both ends of the Canal, resulting in investments approaching $500 million. China is lobbying Panama for a switch in recognition, using this commitment as leverage. But the Panamanians are cautious. Having just ousted the United States, they are not eager to be dominated by another giant. It is more profitable to play the "two Chinas" off against each other.

Panama also feels the effects of its location adjacent to rebel-plagued, drug-infested, and unstable Colombia. For many years Panama City, Central America's only coastal capital city, was a financial center that handled more than just the funds generated by the Canal. The link between the Colón Free Zone and Panama City also had a money-laundering dimension, and the Miami-like skyline of the capital reflected that. In the 1990s, Colombia was the Colón Free Zone's biggest customer, but when Panama began a crackdown on money laundering, its share of trade dropped significantly.

Fortunately for Panama, its border with Colombia in the Darien is short (170 mi/270 km) and densely forested, occasionally providing hideouts for rebel bands and traffickers but creating a buffer between the main theaters of rebel activity and the Panamanian core area. Colombia's campaign against drug production, as we note in Chapter 5, focuses on other parts of the country and plagues its other neighbors far more. But Colombia's chronic difficulties are having another effect: well-to-do Colombians, looking for a relatively safe place to live and invest, are moving to Panama in growing numbers. Panama welcomes them, an ironic revival of linkage to a country on which the Panamanians turned their backs a century ago.

THE CARIBBEAN BASIN

As Figure 4-1 reveals, the Caribbean Basin, Middle America's island region, consists of a broad arc of numerous islands extending from the western tip of Cuba to the southern coast of Trinidad. The four larger islands or *Greater Antilles* (Cuba, Hispaniola,

Jamaica, and Puerto Rico) are clustered in the western half of this arc. The smaller islands or *Lesser Antilles* extend to the east in a crescent-shaped zone from the Virgin Islands to Trinidad and Tobago. (Breaking this tectonic-plate-related regularity are the Bahamas and the Turks and Caicos, north of the Greater Antilles, and numerous other islands too small to appear on a map at the scale of Fig. 4-1.)

On these islands, whose combined land area constitutes only 9 percent of Middle America, lie 33 states and several other entities. (Europe's colonial flags have not totally disappeared from this region, and the U.S. flag flies over Puerto Rico.) The populations of these states and territories, however, comprise 21 percent of the entire geographic realm, making this the most densely peopled part of the Americas.

Economic and Social Patterns

The Caribbean is also one of the poorest regions of the world. Environmental, demographic, political, and economic circumstances combine to impede development in the region at almost every turn. The Caribbean area is scenically beautiful, but it is a difficult place to make a living, and poverty is the norm.

It was not always thus. Untold wealth flowed to the European colonists who invaded this region, enslaved and eventually obliterated the indigenous Amerindian (Carib) population, and brought to these islands in bondage the Africans to work on the sugar plantations that made them rich. By the time competition from elsewhere ended the near-monopoly Caribbean sugar had enjoyed on European markets, the seeds of disaster had been sowed.

As the sugar trade collapsed, millions were pushed into a life of subsistence, malnutrition, and even hunger. Population growth now created ever-greater pressure on the land, but Caribbean islanders had few emigration options in their fragmented, mountainous habitats. The American market for sugar allowed some island countries to revive their sugar exports, including the Dominican Republic and Jamaica (Cuba's exports went to the Soviet Union, but when the USSR disintegrated Cuba's sugar industry was doomed). Some agricultural diversification occurred in the Lesser Antilles where bananas, spices, and sea-island cotton replaced sugar, but farming is a risky profession in the Caribbean. Foreign markets are undependable, and producers are trapped in a disadvantageous international economic system they cannot change.

Not surprisingly, many farmers and their families simply abandon their land and leave for towns and cities, hoping to do better there. Although the Caribbean region is less urbanized, overall, than other parts of the Americas, it is far more urbanized than, say, Africa or the Pacific Realm. Nearly two-thirds of the Caribbean's population now live in cities such as Santo Domingo in the Dominican Republic, Havana in Cuba, Port-au-Prince in Haiti, and San Juan in Puerto Rico. Many of the region's cities reflect the poverty of the citizens who were driven to seek refuge there. Port-au-Prince has some of the world's worst slums, and desolate squatter settlements ring cities such as Kingston, Jamaica, and smaller towns.

Ethnicity and Advantage

The human geography of the Caribbean region still carries imprints of the cultures of Subsaharan Africa. And the legacy of European domination also lingers. The historical geography of Cuba, Hispaniola, and Puerto Rico is suffused with Hispanic culture; Haiti and Jamaica carry stronger African legacies. But the reality of this ethnic diversity is that European lineages still hold the advantage. Hispanics tend to be in the best positions in the Greater Antilles; people who have mixed European-African **19** ancestries, and who are described as **mulatto**, rank next. The largest part of this social pyramid is also the least advantaged: the Afro-Caribbean majority. In virtually all societies of the Caribbean, the minorities hold disproportionate power and exert overriding influence. In Haiti, the mulatto minority accounts for perhaps 5 percent of the population but has long held most of the power. In the adjacent Dominican Republic, the pyramid of power puts Hispanics (16 percent) at the top, the mixed sector (73 percent) in the middle, and the Afro-Caribbean minority (11 percent) at the bottom. Historic advantage has a way of perpetuating itself.

The composition of the population of the islands is further complicated by the presence of Asians from both China and India. During the nineteenth century, the emancipation of slaves and ensuing local labor shortages brought some far-reaching solutions. Some 100,000 Chinese emigrated to Cuba as indentured laborers, and Jamaica, Guadeloupe, and especially Trinidad saw nearly 250,000 South Asians arrive for similar purposes. To the African-modified forms of English and French heard in the Caribbean, therefore, can be added several Asian languages. The ethnic and cultural variety of the plural societies of Caribbean America is indeed endless.

Tourism: Promising Alternative?

Given the Caribbean region's limited economic options, does the tourist industry offer better opportunities? Opinions on this question are divided. The resort areas, scenic treasures, and historic locales of Caribbean America attract well over 20 million visitors annually, with about half of these tourists traveling on Florida-based cruise ships. Certainly, Caribbean tourism is a prospective money-maker for many islands. In Jamaica alone, this industry now accounts for about one-sixth of the gross domestic product and employs more than one-third of the labor force.

But Caribbean tourism also has serious drawbacks. The invasion of poor communities by affluent tourists contributes to rising local resentment, which is further fueled by the glaring contrasts of shiny new hotels towering over substandard housing and luxury liners gliding past poverty-stricken villages. At the same time, tourism can have the effect of debasing local culture, which often is adapted to suit the visitors' tastes at hotel-staged

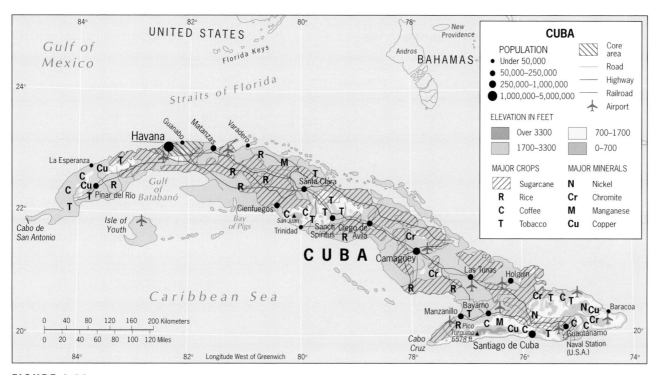

FIGURE 4-14

"culture" shows. And while tourism does generate income in the Caribbean, the intervention of island governments and multinational corporations removes opportunities from local entrepreneurs in favor of large operators and major resorts.

The Greater Antilles

The four islands of the Greater Antilles contain five political entities: Cuba, Haiti, the Dominican Republic, Jamaica, and Puerto Rico (Fig. 4-1). Haiti and the Dominican Republic share the island of Hispaniola.

Cuba the largest Caribbean island-state in terms of both territory (43,000 square miles/ 111,000 sq km) and population (11.4 million), lies only 90 miles (145 km) from the southern tip of Florida (Fig. 4-14). Havana, the now-dilapidated capital, lies almost directly across from the Florida

Keys on the northwest coast of the elongated island. Cuba was a Spanish possession until the late 1890s when, with American help in the Spanish-American War, it attained independence. Fifty years later, a U.S.-backed dictator was in control, and by the 1950s Havana had become an American playground. The island was ripe for revolution, and in 1959 Fidel Castro's insurgents gained control, thereby making Cuba a communist dictatorship and a Soviet client. Castro's rule survived the collapse of the Soviet Empire despite the loss of subsidies and sugar markets on which it had long relied.

As the map shows, sugar was Cuba's economic mainstay for many years; the plantations, once the property of rich landowners, extend all across the territory. But as this map also shows, sugarcane is losing its position as the leading Cuban foreign-exchange earner. Mills are being closed down, and

the cane fields are being cleared for other crops and for pastures. Cuba has other economic opportunities, however, especially in its highlands. There are three mountainous areas, of which the southeastern chain, the Sierra Maestra, is the highest and most extensive. These highlands create considerable environmental diversity as reflected by extensive, timber-producing tropical forests and varied soils on which crops ranging from tobacco to subtropical and tropical fruits are grown. Rice and beans are the staples, but Cuba cannot meet its needs and so must import food. The savannas of the center and west support livestock. Cuba has limited mineral reserves and no oil, but its nickel deposits are extensive and have been mined for a century.

Poverty, crumbling infrastructure, crowded slums, and unemployment mark the Cuban cultural landscape, but Cuba's regime still has support among the general population. During the Castro period, much was done to bring the Afro-Cuban population into the mainstream through education and health provision. Cubans point to Guatemala, Nicaragua, and El Salvador and ask whether those countries are better off than they are under Castro. But in the United States, the view is different: Cuba, exiles and locals agree, could be the shining star of the Caribbean, its people free, its tourist economy booming, its products flowing to American markets. "The future Ireland of the Caribbean," suggested a Cuban geographer recently. Much will have to change for that prediction to come true.

Jamaica lies across the deep Cayman Trench from southern Cuba, and a cultural gulf separates these two countries as well. Jamaica, a former British dependency, has an almost entirely Afro-Caribbean population. As a member of the British Commonwealth, Jamaica still recognizes the British

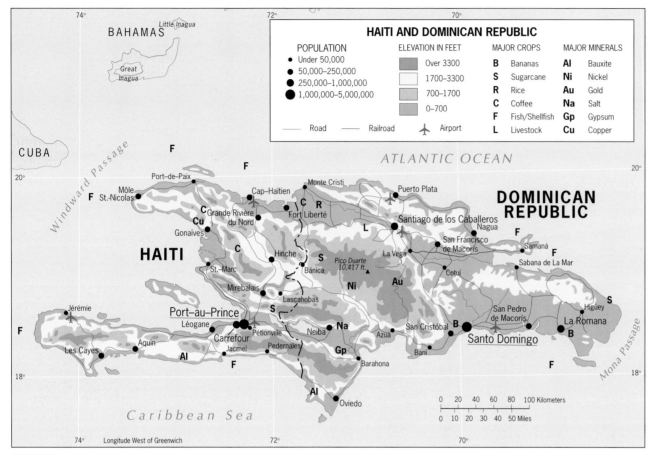

FIGURE 4-15

monarch as the chief of state, represented by a governor-general. The effective head of government in this democratic country, however, is the prime minister. English remains the official language here, and British customs still linger.

Smaller than Connecticut and with 2.7 million people, Jamaica has experienced a steadily declining GNP over the past decades despite its relatively slow population growth. Tourism has become the largest source of income, but the markets for bauxite (aluminum ore), of which Jamaica is a major exporter, have dwindled. And like other Caribbean countries, Jamaica has trouble making money from its sugar

exports. Jamaican farmers also produce crops ranging from bananas to tobacco, but the country faces the disadvantages on world markets common to those in the periphery. Meanwhile, Jamaica must import all of its oil and much of its food because the densely populated coastal flatlands suffer from overuse and shrinking harvests.

The capital, Kingston on the south coast, reflects Jamaica's economic struggle. Almost none of the hundreds of thousands of tourists who visit the country's beaches, explore its Cockpit Country of (karst) limestone towers and caverns, or populate the cruise ships calling at Montego Bay or

other points along the north coast get even a glimpse of what life is like for the ordinary Jamaican.

Haiti, the poorest state in the Western Hemisphere by virtually any measure, occupies the western part of the island of Hispaniola, directly across the Windward Passage from eastern Cuba (Fig. 4-15). Arawaks, not Caribs, formed the dominant indigenous population here, but the Spanish colonists who first took Hispaniola killed many, worked most of the survivors to death on their plantations, and left the others to die of the diseases they brought with them. French pirates established themselves in coastal coves along Hispaniola's west coast even as Spanish activity focused on the east, and just before the eighteenth century opened France formalized the colonial status of "Saint Dominique." During the following century, French colonists established vast plantations in the valleys of the northern mountains and in the central plain, laid out elaborate irrigation systems, built dams, and brought in a large number of Africans in bondage to work the fields. Prosperity made the colonists rich, but the African workers suffered terribly. They rebelled and, in a momentous victory over their European oppressors, established the independent republic of Haiti (resurrecting the original Arawak name for it) in 1804.

Strife among Haitian groups, American intervention, mismanagement, and dictatorship doomed the fortunes of the republic. In the 1990s, the United States attempted to help move Haiti toward more representative government, but the country was in economic and social collapse. By 2003, Haiti's GNP was one-fourth that of Jamaica and lower than that of many of Africa's poorest countries; foreign aid makes possible most of the country's limited public expenditures. Health conditions are dreadful: malnutrition is common, AIDS is rampant, diseases ranging from malaria to tuberculosis are rife, but hospital beds and doctors are in short supply. Conditions in and around the capital, Port-au-Prince, are among the worst in the world—600 miles (less than 1000 km) from the United States.

The **Dominican Republic** has a larger share of the island of Hispaniola than Haiti (Fig. 4-15) in terms of both territory and population. Fly along the north-south border between the two countries, and you see a crucial difference: to the west, Haiti's hills and plains are treeless and gulleyed, its soils eroded and its streams silt-laden. To the east, forests drape the countryside and streams run clear.

Indigenous Caribs inhabited this eastern part of Hispaniola, and when the European colonists arrived on the island, they were in the process of driving the Arawaks westward. Spanish colonists made this a prosperous colony, but then Mexico and Peru diverted Spanish attention from Hispaniola and the territory was ceded to France. But Hispanic culture lingered, and after the revolution in Haiti the French gave eastern Hispaniola back to Spain. After the Dominican Republic declared its independence in 1821, Haitian forces invaded it and occupied the republic until 1844, creating an historic animosity that persists today.

The mountainous Dominican Republic has a wide range of natural environments and a far stronger resource base than Haiti. Nickel, gold, and silver have long been exported along with sugar, tobacco, coffee, and cocoa, but tourism (the great opportunity lost to Haiti) is the leading industry. A long period of dictatorial rule punctuated by revolutions and U.S. military intervention ended in 1978 with the first peaceful transfer of power following a democratic election.

Political stability brought the Dominican Republic rich rewards, and during the late 1990s the economy, based on manufacturing, high-tech industries, and remittances from Dominicans abroad as well as tourism, grew at an average 7 percent per year. But in the early 2000s the economy collapsed, not only because of the downturn in the world economy but also because of bank fraud and corruption in government. Suddenly the Dominican peso collapsed, inflation skyrocketed, jobs were lost, and blackouts prevailed. As the people protested, lives were lost and the self-enriched elite blamed foreign financial institutions that were unwilling to lend the government more money. Once again the hopes of ordinary citizens were dashed by greed and corruption among those in power.

Puerto Rico is the largest U.S. domain in Middle America, the easternmost and smallest island of the Greater Antilles (Fig. 4-16). This 3400-square-mile (8800-sq km) island, with a population of 4 million, is larger than Delaware and more populous than Oregon. It fell to the United States more than a century ago during the Spanish-American War of 1898. Since the Puerto Ricans had been struggling for some time to free themselves from Spanish control, this transfer of power was in their view, only a change from one colonial power to another. As a result, the first half-century of U.S. administration was difficult, and it was

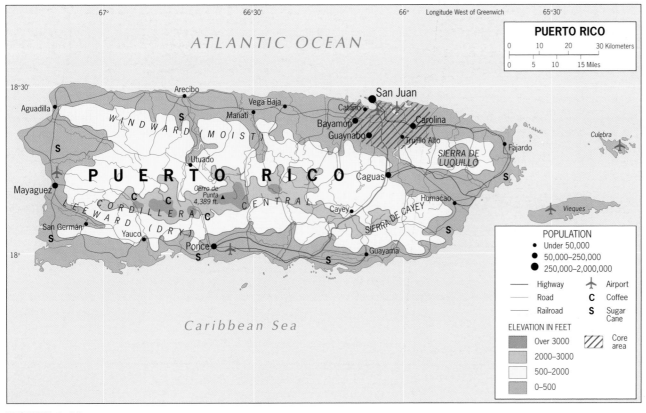

FIGURE 4-16

not until 1948 that Puerto Ricans were permitted to elect their own governor.

When the island's voters approved the creation of a Commonwealth in a 1952 referendum, Washington D.C. and San Juan, the two seats of government, entered into a complicated arrangement. Puerto Ricans are U.S. citizens but pay no federal taxes on local incomes. The Puerto Rican Federal Relations Act governs the island under the terms of its own constitution and awards it considerable autonomy. Puerto Rico also receives a large annual subsidy from Washington, totaling about U.S. $9 billion per year at the end of the 1990s.

Despite these apparent advantages in the poverty-mired Caribbean, Puerto Rico has not thrived under U.S. administration. Long dependent on a single-crop economy (sugar), the island's industrialization during the 1950s and 1960s was based on its comparatively cheap labor, tax breaks for corporations, political stability, and special access to the U.S. market. As a result, pharmaceuticals, electronic equipment, and apparel top today's list of exports, not sugar or bananas. But this industrialization failed to stem a tide of emigration that carried more than 1 million Puerto Ricans to New York City alone. The same wages that favored corporations kept many Puerto Ricans poor or unemployed. Behind the impressive waterfront skyline of San Juan with its high-rise hotels and tourist attractions lies a landscape of economic malaise. Today, approximately 60 percent of all Puerto Ricans continue to live below the poverty line.

Not surprisingly, Puerto Ricans have been vocal in demanding political change. Nonetheless, successive referendums during the 1990s resulted in retention of the status quo—continuation of Commonwealth status rather than either Statehood or independence. The issue will no doubt continue to pose a formidable challenge to American statecraft in the years ahead.

The Lesser Antilles

As Figure 4-1 shows, the Greater Antilles are flanked by two clusters of islands: the extensive Bahamas-Turks/Caicos archipelago to the north and the Lesser Antilles to the east and south. The Lesser Antilles are grouped geographically into the Leeward Islands and the Windward Islands, a climatologically incorrect reference to the prevailing airflows in this tropical area. The Leeward Islands extend from the U.S. Virgin Islands to the French dependencies of Guadeloupe and Martinique, and the Windward Islands from St. Lucia to the Netherlands Antilles off the Venezuelan coast (Fig. 4-1). It should be noted that these countries and territories share the environmental risks of this region: earthquakes, volcanic eruptions, and hurricanes; that they face to varying degrees similar economic problems in the form of limited domestic resources, overpopulation, soil deterioration, land fragmentation, and market limitations; that tourism has become the leading industry for many; that political status ranges from complete sovereignty to continuing dependency; and that cultural diversity is strong not only between but also within islands. In economic terms, the GNP figures given (where available) in Table G-1 may look encouraging, but these figures conceal the social reality in virtually every entity: the gap between the fortunate few who are well-off and the great majority who are poor remains enormous.

Middle America is a physically, culturally, and economically fragmented and diverse geographic realm that defies generalization. Given its strong Amerindian presence, its North American infusions, and its lingering Western European traditions, this certainly is not "Latin" America.

DISCOVERY TOOLS
www.wiley.com/college/deblij

The *Concepts and Regions* Website, featuring GeoDiscoveries, offers many additional resources to enhance your understanding and experience of this chapter. Be sure to explore the following:

GeoDiscoveries Video Clips

A Poor Colonia of Cuidad Juárez
A Rich Neighborhood of Cuidad Juárez
Remotely Controlled
Crossing the Bar
Maquiladoras
Inside a Maquiladora
The Future of Maquiladoras

GeoDiscoveries Interactivities

Overview of the Middle American Realm

Photo Galleries

Haiti/Dominican Republic Border
Point Lisas, Trinidad
Acapulco, Mexico
Panama Canal
Altun Ha, Belize
Belize
Panama Canal, Panama

Virtual Field Trip

Crossroads of the Western Hemisphere:
The Panama Canal

More To Explore

Systematic Essay: Historical Geography
Issues in Geography: Tropical Deforestation
Major Cities: Mexico City

Expanded Regional Coverage

Mexico, Central America, The Greater Antilles;
The Lesser Antilles

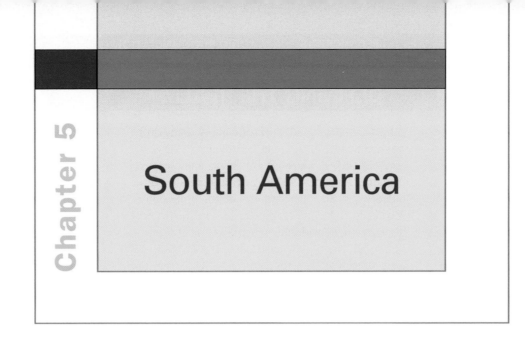

South America

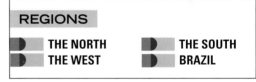

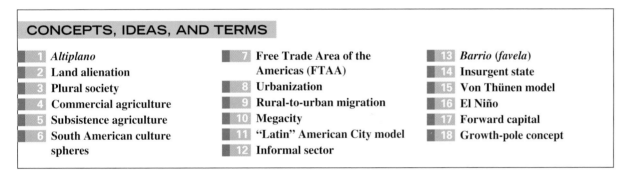

CONCEPTS, IDEAS, AND TERMS

1 *Altiplano*
2 Land alienation
3 Plural society
4 Commercial agriculture
5 Subsistence agriculture
6 South American culture spheres

7 Free Trade Area of the Americas (FTAA)
8 Urbanization
9 Rural-to-urban migration
10 Megacity
11 "Latin" American City model
12 Informal sector

13 *Barrio (favela)*
14 Insurgent state
15 Von Thünen model
16 El Niño
17 Forward capital
18 Growth-pole concept

𝒪F ALL THE continents, South America has the most familiar shape—a giant triangle connected by mainland Middle America's tenuous land bridge to its neighbor in the north. South America also lies not only south but mostly east of its northern counterpart. Lima, the capital of Peru—one of the continent's westernmost cities—lies farther east than Miami, Florida. Thus South America juts out much more prominently into the Atlantic Ocean toward Southern Europe and Africa than does North America. But lying so far eastward means that South America's western flank faces a much wider Pacific Ocean, with the distance from Peru to Australia nearly twice that from California to Japan.

As if to reaffirm South America's northward and eastward orientation, the western margins of the continent are rimmed by one of the world's longest and highest mountain ranges, the Andes, a gigantic wall that extends unbroken from Tierra del Fuego near the continent's southern tip in Chile to northeastern Venezuela in the far north (Fig. 5-1). The other major physiographic feature of South America dominates its central north—the Amazon Basin; this vast humid-tropical amphitheater is drained by the mighty Amazon, which is fed by several major tributaries. Much of the remainder of the continent can be classified as plateau, with the most important components being the Brazilian Highlands that cover most of Brazil southeast of the Amazon Basin, the Guiana Highlands located north of the lower Amazon Basin, and the cold Patagonian Plateau that blankets the southern third of Argentina. Figure 5-1 also reveals two other noteworthy river basins beyond Amazonia: the Paraná-Paraguay Basin of south-central South America, and the Orinoco Basin in the far north that drains interior Colombia and Venezuela.

DEFINING THE REALM

During most of the twentieth century, South America was a realm of political turmoil and dictatorial regimes, strong and damaging regional disparities, poor internal connections and limited international circulation, and development inertia. Toward the end of the century, however, things appeared to be improving. The major countries, previously accustomed to going their separate ways, saw the benefits of forging closer multinational ties. New transport routes crossed international borders and opened settlement frontiers. Democratic governments replaced authoritarian regimes. Economies were growing and an era of stability and progress seemed to have arrived.

But in this first decade of the twenty-first century, South America's promise is again in doubt. Corruption and mismanagement, endemic failures in this part of the world, damaged economies from Argentina, whose economy imploded, to Ecuador, where the political framework faltered. Instability afflicted poverty-stricken Bolivia, where an elected government was ousted by violent protests, and oil-rich Venezuela, where deathly turmoil accompanied a drive to recall the president. A costly drug war continued in Colombia, confidence in the government was at a low ebb in Peru, and isolated Paraguay remained South America's version of a failed state. So what remains of the hopes of a decade ago? A democratic election in Brazil, representing a considerable ideological shift from right to left, was a defining moment for this realm's giant state. The economic success of Chile can be cited as an example for the entire continent. And the stability of Uruguay in a turbulent area (Argentina is its neighbor) is a beacon of hope.

South America thus remains a realm of boom-bust cycles that erode confidence and deter long-term investment. Infrastructures remain weak, inefficiency rampant, and corruption endemic. And the gulf between rich and poor is widening, not narrowing: overall, the richest 20 percent of the population control 70 percent of South America's wealth while the poorest one-fifth own only 2 percent (in some individual countries, the gap is even wider). By some measures this disparity is greater in South America than in any other world realm.

The prospects of this continent, therefore, remain uncertain, but its needs, as we will see, are clear.

THE HUMAN SEQUENCE

Although modern South America's largest populations are situated in the east and north, during the height of the Inca Empire the Andes Mountains contained the most densely peopled and best organized state on the continent. Although the origins of Inca civilization are still shrouded in mystery, it has become generally accepted that the Incas were descendants of ancient peoples who came to South America via the Middle American land bridge. Thus, for thousands of years before the Europeans arrived in the sixteenth century, indigenous Amerindian societies had been developing in South America.

The Inca Empire

About one thousand years ago, a number of regional cultures thrived in Andean valleys and basins and at places along the Pacific coast. By AD 1300, the Incas

MAJOR GEOGRAPHIC QUALITIES OF SOUTH AMERICA

1. South America's physiography is dominated by the Andes Mountains in the west and the Amazon Basin in the central north. Much of the remainder is plateau country.

2. Half of the realm's area and half of its population are concentrated in one country—Brazil.

3. South America's population remains concentrated along the continent's periphery. Most of the interior is sparsely peopled, but sections of it are now undergoing significant development.

4. Interconnections among the states of the realm are improving rapidly. Economic integration has become a major force, particularly in southern South America.

5. Regional economic contrasts and disparities, both in the realm as a whole and within individual countries, are strong.

6. Cultural pluralism exists in almost all of the realm's countries and is often expressed regionally.

7. Rapid urban growth continues to mark much of the South American realm, and the urbanization level overall is today on a par with the levels in the United States and Europe.

FIGURE 5-1

had established themselves in the intermontane basin of Cuzco (Fig. 5-2). With their hearth consolidated, they were now ready to begin forging the greatest pre-European empire in the Americas by steadily conquering and extending their authority over the peoples of coastal Peru and other Andean basins. When the Inca civilization is compared to that of ancient Mesopotamia, Egypt, and the Mexican Aztec Empire, it quickly becomes clear that this civilization was an unusual achievement. Everywhere else, rivers and waterways provided avenues for the circulation of goods and ideas. Here, however, an empire was forged from a **1** series of elongated basins (called *altiplanos*) in the high Andes, created when mountain valleys between parallel and converging ranges filled with erosional materials from surrounding uplands. These *altiplanos* are often separated by some of the world's most rugged terrain, with high snowcapped mountains alternating with precipitous canyons.

More impressive than the Incas' military victories was their subsequent capacity to integrate the peoples and regions of the Andean domain into a stable and efficiently functioning state. The Incas, however, were expert road and bridge builders, colonizers, and administrators, and in an incredibly short time they unified their new territories that stretched from Colombia southward to central Chile (brown zone, Fig. 5-2).

The Incas themselves were always in a minority in this huge state, and their position became one of a ruling elite in a rigidly class-structured society. A bureaucracy of Inca administrators strictly controlled the life of the empire's subjects, and the state was so highly centralized that a takeover at the top was enough to gain power over the entire empire—as the Spaniards quickly proved in the 1530s. The Inca Empire disintegrated abruptly under the impact of the Spanish invaders, but it left behind spectacular structures such as those at Peru's Machu Picchu. It also bequeathed a legacy of social values that have remained a part of Amerindian life in the Andes to this day and still contribute to fundamental divisions between the

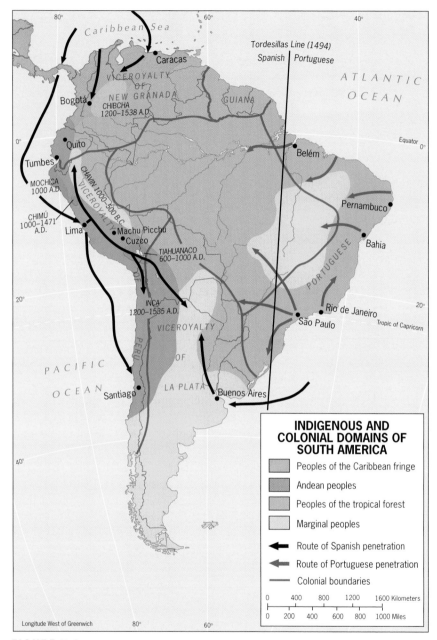

FIGURE 5-2

Hispanic and Amerindian population in this part of South America.

The Iberian Invaders

In South America as in Middle America, the location of indigenous peoples largely determined the direction of the thrusts of European invasion. The Incas, like Mexico's Maya and Aztec peoples, had accumulated gold and silver at their headquarters, possessed productive farmlands, and constituted a ready labor force. Not long after the defeat of the Aztecs in 1521, Francisco Pizarro sailed southward along the continent's northwestern coast, learned of the existence of the Inca Empire, and withdrew to Spain to organize its overthrow. He returned to the Peruvian coast in 1531 with 183 men and two dozen horses, and the events that followed are well known. In 1533, his party rode victorious into Cuzco.

At first, the Spaniards kept the Incan imperial structure intact by permitting the crowning of an emperor who was under their control. But soon the breakdown of the old order began. The new order that gradually emerged in western South America placed the indigenous peoples in serfdom to the Spaniards. Great haciendas were formed by **2** **land alienation** (the takeover of former Amerindian lands), taxes were instituted, and a forced-labor system was introduced to maximize the profits of exploitation.

Lima, the west coast headquarters of the Spanish conquerors, soon became one of the richest cities in the world, its wealth based on the exploitation of vast Andean silver deposits. The city also served as the capital of the viceroyalty of Peru, as the Spanish authorities quickly integrated the new possession into their colonial empire (Fig. 5-2). Subsequently, when Colombia and Venezuela came under Spanish control and, later, when Spanish settlement expanded in what is now Argentina and Uruguay, two additional viceroyalties were added to the map: New Granada and La Plata.

Meanwhile, another vanguard of the Iberian invasion was penetrating the east-central part of the continent, the coastlands of present-day Brazil. This area had become a Portuguese sphere of influence because Spain and Portugal had signed a treaty in 1494 to recognize a north-south line 370 leagues west of the Cape Verde Islands as the boundary between their New World spheres of influence. This border ran approximately along the meridian of 50°W longitude, thereby cutting off a sizeable triangle of eastern South America for Portugal's exploitation (Fig. 5-2). But a brief look at the political map of South America (Fig. 5-3) shows that this treaty did not limit Portuguese colonial territory to the east of the 50th meridian. Instead, Brazil's boundaries were bent far inland to include almost the entire Amazon Basin, and the country came to be only slightly smaller in territorial size than all the other South American countries combined. This westward thrust was the work of many Brazilian elements, particularly the *Paulistas*, the settlers of São Paulo who needed Amerindian slave labor to run their plantations.

The Africans

As Figure 5-2 shows, the Spaniards initially got very much the better of the territorial partitioning of South America—not just in land quality but also in the size of the aboriginal labor force. When the Portuguese began to develop their territory, they turned to the same lucrative activity that their Spanish rivals had pursued in the Caribbean—the plantation cultivation of sugar for the European market. And they, too, found their labor force in the same source region, as millions of Africans were brought in slavery to the tropical Brazilian coast north of Rio de Janeiro (see map, p. 203). Not surprisingly, Brazil now has South America's largest black population, which is still heavily concentrated in the country's poverty-stricken northeastern States. With blacks of "pure" and mixed ancestry today accounting for 44 percent of Brazil's population of 178 million, the Africans decidedly constitute the third major immigration of foreign peoples into South America.

FIGURE 5-3

Longstanding Isolation

Despite their adjacent location on the same continent, their common language and cultural heritage, and their shared national problems, the countries that arose out of South America's Spanish viceroyalties (as well as Brazil) until quite recently existed in a considerable degree of isolation from one another. Distance and physiographic barriers reinforced this separation, and the realm's major population agglomerations still adjoin the coast, mainly the eastern and northern coasts (Fig. G-9). The viceroyalties existed primarily to extract riches and fill Spanish coffers. In Iberia there was little interest in developing the American lands for their own sake. Only after those who had made Spanish and Portuguese America their home and who had a stake there rebelled against Iberian authority did things begin to change, and then very slowly. South America was saddled with the values, economic outlook, and social attitudes of eighteenth-century Iberia—not the best tradition from which to begin the task of forging modern nation-states.

Independence

Certain isolating factors had their effect even during the wars for independence. Spanish military strength was always concentrated at Lima, and those territories that lay farthest from their center of power—Argentina and Chile—were the first to establish their independence from Spain (in 1816 and 1818, respectively). In the north Simón Bolívar led the burgeoning independence movement, and in 1824 two decisive military defeats there spelled the end of Spanish power in South America.

This joint struggle, however, did not produce unity because no fewer than nine countries emerged from the three former viceroyalties. It is not difficult to understand why this fragmentation took place. With the Andes intervening between Argentina and Chile and the Atacama Desert between Chile and Peru, overland distances seem even greater than they really were, and these obstacles to contact proved quite effective. Hence, from their outset the new countries of South America began to grow apart, and friction and even wars were frequent. Only within the past two decades have the countries of this realm finally begun to recognize the mutual advantages of increasing cooperation and to make lasting efforts to steer their relationships in this direction.

CULTURAL FRAGMENTATION

When we speak of the "interaction" of South American countries, it is important to keep in mind just who does the interacting. The fragmentation of colo-

nial South America into ten individual republics, and the subsequent postures of each of these states, was the work of a small minority that constituted the landholding, upper-class elite. Thus in every country a vast majority—be they Amerindians in Peru or people of African descent in Brazil—could only watch as their European masters struggled with each other for supremacy.

3 South America, then, is a continent of **plural societies**, where Amerindians of different cultures, Europeans from Iberia and elsewhere, blacks from western tropical Africa, and Asians from India, Japan, and Indonesia cluster in adjacent areas but do not mix. The result is a cultural kaleidoscope of almost endless variety, whose internal divisions are also reflected in the realm's economic landscape. This is readily visible in the map of South America's dominant livelihood, agriculture **4** (Fig. 5-4). Here **commercial** (for-profit) and **5** **subsistence** (minimum-life-sustaining) farming exist side by side to a greater degree than anywhere else in the world, where one usually dominates the other. The geography of commercial agricultural systems (map categories 1, 2, 3, 5, 6, and 9) is heavily tied to the distribution of landholders of European background, while subsistence farming (categories 4, 7, and 8) is overwhelmingly associated with the spatial patterns of indigenous peoples as well as populations of African and Asian descent.

Certainly, calling this complex human spatial mosaic "Latin" America (as is so often the case) is not very useful. Is there a more meaningful approach to a regional generalization that would better represent and differentiate the continent's cultural and economic spheres? John Augelli, who also developed the Mainland-Rimland concept for Middle America, made such an attempt. His map (Fig. 5-5 shows **6** that five **South American culture spheres**—internal cultural regions—blanket the realm.

- **Tropical-plantation sphere:** Reminiscent of Middle America's Rimland, consists of five Atlantic- or Caribbean-facing coastal strips in northern South America. Location and tropical environmental conditions favored plantation

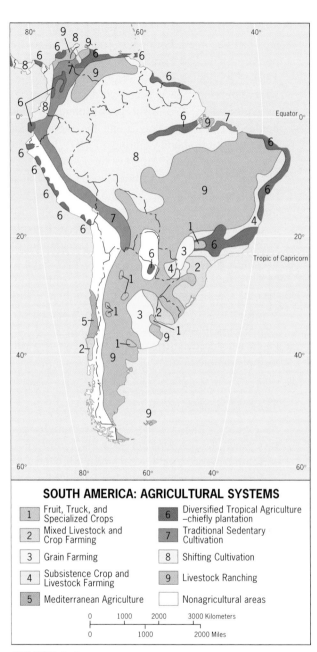

SOUTH AMERICA: AGRICULTURAL SYSTEMS

1	Fruit, Truck, and Specialized Crops	6	Diversified Tropical Agriculture –chiefly plantation
2	Mixed Livestock and Crop Farming	7	Traditional Sedentary Cultivation
3	Grain Farming	8	Shifting Cultivation
4	Subsistence Crop and Livestock Farming	9	Livestock Ranching
5	Mediterranean Agriculture		Nonagricultural areas

0   1000   2000   3000 Kilometers
0   1000   2000 Miles

FIGURE 5-4

crops, especially sugar. Millions of African slave laborers brought here and still strongly influence local cultures. Failure of plantation economy reduced people to the poverty and subsistence that now dominate these five areas.

- **European-commercial sphere:** Covers most of southern, mid-latitude South America, including the core areas of Argentina, Brazil, Chile, and Uruguay. Dominated by populations of European descent with a strong Hispanic cultural imprint. Productive commercial agriculture is the major reason why this region is economically more advanced than the rest of the continent.

- **Amerind-subsistence sphere:** Elongated zone lying astride the central Andes from southern Colombia to northern Chile/northwestern Argentina. The feudal socioeconomic structure established by Spanish conquerors still prevails, with the Amerindian population forming a huge, landless peonage. Much of the population lives in high-altitude environments that are marginal for farming. The region includes some of the realm's poorest areas.

- **Mestizo-transitional sphere:** Surrounds Amerind-subsistence region and covers much of central and northern South America. Zone of mixture between Europeans and Amerindians—or Africans around coastal strips of the tropical-plantation culture sphere. Transitional economy with respect to extremes of European-commercial and Amerind-subsistence regions.

- **Undifferentiated sphere:** The remaining, least accessible areas of the realm that lie in and around the Amazon Basin and in southernmost Chile. Sparsely populated and exhibit only very limited economic development. Augelli's map (Fig. 5-5) depicts this region larger than it is today as penetration of the Amazon proceeds.

SOUTH AMERICA: CULTURE SPHERES

- Tropical-plantation
- European-commercial
- Amerind-subsistence
- Mestizo-transitional
- Undifferentiated

0 400 800 1200 Kilometers
0 200 400 600 Miles

FIGURE 5-5

ECONOMIC INTEGRATION

As noted above, the separatism that has so long characterized international relations is giving way as South American countries discover the benefits of forging new partnerships with one another. With mutually advantageous trade the catalyst, a new continentwide spirit of cooperation is blossoming at every level. Periodic flareups of boundary disputes now rarely escalate into open conflict. Cross-border rail, road, and pipeline projects, stalled for years, are multiplying steadily. In southern South America, five formerly contentious nations are developing the *Hidrovia*, a system of river locks that is opening most of the Paraná-Paraguay Basin to barge transport. Investments today flow freely from one country to another, particularly in the agricultural sector.

Recognizing that free trade can solve many of the realm's economic-geographic problems, governments are pursuing several avenues of economic supranationalism. In 2004, South America's republics were affiliating with the following major trading blocs:

Mercosur Launched in 1995, this Southern Cone Common Market established a free-trade zone and customs union linking Brazil, Argentina, Uruguay, and Paraguay.

Andean Community Formed as the Andean Pact in 1969 but restarted in 1995 as a customs union with common tariffs for imports, its members are Venezuela, Colombia, Peru, Ecuador, and Bolivia.

Group of Three (G-3) Launched in 1995, this free-trade agreement involves Mexico, Venezuela, and Colombia, and aims to eliminate all internal tariffs by 2005.

North American Free Trade Agreement (NAFTA) Launched by the United States, Canada, and Mexico in 1994, NAFTA is seeking to expand into South America to include Chile. Eventually its planners 7 hope to create a **Free Trade Area of the Americas (FTAA)**, encompassing all three geographic realms. In the recent political and economic climate, however, enthusiasm for this project has waned.

URBANIZATION

As in most other less advantaged realms, South Americans are leaving the land and moving to the cities. 8 This **urbanization** process intensified throughout the twentieth century, and it persists so strongly that South America's urban population percentage (79 percent, or 289 mil-

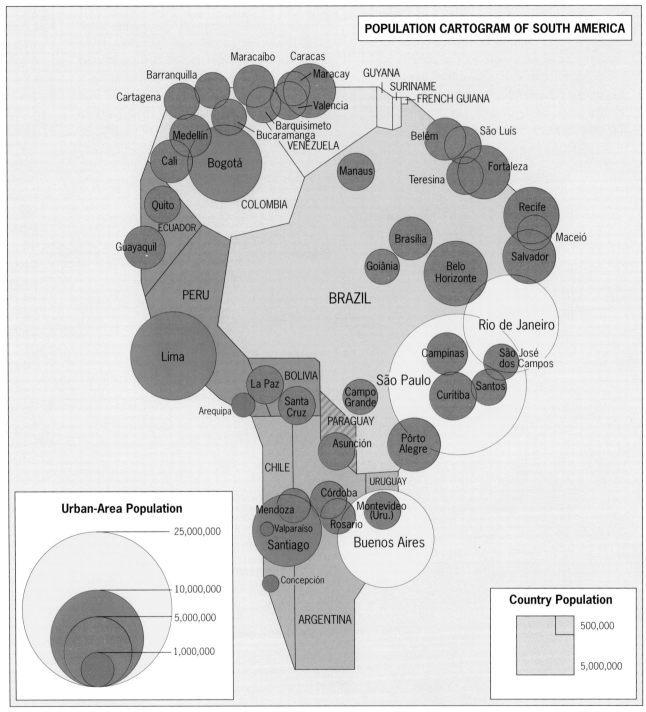

FIGURE 5-6

lion of the continent's 365 million inhabitants) now ranks with those of Europe and the United States. The urban population of South America has grown annually by about 5 percent since 1950, while the increase in rural areas was less than 2 percent. These numbers underscore not only the 9 dimensions but also the durability of the **rural-to-urban migration** from the countryside to the cities.

The generalized spatial pattern of South America's urban transformation is displayed in Figure 5-6, which shows a *cartogram* of the continent's population.* Here we see not only the realm's 13 countries in population-space relative to each other, but also the proportionate sizes of individual large cities within their total national populations.

Regionally, southern South America is most highly urbanized. Today in Argentina, Chile, and Uruguay, at least 86 percent of the people reside in cities and towns (statistically on a par with Europe's most urbanized countries). Ranking next is Brazil at 81 percent. The next highest group of countries, averaging 78 percent urban, border the Caribbean in the north. Not surprisingly, the Amerind-subsistence-dominated Andean countries constitute the realm's least urbanized zone (Peru, whose population is 72 percent urbanized, is that region's leader). Figure 5-6 also tells us a great deal about the relative positions of major metropolises in their countries. Three of them—Brazil's São Paulo and Rio de Janeiro, and Argentina's Buenos 10 Aires—rank among the world's **megacities** (whose populations exceed 10 million).

In South America, as in Middle America, Africa, and Asia, people are attracted to the cities and driven from the poverty of the rural areas. Both *pull* and *push factors* are at work. Rural land reform has been slow in coming, and every year tens

*A cartogram is a specially transformed map in which countries and cities are represented in proportion to their populations. Those containing large numbers are "blown up" in population-space, while those containing lesser numbers are "shrunk" in size accordingly.

of thousands of farmers simply give up and leave, seeing little or no possibility for economic advancement. The cities lure them because they are perceived to provide opportunity—the chance to earn a regular wage. Visions of education for their children, better medical care, upward social mobility, and the excitement of life in a big city draw hordes to places such as São Paulo and Caracas.

But the actual move can be traumatic. Cities in developing countries are surrounded and often invaded by squalid slums, and this is where the urban immigrant most often finds a first—and frequently permanent—abode in a makeshift shack without even the most basic amenities and sanitary facilities. And unemployment is persistently high, often exceeding 25 percent of the available labor force. But still the people come, the overcrowding in the shantytowns worsens, and the threat of epidemic-scale disease (and other disasters) rises.

The "Latin" American City Model

The urban experience in the South and Middle American realms varies because of diverse historical, cultural, and economic influences. Nonetheless, there are many common threads that have prompted geographers to search for useful generalizations. One is the model of the intraurban spatial 11 structure of the **"Latin" American city** proposed by Ernst Griffin and Larry Ford (Fig. 5-7), which may have wider application to cities of developing countries in other geographic realms as well.

The basic spatial framework of city structure, which blends traditional elements of South and Middle American culture with modernization forces now reshaping the urban scene, is a composite of radial sectors and concentric zones. Anchoring the model is the CBD, which is the primary business, employment, and entertainment focus of the surrounding metropolis. The CBD contains many modern high-rise buildings but also mirrors its colonial beginnings. As shown in Figure 4-3, the Spanish colonizers laid out their cities around a central square, or *plaza*, dominated by a church

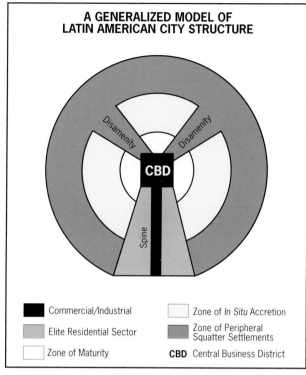

A GENERALIZED MODEL OF LATIN AMERICAN CITY STRUCTURE

CBD

Disamenity   Disamenity

Spine

■ Commercial/Industrial
▦ Elite Residential Sector
□ Zone of Maturity
□ Zone of *In Situ* Accretion
▨ Zone of Peripheral Squatter Settlements
CBD Central Business District

FIGURE 5-7

and government buildings. Santiago's *Plaza de Armas*, Bogotá's *Plaza Bolívar*, and Buenos Aires' *Plaza de Mayo* (photo, p. 182) are classic examples. The plaza was the hub of the city, which later outgrew its old center as new commercial districts formed nearby; but to this day the plaza remains an important link with the past.

Radiating outward from the urban core along the city's most prestigious axis is the commercial *spine*, which is adjoined by the *elite residential sector* (shown in green in Fig. 5-7). This widening corridor is essentially an extension of the CBD, featuring offices, shopping, and housing for the upper and upper-middle classes.

The three remaining concentric zones are home to the less fortunate residents of the city, with income level and housing quality decreasing as distance from the CBD increases. The *zone of maturity* in the inner city contains housing for the middle class, who invest sufficiently to keep their aging dwellings from deteriorating. The adjacent *zone of in situ accretion* is one of much more modest housing interspersed with unkempt areas, representing a transition from inner-ring affluence to outer-ring poverty.

The outermost *zone of peripheral squatter settlements* is home to the impoverished and unskilled hordes that have recently migrated to the city from rural areas. Despite the shantytown living condi-

tions, many newcomers achieve their first economic 12 success by becoming part of the **informal sector**. This primitive form of capitalism is common in many lower-income countries, taking place beyond the control of the government. Participants are unlicensed sellers of homemade goods and services, and their willingness to engage in this hard work has transformed many a shantytown into a beehive of activity that can propel resourceful residents toward a middle-class existence.

A final structural element of many South American cities is the *disamenity sector* that contains 13 relatively unchanging slums, known as ***barrios*** or ***favelas***. The worst of these poverty-stricken areas often include sizeable numbers of people who are so poor that they are forced literally to live in the streets. Thus the realm's cities present stupendous contrasts between poverty and affluence, squalor and comfort—a harsh juxtaposition all too frequently observed in the cityscape (photo, p. 189).

REGIONS OF THE REALM

South America divides geographically into four rather clearly defined regions (Fig. 5-3):

1. *The North* consists of five entities that display a combination of Caribbean and South American features: Colombia, Venezuela, and the three small Guianas representing historic colonial footholds by Britain (Guyana), the Netherlands (Suriname), and France (French Guiana).

2. *The West* is formed by four republics that share a strong Amerindian cultural heritage as well as powerful influences resulting from their Andean physiography: Ecuador, Peru, Bolivia, and, transitionally, Paraguay.

3. *The South*, often called the "Southern Cone," includes three countries that actually conform to the much-misused regional term "Latin" America: Argentina, Chile, and Uruguay, all with strong European imprints and little remaining Amerindian influence, plus aspects of Paraguay.

4. *Brazil* is by itself almost as large as the rest of the continent in terms of territory as well as population, where the dominant Iberian influence is Portuguese, not Spanish, and where Africans, not Amerindians, form a significant element in demography and culture. Brazil in some ways is a bridge between the Americas and Africa, and since we turn to Subsaharan Africa next, we discuss it last in this chapter.

▶ THE NORTH: CARIBBEAN SOUTH AMERICA

As Figure 5-5 reminds us, the countries of South America's northern tier have something in common besides their coastal location: each has a coastal tropical-plantation zone on the Caribbean colonial model. Especially in the three Guianas, early European plantation development entailed the immigration of African laborers and eventually the absorption of this element into the population matrix. Far fewer Africans were brought to South America's northern shores than to Brazil's Atlantic coasts, and tens of thousands of South Asians also arrived as contract laborers and stayed as settlers, so the overall situation here is not comparable to Brazil's. And it is also distinctly different from that of the rest of South America.

Today, the three Guianas still display the coastal orientation and plantation dependency with which the colonial period endowed them, although logging their tropical forests is penetrating and ravaging the interior. In Venezuela and Colombia, however, farming, ranching, and mining drew the population inland, overtaking the coastal-plantation economy and creating diversified economies.

Figure 5-8 shows that not only Venezuela but also Colombia is Caribbean in its orientation: in

Venezuela, coastal oil and natural gas reserves have replaced plantations as the economic mainstay, and Colombia's Andean valleys open toward the north where roads, railways, and pipelines lead toward Caribbean ports such as Cartagena and Barranquilla. Colombia's Pacific coast is hardly a factor in the national economy despite the outlet at Buenaventura. But, as we will see, the North's locational advantages are countered by physical, economic, and political obstacles that continue to prevent the realization of their full potential.

Colombia

Imagine a country more than twice the size of France, not at all burdened by overpopulation, with a physical geography so varied that it can produce crops ranging from the temperate to the tropical, possessing world-class reserves of oil, plus 2000 miles of coastline on Atlantic and Pacific waters. Situated in the northwestern corner of South America, closer than any of its neighbors to the markets of the north, with a single language and one dominant religion, would such a country not assuredly thrive amidst the burgeoning economic geography of its hemisphere? The answer is no. Colombia today is a country as troubled as any in the world, driven by violence, its economy imperiled, its politics destabilized, its future a gigantic

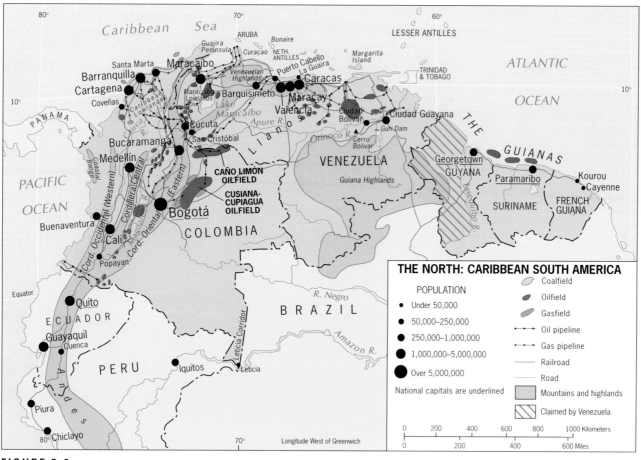

FIGURE 5-8

the politicians. The fabric of Colombian society unraveled.

As Figure 5-8 shows, western Colombia is mountainous while the east lies in the Orinoco and Amazon basins. Most of the country's 45 million people live in the western and northern parts of Colombia. Physiography has contributed to the concentration of this population into more than a dozen separate clusters. Some of them are located in the Caribbean Lowlands; others lie in the Magdalena and Cauca valleys between the parallel ranges of the Andes; and still others concentrate in basins within the mountains themselves, the most important being the cluster around the capital city of Bogotá. The country's renowned coffee crop comes from the *templada* zones on the Andean slopes. Colombia's belt of oil and natural gas reserves flanks the border with Venezuela from Lake Maracaibo south to the Caño Limón oilfield, and then continues southwest along the Andean front through the country's largest deposits, the Cusiana-Cupiagua oilfield (Fig. 5-8). Meanwhile, the sparsely populated east and south have become the production centers for the coca crop that sustains the country's illicit narcotics industry.

With its fuel, mineral, and agricultural productivity, Colombia might have been on the fast track toward prosperity, but the rise of the narcotics industry, coupled with the legacy of *La Violencia*, has dealt the country a devastating blow. Today, Colombia is in combat. In the cities, narco-terrorists repeatedly commit appalling acts of violence; in the countryside, rebel forces and drug-financed armies of the political "left" fight equally vicious paramilitary forces of the "right." Civilians by the thousands die in the crossfire of all these belligerent groups, and countless more are dislocated.

question mark. Colombia's cultural uniformity has not produced social cohesion. The same physical geography that diversifies its environments also divides its population clusters, which are so tenuously interconnected that the country totals less than 250 miles (400 km) of paved four-lane highway. Its proximity to U.S. markets is a curse as well as a blessing: at the root of Colombia's current problems lies its role in the international drug trade.

Colombia's current disorder is not its first. In the past, civil wars between conservatives and liberals (on Roman Catholic religious issues) developed into conflicts pitting rich against poor, elites against workers. In Colombia today, people still refer to the last of these wars as *La Violencia*, a decade of strife beginning in 1948 during which as many as 200,000 people died. In the 1970s, disaster struck again. In remote parts of the country, groups opposed to the political power structure began a campaign of terrorism, damaging the developing infrastructure and destroying confidence in the future. Simultaneously, the U.S. market for narcotics expanded rapidly, and many Colombians got involved in the drug trade. Powerful and wealthy drug cartels formed in major cities such as Medellín and Cali, with networks that influenced all facets of Colombian life from the peasantry to

Meanwhile, U.S. involvement in the war against drug cultivation and production has complicated the situation further.

Parts of Colombia are beyond the control of its government and its national armed forces, and are essentially ruled by rebels who "protect" the local citizenry against Bogotá's efforts to reestablish authority there. In effect, insurgents in several parts of Colombia have created their own domains, and some of their rulers have even negotiated their status with the legitimate government in the capital. What is happening here is a process long studied by political scientists. It involves three stages: (1) *contention*, the period of initial rebellion of the kind that marked the decade of *La Violencia*; (2) *equilibrium*, when the rebels gain effective control over some of the national territory, as has happened in parts of Colombia; and (3) *counteroffensive*, in which the government and the rebels, usually following the breakdown of negotiations, engage in armed conflict that will decide the future of the country. The geographer Robert McColl translated this sequence into spatial terms, suggesting that the equilibrium stage represents the emergence of an **14 insurgent state**. The formation of an insurgent state is the key stage in the process because it creates an informally bounded territory, usually with a distinct core area centered on a captured town, establishes a parallel government, and provides schools and other social services that substitute for those the formal state may have provided.

Colombia today contains several enclaves that were taking on the properties of insurgent states (Fig. 5-9). This is because Colombia confronts not just one but several different guerrilla forces, each operating in a particular area of the country. Thus the M-19 Movement focuses on the zone near the Panamanian border. The Popular Liberation Army is active in the Cauca Valley. The National Liberation Army (ELN) operated in the Arauca area on the Venezuelan border and in the middle Magdalena Valley prior to its recent cease-fire agreement. And the Revolutionary Armed Forces of Colombia (FARC), by far the most powerful insur-

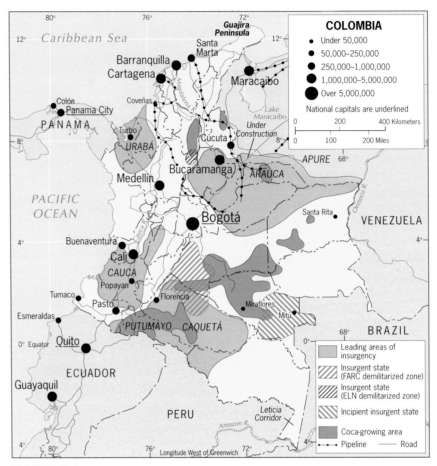

FIGURE 5-9

gent group, has most recently concentrated on the central-south and east.

In 1999, as an incentive to negotiate, FARC compelled the government to demilitarize a large zone south of Bogotá, in effect leaving the insurgents there in control of an area the size of Switzerland (Fig. 5-9). Also important was FARC's simultaneous initiative to take over the far southeastern town of Mitú and its environs, an area whose remoteness made it particularly difficult for the Colombian army to regain control. And although the insurgents claimed they were suppressing drug production in the zones under their control, the ev-

idence indicates that coca cultivation has skyrocketed in the FARC-held insurgent state.

Colombia's twin scourges, narcotics and insurgencies, have set the country back immeasurably and pose a colossal threat to its future as a coherent state. Guerrilla movements—which literally and/or virtually control large parts of Colombia—obstruct foreign investment and destroy infrastructural facilities, organize and tax coca farmers, operate airstrips and charge takeoff and landing fees, run sophisticated communications networks, and control hundreds of towns and even entire provinces. In the process, they have driven well over 1 million Colombian farmers off their lands and into the cities as refugees, as well as convincing an additional million to flee to neighboring countries and the United States. As the rebels' hold over their domains strengthens, local governments, companies, and individuals are forced to pay "protection" money by direct deposit into guerrilla-controlled bank accounts. Such funds are used, among other things, to buy legitimate businesses such as hotels, restaurants, and stores. Thus the drug trade, directly or indirectly, infects every corner of what remains of Colombian society.

The geographic dimensions of the centrifugal forces these activities represent are particularly worrisome. As Figure 5-9 reveals, the activity spheres of the guerrilla movements place extensive parts of the country beyond the reach of law and order. Colombia's international borders continue to be represented on maps (such as those in this chapter), but the reality on the ground is quite different. When any country confronts disruptive centrifugal forces that put portions of the national territory beyond government control, the very foundations of the state are threatened.

Venezuela

Much of what is important in Venezuela is concentrated in the northern and western parts of the country, where the Venezuelan Highlands form the eastern spur of the north end of the Andes system. Most of Venezuela's 26.1 million people are concentrated in these uplands, which include the capital of Caracas. The Venezuelan Highlands are flanked by the Maracaibo Lowlands to the west and by a vast plainland of savanna country, known as the *Llanos*, in the Orinoco Basin to the south and east (Fig. 5-8).

The Maracaibo Lowlands constitute one of the world's leading oil-producing areas. Much of the oil is drawn from reserves that lie beneath the shallow lake at its center. Actually, Lake Maracaibo is a misnomer, for the "lake" is open to the ocean and is in fact a gulf with a very narrow entry. The *Llanos* on the southern side of the Venezuelan Highlands and the Guiana Highlands in the coun-

try's southeast are in an early stage of development. The *Llanos* slope gently from the base of the Andean spur to the Orinoco River. Its mixture of savanna grasses and scrub woodland support cattle grazing on higher ground, but wet-season flooding of the more fertile lower-lying areas has inhibited the *Llanos'* commercial agricultural potential. Economic integration of this interior zone with the rest of Venezuela has so far been spearheaded by the exploitation of rich iron ores on the northern flanks of the Guiana Highlands.

Despite these assets, since 1998 Venezuela has been in upheaval as longstanding economic and social problems finally intensified to push the country into a new era of radical political change. A major reason was that oil had not bettered the lives of most Venezuelans because the government had acquired the habit of living off oil profits, forcing the country to suffer the consequences of the long global oil depression that began in the early 1980s. With more and more Venezuelans enraged at the way their oil-rich country was approaching bankruptcy without making progress toward the more equitable distribution of the national wealth, voters turned in an extreme direction in the 1998 presidential election. They elected Hugo Chávez, a former colonel who had led a failed military coup, and gave him a mandate to be a strongman type of leader.

Chávez has indeed pursued this course since entering office in 1999, sweeping aside Congress and the Supreme Court, supervising the rewriting of the Venezuelan constitution in his own image, and proclaiming himself the leader of a "peaceful left-ist revolution" that will transform the country. Although he professes that social equality ranks highest on his agenda, Chávez has stirred up racial divisions by actively promoting mestizos (67 percent of the population) over those of European background (20 percent). And in the international arena, Chávez sparks controversy at every turn: angering the government of neighboring Colombia by expressing his "neutrality" in its confrontation with cocaine-producing insurgent forces; unsettling another neighbor, Guyana, by aggressively

reviving a century-old territorial claim to the western zone of that country (the striped area in Fig. 5-8); and harassing the United States—which imports more oil from Venezuela than from any other country—by pressing for higher fuel prices within the OPEC petroleum cartel.

Venezuela is a key country in the Americas, and not only because its oil reserves rank among the world's top ten. Mismanagement of its economy led to popular dissatisfaction and the rise of a leader whose earlier actions evinced a disdain for democracy. This was a serious setback for a continent just emerging from an era of disastrous dictatorial rule.

The Guianas

The three small entities that form the remainder of the region remind us again why the term "Latin" America constitutes an excessive generalization for South as well as Middle America. Formerly known as British, Dutch, and French Guiana, two are now independent (Guyana since 1966, Suriname since 1975) and the third, French Guiana, has representation in the French Assembly. None has a population exceeding 1 million (see Table G-1), and per-capita incomes resemble Caribbean rather than South American norms.

British-influenced **Guyana**'s population of just under 800,000 is divided about equally between people of African and South Asian descent, which has caused social problems since before independence and political crises since then. For the people living in the small coastal villages along the coast, economic options are few; for the government, Venezuela's claims to a large part of the national territory (Fig. 5-8) is a constant worry. Guyana, however, has at times claimed parts of its neighbor **Suriname**, where political instability and joblessness drove more than 100,000 residents—more than one-fifth of the population—to emigrate to the Netherlands. Suriname is somewhat better off economically, with a large bauxite mine, self-sufficiency in rice, and some banana and fish exports. But ethnic problems recur here too, with people of African descent (including those who escaped slavery in the deep interior forests) slightly in the majority over those of South Asian and Indonesian descent. In the early 2000s, **French Guiana** remains officially a *département* of the French Republic, with the space center at Kourou its main asset but the lowest per-capita income among ordinary inhabitants.

As a South American region, the North continues to be volatile and rife with actual and potential conflict on the doorstep of the Caribbean. Legal (oil) and illicit (cocaine) exports move continuously to North American markets. Indigenous peoples, confined to the mountains of southern Colombia and to the Amazonian interiors of Colombia and Venezuela, are not part of the national fabric.

▶ THE WEST: ANDEAN SOUTH AMERICA

The second regional grouping of South American states—the Andean West (Fig. 5-10)—encompasses Peru, Ecuador, Bolivia, and transitional Paraguay. The map of culture spheres (Fig. 5-5) shows the Amerind-subsistence region extending along the Andes Mountains, indicating that these countries have large Amerindian components in their populations (Bolivia, about 55 percent; Peru, about 45 percent; and Ecuador, about 25 percent). Paraguay, not an Andean country but strongly Amerindian-infused, is transitional to the next region, the South. These countries also exhibit other similarities: their incomes are low, they are comparatively unproductive, and they exemplify the grinding poverty of the landless peonage. Today, these desperate conditions are no longer passively tolerated, and uprisings continue to emerge throughout the Andean West.

Peru

Peru straddles the Andean spine for more than 1000 miles (1600 km) and is the largest of the re-gion's four republics in both territory and population (27.7 million). Physiographically and culturally, Peru divides into three subregions: (1) the desert coast, the European-mestizo region; (2) the Andean highlands or *Sierra*, the Amerindian region; and (3) the eastern slopes, the sparsely populated Amerindian-mestizo interior (Fig. 5-10).

Lima and its port, Callao, lie at the center of the desert coastal strip, and it is symptomatic of the cultural division prevailing in Peru that for nearly 500 years the capital city has been positioned on the periphery, not in a central location in a basin of the Andes. From an economic point of view, however, the Spaniards' choice of a headquarters on the Pacific coast proved to be sound, for the coastal subregion has become commercially the most productive part of the country. A thriving fishing industry contributes significantly to the export trade; so do the products of irrigated agriculture in some 40 oases distributed all along the arid coast, which include cotton, sugar, and fruits and vegetables.

The Sierra (Andean) subregion occupies about one-third of the country and is the ancestral home of the largest component in the total population, the Quechua-speakers who had been subjugated by the Inca rulers when the Spanish conquerors arrived. Their survival during the harsh colonial regime was made possible by their adaptation to the high-altitude environments they inhabited, but their social fabric was ripped apart by communalization, forced cropping, religious persecution, and outward migration to towns and haciendas where many became serfs. Another Amerindian people, the Aymara, live in the area of Lake Titicaca, making up about 6 percent of Peru's population but larger numbers in neighboring Bolivia.

Although these Amerindian people make up about half of the population of Peru, their political influence remains slight—despite the recent election of a president, Alejandro Toledo, whose mestizo ancestry made him popular in the Andean districts. Nor is the Andean subregion a major factor in Peru's commercial economy—except, of course, for its mineral storehouse, which yields copper, zinc, and lead from mining centers, the largest of which is Cerro de Pasco. In the high valleys and intermontane basins, the Amerindian population is clustered either in isolated villages, around which people practice a precarious subsistence agriculture, or in the more favorably located and fertile areas where they are tenants, peons on white- or mestizo-owned haciendas. Most of these people never receive an adequate daily caloric intake or balanced diet of any sort. The wheat produced around Huancayo, for example, is mainly exported and would in any case be too expensive for the Amerindians themselves to buy. Potatoes, barley, and corn are among the subsistence crops grown here in the *tierra fría* zone, and in the *tierra helada* of the higher basins the Amerindians graze their llamas, alpacas, cattle, and sheep.

Of Peru's three subregions, the *Oriente*, or East—the inland slopes of the Andes and the Amazon-drained, rainforest-covered *montaña*—is the most isolated. The focus of the eastern subregion, in fact, is Iquitos, a city that looks east rather than west and can be reached by oceangoing vessels sailing 2300 miles (3700 km) up the Amazon River across northern Brazil. Iquitos grew rapidly during the Amazon wild-rubber boom of a century ago and then declined; now it is finally growing again and reflects Peruvian plans to open up the eastern interior. Petroleum was discovered west of Iquitos in the 1970s, and since 1977, oil has flowed through a pipeline built across the Andes to the Pacific port of Bayóvar.

Peruvian planners see their country on the verge of an energy age as interior exploration is proving large reserves of gas and oil east of the Andes. Gas will flow from the Camisea reserve (north of Cuzco) across the Andes to a conversion plant on the Paracas Peninsula south of Lima, from where a 2-mile (3-km) ocean-floor pipeline will carry it to an offshore loading platform for tankers that will transport it to the U.S. market (Fig. 5-10).

Peru, at the turn of this century, appeared to have put behind it the lengthy period of instability during which its government was threatened by well-

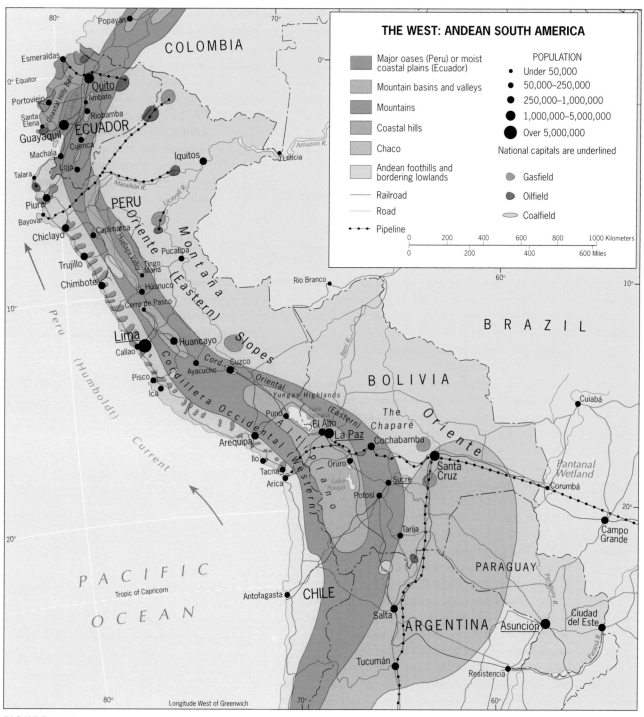

FIGURE 5-10

organized guerrilla movements. The most serious threat, which for a time during the 1980s seemed on the verge of achieving an insurgent state in its Andean base, was the so-called *Sendero Luminoso* (Shining Path) movement. But forceful action under the later-disgraced president of Japanese ancestry, Alberto Fujimori, defeated the rebel movement and returned Peru to political stability and economic growth.

The social cost of the Fujimori regime, however, was high. Democratic principles were violated and corruption was endemic. When the succeeding Toledo administration soon ran into trouble on economic and political fronts, the specter of Peru's chronic instability loomed again.

Ecuador

On the map, Ecuador, smallest of the Andean West republics, appears to be just a corner of Peru. But that would be a misrepresentation because Ecuador possesses a full range of regional contrasts (Fig. 5-10). It has a coastal belt; an Andean zone that may be narrow (under 150 miles or 250 km) but by no means of lower elevation than elsewhere; and an *Oriente*—an eastern subregion that is as sparsely populated and long was as economically marginalized as that of Peru.

Ecuador's Pacific coastal zone consists of a belt of hills interrupted by lowlands, of which the most important lies in the south between the hills and the Andes, drained by the Guayas River. Guayaquil—the country's largest city, main port, and leading commercial center—forms the focus of this subregion. Unlike Peru's coastal strip, Ecuador's is not a desert: it consists of fertile tropical plains. Seafood (especially shrimp) is a leading product, and these lowlands support a thriving commercial agricultural economy built around bananas, cacao, cattle-raising, and coffee on the hillsides. The coastal zone is home to most of the 40 percent of Ecuadorians who are mestizos, and from time to time they express their displeasure with the government based in the Andean city of Quito by proclaiming a desire for autonomy or even secession.

The great majority of the Amerindians (who constitute 25 percent of Ecuador's population of 13.6 million) and the 10 percent who are of European ancestry live in the central Andean zone, where landtenure reform is an explosive issue. The differing interests of the Guayaquil-dominated coastal lowland and the Andean-highland subregion focused on the capital (Quito) have long fostered a deep regional cleavage between the two.

In the early 2000s, Ecuador, like Peru, is rapidly expanding its energy industry based on a growing volume of proven reserves not only in the *Oriente* but also in the coastal zone. So large are these reserves that experts compare Ecuador's energy resources to those Mexico and Nigeria, which would rank this country among South America's leaders in this respect. Already, oil flows from beneath the rainforests of the *Oriente* via a Transandean pipeline to the port of Esmeraldas; another 300-mile (500-km) pipeline was put into operation in 2003. Oil and gas in 2004 yielded about 50 percent of Ecuador's export revenues, and the government expects this portion to grow rapidly in the years ahead.

Bolivia

From Ecuador southward through Peru, the Andes broaden until in Bolivia they reach a width of some 450 miles (720 km). Here, between the eastern and western Andean cordilleras (ranges), lies the *Altiplano* proper (Fig. 5-10; as noted earlier, an altiplano is an elongated, high-altitude basin). On the boundary between Peru and Bolivia, freshwater Lake Titicaca lies at 12,507 feet (3700 m) above sea level and helps make the adjacent *Altiplano* livable by ameliorating the coldness in its vicinity, where the snow line lies just above the plateau surface. On the surrounding cultivable land, grains have been raised for centuries dating back to pre-Inca times, and the Titicaca Basin still supports a major cluster of Aymara subsistence farmers. This portion of the *Altiplano* is the heart of modern Bolivia and also contains the capital city, La Paz.

Modern Bolivia is the product of the European impact, and the country's Amerindians (who still make up 55 percent of the national population of 9.2 million) no more escaped the loss of their land than did their Peruvian or Ecuadorian counterparts. What made the richest Europeans in Bolivia wealthy, however, was not land but minerals. The town of Potosí in the eastern cordillera became a legend for the immense deposits of silver in its vicinity; tin, zinc, copper, and several ferroalloys were also discovered there.

Today oil and natural gas, exported to Argentina and Brazil, are major sources of foreign revenues, and zinc has replaced tin as the leading metal export. But Bolivia's economic prospects will always be impeded by the loss of its outlet to the Pacific Ocean during its war with Chile in the 1880s despite its transit rights and dedicated port facilities at Antofagasta.

More critical than its economic limitations or its landlocked situation is Bolivia's social predicament. The government's history of mistreatment of indigenous people and the harsh exploitation of its labor force hangs heavily over a society where about 70 percent of the people, almost all Amerindian, live in dire poverty, landless, and devoid of hope. In recent years, however, this underrepresented majority has been making an impact on national affairs. In 2000, the bitter feelings of indigenous peoples burst into the open all across the *Altiplano* during a week-long protest that began as a riot against hikes in water rates in the city of Cochabamba: the government was forced to retreat. In 2003, violent opposition to a government plan to export natural gas to the United States via a new pipeline to the Chilean coast led to chaos and the government's resignation.

None of this, however, alters the fundamental social geography of Bolivia, where the privileged and powerful in La Paz and Sucre are seen as adversaries by millions of Amerindian citizens. Less than half of Bolivians are able to speak Spanish in this "Latin" American country where the government is viewed by most as unwilling to integrate the impoverished masses into the mainstream of national life.

Paraguay

Earlier we noted the transitional situation of Paraguay (see Figs. 5-3 and 5-10). In terms of economic development, Paraguay is more typical of the Andean West than of the richer South. Ethnically, too, the pattern is more characteristic of the West: about 95 percent of Paraguay's 6.3 million people are mestizo, but with so pervasive an Amerindian influence that any white ancestry is almost totally submerged. As for languages, Amerindian Guaraní is so widely spoken alongside Spanish that the country is surely one of the world's most completely bilingual. The physiography, however, is decidedly non-Andean because all of Paraguay lies to the east of the mountains in the center of the Paraná-Paraguay Basin.

And note that two major rivers link Paraguay to the South: the Paraguay River, on which the capital of Asunción is located, and the Paraná River, along Paraguay's eastern border, where three countries—Paraguay, Brazil, and Argentina—meet in what is known as the *Triple Frontier*.

Apart from the Guianas, Paraguay long has been South America's least-developed country by most measures, the result of a combination of factors including not only its landlocked situation but also land expropriation, subjugation of the poor, power struggles, mismanagement, and corruption. In 2004 about 60 percent of the people, notably in

Asunción's dreadful slums but also in the dry, scrub-covered Chaco of the west, lived in abject poverty.

Remote and dictatorially-run countries ruled by corrupt oligarchies make attractive targets for terrorist organizations. The Triple Frontier area, notably the lawless town of Ciudad del Este, became a place of concern during the current global War on Terror, when a map of it was found in an al-Qaeda hideout in Kabul, Afghanistan. An investigation exposed financial ties involving millions of dollars between certain local businesspeople and terrorist-linked organizations in the Middle East. Bringing Paraguay into the mainstream of South American development is more than an economic objective.

THE SOUTH: MID-LATITUDE SOUTH AMERICA

South America's three southern countries—Argentina, Chile, and Uruguay—constitute a region sometimes referred to as the *Southern Cone* because of its pointed, ice-cream-cone shape (Fig. 5-11). As noted earlier, Paraguay has strong links to this region and in some ways forms part of it, although social contrasts between Paraguay and those in the Southern Cone remain sharp.

Since 1995, the countries of this region have been drawing closer together in an economic union named Mercosur, the hemisphere's second-largest trading bloc after NAFTA. Despite setbacks and disputes, Mercosur has expanded and today encompasses Argentina, Uruguay, Paraguay, and Brazil, with Chile and Bolivia participating as associate members.

Argentina

The largest Southern Cone country by far is Argentina, whose territorial size ranks second only to Brazil in this geographic realm; its population of 37.3 million ranks third after Brazil and Colombia. Argentina exhibits a great deal of physical-environmental variety within its boundaries, and

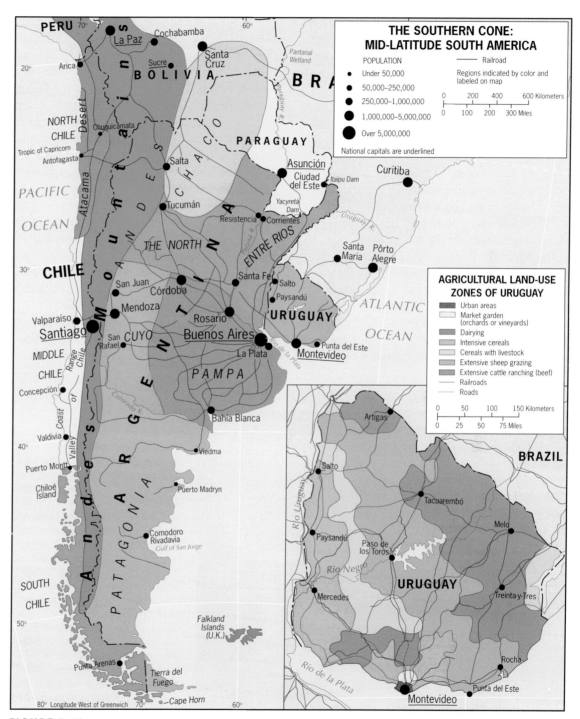

FIGURE 5-11

the vast majority of the Argentines are concentrated in the physiographic subregion known as the *Pampa* (a word meaning "plain"). Figure G-9 underscores the degree of clustering of Argentina's inhabitants on the land and in the cities of the Pampa. It also shows the relative emptiness of the other six subregions (mapped in Fig. 5-11): the scrub-forest *Chaco* in the northwest; the mountainous *Andes* in the west, along whose crestline lies the boundary with Chile; the arid plateaus of *Patagonia* south of the Rio Colorado; and the undulating transitional terrain of intermediate *Cuyo*, *Entre Rios* (also known as "Mesopotamia" because it lies between the Paraná and Uruguay rivers), and the *North*.

The Argentine Pampa is the product of the past 150 years. During the second half of the nineteenth century, when the great grasslands of the world were being opened up (including those of the interior United States, Russia, and Australia), the economy of the long-dormant Pampa began to emerge. The food needs of industrializing Europe grew by leaps and bounds, and the advances of the Industrial Revolution—railroads, more efficient ocean transport, refrigerated ships, and agricultural machinery—helped make large-scale commercial meat and grain production in the Pampa not only feasible but also highly profitable. Large haciendas were laid out and farmed by tenant workers; railroads radiated ever farther outward from the booming capital of Buenos Aires and brought the entire Pampa into production.

Argentina, as we noted in our introductory chapter, once was one of the richest countries in the world. Its historic affluence is still reflected in its architecturally splendid cities whose plazas and avenues are flanked by ornate public buildings and private mansions. This is true not only of the capital, Buenos Aires, at the head of the Rio de la Plata estuary; it also applies to interior cities such as Mendoza and Córdoba. The cultural imprint is dominantly Spanish, but the cultural landscape was diversified by a massive influx of Italians and smaller but influential numbers of British, French, and German immigrants. A sizeable immigration from Lebanon resulted in the diffusion of Arab ancestry to more than 12 percent of the Argentinian population.

Argentina has long been one of the realm's most urbanized countries: 90 percent of its population is concentrated in cities and towns, a higher percentage even than Western Europe or the United States. About one-third of all Argentinians live in metropolitan Buenos Aires, also by far the leading industrial complex where processing Pampa products dominates. Córdoba has become the second-ranking industrial center and was chosen by foreign automobile manufacturers as the car-assembly center for the expanding Mercosur market. One in three Argentinian wage earners is engaged in manufacturing, another indication of the country's economic progress. But what concentrates the urban populations is the processing of products from the vast, sparsely peopled interior: Tucumán (sugar), Mendoza (wines), Santa Fe (forest products), Salta (livestock). Argentina's product range is enormous. There is even an oil reserve near Comodoro Rivadávia on the coast of Patagonia.

Despite all these riches, Argentina's economic history is one of boom and bust. With fewer than 40 million inhabitants, a vast territory with diverse natural resources, adequate infrastructure, and good international linkages, Argentina should still be one of the world's wealthiest countries, as it once was. But political infighting and economic mismanagement have combined to ruin a vibrant and varied economy. What began as a severe recession toward the end of the 1990s became an economic collapse in the first years of the new century.

To understand how Argentina finds itself in this situation, a map of its administrative structure is useful. On paper, Argentina is a federal state consisting of the Buenos Aires Federal District and 23 provinces (Fig. 5-12). As would be expected from what we have just learned, the urbanized provinces are populous while the mainly rural ones have smaller populations—but the gap between Buenos Aires Province (whose capital is La Plata), with nearly 15 million, and Tierra del Fuego (capital:

FROM THE FIELDNOTES

"At the heart of Buenos Aires lies the Plaza de Mayo, flanked by impressive buildings but, unusual for such squares in Iberian America, carpeted with extensive lawns shaded by century-old trees. Getting a perspective of the plaza was difficult until I realized that you could get to the top of the 'English Tower' across the avenue you see in the foreground. From there, one can observe the prominent location occupied by the monument to the approximately 700 Argentinian military casualties of the Falklands War of 1982 (center), where an eternal flame behind a brass map of the islands symbolizes Argentina's undiminished determination to wrest the islands from British control."

PROVINCES OF ARGENTINA

- – - – International boundary
- – – – Provincial boundary
- • Location of Provincial capital

Provincial population

- Greater than 5 million
- 3–5 million
- 1–3 million
- 500,000–1 million
- Less than 500,000

| 0 | 200 | 400 | 600 Kilometers |

| 0 | 100 | 200 | 300 Miles |

Falkland Islands (U.K.)

FIGURE 5-12

Ushuaia), with barely over 100,000, is wide indeed. Several other provinces have under 500,000 inhabitants, so that the larger ones in addition to dominant Buenos Aires are also disproportionately influential in domestic politics, especially Córdoba and Santa Fe, both with capitals of the same name.

The never-ending problem for Argentina has been corrupt politics and associated mismanagement. Following an army coup in 1946, Juan Perón got himself elected president, bankrupted the country, and was succeeded by a military junta that plunged the country into its darkest days, culminating in the "Dirty War" of 1976–1983 which saw more than ten thousand Argentinians disappear without a trace. In 1982, this ruthless military clique launched an invasion of the British-held Malvinas (Falkland Islands), resulting in a major defeat for Argentina. By the time civilian government replaced the discredited junta, inflation was soaring and the national debt had become staggering. Economic revival during the 1990s was followed by another severe downturn that exposed the flaws in Argentina's fiscal system, including scandalously inefficient tax collection and unconditional federal handouts to the politically powerful provinces.

In 2003, a new administration took office led by President Néstor Kirchner who hails from a small Patagonian province, not from one the country's traditional power centers. He vowed to revive Argentina's fortunes and deal with corruption, military immunity from prosecution, and foreign economic intervention. The question is how long his indispensable initiatives will last.

Chile

For 2500 miles (4000 km) between the crestline of the Andes and the coastline of the Pacific lies the narrow strip of land that is the Republic of Chile. On average just 90 miles (150 km) wide (and only rarely over 150 miles or 250 km in width), Chile is the world's quintessential example of what *elongation* means to the functioning of a state. Accentuated by its north-south orientation, this severe territorial attentuation not only results in Chile ex-

tending across numerous environmental zones; it has also contributed to the country's external political, internal administrative, and general economic problems. Nonetheless, throughout most of their modern history, the Chileans have made the best of this potentially disastrous centrifugal force: from the beginning, the sea has constituted an avenue of longitudinal communication; the Andes Mountains continue to form a barrier to encroachment from the east; and when confrontations loomed at the far ends of the country, Chile proved to be quite capable of coping with its northern rivals, Bolivia and Peru, as well as Argentina in the extreme south.

As Figures G-8 and 5-11 indicate, Chile is a three-subregion country. About 90 percent of its 16 million people are concentrated in what is called Middle Chile, where Santiago, the capital and largest city, and Valparaíso, the chief port, are located. North of Middle Chile lies the Atacama Desert, which is wider, drier, and colder than the coastal desert of Peru. South of Middle Chile, the coast is broken by a plethora of fjords and islands, the topography is mountainous, and the climate—wet and cool near the Pacific—soon turns drier and colder against the Andean interior. South of the latitude of Chiloé Island, there are no permanent overland routes and hardly any settlements. These three subregions are also clearly apparent on the map of culture spheres (Fig. 5-5), which displays a mestizo north, a European-commercial zone in Middle Chile, and an undifferentiated south. In addition, a small Amerind-subsistence zone in northern Chile's Andes is shared with Argentina and Bolivia.

Some intraregional differences exist between northern and southern Middle Chile, the country's core area. Northern Middle Chile, the land of the hacienda and of Mediterranean climate with its dry summer season, is an area of (usually irrigated) crops that include wheat, corn, grapes, and other Mediterranean products. Livestock raising and fodder crops also take up much of the productive land, but continue to give way to the more efficient and profitable cultivation of fruits for export. Southern Middle Chile, into which immigrants from both the north and Europe (especially Germany) have pushed, is a

better-watered area where raising cattle has predominated. But here, too, more lucrative fruits, vegetables, and grains are changing the area's agricultural specializations.

Prior to the 1990s, the arid Atacama region in the north accounted for more than half of Chile's foreign revenues. The Atacama Desert contains the world's largest exploitable deposits of nitrates, which was the country's economic mainstay before the discovery of methods of synthetic nitrate production a century ago. Subsequently, copper became the chief export (Chile again possesses the world's largest reserves). It is found in several places, but the main concentration lies on the eastern margin of the Atacama near Chuquicamata, not far from the port of Antofagasta.

Chile today is emerging from a development boom that transformed its economic geography during the 1990s, a growth spurt that established its reputation as South America's greatest success story. Following the withdrawal of its brutal military dictatorship in 1990, Chile embarked on a program of free-market economic reform that brought stable growth, lowered inflation and unemployment, reduced poverty, and attracted massive foreign investment. The last is of particular significance because these new international connections enabled the export-led Chilean economy to diversify and develop in some badly needed new directions. Copper remains the single leading export, but many other mining ventures have been launched. In the agricultural sphere, fruit and vegetable production for export has soared because Chile's harvests coincide with the winter farming lull in the affluent countries of the Northern Hemisphere. Industrial expansion is occurring as well, though at a more leisurely pace, and new manufactures include a modest array of goods that range from basic chemicals to computer software.

Chile's newly internationalized economy has propelled the country to forge a prominent role for itself on the global trading scene. Even the Japanese expressed interest in building ties, and for a time during the 1990s Japan was its leading trading partner. This foothold in the Pacific Rim arena notwithstanding, Chile's trade linkages are now more heavily oriented toward its own hemisphere. The Chileans quickly realized that regional economic integration offered the best opportunities for sustained progress. Chile has affiliated locally with Mercosur as an associate member; however, despite exhortations from its Southern Cone partners, Chile in 2000 decided against seeking full membership.

Chile ended the twentieth century on a less encouraging note as its still-dominant copper sector dragged the national economy down to a growth standstill. With this commodity earning more than a third of all foreign revenues, there was little the Chileans could do in 1998 when the Asian economic crisis stifled demand and caused a worldwide collapse in the already low price of copper. This was followed by a period of painful readjustment as Chile's planners discovered that the country's economic diversification away from its reliance on copper had been neither as comprehensive nor as rapid as was first believed. Among the other lessons learned were that Chile remains too dependent on its natural resources, needs to expand the skills of its labor force, and must work harder to develop its manufacturing, services, and quaternary sectors.

Chile's recent transformation is widely touted as the "economic model" for all of Middle and South America to emulate. That is almost certainly an overstatement because few countries in these geographic realms can match the Chilean combination of natural and human resources and development circumstances. Yet all of those countries strongly aspire to follow in Chile's footsteps and for the first time in their modern histories are cooperating toward making that a reality. Their goal is the economic integration of all the Americas within the next few years—an accomplishment that could lead to a better life for every inhabitant of the Western Hemisphere.

Uruguay

Uruguay, unlike Argentina or Chile, is compact, small, and rather densely populated. This buffer state of old became a fairly prosperous agricultural country, in effect a smaller-scale Pampa (though possessing less favorable soils and topography). Figures 5-1 and G-8 shows the similarity of physical conditions on the two sides of the Plata estuary. Montevideo, the coastal capital, contains more than 35 percent of the country's population of 3.4 million; from here, railroads and roads radiate outward into the productive agricultural interior (Fig. 5-11). In the immediate vicinity of Montevideo lies Uruguay's major farming area, which produces vegetables and fruits for the metropolis as well as wheat and fodder crops. Most of the rest of the country is used for grazing cattle and sheep, with beef products, wool and textile manufactures, and hides dominating the export trade. Tourism is another major economic activity as Argentines, Brazilians, and other visitors increasingly flock to the Atlantic beaches at Punta del Este and other thriving resort towns.

15 In Chapter 1, we noted the work of **von Thünen** and his model layout of agricultural zones. Here in Uruguay we can discern a real-world example of that scheme (inset map, Fig. 5-11). Market gardens and dairying cluster nearest the urban area of Montevideo, with increasingly extensive agriculture ringing this national market in concentric land-use zones.

Uruguay's area of 68,500 square miles (177,000 sp km)—less territory even than Guyana—does not leave much room for population clustering. Nevertheless, a special quality of the land area of Uruguay is that it is rather evenly peopled right up to the boundaries with Brazil and Argentina. And of all the countries in South America, Uruguay is the most truly European. As for future development prospects as Mercosur matures (Montevideo is the bloc's administrative capital), Uruguay is in an excellent position to capitalize on its location between the Pampa to the south and Brazil's most dynamic areas to the north.

BRAZIL: GIANT OF SOUTH AMERICA

By any measure, Brazil is South America's giant. It is so large that it has common boundaries with all the realm's other countries except Ecuador and Chile (Fig. 5-3). Its tropical and subtropical environments range from the equatorial rainforest of the Amazon Basin to the humid temperate climate of the far south. Territorially, Brazil occupies just under 50 percent of South America and is exceeded on the global stage only by Russia, Canada, the United States, and China. In population size, it also accounts for just below half the continental total and is again the world's fifth-largest country (surpassed only by China, India, the United States, and Indonesia). The Brazilian economy is now the eighth largest on Earth, and its modern industrial base is the seventh largest. Moreover, the country is likely to continue advancing in the international ranks and is poised to become a world force as the twenty-first century unfolds.

Population Patterns

Brazil's population of 178 million is as diverse as that of the United States. In a pattern quite familiar in the Americas, the indigenous inhabitants of the country were decimated following the European invasion (fewer than 200,000 Amerindians now survive deep in the Amazonian interior). Africans came in great numbers, too, and today there are more than 10 million Afro-Brazilians in Brazil. Significantly, however, there was also much racial mixing, and 70 million Brazilians have combined European, African, and minor Amerindian ancestries. The remaining 97 million—now barely in the majority at 54 percent—are mainly of European origin, the descendants of immigrants from Portugal, Italy, Germany, and Eastern Europe. The complexion of the population was further diversified by the arrival of Lebanese and Syrian Muslims following the collapse of the Ottoman Empire nearly a century ago. Estimates of the Muslim component in the Brazilian population suggest that about 10 million can claim to be of Arab descent, of whom about 1 million are nominally Muslims. But practicing Muslims are far fewer in number: in 2003 Brazil had only about 35 functioning mosques.

Another significant, though small, minority began arriving in Brazil in 1908; the Japanese, who today are concentrated in São Paulo State. The more than 1 million Japanese-Brazilians form the largest ethnic Japanese community outside Japan, and in their multicultural environment they have risen to the top ranks of Brazilian society as business leaders, urban professionals, farmers—and even as politicians in the city of São Paulo. Committed to their Brazilian homeland as they are, the Japanese community also retains its contacts with Japan, resulting in many a trade connection.

Brazilian society, to a greater degree than is true elsewhere in the Americas, has made progress in dealing with its racial divisions. To be sure, blacks are still the least advantaged among the country's major population groups, and community leaders continue to complain about discrimination. But ethnic mixing in Brazil is so pervasive that hardly any group is unaffected, and official census statistics about "blacks" and "Europeans" are meaningless. What the Brazilians do have is a true national culture, expressed in an historic adherence to the Catholic faith (now eroding under Protestant-evangelical and secular pressures), in the universal use of a modified form of Portuguese as the common language, and in a set of lifestyles in which vivid colors, distinctive music, and a growing national consciousness and pride are fundamental ingredients.

Brazil's population grew rapidly during the world's twentieth-century population explosion. But over the past three decades, the rate of natural increase has slowed from nearly 3.0 percent to 1.2 percent, and the average number of children born to a Brazilian woman has been halved from 4.4 in 1980 to 2.2 in 2004. These dramatic reductions occurred in the absence of any active family-planning policies by Brazil's federal or State governments and in direct contradiction to the teachings of the Catholic Church, reflecting religion's changing role in this former bastion of Roman Catholicism. Brazil's rapid urbanization, its prevalent economic uncertainties, and the widespread use of contraceptives are among factors influencing this significant change.

African Heritage

As we noted earlier, Afro-Brazilians still suffer disproportionately from the social ills of this relatively integrated nation. And yet Brazil's culture is infused with African themes, a quality that has marked it from the very beginning. Three centuries ago, the African sculptor and architect affectionately called Aleijadinho was Brazil's most famous artist. The world-famous composer Heitor Villa-Lobos used numerous Afro-Brazilian folk themes in his music. So many Africans were brought in bondage to the city and hinterland of Salvador (Bahia State) from what is today Benin (formerly Dahomey) in West Africa that Bahia has become a veritable outpost of African culture. Candomblé, one of the Macumba religious sects arising from the arrival of African cultures in Brazil, is concentrated in Salvador and Bahia. Its presence there has led to a reconnection with modern Benin as Afro-Brazilians travel to West Africa to search for their roots and West Africans come to Bahia to experience the cultural landscape to which their ancestors gave rise.

In some parts of Bahia, the rural Northeast, and urban areas in Brazil's core you can almost imagine being in Africa. As a nation, Brazil is taking an increased interest in African linkages, not least because the Lusitanian world incorporates not only Portugal and Brazil but also Angola and Moçambiuqe, where the Portuguese colonial experiment failed but where the Lusitanian imprint also survives.

Inequality in Brazil

For all its accomplishments in multiculturalism, Brazil remains a country of stark, appalling social inequalities. Although such inequality is hard to measure precisely, South America is often cited as the geographic realm exhibiting the world's sharpest division between affluence and poverty. And in South America, Brazil is reputed to have the widest gap of all.

Data that underscore this situation include the following: today the richest 10 percent of the population own two-thirds of all the land and control more than half of Brazil's wealth. The poorest 20 percent of the people live in the most squalid conditions prevailing anywhere on the planet, even including the megacities of Africa and Asia. According to UN estimates, in this age of adequate available (but not everywhere affordable) food, about half the population of Brazil suffers from some form of malnutrition and, in the northeastern States, even hunger. Packs of young orphans and abandoned children roam the cities, sleeping where they can, stealing or robbing when they must. Some of the world's most magnificent central cities are ringed and sectored by the most wretched *favelas* where poverty, misery, and crime converge (see Fig. 5-7).

Most Brazilian voters had long believed that these conditions were worsening, that what economic development was taking place benefited those already well off, and that powerful corporations and foreign investors were to blame for the spiral of rising and growing unemployment (about 10 percent of the workforce was jobless in 2001). When the latest official reports announced that the number of Brazilians afflicted by poverty had increased by 50 percent since 1980, the people demanded change and, as we will note, they achieved that in 2002.

Development Prospects

Brazil is richly endowed with mineral resources, including enormous iron and aluminum ore reserves, extensive tin and manganese deposits, and sizeable oil- and gasfields (Fig. 5-13). Other significant energy developments involve massive new hydroelectric facilities and the successful substitution of sugarcane-based alcohol (gasohol) for gasoline— allowing well over half of Brazil's cars to use this fuel instead of costly imported petroleum. Besides these natural endowments, Brazilian soils sustain a bountiful agricultural output that makes the country a global leader in the production and export of coffee, soybeans, and orange juice concentrate. Commercial agriculture, in fact, is now the fastest growing economic sector, propelled by mechanization and the opening of a major new farming frontier in the fertile grasslands of southwestern Brazil.

As already noted, Brazil ranks seventh among the world's industrial powers. Much of the momentum for this continuing development was unleashed in the early 1990s after the government opened the country's long-protected industries to international competition and foreign investment. These new policies are proving to be effective because productivity has risen by over a third since 1990, as Brazilian manufacturers attain world-class quality (you may very well fly in a Brazil-made passenger jet the next time you travel by air within the United States). In the mid-1990s the revenues from industrial exports surpassed those from agriculture. Commerce with Argentina made that member of *Mercosul* (as the Portuguese-speaking Brazilians call Mercosur) one of Brazil's leading trade partners. On the global stage, Brazil became a formidable presence in other ways. The country's enormous and easily accessible iron-ore deposits, the relatively low wages of its workers, and the mechanized efficiency of its steelmakers enable Brazil to produce that commodity at half the cost of steel made in the United States. This causes American steel producers to demand government protection through tariffs, which goes against purported U.S. principles of free trade.

Even though Brazil has become a major player on the world's economic stage, it has not escaped the cycles of boom and bust that affect transforming economies. Needing loans from international lending agencies to weather the downturns, Brazil's government had to agree to terms that included the privatization of public companies ranging from telephones to utilities. This often meant layoffs, rising prices to consumers, and even civil unrest. Brazil's currency, the *real*, was devalued in 1999, resulting in the outflow of billions of dollars withdrawn from Brazil's banks by panicked depositors. Strikes and marches against increased gas prices, highway tolls, and utility charges paralyzed the country for days. Newspapers reported scandals and corruption involving leading public figures and hundreds of millions of *reals*. In the political campaign that led up to the presidential election of 2002, these

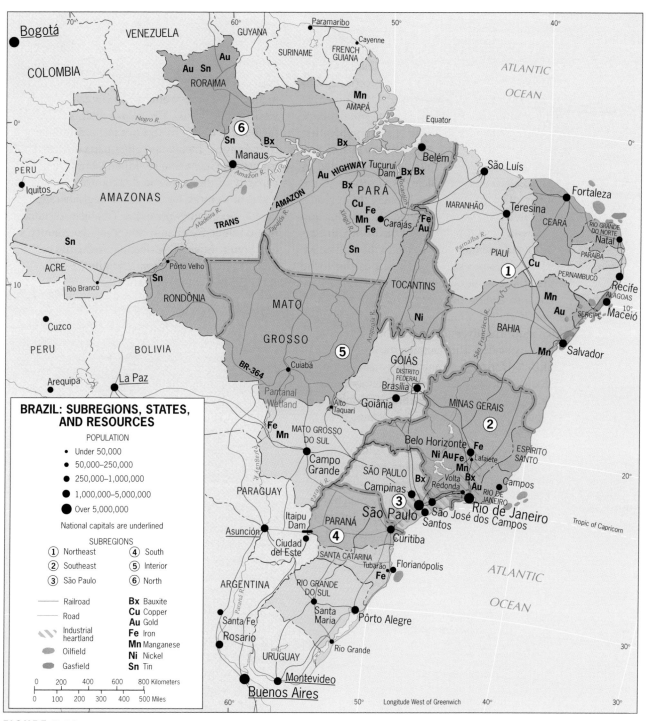

FIGURE 5-13

issues guaranteed the defeat of the right-of-center government held responsible for it all. In that election a leftist candidate, Luiz Inácio Lula da Silva, won in a landslide victory. He faced the daunting task of adhering to Brazil's existing financial commitments while steering a new course toward more fairness, openness, and honesty in government.

Brazil's Subregions

Brazil is a federal republic consisting of 26 States and the federal district of the capital, Brasília (Fig. 5-13). As in the United States, the smallest States lie in the northeast and the larger ones farther west; their populations range from 3 million in the huge, sparsely populated State of Amazonas to 37 million in burgeoning São Paulo State. Although Brazil is about as large as the 48 contiguous United States, it does not exhibit a clear physiographic regionalism. Even the Amazon Basin, which covers almost 60 percent of the country, is not entirely a plain: between the tributaries of the great river lie low but extensive tablelands. Given this physiographic ambiguity, the six Brazilian subregions discussed next have no absolute or even generally accepted boundaries. In Figure 5-13, those boundaries have been drawn to coincide with the borders of States, making identifications easier.

The **Northeast** was Brazil's source area, its culture hearth. The plantation economy took root here at an early date, attracting Portuguese planters who soon imported the country's largest group of African slaves to work in the sugar fields. But the ample rainfall occurring along the coast soon gives way to lower and more variable patterns in the interior, which is home to about half of the region's 50 million people. This drier inland backcountry—called the *sertão*—is not only seriously overpopulated but also contains some of the worst poverty to be found anywhere in the Americas. The Northeast produces less than one-sixth of Brazil's gross domestic product, but its inhabitants constitute almost one-third of the national population. Given this staggering imbalance, it is not surprising that the region contains half of the country's poor, a literacy rate 20 percent below Brazil's mean, and an infant mortality rate twice the national average.

Much of the Northeast's misery is rooted in its unequal system of land tenure. Farms must be at least 250 acres to be profitable in the hardscrabble *sertão*, a size that only large landowners can afford. Moreover, the Northeast is plagued by a monumental environmental problem: the recurrence of devastating 16 droughts at least partly attributable to **El Niño** (periodic sea-surface-warming events off the continent's northwestern coast). Indeed, extended dry periods occur so regularly that the region has become known as the *Polygon of Drought*.

Understandably, these conditions propel substantial emigration toward the coastal cities and, increasingly, out of the Northeast to the more prosperous Brazilian subregions that line the Atlantic seaboard to the south. To stem that human tide, Brazil's government has pursued the purchase of underutilized farmland for the settling of landless peasants; although tens of thousands of families have benefited, overall barely a dent has been made in the Northeast's massive rural poverty crisis.

The Northeast today is Brazil's great contradiction. In cities such as Recife and Salvador, hordes of peasants driven from the land constantly arrive to expand the surrounding shantytowns. As yet, few of the generalizations about emerging Brazil apply here, but there are some bright spots. A petrochemical complex has been built near Salvador, creating thousands of jobs and luring foreign investment. Irrigation projects have nurtured a number of productive new commercial agricultural ventures. Tourism is booming along the entire Northeast coast, whose thriving beachside resorts attract thousands of vacationing Europeans. Recife has spawned a budding software industry and a major medical complex. And Fortaleza is the center of new clothing and shoe industries that have already put the city on the global economic map.

The **Southeast** has been modern Brazil's *core area*, with its major cities and leading population clusters. Gold first drew many thousands of settlers, and other mineral finds also contributed to the influx—with Rio de Janeiro itself serving as the terminus of the "Gold Trail." Later, the region's agricultural possibilities ensured its stabilization and growth. The third quarter of the twentieth century brought another mineral age to the Southeast, based on the iron ores around Lafaiete carried to the steelmaking complex at Volta Redonda (Fig. 5-13).

The surrounding State of Minas Gerais (the name means "General Mines") formed the base from which industrial diversification in the Southeast has steadily expanded. The booming metallurgical center of Belo Horizonte paved the way and is now the endpoint of a rapidly developing, ultramodern manufacturing corridor that stretches 300 miles (500 km) southwest to metropolitan São Paulo.

São Paulo State is the leading industrial producer and primary focus of ongoing Brazilian development. This economic-geographic powerhouse accounts for nearly half of the country's gross domestic product, with an economy that today matches Argentina's in overall size. Not surprisingly, this region is growing phenomenally (it already contains 21 percent of Brazil's population) as a magnet for migrants, especially from the Northeast.

The wealth of São Paulo State was built on its coffee plantations (known as *fazendas*), and Brazil is still the world's leading producer. But coffee today has been eclipsed by other farm commodities. One of them is orange juice concentrate (here, too, Brazil leads the world). São Paulo State now produces more than double the annual output of Florida, thanks to a climate all but devoid of winter freezes, to ultramodern processing plants, and to a fleet of specially equipped tankers that ship the concentrate to foreign markets. Another leading pursuit is soybeans, in which Brazil ranks second among the world's producers.

Matching this agricultural prowess is the State's industrial strength. The revenues derived from the coffee plantations provided the necessary investment capital, ores from Minas Gerais supplied the vital raw materials, the nearby outport of Santos facilitated access to the ocean, and immigration from Europe, Japan, and other parts of Brazil contributed the increasingly skilled labor force. As the capacity of the domestic market grew, the advantages of central location and agglomeration secured São Paulo's primacy. This also resulted in metropolitan São Paulo becoming the country's—and South America's—leading industrial complex and megacity (population: 25 million).

The **South** consists of three States, whose combined population exceeds 25 million: Paraná, Santa Catarina, and Rio Grande do Sul. Southernmost Brazil's excellent agricultural potential has long attracted large numbers of European immigrants. Here the newcomers introduced their advanced farming methods to several areas. Portuguese rice farmers clustered in the valleys of Rio Grande do Sul, where tobacco production has now propelled Brazil to become the world's leading exporter. The Germans, specialists in raising grain and cattle, occupied the somewhat higher areas to the north and in Santa Catarina. The Italians selected the highest slopes, where they established thriving vineyards. All of these fertile lands proved highly productive, and with growing markets in the large urban areas to the north, this tristate subregion became Brazil's most affluent corner.

With the South firmly rooted in the European-commercial culture sphere (Fig. 5-5), European-style standards of living match the diverse Old World heritage that is reflected in the towns and countryside (where German and Italian are spoken alongside Portuguese). This has led to hostility against non-European Brazilians, and many communities actively discourage poor, job-seeking migrants from the north by offering to pay return bus fares or even blocking their household-goods-laden vehicles. Moreover, extremist groups have

FROM THE FIELDNOTES

"Having seen Rio de Janeiro from the air, from the sea, from its avenues, and from the alleys of two of its hillside *favelas*, I can attest to the searing inequities in this great city's social geography. With nearly 12 million inhabitants, Rio lost its capital functions to Brasília and its demographic lead to São Paulo, but it remains Brazil's cultural heart, its educational core, and its international gateway. Fortunate *cariocas* (as Rio de Janeiro's residents call themselves) live in modern highrises overlooking Ipanema Beach while the poor have million-dollar views from the shacks of Rocinho."

arisen to openly espouse the secession of the South from Brazil. The largest, calling itself "The South Is My Country," promotes the creation of a *Republic of the Pampas*. Accusing the separatist movement of subversion and racism, State governments respond with the claim that the issue is an economic one, complaining to Brasília that the South receives only about a 60 percent return on the federal taxes it pays. But with polls showing at least one-third of the population sympathetic to the idea of secession, officialdom may well have a longer-term devolutionary struggle on its hands.

Economic development in the South is not limited to the agricultural sector. Coal from Santa Catarina and Rio Grande do Sul is shipped north to the steel plants of Minas Gerais. Local manufacturing is growing as well, especially in Pôrto Alegre and Tubarão. During the 1990s, a major center of the computer software industry was established in Florianópolis, the island city and State capital just off Santa Catarina's coast. Known as *Tecnópolis*, this budding technopole continues to grow by capitalizing on its seaside amenities, skilled labor force, superior air-travel and global communications linkages, and government and private-sector incentives to support new companies. Only the sparsely populated interior portion of the South has lagged behind the rest of the region, despite the recent completion of massive Itaipu Dam in westernmost Paraná State.

The **Interior** subregion—constituted by the States of Goiás, Mato Grosso, and Mato Grosso do Sul—is also known as the Central-West. This is the region that Brazil's developers have long sought to make a part of the country's productive heartland, and in 1960 the new capital of Brasília was deliberately situated on its margins (Fig. 5-13).

By locating the new capital city in the wilderness 400 miles (650 km) inland from its predecessor, Rio de Janeiro, the nation's leaders dramatically signaled the opening of Brazil's development thrust toward the west. Brasília is also noteworthy in another regard because it represents what political geographers [17] call a **forward capital**. A state will sometimes relocate its capital to a sensitive area, perhaps near a peripheral zone under dispute with an unfriendly neighbor, in part to confirm its determination to sustain its position in that contested zone. Brasília does not lie near a contested area, but Brazil's interior was an internal frontier to be conquered by a growing nation. Spearheading that drive, the new capital occupied a decidedly forward position.

Despite the subsequent growth of Brasília to 2.5 million inhabitants today, it was not until the 1990s that the Interior began its economic integration with the rest of Brazil. The catalyst was the exploitation of the vast *cerrado*—the fertile savannas that blanket the Central-West and make it one of the world's most promising agricultural frontiers (at least two-thirds of its arable land still awaits development). As with the U.S. Great Plains, the flat terrain of the *cerrado* is one of its main advantages because it facilitates the large-scale mechanization of farming with a minimal labor force. Another advantage is rainfall, which is more prevalent than in the Great Plains or Argentine Pampa (Fig. G-7).

The leading crop is soybeans, whose output per acre here exceeds even that of the U.S. Corn Belt. Other grains and cotton are also expanding across the farmscape of the *cerrado*, but the current pace of regional development is inhibited by a serious accessibility problem. Unlike the Great Plains and Pampa, the growth of an efficient transportation network did not accompany the opening of this farming frontier. Thus the Interior's products must travel along poor roads and intermittent railroads to reach the markets and ports of the Atlantic seaboard. Today several projects are finally underway to alleviate these bottlenecks, including the privately financed *Ferronorte* railway that links Santos to the southeastern corner of Mato Grosso State.

The **North** is Brazil's territorially largest and most rapidly developing subregion, which consists of the seven States of the Amazon Basin (Fig. 5-13). This was the scene of the great rubber boom a century ago, when the wild rubber trees in the *selvas* (tropical rainforests) produced huge profits and the central Amazon city of Manaus enjoyed a brief period of wealth and splendor. But the rubber boom ended in 1910, and for most of the seven decades that followed Amazonia was a stagnant hinterland lying remote from the centers of Brazilian settlement. All that changed dramatically during the 1980s as new development began to stir throughout this awakening region, which currently is the scene of the world's largest migration into virgin territory as over 200,000 new settlers arrive each year. Most of this influx is occurring south of the Amazon River, in the tablelands between the major waterways and along the Basin's wide rim.

Two ongoing development schemes are especially worth noting. The first is the *Grande Carajás Project* in southeastern Pará State, a huge multifaceted scheme centered on one of the world's largest known deposits of iron ore in the hills around Carajás (Fig. 5-13). In addition to a vast mining complex, other new construction here includes the Tucuruí Dam on the nearby Tocantins River and a 535-mile (850-km) railroad to the Atlantic port of São Luis. This ambitious development project also emphasizes the exploitation of additional minerals, cattle raising, crop farming, and forestry. What is occurring is an application of 18 the **growth-pole concept**. A growth pole is a location where a set of activities, given a start, will expand and generate widening ripples of development in the surrounding area. In this case, the stimulated hinterland could one day cover one-sixth of all Amazonia.

Understandably, tens of thousands of settlers have descended on this part of the Amazon Basin. Those seeking business opportunities have been in the vanguard, but they have been followed by masses of lower-income laborers and peasant farmers in search of jobs and land ownership. The initial stage of this colossal enterprise has boosted the fortunes of many towns, particularly the revival of Manaus northwest of Carajás. Here, a thriving industrial complex (specializing in the production of electronic goods) has emerged in the free-trade zone adjoining the city thanks to the outstanding air-freight operations at Manaus' ultramodern airport. But many problems have also arisen as the tide of pioneers rolled across central Amazonia. One of the most tragic involved the Yanomami people, whose homeland in Roraima State was overrun by thousands of claim-stakers (in search of newly discovered gold) who triggered

FROM THE FIELDNOTES

"Near the waterfront in Belém lie the now-deteriorating, once-elegant streets of the colonial city built at the mouth of the Amazon. Narrow, cobblestoned, flanked by tiled frontages and arched entrances, this area evinces the time of Dutch and Portuguese hegemony here. Mapping the functions and services here, we recorded the enormous diversity of activities ranging from carpentry shops to storefront restaurants and from bakeries to clothing stores. Dilapidated sidewalks were crowded with shoppers, workers, and people looking for jobs (some newly arrived attracted by perceived employment opportunities in this growing city of 1.9 million). The diversity of population in this and other tropical South American cities reflects the varied background of the region's peoples and the wide hinterland from which these urban magnets have drawn their inhabitants."

violent confrontations that ravaged the fragile aboriginal way of life.

The second leading development scheme, known as the *Polonoroeste Plan*, is located about 500 miles (800 km) to the southwest of Grande Carajás in the 1500-mile-long Highway BR-364 corridor that parallels the Bolivian border and connects the western Brazilian towns of Cuiabá, Pôrto Velho, and Rio Branco (Fig. 5-13). Although the government had planned for the penetration of western Amazonia to proceed via the east-west Trans-Amazon Highway, the migrants of the 1980s and 1990s preferred to follow BR-364 and settle within the Basin's southwestern rim zone, mostly in Rondônia State. Agriculture has been the dominant activity here, but in the quest for land, bitter conflicts continue to break out between peasants and landholders as the Brazilian government pursues the volatile issue of land reform.

The usual pattern of settlement in this part of the Brazilian North is something like this. As main and branch highways are cut through the wilderness, settlers, enticed by cheap land, follow and move out laterally to clear spaces for farming. Crops are planted, but within three years the heavy equatorial-zone rains leach out soil nutrients and accelerate surface erosion. As soil fertility declines, pasture grasses are planted, and the plot of land is soon sold to cattle ranchers. The peasant farmers then move on to newly opened areas, clear more land for planting, and the cycle repeats itself. Unfortunately, this not only establishes low-grade land uses across widespread areas but also requires the burning and clearing of enormous stands of tropical woodland. Since the 1980s, an area of rainforest almost the size of Ohio has been disappearing *annually* in Amazonia—accounting for over half of all tropical deforestation and exacerbating an environmental crisis of global proportions.

Brazil is the cornerstone of South America, the dominant economic force in Mercosur/Mercosul, the only dimensional counterweight to the United States in the Western Hemisphere, a maturing democracy, and an emerging global giant. The future of the entire South American realm depends on Brazil's stability and social as well as economic progress.

DISCOVERY TOOLS

www.wiley.com/college/deblij

The *Concepts and Regions* Website, featuring *GeoDiscoveries*, offers many additional resources to enhance your understanding and experience of this chapter. Be sure to explore the following:

GeoDiscoveries Video Clips
Urbanization of South America
The Central Business District

GeoDiscoveries Animations
Exploring the Griffin-Ford Model
Comparing Land-Use Patterns

GeoDiscoveries Interactivities
The Griffin-Ford Model
Major South American Cities
Rank the Cities by Population

Photo Galleries
Salvador, Brazil
Punta Arenas, Chile
Rio de Janeiro, Brazil
Pisac, Peru
Lima, Peru
Belém, Brazil
El Cojo, Venezuela
Buenos Aires, Argentina
Callao, Peru
Avenida Florída, Buenos Aires, Argentina
Otavalo, Ecuador
Mendoza, Argentina
Pan-American Highway, Chile

Virtual Field Trip
Brazil's Carnaval

More To Explore
Systematic Essay: Economic Geography
Issues in Geography: The Guianas, The Geography of Cocaine
Major Cities: Lima, Buenos Aires, Rio de Janeiro, Sao Paulo

Expanded Regional Coverage
The North, The West, The South, Brazil

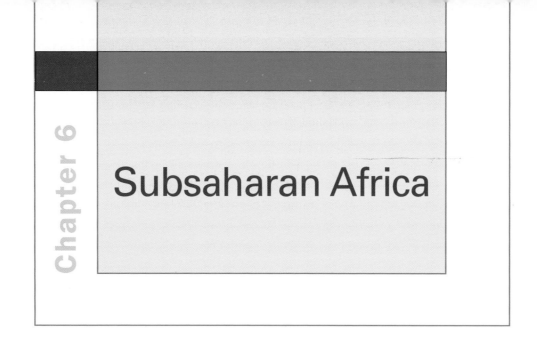

Chapter 6

Subsaharan Africa

CONCEPTS, IDEAS, AND TERMS

1 **Human evolution**
2 **Rift valley**
3 **Continental drift**
4 **Medical geography**
5 **Endemic**
6 **Epidemic**
7 **Pandemic**

8 **Land tenure**
9 **Land alienation**
10 **Green Revolution**
11 **State formation**
12 **Colonialism**
13 **Multilingualism**

14 **Apartheid**
15 **Separate development**
16 **Landlocked state**
17 **Exclave**
18 **Periodic market**
19 **Islamic Front**

REGIONS

▶ **SOUTHERN AFRICA**
▶ **EAST AFRICA**
▶ **EQUATORIAL AFRICA**

▶ **WEST AFRICA**
▶ **AFRICAN TRANSITION ZONE**

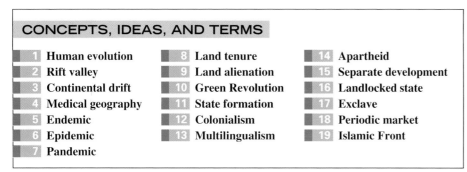

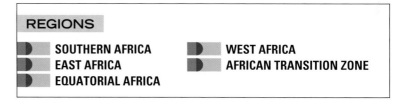

THE AFRICAN CONTINENT occupies a special place in the physical as well as the human world. You need a globe to confirm the former: the Earth has a Land Hemisphere and a Water Hemisphere, and Africa lies at the center of the Land Hemisphere, surrounded in all directions by other land masses. As for Africa's human significance, this is where the great saga 1 of **human evolution** began. In Africa we formed our first communities, spoke our first words, created our first art and our first weapons.

From Africa our ancestor hominids spread outward into Eurasia more than two million years ago. From Africa our own species emigrated, beginning perhaps 130,000 years ago, northward into present-day Europe and eastward via southern Asia into Australia and, much later, farther afield into the Americas. Disperse our forebears did, but we should remember that, at the source, we are all Africans.

For millions of years, therefore, Africa served as the cradle for humanity's emergence. For tens of

thousands of years, Africa was the source of human cultures. For thousands of years, Africa led the world in countless spheres ranging from tool manufacture to plant domestication.

But in this chapter we will encounter an Africa that has been struck by a series of disasters ranging from environmental deterioration to human dislocation on a scale unmatched anywhere in the world. When we assess Africa's misfortunes, though, we should remember that they have lasted hundreds, not thousands or millions of years. Africa's catastrophic

193

STUDY TOOLS

3D GLOBE
AREA AND DEMOGRAPHIC DATA

FLASHCARDS
MAP QUIZZES

AUDIO PRONUNCIATION GUIDE
CHAPTER QUIZ

ANNOTATED WEBLINKS
LONELY PLANET WEBLINKS

interlude will end, and Africa's time and turn will come again.

The focus in this chapter will be on Africa south of the Sahara, for which the unsatisfactory but convenient name *Subsaharan Africa* has come into use to signify not physically "under" the great desert but directionally "below" it. The African continent contains two geographic realms: the African, extending from the southern margins of the Sahara to the Cape of Good Hope, and the western flank of the realm dominated by the Muslim faith and Islamic culture whose heartland lies in the Middle East and the Arabian Peninsula. The great desert forms a formidable barrier between the two, but

the powerful influences of Islam crossed it centuries before the first Europeans set foot in West Africa. By that time, the African kingdoms in what is known today as the Sahel had been converted, creating an Islamic foothold all along the northern periphery of the African realm (see Fig. G-2, p. 4). As we note later, this cultural and ideological penetration had momentous consequences for Subsaharan Africa.

In the three previous chapters on the Americas, we made frequent reference to the forced migration of Africans to Brazil, the Caribbean region, and the United States. The slave trade was one of those African disasters alluded to above, and it was

facilitated in part by what we may call the peril of proximity. The northeastern tip of Brazil, by far the largest single destination for the millions of Africans forced from their homes in bondage, lies about as far from the nearest West African coast as South Carolina lies from Venezuela, a short maritime intercontinental journey indeed (it is more than twice as far from West Africa to South Carolina). That proximity facilitated the forced migration of millions of West Africans to Brazil, which in turn contributed to the emergence of an African cultural diaspora in Brazil that is without equal in the New World. It is therefore logical to focus next on the African realm.

DEFINING THE REALM

The African continent may be partitioned into two human-geographic realms, but the landmass is indivisible. Before we investigate the human geography of Subsaharan Africa, therefore, we should take note of the entire continent's unique physical geography (Fig. 6-1). We have already noted Africa's situation at the center of the planet's Land Hemisphere; moreover, no other landmass is positioned so squarely astride the equator, reaching almost as far to the south as to the north. This location has much to do with the distribution of Africa's climates, soils, vegetation, agricultural potential, and human population.

AFRICA'S PHYSIOGRAPHY

Africa accounts for about one-fifth of the Earth's entire land surface. The north coast of Tunisia lies 4800 miles (7700 km) from the south coast of South Africa. Coastal Senegal, on the extreme western *Bulge* of Africa, lies 4500 miles (7200 km) from the tip of the *Horn* in easternmost Somalia. These distances have environmental implications. Much of Africa is far from maritime sources of moisture. In addition, as Figure G-8 shows, large parts of the

landmass lie in latitudes where global atmospheric circulation systems produce arid conditions. The Sahara in the north and the Kalahari in the south form part of this globe-girdling desert zone. Water supply is one of Africa's great problems.

Africa's topography reveals several properties that are not replicated on other landmasses. Alone among the continents, Africa does not have an Andes-like linear mountain backbone; neither the northern Atlas nor the southern Cape Ranges are in the same league. Where Africa does have high mountains, as in Ethiopia and South Africa, these are really deeply eroded plateaus or, as in East Africa, high, snowcapped volcanoes. Furthermore, Africa is one of only two continents containing a cluster of Great Lakes, and the only one whose lakes result from powerful tectonic forces in the Earth's crust. These lakes (with the exception of Lake Victoria) lie **2** in deep trenches called **rift valleys**, which form when huge parallel cracks or faults appear in the Earth's crust and the strips of crust between them sink, or are pushed down, to form great, steep-sided, linear valleys. In Figure 6-1 these rift valleys, which stretch more than 6000 miles (9600 km) from the Red Sea to Swaziland, are marked by red lines.

Africa's rivers, too, are unusual: their upper courses often bear landward, seemingly unrelated to the coast toward which they eventually flow. Several rivers, such as the Nile and the Niger, have inland as well as coastal deltas. Major waterfalls, notably Victoria Falls on the Zambezi, or lengthy systems of cataracts, separate the upper from the lower courses.

Finally, Africa may be described as the "plateau continent." Except for some comparatively limited coastal plains, almost the entire continent lies above 1000 feet (300 m) in elevation, and fully half of it lies over 2500 feet (800 m) high. As Figure 6-1 shows, the plateau surface has sagged under the weight of accumulating sediments into a half dozen major basins (three of them in the Sahara). The margins of Africa's plateau are marked by escarpments, often steep and step-like. Most notable among these is the Great Escarpment of South Africa, marking the eastern edge of the Drakensberg Mountains.

Continental Drift and Plate Tectonics

Africa's remarkable and unusual physiography was one piece of evidence that geographer Al-

MAJOR GEOGRAPHIC QUALITIES OF **SUBSAHARAN AFRICA**

1. Physiographically, Africa is a plateau continent without a linear mountain backbone, with a set of Great Lakes, variable rainfall, generally low fertility soils, and mainly savanna and steppe vegetation.

2. Dozens of nations, hundreds of ethnic groups, and many smaller entities make up Subsaharan Africa's culturally rich and varied population.

3. Most of Subsaharan Africa's peoples depend on farming for their livelihood.

4. Health and nutritional conditions in Subsaharan Africa need improvement as the incidence of disease remains high and diets are often unbalanced. The AIDS pandemic began in Africa and has become a major health crisis in this realm.

5. Africa's boundary framework is a colonial legacy; many boundaries were drawn without adequate knowledge of or regard for the human and physical geography they divided.

6. The realm is rich in raw materials vital to industrialized countries, but much of Subsaharan Africa's population has little access to the goods and services of the world economy.

7. Patterns of raw-material exploitation and export routes set up in the colonial period still prevail in most of Subsaharan Africa. Interregional and international connections are poor.

8. During the Cold War, great-power competition magnified conflicts in several Subsaharan African countries, with results that will be felt for generations.

9. Severe dislocation affects many Subsaharan African countries, from Liberia to Rwanda. This realm has the largest refugee population in the world today.

10. Government mismanagement and poor leadership afflict the economies of many Subsaharan African countries.

---

fred Wegener used to construct his hypothesis of **3** **continental drift**. The present continents, Wegener reasoned, lay assembled as one giant landmass called *Pangaea* not very long ago in geologic time (220 million years ago). The southern part of this supercontinent was *Gondwana*, of which Africa formed the core (Fig. 6-2). When, about 200 million years ago, tectonic forces began to split Pangaea apart, Africa (and the other landmasses) acquired their present configurations. That process, now known as *plate tectonics*, continues, marked by earthquakes and volcanic eruptions. By the time it started, however, Africa's land surface had begun to acquire some of the features that mark it today— and make it unique. The rift valleys, for example, demarcate the zones where plate movement continues—hence the linear shape of the Red Sea, where the Arabian Plate is separating from the African

Plate (Fig. G-4). And yes, the rift valleys of East Africa probably mark the further fragmentation of the African Plate (some geophysicists have already referred to a "Somali Plate" which will separate, Madagascar-like, from the rest of Africa).

So Africa's ring of escarpments, its rifts, its river systems, its interior basins, and its lack of Andes-like mountains, all relate to the continent's central location in Pangaea, all pieces of the puzzle that led to the plate-tectonics solution. Once again, geography was the key.

NATURAL ENVIRONMENTS

Only the southernmost tip of Subsaharan Africa lies outside the tropics. Although African elevations are comparatively high, they are not high enough to

ward off the heat that comes with tropical location except in especially favored locales such as the Kenya Highlands and parts of Ethiopia. And, as we have noted, Africa's bulky shape means that much of the continent lies far from maritime moisture sources. Variable weather and frequent droughts are among Africa's environmental problems.

It is useful at this point to refer back to Figure G-8 on page 12. As that map shows, Africa's climatic regions are distributed almost symmetrically about the equator, though more so in the center of the landmass than in the east, where elevation changes the picture. The hot, rainy climate of the Congo Basin merges gradually, both northward and southward, into climates with distinct winter-dry seasons. "Winter," however, is marked more by drought than by cold. In parts of the area mapped *Aw* (savanna), the annual seasonal cycle

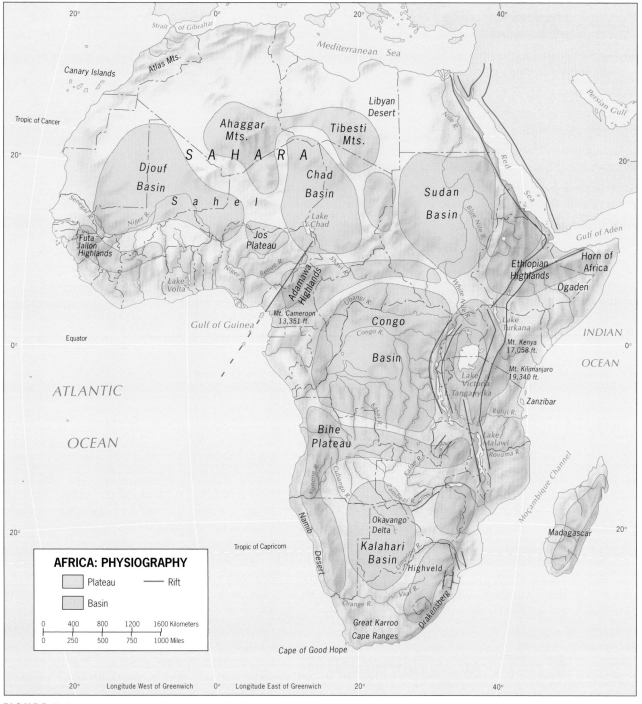

FIGURE 6-1

produces two rainy seasons, often referred to locally as the "long rains" and the "short rains," separated by two "winter" dry periods. As you go farther north and south, away from the moist Congo Basin, the dry season(s) grow longer and the rainfall diminishes and becomes less and less dependable.

Most Africans make their living by farming, and many grow crops in marginal areas where rainfall variability can have catastrophic consequences. Compare Figures G-7 and G-8 and note the steep decline in annual rainfall from more than 80 inches (200 cm) around the equator in the Congo Basin to a mere 4 inches (10 cm) in parts of Chad to the north and Namibia to the south. Millions of farmers till the often unproductive soils of the *savanna*, and many mix livestock herding with agriculture to reduce their risk in this difficult environment. But the savanna's wildlife carries diseases that also infect livestock, which makes herding a risky proposition as well. Even where the savanna gives way to the still-drier *steppe*, human pressure continues to grow and people as well as animals trample fragile, desert-margin ecologies.

Africa's shrinking rainforests and vast savannas form the world's last refuges for wildlife ranging from primates to wildebeests. Gorillas and chimpanzees survive in dwindling numbers in threatened forest habitats, while millions of herbivores range in great herds across the savanna plains where people compete with them for space. European colonizers, who introduced hunting as a "sport" (a practice that was not part of African cul-

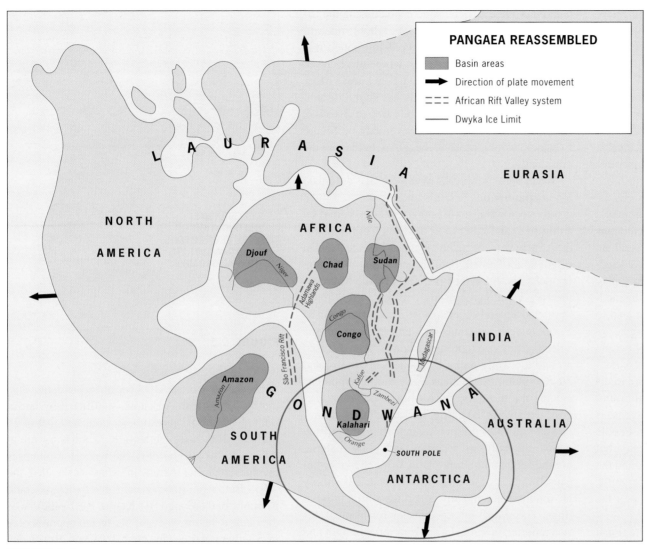

FIGURE 6-2

and animals in Africa has taken a new turn. It is the end of an era.

And yet it would seem that there could be room for humans as well as wildlife in Subsaharan Africa. As Table G-1 confirms, all the countries of this realm *combined* have a population only about half that of China alone. Indeed, Africa does have some areas of good soils, ample water, and high productivity: the volcanic soils of Mount Kilimanjaro and those of the highlands around the Western Rift Valley, the soils in the Ethiopian Highlands, those in the moister areas of higher-latitude South Africa and parts of West Africa yield good crops when social conditions do not disrupt the farming communities. But such areas are small in the vastness of Africa. This realm has nothing to compare to the huge alluvial basins of India and China, not even a Nile Valley and Delta, where comparatively small areas of fertile, irrigated soils can support tens of millions of people. Except for small (in some cases experimental) patches of rice and wheat, Africa is the land of corn (maize), millet, and root crops, far less able to provide high per-acre yields. Africa's natural environment poses a formidable challenge to the millions who depend directly on it.

tural traditions) and who brought their capacities for mass destruction to animals as well as people in Africa, helped clear vast areas of wildlife and push species to near-extinction. Later, they laid out game reserves and other types of conservation areas, but these were not sufficiently large or well enough connected to allow herd animals to follow their seasonal and annual migration routes. The same climatic variability that affects farmers also affects wildlife, and when the rangelands wither, the animals seek better pastures. When the fences of a game reserve wall them off, they cannot survive. When there are no fences, the wildlife invades neighboring farmlands and destroys crops, and the farmers retaliate. After thousands of years of equilibrium, the competition between humans

ENVIRONMENT AND HEALTH

From birth, Africans (especially rural people living in *A* climates) are exposed to a wide range of diseases spread by insects and other organisms. The study of human health in spatial context is the field of **4** **medical geography**, and medical geographers employ modern methods of analysis (including geographic information systems) to track disease outbreaks, identify their sources, detect their carriers, and prevent their repetition. Alliances between doctors and geographers have already yielded significant results. Doctors know how a disease ravages the body; geographers know how climatic conditions such as wind direction or variations in river flow can affect the distribution and effectiveness of disease carriers. This collaboration helps protect vulnerable populations.

Tropical Africa, the source of many serious illnesses, is the focus of much of medical geography's work. Not only the carriers (*vectors*) of infectious diseases but also cultural traditions that facilitate transmission, such as sexual practices, food selection and preparation, and personal hygiene, play their role—and all can be mapped. Comparing medical, environmental, and cultural maps can lead to crucial evidence that helps combat the scourge.

In Africa today, hundreds of millions of people carry one or more maladies, often without knowing exactly what ails them. A disease that infects many people (the *hosts*) in a kind of equilibrium, without causing rapid and widespread deaths, is **5** said to be **endemic** to the population. People affected may not die suddenly or dramatically, but their health deteriorates, energy levels fall, and the quality of life declines. In tropical Africa hepatitis, venereal diseases, and hookworm are among the public health threats in this category.

When a disease outbreak has local or regional **6** dimensions, it is called **epidemic**. It may claim thousands, even tens of thousands of lives, but it remains confined to a certain area, perhaps one defined by the range of its vector. In tropical Africa, trypanosomiasis, the disease known as sleeping sickness and vectored by the tsetse fly, has regional proportions. The great herds of savanna wildlife form the *reservoir* of this disease, and the tsetse fly transmits it to livestock and people. It is endemic to the wildlife, but it kills cattle, so Africa's herders try to keep their animals in tsetse-free zones. African sleeping sickness appears to have origi-

nated in a West African source area during the fifteenth century, and it spread throughout much of tropical Africa (Fig. 6-3). Its epidemic range was limited by that of the tsetse fly: where there are no tsetse flies, there is no sleeping sickness. More than anything else, the tsetse fly has kept Subsaharan Africa's savannalands largely free of livestock and open to wildlife. Should a remedy be found, livestock would replace the great herds on the grasslands.

When a disease spreads worldwide, it is described **7** as **pandemic**. Africa's and the world's most deadly vectored disease is malaria, transmitted by a mosquito and killer of as many as 1 million children each year. Whether malaria has an African origin is not known, but it is an ancient affliction. Hippocrates, the Greek physician of the fifth century BC, mentions it in his writings. Apes, monkeys, and several other species also suffer from it. Fever attacks, anemia, and enlargement of the spleen are its symptoms. Malaria has diffused around the world and prevails not only in tropical but also in temperate areas. Eradication campaigns against the mosquito vector have had some success, but always the carrier has come back with renewed vigor. At present, as many as 300 million people annually are recorded as having malaria, but the actual number probably is much higher because in much of tropical Africa malaria is simply an omnipresent malady affecting entire populations. The short life expectancies for tropical Africa reported in Table G-1 in part reflect infant and child mortality from malarial infection.

Another pandemic disease, also vectored by a mosquito and with African origins, is yellow fever. It crossed the Atlantic, reached South and Middle America, and even penetrated the United States during the nineteenth century; its defeat in Panama nearly a century ago made possible the construction of the Panama Canal.

But all these (and many other) maladies are overshadowed today by Africa's latest scourge: AIDS. Monkeys may have activated this disease in humans (monkeys are part of the diet in parts of Africa), but

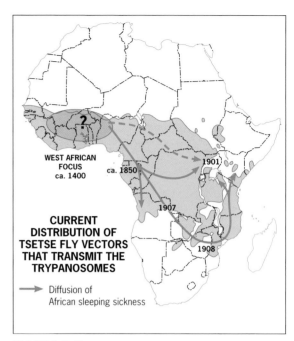

CURRENT DISTRIBUTION OF TSETSE FLY VECTORS THAT TRANSMIT THE TRYPANOSOMES

WEST AFRICAN FOCUS ca. 1400

ca. 1850

1901

1907

1908

→ Diffusion of African sleeping sickness

FIGURE 6-3

whatever the source, AIDS became a pandemic during the 1990s with Africa worst afflicted by far. In 2004, United Nations estimates indicated that of 38 million AIDS cases worldwide, 32 million were in Africa. In the early 1990s, the worst-hit countries lay in Equatorial Africa in what was called the AIDS Belt, from (then) Zaïre (now the Democratic Republic of the Congo*) to Kenya. But by the end of the decade the most severely affected countries lay in Southern Africa. In Zimbabwe and Botswana, more than 25 percent of persons aged 15 to 49 were infected with HIV; in Zambia, nearly 20 percent; and in South Africa, the region's most populous country, 13 percent. These are the official numbers; medical geographers estimate that between 20 and 25 percent of the *entire* population of several tropi-

---

*Two countries in Africa have the same short-form name: Congo. In this book, we use *The Congo* for the larger Democratic Republic of the Congo and *Congo* for the smaller Republic of Congo.

cal African countries is infected. Africa's vital statistics are showing the demographic impact. In South Africa, life expectancy declined from 66 to 47 in just ten years. In Zimbabwe, population growth in the early 1980s was 3.3 percent; today it is less than 1 percent. UN projections suggest that by 2010, AIDS will be responsible for 20 million orphans in Africa. It is another health-related disaster that is wrecking families and communities, ruining budgets, and devastating national and local economies.

LAND AND FARMING

In their penetrating book, *Geography of Sub-Saharan Africa*, editor Samuel Aryeetey-Attoh and his colleagues focus on the issue of land tenure, crucial in 8 Africa because most Africans are farmers. **Land tenure** refers to the way people own, occupy, and use land. African traditions of land tenure are different from those of Europe or the Americas. In most of Subsaharan Africa, communities, not individuals, customarily hold land. Occupants of the land have temporary, custodial rights to it and cannot sell it. Land may be held by large (extended) families, by a village community, or even by a traditional chief who holds the land in trust for the people. His subjects may house themselves on it and farm it, but in return they must follow his rules.

When the European colonizers took control of much of Subsaharan Africa, their land ownership practices clashed head-on with those of Africa. Africans believed that their land belonged to their ancestors, the living, and the yet-unborn; Europeans saw unclaimed space and felt justified in claiming it. 9 What Africans called **land alienation**—the expropriation of (often the best) land by Europeans— changed the pattern of land tenure in Africa. By the time the Europeans withdrew, private land ownership was widely dispersed and could not be reversed. Postcolonial African states tried to deal with this legacy by nationalizing all the land and doing away with private ownership, reverting in theory to the role of the traditional chief. But this policy has not worked well. In the rural areas, the government in the capital is a remote authority often seen as unsympathetic to the plight of farmers. Governments keep the price of farm products low on urban markets, pleasing their supporters but frustrating farmers. Large landholdings once owned by Europeans seem now to be occupied by officials or those the government favors. The colonial period's approach to land tenure left Africa with a huge social problem.

Rapid population growth, as Africa experienced early in the postcolonial period, makes the problem even worse. Traditional systems of land tenure, which involve subsistence farming in various forms ranging from shifting cultivation to pastoralism, work best when population is fairly stable. Land must be left fallow to recover from cultivation and pastures kept free of livestock so the grasses can revive. A population explosion of the kind Africa experienced during the mid-twentieth century destroys this equilibrium. Soils cannot rest, and pastures are overgrazed. As land becomes degraded, yields decline.

Although there is considerable commercial farming in Africa (as we will note in the regional section of this chapter), most African farmers remain subsistence farmers who grow grain crops (corn, millet, sorghum) in drier areas and root crops (yams, cassava, sweet potatoes) where moisture is ample. Others herd livestock (mostly cattle and goats) in constant search of water and pasture, sometimes clashing with sedentary farmers whose fields their animals invade. Although subsistence is the primary objective, these farmers will take some of their yields to market in good years. But African governments, eager to please their city-dwellers, have tended to keep produce prices low, thereby discouraging their rural, less influential citizens.

African farmers adapt to the problems they face in remarkable ways, but farm production in Africa has been declining for many years. Governments bent on high-profile industrial projects neglected agriculture (Nigeria is a tragic example); rural needs (roads, dams, electricity) were ignored; and banks would not lend to needy farmers. All this has been especially hard on Africa's women who, according to current estimates, produce 75 percent of local food in Subsaharan Africa.

10 Even the **Green Revolution**—the development of more productive, drought-tolerant, pest-resistant,

FROM THE FIELDNOTES

"Visiting the village in the Kenya Highlands where a graduate student was doing fieldwork on land reform, I took the long way and drove along the top of the eastern wall of the Eastern Rift Valley. Often the valley wall is not sheer but terraced, and soils on those terraces are quite fertile; also the west-facing slopes tend to be well-watered. Here African farmers built villages and laid out communal plots, farming these lands in a well-organized way long before the European intrusion."

higher-yielding types of grain—has had less impact in Africa than elsewhere. Where people depend mainly on rice and wheat, the Green Revolution pushed back the specter of hunger. But Africa's crops proved more difficult to modify genetically, although lately there have been some positive signs with higher-yielding strains of corn. Nor is the Green Revolution an unqualified remedy: the poorest farmers, who need help the most, can least afford the more expensive, higher-yielding seeds or the pesticides that may be required. As overall productivity continues to decline and dependence on food imports grows, it is clear that the battle for food sufficiency in Africa is far from won.

AFRICA'S HISTORICAL GEOGRAPHY

Africa is the cradle of humanity. Archeological research has chronicled 7 million years of transition from Australopithecenes to Hominids to *Homo sapiens*. It is therefore ironic that we know comparatively little about Subsaharan Africa from 5000 to 500 years ago—that is, before the onset of European colonialism. This is partly due to the colonial period itself, during which African history was neglected, many African traditions and artifacts were destroyed, and many misconceptions about African cultures and institutions became entrenched. It is also a result of the absence of a written history over most of Africa south of the Sahara until the sixteenth century—and over a large part of it until much later than that.

African Genesis

Africa on the eve of the colonial period was a continent in transition. For several centuries, the habitat in and near one of the continent's most culturally and economically productive areas—West Africa—had been changing. For 2000 years, probably more, Africa had been innovating as well as adopting ideas from outside. In West Africa, cities were developing on an impressive scale; in central and Southern Africa, peoples were moving, readjusting, sometimes struggling with each other for territorial supremacy. The Romans had penetrated to southern Sudan, North African peoples were trading with West Africans, and Arab *dhows* were sailing the waters along the eastern coasts, bringing Asian goods in exchange for gold, copper, and a comparatively small number of slaves.

It is known that African cultures had been established in all the environmental settings shown in Figure G-8 for thousands of years and thus long before Islamic or European contact. One of these, the Nok culture, endured for over eight centuries on the Benue Plateau (north of the Niger-Benue confluence in modern Nigeria) from about 500 BC to the third century AD. The Nok people made stone as well as iron tools, and they left behind a treasure of art in the form of clay figurines representing humans and animals. But we have no evidence that they traded with distant peoples. The opportunities created by environments and technologies still lay ahead.

Early Trade

West Africa, over a north-south span of a few hundred miles, displayed an enormous contrast in environments, economic opportunities, modes of life, and products. The peoples of the tropical forest produced and needed goods that were different from the products and requirements of the peoples of the dry, distant north. For example, salt is a prized commodity in the forest, where the humidity precludes its formation, but it is plentiful in the desert and steppe. This enabled the desert peoples to sell salt to the forest peoples in exchange for ivory, spices, and dried foods. Thus there evolved a degree of *regional complementarity* between the peoples of the forest and those of the drylands. And the savanna peoples—those located in between—found themselves in a position to channel and handle the trade (which is always economically profitable).

The markets in which these goods were exchanged prospered and grew, and cities arose in the savanna belt of West Africa. One of these old cities, now an epitome of isolation, was once a thriving center of commerce and learning and one of the leading urban places in the world—Timbuktu. Others, predecessors as well as successors of Timbuktu, have declined, some of them into oblivion. Still other savanna cities, such as Kano in the northern part of Nigeria, are still important.

Early States

Strong and durable states arose in West Africa. The oldest state we know anything about is ancient Ghana, located to the northwest of the modern country of Ghana. It covered parts of present-day Mali, Mauritania, and adjacent territory. Ghana lay astride the upper Niger River and included gold-rich streams flowing off the Futa Jallon Highlands, where the Niger has its origins. For a thousand years, perhaps longer, old Ghana managed to weld various groups of people into a stable state. The country had a large capital city complete with markets, suburbs for foreign merchants, religious shrines, and, some distance from the city center, a fortified royal retreat. Taxes were collected from the citizens, and tribute was extracted from subjugated peoples on Ghana's periphery; tolls were levied on goods entering Ghana, and an army maintained control. Muslims from the northern drylands invaded Ghana in the mid-eleventh century, when it may already have been in decline. Even so, the Ghanaians managed to protect their capital for 14 years. However, the invaders had ruined the farmlands and destroyed the trade links with the north. Ghana could not survive. It finally broke into smaller units.

In the centuries that followed, the focus of politico-territorial organization in the West African culture hearth shifted almost continuously eastward—first to ancient Ghana's successor state of Mali, which was centered on Timbuktu and the middle Niger River Valley, and then to the state of Songhai, whose focus was Gao, a city on the Niger that still exists. This eastward movement may have been the result of the growing influence and power of Islam. Traditional religions prevailed in ancient Ghana, but Mali and its successor states sent huge, gold-laden pilgrimages to Mecca along the savanna corridor south of the Sahara, passing through present-day Khartoum and Cairo. Of the tens of thousands who participated in these pilgrimages, some remained behind. Today, many Sudanese trace their ancestry to the West African savanna kingdoms.

West Africa's savanna region undoubtedly witnessed momentous cultural, technological, and economic developments, but other parts of Africa also progressed. Early states emerged in present-day Sudan, Eritrea, and Ethiopia. Influenced by innovations from the Egyptian culture hearth, these kingdoms were stable and durable: the oldest, Kush, lasted 23 centuries (Fig. 6-4). The Kushites built elaborate irrigation systems, forged iron tools, and built impressive structures as the ruins of their long-term capital and industrial center, Meroe, reveal. Nubia, to the southeast of Kush, was Christianized until the Muslim wave overtook it in the eighth century. And Axum was the richest market in northeastern Africa, a powerful kingdom that controlled Red Sea trade and endured for six centuries. Axum, too, was a Christian state that confronted Islam, but Axum's rulers deflected the Muslim advance and gave rise to the Christian dynasty that eventually shaped modern Ethiopia.

11 The process of **state formation** spread throughout Africa and was still in progress when the first European contacts occurred in the late fifteenth century. Large and effectively organized states developed on the equatorial west coast (notably Kongo) and on the southern plateau from the southern part of The Congo to Zimbabwe. East Africa had several city-states, including Mogadishu, Kilwa, Mombasa, and Sofala.

A crucial event affected virtually all of Equatorial, West, and Southern Africa: the great Bantu migration from present-day Nigeria-Cameroon southward and eastward across the continent. This migration appears to have occurred in waves starting

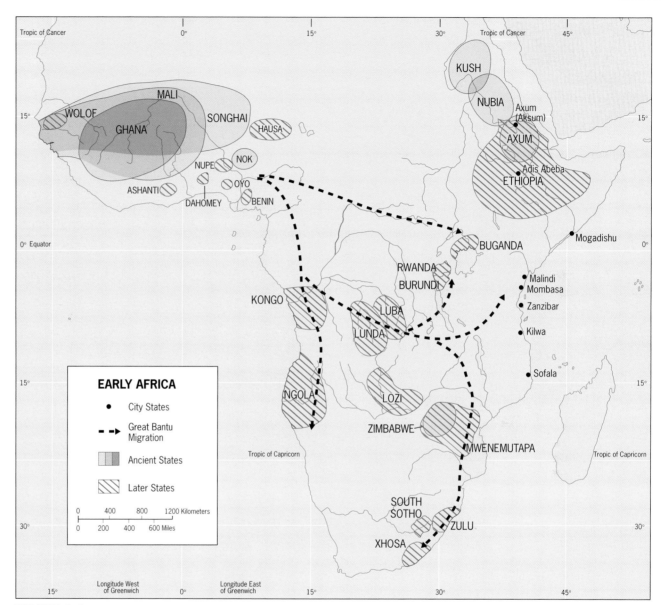

FIGURE 6-4

as long as 5000 years ago, populating the Great Lakes area and penetrating South Africa, where it resulted in the formation of the powerful Zulu Empire in the nineteenth century (Fig. 6-4).

All this reminds us that, before European colonization, Africa was a realm of rich and varied cultures, diverse lifestyles, technological progress, and external trade. It was, however, also a highly fragmented realm, its cultural mosaic (Fig. 6-5) spelling weakness when European intervention came to change the social and political map.

The Colonial Transformation

European involvement in Subsaharan Africa began in the fifteenth century. It would interrupt the path of indigenous African development and irreversibly alter the entire cultural, economic, political, and social makeup of the continent. It started quietly in the late fifteenth century, with Portuguese ships groping their way along the west coast and rounding the Cape of Good Hope. Their goal was to find a sea route to the spices and riches of the Orient. Soon other European countries were sending their vessels to African waters, and a string of coastal stations and forts sprang up. In West Africa, the nearest part of the continent to European spheres in Middle and South America, the initial impact was strongest. At their coastal control points, the Europeans traded with African middlemen for the slaves who were needed to work New World plantations, for the gold that had been flowing northward across the desert, and for ivory and spices.

Suddenly, the centers of activity lay not with the cities of the savanna but in the foreign stations on the Atlantic coast. As the interior declined, the coastal peoples thrived. Small forest states gained unprecedented wealth, transferring and selling slaves captured in the interior to the European traders on the coast. Dahomey (now called Benin) and Benin (now part of neighboring Nigeria) were states built on the slave trade. When slavery eventually came under attack in Europe, those who had inherited the power and riches it had brought opposed abolition vigorously in both continents.

Although slavery was not new to West Africa, the *kind* of slave raiding and trading the Europeans introduced certainly was. In the savanna, kings, chiefs, and prominent families traditionally took a few slaves, but the status of those slaves was unlike anything that lay in store for those who were shipped across the Atlantic. In fact, large-scale slave trading had been introduced in East Africa long before the Europeans brought it to West Africa. African middlemen from the coast raided the interior for able-bodied men and women and marched them in chains to the Arab markets on the coast (Zanzibar was a notorious market). There, packed in specially built *dhows*, they were carried off to Arabia, Persia, and India. When the European slave trade took hold in West Africa, however, its volume was far greater. Europeans, Arabs, and collaborating Africans ravaged the continent, forcing perhaps as many as 30 million persons away from their homelands in bondage (Fig. 6-6). Families were destroyed, as were whole villages and cultures; those who survived their exile suffered unfathomable misery.

The European presence on the West African coast completely reoriented its trade routes, for it initiated the decline of the interior savanna states and strengthened the coastal forest states. Moreover, the Europeans' insatiable demand for slaves ravaged the population of the interior. But it did not lead to any major European thrust toward the interior or produce colonies overnight. The African middlemen were well organized and strong, and they held off their European competitors, not just for decades but for centuries. Although the

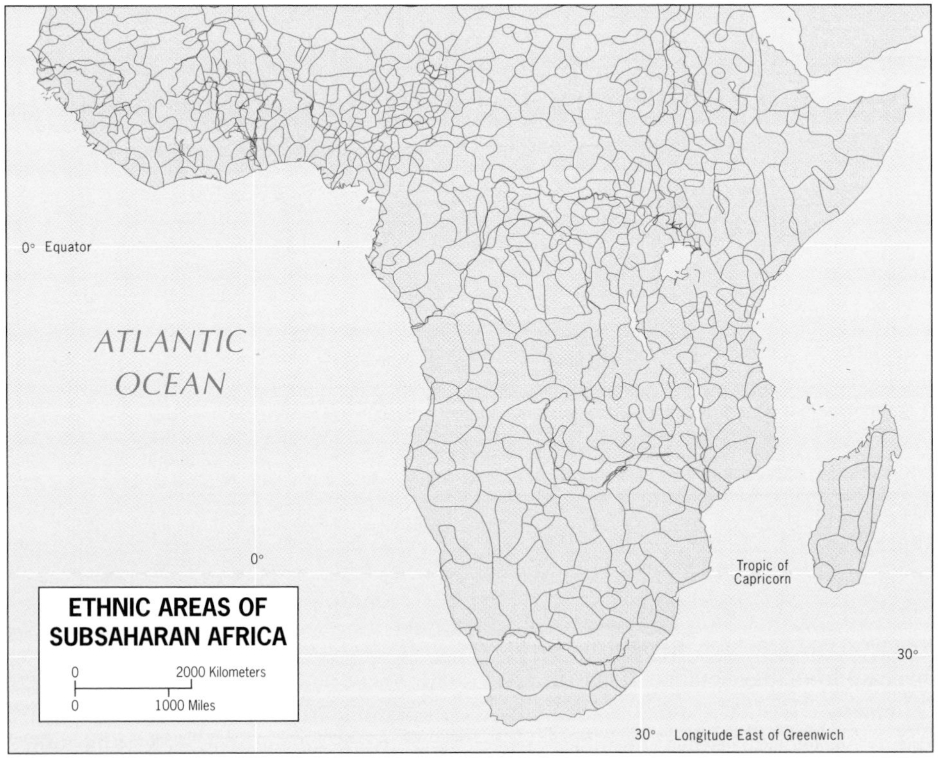

ETHNIC AREAS OF SUBSAHARAN AFRICA

ATLANTIC OCEAN

0° Equator

0°

Tropic of Capricorn

30°

30°   Longitude East of Greenwich

0   2000 Kilometers
0   1000 Miles

FIGURE 6-5

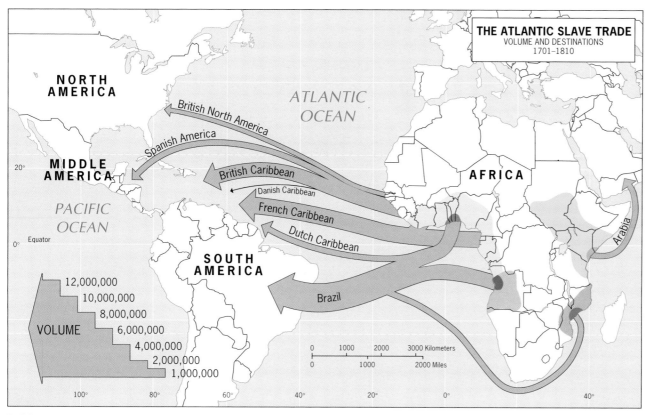

FIGURE 6-6

Europeans first appeared in the fifteenth century, they did not carve West Africa up until nearly 400 years later, and they did not conquer many other areas until after the beginning of the twentieth century.

Colonization

In the second half of the nineteenth century, whether or not they had control, the European powers finally laid claim to virtually all of Africa. Colonial competition was intense; spheres of influence began to overlap. It was time for negotiation among the powerful, and in 1884 a conference was convened in Berlin to sort things out. Fourteen states participated (including the United States, which had no claims

on Africa). The major colonial contestants were the British, French, Portuguese, King Leopold II of Belgium, and Germany itself. On maps spread on a large table, representatives from these powers drew boundaries, exchanged real estate, and forged a map that would become a permanent African liability decades later. As Figure 6-7 indicates, when the three-month conference was in progress most of Africa remained under traditional African rule. Not until 1910 did the colonial powers manage to control all the areas they had marked off on their maps.

It is important to examine Figure 6-7 carefully because the colonial powers governed their new dependencies in very different ways, and their contrasting legacies remain in evidence to this day in the countries their colonies spawned. Some colonial

powers were democracies (the United Kingdom and France); others were dictatorships (Portugal and Spain). The British established a system of indirect rule over much of their domain, leaving indigenous power structures in place and making local rulers representatives of the British Crown. This was unthinkable in the Portuguese colonies, where harsh, direct control was the rule. The French sought to create culturally assimilated elites that would represent French ideals in the colonies. In the Belgian Congo, however, King Leopold II, who had financed the expeditions that staked Belgium's claim in Berlin, embarked on a campaign of ruthless exploitation. His enforcers mobilized almost the entire Congolese population to gather rubber, kill elephants for their ivory, and build public works to improve export routes. For failing to meet production quotas, entire communities were massacred. Killing and maiming became routine in a colony in which horror was the only common denominator. After the impact of the slave trade, King Leopold's reign of terror was Africa's most severe demographic disaster. By the time it ended, after a growing outcry around the world, as many as 10 million Congolese had been murdered. In 1908, the Belgian government took over, and slowly its Congo began to mirror Belgium's own internal divisions: corporations, government administrators, and the Roman Catholic Church each pursued their sometimes competing interests. But no one thought to change the name of the colonial capital: it was Leopoldville until the Belgian Congo achieved independence.

12 Colonialism transformed Africa, but in its post-Berlin form it lasted less than a century. In Ghana, for example, the Ashanti (Asante) Kingdom still was fighting the British in the early years

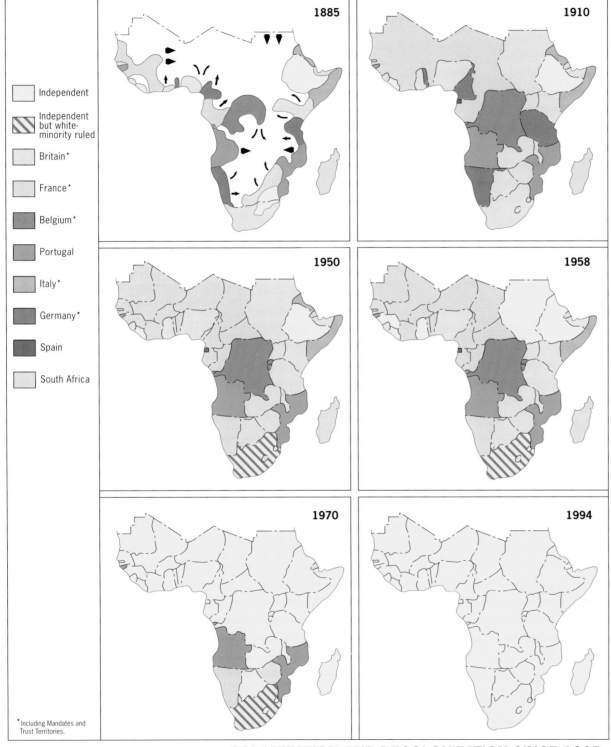

COLONIZATION AND DECOLONIZATION SINCE 1885

of the twentieth century; by 1957, Ghana was independent again. In a few years, much of Subsaharan Africa will have been independent for half a century, and the colonial period is becoming an interlude rather than a paramount chapter in modern African history.

CULTURAL PATTERNS

We may tend to think of Africa in terms of its prominent countries and famous cities, its development problems and political dilemmas, but Africans themselves have another perspective. The colonial period created states and capitals, introduced foreign languages to serve as the *linguae francae*, and brought railroads and roads. The colonizers stimulated labor movements to the mines they opened, and they disrupted other migrations that had been part of African life for many centuries. But they did not change the ways of life of most of the people. More than 70 percent of the realm's population still live in, and work near, Africa's hundreds of thousands of villages. They speak one of more than a thousand languages in use in the realm. The villagers' concerns are local; they focus on subsistence, health, and safety. They worry that the conflicts over regional power or political ideology will engulf them, as has happened to millions in Liberia, Sierra Leone, Ethiopia, Rwanda, The Congo, Moçambique, and Angola since the 1970s. Africa's largest peoples are major nations, such as the Yoruba of Nigeria and the Zulu of South Africa. Africa's smallest peoples number just a few thousand. As a geographic realm, Sub-

FIGURE 6-7

saharan Africa has the most complex cultural mosaic on Earth.

African Languages

Africa's linguistic geography is a key component of that cultural intricacy. Most of Subsaharan Africa's more than one thousand languages do not have a written tradition, making classification and mapping difficult. Scholars have attempted to delimit an African language map, and Figure 6-8 is a composite of their efforts. One feature is common to all language maps of Africa: the geographic realm begins approximately where the Afro-Asiatic language family (mapped in yellow in Fig. 6-8) ends, although the correlation is sharper in West Africa than to the east.

In Subsaharan Africa, the dominant language family is the Niger-Kordofanian family. It consists of two subfamilies, the tiny Kordofanian concentration in northeastern Sudan and the pervasive Niger-Congo subfamily that extends across most of the realm from West Africa to East and Southern Africa. The Bantu language forms the largest branch in this subfamily, but Niger-Congo languages in West Africa, such as Yoruba and Akan, also have millions of speakers. Another important language family is the Nilo-Saharan family, extending from Maasai in Kenya northwest to Teda in Chad. No other language families are of similar extent or importance: the Khoisan family, of ancient origins, now survives among the dwindling Khoi and San peoples of the Kalahari; the small white minority in South Africa speak Indo-European languages; and Malay-Polynesian languages prevail in Madagascar, which was peopled from Southeast Asia before Africans reached it.

About 40 African languages are spoken by 1 million people or more, and a half-dozen by about 10 million or more: Hausa (50 million), Yoruba (23 million), Ibo, Swahili, Lingala, and Zulu. Although English and French have become important *linguae francae* in multilingual countries such as Nigeria and Côte d'Ivoire (where officials even insist on spelling the name of their country—Ivory Coast—in the Francophone way), African languages also serve this purpose. Hausa is a common language across the West African savanna; Swahili is widely used in East Africa. And pidgin languages, mixtures of African and European tongues, are spreading along West Africa's coast. Millions of Pidgin English (called *Wes Kos*) speakers use this medium in Nigeria and Ghana.

13 **Multilingualism** can be a powerful centrifugal force in society, and African governments have tried with varying success to establish "national" alongside local languages. Nigeria, for example, made English its official language because none of its 250 languages, not even Hausa, had sufficient internal interregional use. But using a European,

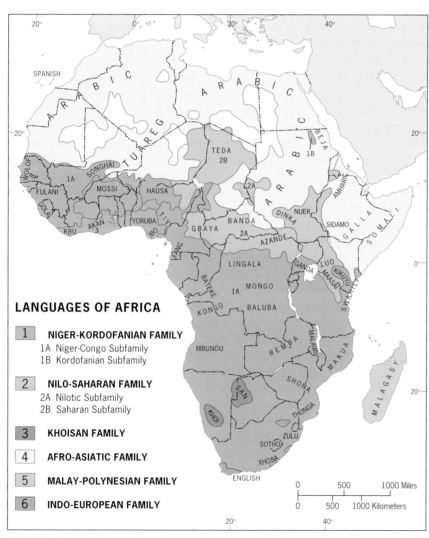

LANGUAGES OF AFRICA

1   **NIGER-KORDOFANIAN FAMILY**
1A Niger-Congo Subfamily
1B Kordofanian Subfamily

2   **NILO-SAHARAN FAMILY**
2A Nilotic Subfamily
2B Saharan Subfamily

3   **KHOISAN FAMILY**

4   **AFRO-ASIATIC FAMILY**

5   **MALAY-POLYNESIAN FAMILY**

6   **INDO-EUROPEAN FAMILY**

FIGURE 6-8

colonial language as an official medium invites criticism, and Nigeria remains divided on the issue. On the other hand, making a dominant local language official would invite negative reactions from ethnic minorities. Language remains a potent force in Africa's cultural life.

Religion in Africa

Africans had their own religious belief systems long before Christians and Muslims arrived to convert them. And for all of Subsaharan Africa's cultural diversity, Africans had a consistent view of their place in nature. Spiritual forces, according to African tradition, are manifest everywhere in the natural environment, not in a supreme deity that exists in some remote place. Thus gods and spirits affect people's daily lives, witnessing every move, rewarding the virtuous, and punishing (through injury or crop failure, for example) those who misbehave. Ancestral spirits can inflict misfortune on the living. They are everywhere: in the forest, rivers, and mountains.

As with land tenure, the religious views of Africans clashed fundamentally with those of outsiders. Monotheistic Christianity first touched Africa in the northeast when Nubia and Axum were converted, and Ethiopia has been a Coptic Christian stronghold since the fourth century AD. But the Christian churches' real invasion did not begin until the onset of colonialism after the turn of the sixteenth century. Christianity's various denominations made inroads in different areas: Roman Catholicism in much of Equatorial Africa mainly at the behest of the Belgians, the Anglican Church in British colonies, and Presbyterians and others elsewhere. But almost everywhere, Christianity's penetration led to a blending of traditional and Christian beliefs, so that much of Subsaharan Africa is nominally, though not exclusively, Christian. Go to a church in Gabon or Uganda or Zambia, and you may hear drums instead of church bells, sing African music rather than hymns, and see African carvings alongside the usual statuary.

Islam had a different arrival and impact. Long before the colonial invasion, Islam advanced out of Arabia, across the desert, and down the east coast. Muslim clerics converted the rulers of African states and commanded them to convert their subjects. They Islamized the savanna states and penetrated into present-day northern Nigeria, Ghana, and Ivory Coast. They encircled and isolated Ethiopia's Coptic Christians and Islamized the Somali people in Africa's Horn. They established beachheads on the Kenya coast and took over Zanzibar. Arabizing Islam and European Christianity competed for African minds, and Islam proved to be a far more pervasive force. From Senegal to Somalia, the population is virtually 100 percent Muslim, and Islam's rules dominate everyday life. The Sunni *mullahs* would never allow the kind of marriage between traditional and Christian beliefs seen in much of formerly colonial Africa. This fundamental contradiction between Islamic dogma and Christian accommodation creates a potential for conflict in countries where both religions have adherents.

MODERN MAP AND TRADITIONAL SOCIETY

The political map of Subsaharan Africa has 45 states but no nation-states (apart from some microstates and ministates in the islands and in the south). Centrifugal forces are powerful, and outside interventions during the Cold War, when communist and anticommunist foreigners took sides in local civil wars, worsened conflict within African states. Colonialism's economic legacy was not much better. In tropical Africa, capitals, core areas, port cities, and transport systems were laid out to maximize profit and facilitate exploitation of minerals and soils; the colonial mosaic inhibited interregional communications except where cooperation enhanced efficiency. Colonial Zambia and Zimbabwe, for example (then called Northern and

Southern Rhodesia), were landlocked and needed outlets, so railroads were built to Portuguese-owned ports. But such routes did little to create intra-African linkages. The modern map reveals the results: in West Africa you can travel from the coast into the interior of all the coastal states along railways or adequate roads. But no high-standard roadway was ever built to link these coastal neighbors to each other.

To overcome such disadvantages, African states must cooperate internationally, continentwide as well as regionally. The Organization of African Unity (OAU) was established for this purpose in 1963 and in 2001 was superseded by the African Union. In 1975 the Economic Community of West African States (ECOWAS) was founded by 15 countries to promote trade, transportation, industry, and social affairs in the region. And in the early 1990s another important step was taken when 12 countries joined in the Southern African Development Community (SADC), organized to facilitate regional commerce, intercountry transport networks, and political interaction.

Population and Urbanization

Subsaharan Africa remains the least urbanized world realm, but it is urbanizing at a fast pace. By the time you read this, the percentage of urban dwellers will have passed 30. This means that nearly 200 million people now live in cities and towns, of which many were founded and developed by the colonial powers.

African cities became centers of embryonic national core areas, and of course they served as government headquarters. This *formal sector* of the city used to be the dominant one, with government control and regulations affecting civil service, business, industry, and workers. Today, however, African cities look different. From a distance, the skyline still resembles that of a modern center. But in the streets, on the sidewalks right below the shopwindows, there are hawkers, basket weavers, jewelry sellers, garment makers, wood carvers—a second economy, most of it beyond government control. This *informal sector* now dominates many African cities. It is peopled by the rural immigrants, who also work as servants, apprentices, construction workers, and in countless other menial jobs.

Millions of urban immigrants, however, cannot find work, at least not for months or even years at a time. They live in squalid circumstances, in desperate poverty, and governments cannot assist them. As a result, the squatter rings around (and also within) many of Africa's cities are unsafe— uncomfortable, unhealthy slums without adequate shelter, water supply, or basic sanitation. Garbage-strewn (no solid-waste removal here), muddy and insect-infested during the rainy season, and stifling and smelly during the dry period, they are incubators of disease. Yet few of its residents return to their villages. Every new day brings hope.

In our regional discussion we refer to some of Subsaharan Africa's cities, all of which, to varying degrees, are stressed by the rate of population influx. Despite the plight of the urban poor and the poverty of Africa's rural areas, some of Africa's capitals remain the strongholds of privileged elites

FROM THE FIELDNOTES

"Looking down on this enormous railroad complex, we were reminded of the fact that almost an entire continent was turned into a wellspring of raw materials carried from interior to coast and shipped to Europe and other parts of the world. This complex lies near Witbank in the eastern Rand, a huge inventory of freight trains ready to transport ores from the plateau to Durban and Maputo. But at least South Africa acquired a true transport network in the process, ensuring regional interconnections; in most African countries, railroads serve almost entirely to link resources to coastal outlets."

who, dominant in governments, fail to address the needs of other ethnic groups. Discriminatory policies and artificially low food prices disadvantage farmers and create even greater urban-rural disparities than the colonial period saw. But today the prospect of democracy brings hope that Africa's rural majorities will be heard and heeded in the capitals.

REGIONS OF THE REALM

On the face of it, Africa seems to be so massive, compact, and unbroken that any attempt to justify a contemporary regional breakdown is doomed to fail. No deeply penetrating bays or seas create peninsular fragments as in Europe. No major islands (other than Madagascar) provide the broad regional contrasts we see in Middle America. Nor does Africa really taper southward to the peninsular proportions of South America. And Africa is not cut by an Andean or a Himalayan mountain barrier. Given Africa's colonial fragmentation and cultural mosaic, is regionalization possible? Indeed it is.

Maps of environmental distributions, ethnic patterns, cultural landscapes, historic culture hearths, and other spatial data yield a four-region structure complicated by a fifth, overlapping zone as shown in Figure 6-9. Beginning in the south, we identify the following regions:

1. *Southern Africa*, extending from the southern tip of the continent to the northern borders of Angola, Zambia, Malawi, and Moçambique. Ten countries constitute this region, which extends beyond the tropics and whose giant is South Africa.

2. *East Africa*, where natural (equatorial) environments are moderated by elevation and where plateaus, lakes, and mountains, some carrying permanent snow, define the countryside. Six countries, including the highland part of Ethiopia, comprise this region. The island of Madagascar, with Southeast Asian influences, is neither Eastern nor Southern African.

3. *Equatorial Africa*, much of it defined by the basin of the Congo River, where elevations are lower than in East Africa, temperatures are higher and moisture more ample, and where most of Africa's

surviving rainforests remain. Among the eight countries that form this region, The Congo dominates territorially and demographically, but others are much better off economically and politically.

4. *West Africa*, which includes the countries of the western coast and those on the margins of the Sahara in the interior, a populous region anchored in the southeast by Africa's demo-

graphic giant, Nigeria. Fifteen countries form this crucial African region.

5. *The African Transition Zone*, the complicating factor on the regional map of Africa. In Figure 6-9, note that this zone of increasing Islamic influence completely dominates some countries (e.g., Somalia in the east and Senegal in the west) while cutting across others, thereby creat-

ing Islamized northern areas and non-Islamic southern zones (Nigeria, Chad, Sudan).

SOUTHERN AFRICA

Southern Africa, as a geographic region, consists of all the countries and territories lying south of Equatorial Africa's The Congo and East Africa's Tanzania (Fig. 6-10). Thus defined, the region extends from Angola and Moçambique (on the Atlantic and Indian Ocean coasts, respectively) to South Africa and includes a half-dozen landlocked states. Also marking the northern limit of the region are Zambia and Malawi. Zambia is nearly cut in half by a long land extension from The Congo, and Malawi penetrates deeply into Moçambique. The colonial boundary framework, here as elsewhere, produced many liabilities.

Southern Africa constitutes a geographic region in both physiographic and human terms. Its northern zone marks the southern limit of the Congo Basin in a broad upland that stretches across Angola and into Zambia (the tan corridor extending eastward from the Bihe Plateau in Fig. 6-1). Lake Malawi is the southernmost of the East African rift-valley lakes; Southern Africa has none of East Africa's volcanic and earthquake activity. Most of the region is plateau country, and the Great Escarpment is much in evidence here. There are two pivotal river systems: the Zambezi (which forms the border between Zambia and Zimbabwe) and the Orange-Vaal (South African rivers that combine to demarcate southern Namibia from South Africa).

Southern Africa is the continent's richest region materially. A great zone

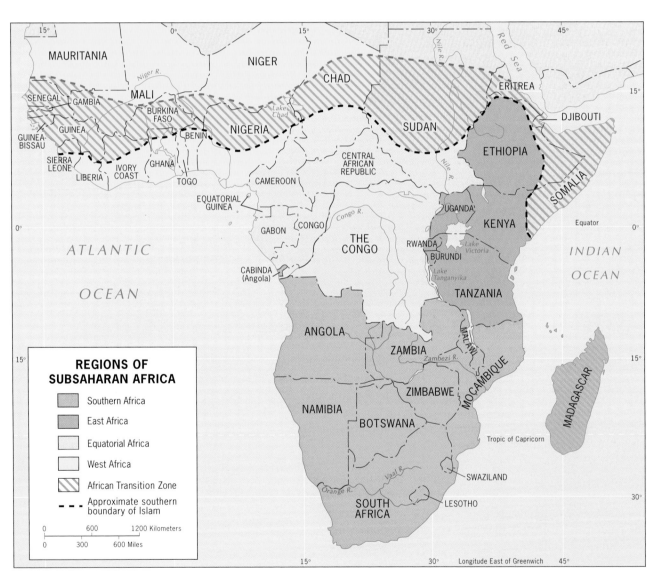

REGIONS OF SUBSAHARAN AFRICA

- Southern Africa
- East Africa
- Equatorial Africa
- West Africa
- African Transition Zone
- - - - Approximate southern boundary of Islam

FIGURE 6-9

FIGURE 6-10

Map legend:

SOUTHERN AFRICA

POPULATION

- Under 50,000
- 50,000–250,000
- 250,000–1,000,000
- 1,000,000–5,000,000

National capitals are underlined

— Railroad — Road

National park

Johannesburg–Maputo Corridor

0 200 400 Kilometers
0 100 200 Miles

Longitude East of Greenwich

of mineral deposits extends through the heart of the region from Zambia's Copperbelt through Zimbabwe's Great Dyke and South Africa's Bushveld Basin and Witwatersrand to the goldfields and diamond mines of the Orange Free State and northern Cape Province in the heart of South Africa. Ever since these minerals began to be exploited in colonial times, many migrant laborers have come to work in the mines.

Southern Africa's agricultural diversity matches its mineral wealth. Vineyards drape the slopes of South Africa's Cape Ranges; tea plantations hug the eastern escarpment slopes of Zimbabwe. Before civil war destroyed its economy, Angola was one of the world's leading coffee producers. South Africa's relatively high latitudes and its range of altitudes create environments for apple orchards, citrus groves, banana plantations, pineapple farms, and many other crops.

Despite this considerable wealth and potential, the countries of Southern Africa have not prospered. As Figure G-10 shows, most remain mired in the low-income category (Malawi, with a per-capita GNP of only $600, is one of the world's poorest states); only South Africa and neighboring Botswana are in the upper-middle-income rank, the latter desert state being the realm's second most sparsely populated country. A period of rapid population growth, followed by the devastating onslaught of AIDS, civil conflict, political instability, incompetent government, corruption, unfair competition on foreign markets, local corruption,

FROM THE FIELDNOTES

"My second visit to the Kariba Dam on the Zambezi River between Northern and Southern Rhodesia (now renamed Zambia and Zimbabwe) showed the huge dam wall complete and the lake filling up. The four-lane highway across the top of the dam was not yet open for traffic, and construction was still in progress on the turbine housing and other buildings. This was the time when megadams were thought to hold the solution to development problems: the Kariba Dam would provide electricity to places as far away as Salisbury (now Harare) and the Copperbelt, would support a fishing industry, would enable farming along its shores, would stimulate the tourist industry. It was therefore worth dislocating tens of thousands of people living in the valley and eliminating wildlife refuges (a massive effort was made to rescue wildlife caught on islands as the water rose). But, like other megaprojects of this kind in Africa and elsewhere in the less advantaged world, the Kariba Dam failed to live up to expectations and generated unanticipated problems, medical and otherwise. A hint of the latter can be seen in the extreme left of this photograph: the floating vegetation at the turbine-tunnel entrance. Brazilian water hyacinth, reportedly introduced by two Brazilian monks who built a small church where the new shoreline would be, clogged the upper reaches of Lake Kariba and spread into Equatorial Africa's rivers."

and environmental problems have constrained economic development. In 2003, countries in this region needed emergency food supplies to stave off malnutrition and hunger; one of them (Zimbabwe) failed to distribute the provisions where they were most needed, and another (Zambia) refused genetically modified (GM) grain because it was thought to be unhealthy.

Still, Southern Africa as a region is better off than any other in Subsaharan Africa: as Figure G-10 shows, three countries here have risen above the lowest-income rank among world states. Some international cooperation including a regional association and a customs union are emerging. And

South Africa, the realm's most important country by many measures, gives hope for a better future.

South Africa

The Republic of South Africa is the giant of Southern Africa, an African country at the center of world attention, a bright ray of hope not only for Africa but for all humankind.

Long in the grip of one of the world's most notorious racial policies (**apartheid**, or "apartness," and its derivative, **separate development**), South Africa today is shedding its past and is building a new future. That virtually all

parties to the earlier debacle are now working cooperatively to restructure the country under a new flag, a new national anthem, and a new leadership was one of the great events of the late twentieth century. Now, with a new century opened, South Africa is poised to take its long-awaited role as the economic engine for the region—and perhaps beyond.

South Africa stretches from the warm subtropics in the north to Antarctic-chilled waters in the south. With a land area in excess of 470,000 square miles (1.2 million sq km) and a heterogeneous population of 44.5 million, South Africa is the dominant state in Southern Africa. It contains the bulk of the region's minerals, most of its good farmlands, its largest cities, best ports, most productive factories, and most developed transport networks. Mineral exports from Zambia and Zimbabwe move through South African ports. Workers from as far away as Malawi and as nearby as Lesotho work in South Africa's mines, factories, and fields.

Historical Geography

South Africa's location has much to do with its human geography. On the African continent, peoples migrated southward—first the Khoisan-speakers and then the Bantu peoples—into the South African cul-de-sac. On the oceans, the Europeans arrived to claim the southernmost Cape as one of the most strategic places on Earth, the gateway from the Atlantic to the Indian Ocean, a waystation on the route to Asia's riches. The Dutch East India Company founded Cape Town as early as 1652, and soon the Hollanders began to bring Southeast Asians to the Cape to serve as domestics and laborers. When the British took over about 150 years later, Cape Town had a substantial population of mixed ancestry, the source of today's so-called Coloured sector of the country's citizenry.

The British also altered the demographic mosaic by bringing tens of thousands of indentured laborers from their South Asian domain to work on the sugar plantations of east-coast Natal. Most of these laborers stayed after their period of indenture was over, and today South Africa counts 1 million Indians among its people. Most are still concentrated in Natal and prominently so in metropolitan Durban.

As we noted earlier, South Africa had been occupied by Europeans long before the colonial "scramble for Africa" gained momentum. The Dutch, after the British took control of the Cape, trekked into the South African interior and, on the high plateau they called the *highveld*, founded their own republics. When diamonds and gold were discovered there, the British challenged the Boers (descendants of the Dutch settlers) for these prizes. In 1899–1902, the British and the Boers fought the Boer War. The British won, and British capitalists took control of South Africa's economic and political life. But the Boers negotiated a power-sharing arrangement and eventually achieved hegemony. Having long since shed their European links, they now called themselves *Afrikaners*, their word for Africans, and proceeded to erect the system known as apartheid.

These foreign immigrations and struggles took place on lands that Africans had already entered and fought over. When the Europeans reached the Cape, Bantu nations were driving the weaker Khoi and San peoples into less hospitable territory or forcing them to work in bondage. One great contest was taking place in the east and southeast, below the Great Escarpment. The Xhosa nation was moving toward the Cape along this natural corridor. Behind them, in Natal (Fig. 6-10), the Zulu Empire became the region's most powerful entity in the nineteenth century. On the highveld, the North and South Sotho, the Tswana, and other peoples could not stem the tide of European aggrandizement, but their numbers ensured survival.

In the process, South Africa became Africa's most pluralistic and heterogeneous society. People had converged on the country from Western Europe, Southeast Asia, South Asia, and other parts of Africa itself. At the end of the twentieth century, Africans outnumbered non-Africans by about 4 to 1.

Social Geography

Heterogeneity also marks the spatial demography of South Africa. Despite centuries of migration and (at the Cape) intermarriage, labor movement (to the mines, farms, and factories), and massive urbanization, regionalism pervades the human mosaic. The Zulu nation still is largely concentrated in the province the Europeans called Natal. The Xhosa still cluster in the Eastern Cape, from the city of East London to the Natal border and below the Great Escarpment. The Tswana still occupy ancestral lands along the Botswana border. Cape Town still is the core area of the Coloured population; Durban still has the strongest Indian imprint. Travel through South Africa, and you will recognize the diversity of rural cultural landscapes as they change from Swazi to Ndebele to Venda.

This historic regionalism was among the factors that led the Afrikaner government to institute its "separate development" scheme, but it could not stem the tide of urbanization. Millions of workers, job-seekers, and illegal migrants converged on the cities, creating vast shantytowns on their margins. In the legal African "townships" such as Johannesburg's Soweto ("SOuth WEstern TOwnships") and in these squatter settlements the anti-apartheid movement burgeoned, and the strength of the African National Congress (ANC) movement grew. In February 1990 Nelson Mandela, imprisoned for 28 years on Robben Island, South Africa's Alcatraz, became a free man, and following the momentous first democratic election of 1994 he became president of an ANC-dominated government in Cape Town.

Before this election could take place, however, South Africa's administrative geography had to be radically changed. The republic's political geography was a legacy of the Union period, when it was divided into four provinces: the Cape, by far the largest and centered on the legislative capital of Cape Town; Natal, anchored by the port city of Durban; the Transvaal ("across the Vaal River"), focused on the administrative capital of Pretoria and including the great Johannesburg metropolis; and the Orange Free State, the Boer stronghold with Bloemfontein as its headquarters.

This structure was replaced by a new map of nine provinces (Fig. 6-11), creating a federal arrangement in which each province would have its own administration while being represented in the central government. The boundaries of Natal and the Orange Free State remained essentially unchanged, but Natal's name was changed to Kwazulu-Natal, recognizing the Zulu presence there, and the Orange Free State became simply the Free State. But the Cape Province was divided into four new provinces, one of them overlapping into the Transvaal, and the

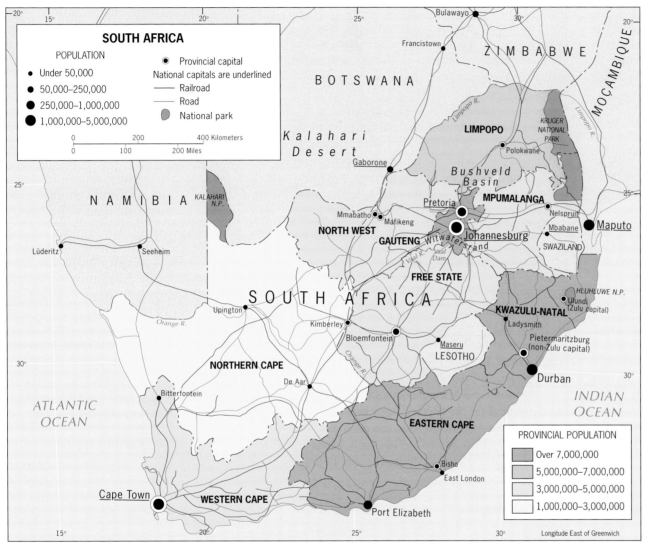

FIGURE 6-11

white opposition that most threatened majority rule. The white extremist political parties collapsed relatively soon, and a brief terrorist episode was quickly halted. More consequential was the uncertain role of the largest single nation in South Africa, the Zulu, whose historic domain lies in Kwazulu-Natal, and whose prominent leader, Chief Buthelezi, at times hinted at secession if the new South Africa was not acceptable. In the runup to the election, violence killed thousands and threatened to precipitate a civil war. But that danger, too, was overcome, and the Zulu nation and its leaders remained in the fold.

In June 1999, ANC leader Thabo Mbeki, who had served as President Mandela's deputy, became the country's second popularly elected president. In other African states, the succession from heroic founder-of-the-nation to political inheritor-of-the-presidency often has not gone well, but South Africa's new constitution proved its worth. The new president faced several disadvantages: inevitable comparisons to the incomparable Mandela; the rising tide of AIDS, which Mbeki controversially attributed to causes other then HIV; and the farm invasions in Zimbabwe (see p. 214) on which his leadership seemed to falter. Nevertheless, Mbeki's economic policies, social programs, and foreign initiatives (The Congo, Rwanda) have served his country well.

Economic Geography

Undoubtedly the most serious immediate problems for the new South Africa are economic. Ever since diamonds were discovered at Kimberley in

rest of the Transvaal was divided into three, one of which is Gauteng, which includes the Johannesburg-Witwatersrand-Pretoria megalopolis.

The 1994 election, based on this new layout, produced an ANC victory in seven of these nine provinces, excepting only Kwazulu-Natal, where the Zulu vote went to a local political movement called Inkatha, and the Western Cape, where a combination of white and Coloured voters outpolled the ANC. This was a very fortunate result, proving that the ANC was not all-powerful and giving minorities, including Coloureds and whites, reason to trust and participate in the new system.

The transition from Afrikaner domination to democratic government still was complicated by numerous circumstances. As it turned out, it was not

the 1860s, South Africa has been synonymous with minerals. The Kimberley finds, made in a remote corner of what was then the Orange Free State (the British soon annexed it to the Cape), set into motion a new economic geography. Rail lines were laid from the coast to the "diamond capital" even as fortune seekers, capitalists, and tens of thousands of African workers, many from as far afield as Lesotho, streamed to the site. One of the capitalists was Cecil Rhodes, of Rhodes Scholarship fame, who used his fortune to help Britain dominate Southern Africa.

Just 25 years after the diamond discoveries, prospectors found what was long to be the world's greatest goldfield on a ridge called the Witwatersrand (Fig. 6-11). This time the site lay in the so-called South African Republic (the Transvaal), and again the Boers were unable to hold the prize. Johannesburg became the gold capital of the world, and a new and even larger stream of foreigners arrived, along with a huge influx of African workers. Cheap labor enlarged the profits. Johannesburg grew explosively, satellite towns developed, and black townships mushroomed. The Boer War was only an interlude here on the mineral-rich Witwatersrand.

During the twentieth century, South Africa proved to be richer than had been foreseen. Additional goldfields were discovered in the Orange Free State. Coal and iron ore were found in abundance, which gave rise to a major iron and steel industry. Other metallic minerals, including chromium and platinum, yielded large revenues on world markets. Asbestos, manganese, copper, nickel, antimony, and tin were mined and sold; a thriving metallurgical industry developed in South Africa itself. Capital flowed into the country, white immigration grew, farms and ranches were laid out, and markets multiplied.

South Africa's cities grew apace. Johannesburg was no longer just a mining town: it became an industrial complex and a financial center as well. The old Boer capital, Pretoria, just 30 miles (50 km) north of the Witwatersrand, became the country's administrative center during apartheid's days. In the

Orange Free State, major industrial growth (including oil-from-coal technology) matched the expansion of mining. While the core area developed megalopolitan characteristics, coastal cities expanded as well. Durban's port served not only the Witwatersrand but a wider regional hinterland as well. Cape Town was becoming South Africa's largest city; its port, industries, and productive agricultural hinterland gave it primacy over a wide area.

Apartheid ruined these prospects and exposed the economy's weaknesses. The Afrikaner government made huge investments in its separate development policy, which contradicted fundamental principles of economic geography. International sanctions against apartheid hurt the economy, labor unrest weakened it further, and lower prices for those products that did get sold hurt even more. In the black townships, the school systems fell apart under the cry of "liberation before education," creating a huge undereducated (later unemployable) mass of young people.

For South Africa later is now, and the problems these policies created have arrived. As the twenty-first century dawned, South Africa's economy still depended heavily on the export of metals and minerals, but mineworkers demanding higher wages cut into the profits. Manufactured goods figured hardly at all in the export picture. Unemployment stood at high levels even as employed workers sought to raise their wages. In the cities, the whites fled to the suburbs, transforming the downtowns into commuter workplaces and the streets into bazaars. Violent robberies replaced the political crime of the pre-transfer period. Most ominous is the revolution of rising expectations, a phenomenon many African countries experienced after decolonization. Land pressure and housing needs cannot be accommodated overnight, but the newly empowered majority will expect its government to act expeditiously. Joblessness, housing shortages, and land pressure can form a potent mix to destabilize a society, and the South African government faces a daunting challenge.

On maps of development indices, South Africa is portrayed as an upper-middle-income economy,

but averages mean little in this country of strong internal core-periphery contrasts. In its great cities, industrial complexes, mechanized farms, and huge ranches, South Africa resembles a high-income economy, much like Australia or Canada. But outside the primary core area (centered on Johannesburg) and beyond the secondary cores and their linking corridors lies a different South Africa, where conditions are more like those of rural Zambia or Zimbabwe. In terms of such indices as life expectancy, infant and child mortality, overall health, nutrition, education, and many others, a wide range marks the country's population sectors. South Africa has been described as a microcosm of the world, exhibiting in a single state not only a diversity of cultures but also a wide range of human conditions. If South Africa can keep on course, the one-time pariah of apartheid will become a guidepost to a better world.

The Middle Tier

Between South Africa's northern border and the region's northern limit lie two groups of states: those with borders with South Africa and those beyond.

Five countries constitute the Middle Tier, neighboring South Africa: Zimbabwe, Namibia, Botswana, and the ministates of Lesotho and Swaziland (Fig. 6-10). As the map shows, four of these five are landlocked. Diamond-exporting (and upper-middle-income) **Botswana** occupies the heart of the Kalahari Desert and surrounding steppe; despite its lucrative diamonds, most of its 1.6 million inhabitants are subsistence farmers. In 2004 Botswana was the most severely AIDS-afflicted country in all of tropical Africa. **Lesotho** and **Swaziland**, both traditional kingdoms, depend heavily on remittances from their workers in South African mines, fields, and factories.

The most important state in the middle tier undoubtedly is **Zimbabwe** (12.5 million), landlocked but well endowed with mineral and agricultural resources. Zimbabwe (the country is named after stone ruins in its interior [photo, p. 192]) is mostly an elevated plateau between the Zambezi and Limpopo rivers, with the desert to the west and the Great Escarpment to the east. Its core area is defined by the mineral-rich Great Dyke and its environs, extending southwest from the vicinity of the capital, Harare, to the country's second city, Bulawayo. Copper, asbestos, and chromium (of which Zimbabwe is one of the world's leading sources) are among its major mineral exports, but Zimbabwe is not just an ore-exporting country. Farms produce tobacco, tea, sugar, cotton, and other crops (corn is the staple).

Two nations form most of Zimbabwe's population: the Shona (71 percent) and the Ndebele (16 percent). A tiny minority of whites owned the best farmland and organized the agricultural economy. Mounting environmental and economic problems beginning in the 1980s led to rising social and political tensions. President Mugabe and his dominant party allowed white farms to be invaded by squatters who sometimes killed the owners; corruption rose and human rights were curbed. Meanwhile, Mugabe used Zimbabwe's armed forces to intervene in The Congo. Once-promising Zimbabwe is the tragedy of the region today.

Southern Africa's youngest independent state, **Namibia** (1.9 million), is a former German colony with a territorial peculiarity: the so-called Caprivi Strip linking it to the right bank of the Zambezi River (Fig. 6-10), another product of colonial partitioning. Administered by South Africa from 1919 to 1990, Namibia is named after one of the world's driest deserts. This state is about as large as Texas and Oklahoma, but only its far north receives enough moisture to permit subsistence farming, which is why most of the people live near the Angolan border. Mining in the Tsumeb area and ranching in the vast steppe country of the south form the main commercial activities. The capital, Windhoek, is centrally situated opposite Walvis Bay, the main port. German influence still lingers in what used to be called South West Africa, as does an Afrikaner presence from the apartheid period. Look for land and other social issues to arise here.

The Northern Tier

In the four countries that extend across the northern tier of the region—Angola, Zambia, Malawi, and Moçambique—problems abound. **Angola** (13.4 million), formerly a Portuguese dependency, with its exclave of Cabinda had a thriving economy based on a wide range of mineral and agricultural exports at the time of independence in 1975. But then Angola fell victim to the Cold War, with northern peoples choosing a communist course and southerners falling under the sway of a rebel movement backed by South Africa and the United States. The results included a devastated infrastructure, idle farms, looting of diamonds, hundreds of thousands of casualties, and millions of landmines that continue to kill and maim. But Angola's oil wealth yields about U.S. $3 billion per year, and stability, if sustained, may attract investors to begin rebuilding this ruined country.

On the opposite coast, the other major former Portuguese colony, **Moçambique** (20.4 million), fared poorly in a different way. Without Angola's mineral base and with limited commercial agriculture, Moçambique's chief asset was its relative location. Its two major ports, Maputo and Beira, handled large volumes of exports and imports for South Africa, Zimbabwe, and Zambia. But upon independence Moçambique, too, chose a Marxist course with dire economic and dreadful political consequences. Another rebel movement supported by South Africa caused civil conflict, created famines, and generated a stream of more than a million refugees toward Malawi. Rail and port facilities lay idle, and Moçambique at one time was ranked by the United Nations as the world's poorest country. In recent years, the port traffic has been somewhat revived and Moçambique and South Africa are working on a joint Maputo Development Corridor (Fig. 6-10), but it will take generations for Moçambique to recover.

Landlocked **Zambia** (10.4 million), the product of British colonialism, shares the riches of the Copperbelt with The Congo's Katanga Province. Not only have commodity prices on which Zambia depends declined severely, but Zambia's outlets—Lobito in Angola and Beira in Moçambique—and the railroads leading there were made inoperative by Cold War conflicts. Neighboring **Malawi** (11.4 million) has an almost totally agricultural economic base, sufficiently diversified to cushion its economy against market swings. But Malawi, too, suffers from its landlocked situation.

▶ EAST AFRICA

East of the row of Great Lakes that marks the eastern border of The Congo (Lakes Albert, Edward, Kivu, and Tanganyika), the land rises from the Congo Basin to the East African Plateau. Hills and valleys, fertile soils, and copious rains mark the transition in Rwanda and Burundi. Eastward the rainforest disappears and open savanna cloaks the countryside. Great volcanoes rise above a rift-valley-dissected highland. At the heart of the region lies Lake Victoria. In the north the surface rises above 10,000 feet (3300 m), and so deep are the trenches cut by faults and rivers there that the land was called, appropriately, Abyssinia (now Ethiopia).

Five countries, as well as the highland part of Ethiopia, form this East African region: Kenya, Tanzania, Uganda, Rwanda, and Burundi (Fig. 6-12). Here the Bantu peoples that make up most of the population met Nilotic peoples from the north.

Kenya is neither the largest nor the most populous country in East Africa, but over the past half-century it has been the dominant state in the region. Its skyscrapered capital at the heart of its core area, Nairobi, is the region's largest city; its port, Mombasa, is the region's busiest.

After independence, Kenya chose a capitalist path of development, aligning itself with Western interests. Without major known mineral deposits, Kenya depended on coffee and tea exports and on a

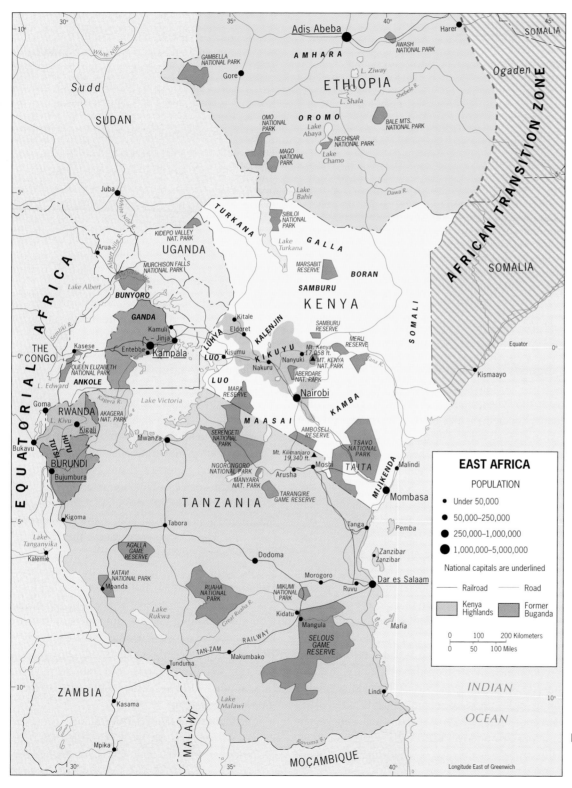

FIGURE 6-12

tourist industry based on its magnificent national parks (Fig. 6-12). Tourism became its largest single earner of foreign exchange, and Kenya prospered, apparently proving the wisdom of its capitalist course.

But serious problems arose. Kenya during the 1980s had the highest rate of population growth in the world, and population pressure on farmlands and on the fringes of the wildlife reserves mounted. Poaching became worrisome, and tourism declined. During the late 1990s, violent weather buffeted Kenya, causing landslides and washing away large segments of the crucial Nairobi-Mombasa Highway. This was followed by a severe drought lasting several years, bringing famine to the interior. Meanwhile, government corruption siphoned off funds that should have been invested. Democratic principles were violated, and relationships with Western allies were strained. The AIDS epidemic brought another setback. And then Kenya sustained damage from terrorist attacks in Nairobi and Mombasa, further ravaging the tourist industry.

Today, Kenya's prospects are uncertain. Geography, history, and politics have placed the Kikuyu (22 percent of the population of 32.4 million) in a position of power. But there are other major peoples (see Fig. 6-12) and several smaller ones. The Luhya, Luo, Kalenjin, and Kamba together constitute about 50 percent of the population, and on the territorial margins of the country there are peoples such as the Maasai, Turkana, Boran, and Galla. Creating and sustaining a political system that ensures democracy and represents the interests of these disparate peoples is Kenya's unmet challenge.

Tanzania (a name derived from Tanganyika plus Zanzibar) is the biggest and most populous East African country (39.2 million). Its total area exceeds that of the other four countries combined. Tanzania has been described as a country without a core because its clusters of population and zones of productive capacity lie dispersed—mostly on

"It was 108 in the shade, but the narrow alleys of Zanzibar's Stone City felt even hotter than that. I spent some time here when I was working on my monograph on Dar es Salaam in the 1960s and had not been back. In those days, African socialism and *uhuru* were the watchwords; since then Tanzania has not done well economically. But here were signs of a new era: a People's Bank in a former government building, and, on the old fort's tower, a poster saying 'Think Digital Go Tritel.' Nearby, on the sandy beach where I relaxed 35 years ago, was evidence that Zanzibar has not escaped the ravages of AIDS. Now the sand served as a refuge for the sick, who were resting there. 'It's better, bwana, than the corridor of the clinic,' said a young man who could walk only a few steps at a time with the help of a cane, and who breathed with difficulty as he spoke. Here as everywhere in Subsaharan Africa, AIDS has severely strained already-limited medical facilities."

its margins on the east coast (where the capital, Dar es Salaam, is located), near the shores of Lake Victoria in the northwest, near Lake Tanganyika in the far west, and near Lake Malawi in the interior south. This is in sharp contrast to Kenya, which has a well-defined core area in the Kenya Highlands (centered on Nairobi in the heart of the country). Moreover, Tanzania is a country of many peoples, none numerous enough to dominate the state. About 100 ethnic groups coexist; one-third of the population, mainly those on the coast, are Muslims.

After independence Tanzania embarked on a socialist course toward development, including a massive and disastrous farm collectivization program. The tourist industry declined sharply, and Tanzania became one of the world's poorest countries. But Tanzania did achieve remarkable political stability and a degree of democracy that none of the other East African states attained. Since 1990, the government has changed course, but the AIDS crisis, problems with the Zanzibar merger, and involvement in the troubles of neighboring countries have set Tanzania back.

Uganda contained this region's most important African state when the British colonialists arrived: the Kingdom of Buganda, peopled by the Ganda, located on the northwest shore of Lake Victoria (Fig. 6-12, the dark brown area). The British established their headquarters near the Ganda capital of Kampala and used the Ganda to control Uganda through indirect rule. Thus the Ganda became the dominant nation in multicultural Uganda, and when the British left, they bequeathed Uganda a complicated federal system designed to perpetuate Ganda supremacy.

The system failed, bringing to power one of Africa's most brutal dictators, Idi Amin. Uganda had a strong economy based on coffee, cotton, and other farm exports, and on copper mining; its Asian minority of about 75,000 dominated local commerce. Amin ousted all the Asians, exterminated his opponents, and destroyed the economy. In addition, the AIDS epidemic struck Uganda severely. Following Amin's expulsion, Uganda moved toward more representative government and, under President Yoweri Museveni, made significant gains in the struggle against AIDS. Unfortunately, Ugan-

dan forces played a less constructive role in the current conflicts involving Rwanda and The Congo.

16 As the map shows, Uganda is a **landlocked state** and depends on Kenya for an outlet to the ocean. Its relative location adjacent to unstable Sudan, Rwanda, and The Congo constitutes a formidable challenge.

Rwanda and **Burundi** would seem to occupy Tanzania's northwest corner, and indeed they were part of the German colonial domain conquered before World War I. But during that war Belgian forces attacked the Germans from their Congo bases and were awarded these territories when the conflict was over in 1918. The Belgians used them as labor sources for their Katanga mines.

Rwanda (7.7 million) and Burundi (7 million) are physiographically part of East Africa, but their cultural geography is linked to the north and west. Here, Tutsi pastoralists from the north subjugated Hutu farmers (who had themselves made serfs of the local Twa [pygmy] population), setting up a conflict that was originally ethnic but became cultural. Certain Hutu were able to advance in the Tutsi-dominated society, becoming to some extent converted to Tutsi ways, leaving subsistence farming behind, and rising in the social hierarchy. These so-called moderate Hutu were—and are—often targeted by other Hutus, who resent their position in society. This longstanding discord, worsened by colonial policies, had repeatedly devastated both countries and, in the 1990s, spilled over into The Congo, generating the first interregional war in Subsaharan Africa.

As many as 3 million people have perished as Hutu, Tutsi, Ugandan, and Congolese rebel forces have fought for control over areas of the eastern Congo, unleashing longstanding local animosities (such as those between Hema and Lendu around Bunia) that worsened the death toll. Only massive international intervention could stabilize the situation, but the world has turned a blind eye to the region's woes—again.

The highland zone of **Ethiopia** also forms part of East Africa. Adis Abeba, the historic capital, was the headquarters of a Coptic-Christian, Amharic empire that held its own against the colonial intrusion except for a brief period from 1935 to 1941, when the Italians defeated it. Indeed, the Ethiopians in their mountain fortress (Adis Abeba lies about 10,000 feet above sea level) became colonizers themselves, tak-

ing control of much of the Islamic part of the African Horn to the east.

Ethiopia's natural outlets are toward the Gulf of Adan and the Red Sea, but its government was forced to yield independence to Eritrea and the country is now effectively landlocked. Physiographically and culturally, however, it is part of East Africa, and the Amhara and Oromo peoples are neither Arabized nor Muslim: they are Africans. The map shows that functional linkages between Ethiopia and East Africa remain weak, but this is likely to change in the future.

Madagascar

Off Africa's east coast lies the world's fourth-largest island, Madagascar (Fig. 6-13). About 2000 years ago the first human settlers arrived here—not from Africa but from distant Southeast Asia. A powerful Malay kingdom of the Merina flourished in the highlands, whose language, Malagasy, became Madagascar's tongue (Fig. 6-8).

The Malay immigrants eventually brought Africans to their island, but today the Merina (4 million) and the Betsimisaraka (2 million) remain the largest of about 20 ethnic groups in the population of 17.6 million. After successfully resisting colonial invasion, the locals eventually yielded to France, and French became the *lingua franca* of the educated elite.

Madagascar is not part of either East Africa or Southern Africa. Its human as well as wildlife population is unique; its cultural landscapes still carry Southeast Asian imprints. Here the people eat rice, not corn, grown on terraced paddies. Rapid population growth is shrinking forests and animal refuges, the economy is weak, and poverty reigns in one of the world's most scenic outposts.

◗ EQUATORIAL AFRICA

The term *equatorial* is not just locational but also environmental. The equator bisects Africa, but only the western part of central Africa features the conditions

associated with the low-elevation tropics: intense heat, high rainfall and extreme humidity, little seasonal variation, rainforest and monsoon-forest vegetation, and enormous biodiversity. To the east, beyond the Western Rift Valley, elevations rise, and cooler, more seasonal climatic regimes prevail. As a result, we recognize two regions in these lowest latitudes: Equatorial Africa to the west and East Africa to the east.

Equatorial Africa is physiographically dominated by the gigantic Congo Basin. The Adamawa Highlands separate this region from West Africa; rising elevations and climatic change mark its southern limits (see the *Cwa* boundary in Fig. G-8). Its political geography consists of eight states, of which The Congo (formerly Zaïre) is by far the largest in both territory and population (Fig. 6-14).

Five of the other seven states—Gabon, Cameroon, São Tomé and Príncipe, Congo, and Equatorial Guinea—all have coastlines on the Atlantic Ocean. The Central African Republic and Chad, the south of which is part of this region, are landlocked. In certain respects, the physical and human characteristics of Equatorial Africa extend even into southern Sudan. This vast and complex region is in many ways the most troubled region in the entire Subsaharan Africa realm.

The Congo

As the map shows, The Congo has but a tiny window (23 miles; 37 km) on the Atlantic Ocean, just enough to accommodate the mouth of the Congo River. Oceangoing ships can reach the port of Matadi, inland from which falls and rapids make it necessary to move goods by road or rail to the capital, Kinshasa. This is not the only place where the Congo River fails as a transport route. Follow it upstream on Figure 6-14, and you note that other transshipments are necessary between Kisangani and Ubundu, and at Kindu. Follow the railroad south from Kindu, and you reach another narrow corridor of territory at the city of Lubumbashi. That vital part of Katanga Province contains most of The Congo's major mineral resources, including copper and cobalt.

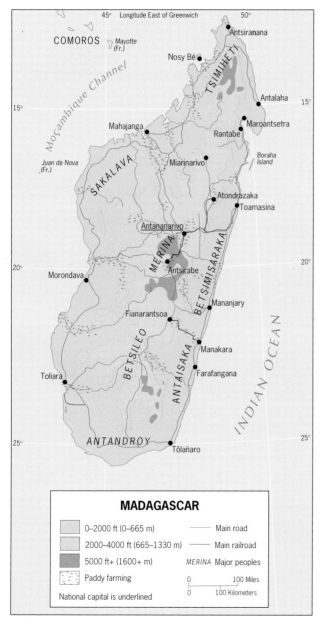

MADAGASCAR

☐ 0–2000 ft (0–665 m)	—— Main road
☐ 2000–4000 ft (665–1330 m)	—— Main railroad
☐ 5000 ft+ (1600+ m)	*MERINA* Major peoples
☐ Paddy farming	0 100 Miles
National capital is underlined	0 100 Kilometers

FIGURE 6-13

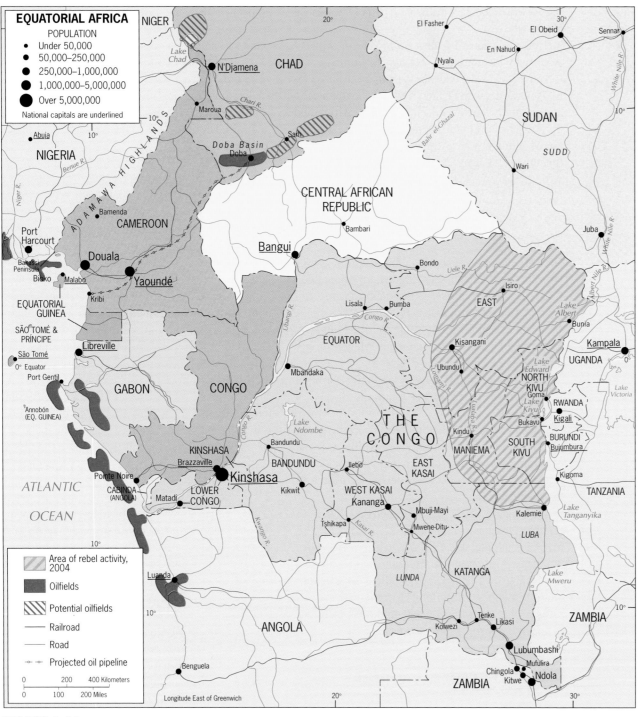

FIGURE 6-14

With a territory not much smaller than the United States east of the Mississippi, a population of 58.7 million, a rich and varied mineral base, and much good agricultural land, The Congo would seem to have all the ingredients needed to lead this region and, indeed, Africa. But strong centrifugal forces, arising from its physiography and cultural geography, pull The Congo apart. The immense forested heart of the basin-shaped country creates communication barriers between east and west, north and south. Many of The Congo's productive areas lie along its periphery, separated by enormous distances. These areas tend to look across the border, to one or more of The Congo's nine neighbors, for outlets, markets, and often ethnic kinship as well.

The Congo's civil wars of the 1990s started in one such neighbor, Rwanda, and spilled over into what was then still known as Zaïre. Rwanda, Africa's most densely populated country, has for centuries been the scene of conflict between sedentary Hutu farmers and invading Tutsi pastoralists. Colonial borders and practices worsened the situation, and after independence a series of terrible crises followed. In the mid-1990s the latest of these crises generated one of the largest refugee streams ever seen in the world, and the conflict engulfed eastern (and later northern and western) Congo. The death toll will never be known, but estimates range from 3 to 4 million, a calamity that was not enough to propel the "international community" into concerted peacemaking action. By the beginning of 2004 a combination of power transfer in the capital, Kinshasa, negotiation among the rebel groups and the African states involved in various ways in the conflict, UN assistance, and exhaustion had produced a semblance of stability in all but some eastern areas of The Congo (Fig. 6-14).

Across the River

To the west and north of the Congo and Ubangi rivers lie Equatorial Africa's other seven countries (Fig. 6-14). Two of these are landlocked. **Chad**, straddling the African Transition Zone as well as the regional boundary with West Africa, is one of Africa's most remote countries, although recent oil discoveries in the south are likely to change this. The **Central African Republic**, chronically unstable and poverty-stricken, never was able to convert its agricultural potential and mineral resources (diamonds, uranium) into real progress. And one country consists of two small, densely forested volcanic islands: **São Tomé and Príncipe**, a ministate with a population of only 220,000 and a few exports derived from its cocoa plantations and coconut trees.

The four coastal states present a different picture. All four possess oil reserves and share the Congo Basin's equatorial forests; oil and timber, therefore, rank prominently among their exports. In **Gabon**, this combination has produced Equatorial Africa's only upper-middle-income economy. Of the four coastal states, Gabon also has the largest proven mineral resources, including manganese, uranium, and iron ore. Its capital, Libreville (the only coastal capital in the region), reflects all this in its high-rise downtown, bustling port, and fast-growing squatter settlements.

Cameroon, less well endowed with oil or other raw materials, has the region's strongest agricultural sector by virtue of its higher-latitude location and high-relief topography. Western Cameroon is one of the more developed parts of Equatorial Africa and includes the capital, Yaoundé, and the port of Douala.

With five neighbors, **Congo** could be a major transit hub for this region, especially for The Congo if it recovers from civil war. Its capital, Brazzaville, lies across the Congo River from Kinshasa and is linked to the port of Pointe Noire by road and rail. But devastating power struggles have negated Congo's geographic advantages.

As Figure 6-14 shows, **Equatorial Guinea** consists of a rectangle of mainland territory and the island of Bioko, where the capital of Malabo is located. A former Spanish colony that remained one of Africa's least-developed territories, Equatorial Guinea, too, has been affected by the oil business in this area. Petroleum products now dominate its exports, but, as in so many other oil-rich countries, this bounty has not significantly raised incomes for most of the people.

One other territory would seem to be a part of Equatorial Africa: *Cabinda*, wedged between the two Congos just to the north of the Congo River's mouth. But *Cabinda* is one of those colonial legacies on the African map—it belonged to the Portuguese and was administered as part of Angola. **17** Today it is an **exclave** of independent Angola, and a valuable one: it contains major oil reserves.

▶ WEST AFRICA

West Africa occupies most of Africa's Bulge, extending south from the margins of the Sahara to the Gulf of Guinea coast and from Lake Chad west to Senegal (Fig. 6-15). Politically, the broadest definition of this region includes all those states that lie to the south of Western Sahara, Algeria, and Libya and to the west of Chad (itself sometimes included) and Cameroon. Within West Africa, a rough division is sometimes made between the large, mostly steppe and desert states that extend across the southern Sahara (Chad included) and the smaller, better-watered coastal states.

France and Britain dominated the colonial map of West Africa, and to this day the interaction among West Africa's states is limited. But West Africa's cultural vitality, historic legacies, populous cities, crowded countrysides, and bustling markets combine to create a regional imprint that is distinct and pervasive.

In the 1990s and into the 2000s, West Africa also was the scene of violent rebellions, collapsing governments, cultural strife and anarchy. Sierra Leone, Liberia, and Ivory Coast (Côte d'Ivoire) were most severely afflicted, but the mayhem spilled over into Guinea and erupted even in Nigeria over oil and land issues in the Niger Delta and religious issues in the north. Millions of ordinary citizens, caught up in conflicts caused by outsiders, have paid a dreadful price.

Nigeria

When Nigeria achieved full independence from Britain in 1960, its new government faced the task of administering a European political creation containing three major nations and nearly 250 other peoples ranging from several million to a few thousand in number.

For reasons obvious from the map, Britain's colonial imprint always was stronger in the two southern regions than in the north. Christianity became the dominant faith in the south, and southerners, especially the Yoruba, took a lead role in the transition from colony to independent state. The choice of Lagos, the port of the Yoruba-dominated southwest, as the capital of a federal Nigeria (and not one of the cities in the more populous north) reflected British desires for the country's future. A three-region federation, two of which lay in the south, would ensure the primacy of the non-Islamic part of the state. But this framework did not last long. In 1967, the Ibo-dominated Eastern Region declared its independence as the Republic of Biafra, leading to a three-year civil war at a cost of 1 million lives. Since then, Nigeria's federal system has been modified repeatedly; today there are 36 States, and the capital has been moved from Lagos to centrally-located Abuja (Fig. 6-16).

Large oilfields were discovered beneath the Niger Delta during the 1950s, when Nigeria's agricultural sector produced most of its exports

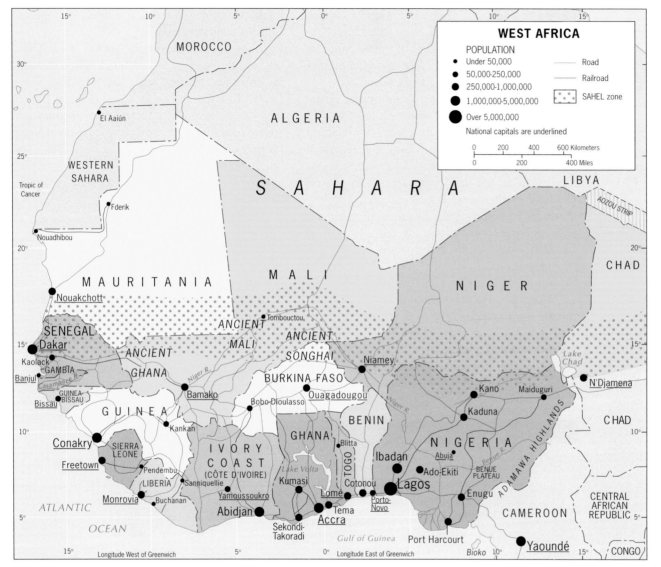

FIGURE 6-15

enues during military misrule, and excessive borrowing against future oil income led to economic disaster. The country's infrastructure collapsed. In the cities, basic services broke down. In the rural areas, clinics, schools, water supplies, and roads to markets crumbled. In the Niger Delta area, local people beneath whose land the oil was being exploited demanded a share of the revenues and reparations for ecological damage; the military regime under General Abacha responded by arresting and executing nine of their leaders. On global indices of national well-being, Nigeria sank to the lowest rungs even as its production ranked it as high as the world's tenth-largest oil producer, with the United States its chief customer.

In 1999, Nigeria's hopes were raised when, for the first time since 1983, a democratically elected president was sworn into office. But Nigeria's problems (now also including a deepening AIDS crisis) worsened when northern States, beginning with Zamfara, proclaimed Sharia law. When Kaduna State followed suit, riots between Christians and Muslims devastated the old capital city of Kaduna. There, and in ten other northern States (Fig. 6-16), the imposition of Sharia law led to the departure of thousands of Christians, intensifying the cultural fault line that threatens the cohesion of the country. Although Kaduna State temporarily repealed its decision, the Islamic revivalism now exhibited in the north raises the prospect that Nigeria, West Africa's cornerstone and one of Africa's most important states, may succumb to devolutionary forces arising from its loca-

(peanuts, palm oil, cocoa, cotton) and farming still had priority in national and State development plans. Soon, revenues from oil production dwarfed all other sources, bringing the country a brief period of prosperity and promise. But before long Nigeria's oil wealth brought more bust than boom. Misguided development plans now focused on grand, ill-founded industrial schemes and costly luxuries such as a national airline; the continuing mainstay of the vast majority of Nigerians, agriculture, fell into neglect. Worse, poor management, corruption, outright theft of oil rev-

THE STATES OF FEDERAL NIGERIA

Legend	
Road	
Railroad	
Pipeline	
Oilfield	
Sharia law proclaimed	
Core area	
TIV	Major ethnic group

POPULATION
- Under 50,000
- 50,000–250,000
- 250,000–1,000,000
- 1,000,000–5,000,000
- Over 5,000,000

National capitals are underlined

Longitude East of Greenwich

FIGURE 6-16

tion in the African Transition Zone. For Africa, this would be a calamity.

Coast and Interior

Nigeria is one of 17 states (counting Chad and offshore Cape Verde) that form the region of West Africa. Four of these countries, comprising a huge territory under desert and steppe environments but containing small populations, are landlocked: Mali, Burkina Faso, Niger, and Chad (Fig. 6-15).

Among the coastal states, **Ghana**, once known as the Gold Coast, was the first West African state to achieve independence, with a democratic government and a sound economy based on cocoa exports. Two grandiose post-independence schemes can be seen on the map: the port of Tema, intended to serve a vast West African hinterland, and Lake Volta, which resulted from the region's largest dam project. Mismanagement contributed to Ghana's economic collapse, but in the 1990s, a military regime was replaced by stable and democratic government and recovery began. **Ivory Coast** (officially Côte d'Ivoire) translated three decades of autocratic but stable rule into economic progress, but excesses by the country's president-for-life cost it dearly. One of these excesses involved the construction of a Roman Catholic basilica to rival St. Peter's in Rome in the president's home village, Yamoussoukro, also designated to replace Abidjan as the country's new capital. At the turn of this century, the political succession became entangled in a north-south, Muslim-Christian schism that threatened the country's stability. Democratic **Senegal**, on the far west coast, demonstrates what stability can achieve: without oil, diamonds, or other valuable income sources and with an overwhelmingly subsistence-farming population, Senegal nevertheless managed to achieve some of the region's highest GNP levels. Over 90 percent Muslim and dominated by the Wolof, the ethnic group concentrated in the capital of Dakar, and with continuing close ties to France, Senegal has even been able to overcome a failed effort to unify with its English-speaking enclave, Gambia, and a secession movement in its southern Casamance District.

Other parts of West Africa have been afflicted by civil war and horror. **Liberia**, founded in 1822 by freed American slaves who returned to Africa with the help of American colonization societies, was ruled by "Americo-Liberians" for six generations and sold rubber and iron ore abroad. But a military coup in 1980 ended that era, and full-scale civil war beginning in 1989 embroiled virtually every ethnic group in the country. About a quarter of a million people, one-tenth of the population, perished; hundreds of thousands of others fled as refugees, many to **Sierra Leone**, which also was originally founded as a haven for freed slaves, in this case by the British in 1787. Sierra Leone went the all-too-familiar route from self-governing Commonwealth member to republic to one-party state to military dictatorship. In the 1990s worse was to follow: a rebel movement, funded by diamond sales, inflicted dreadful punishment on the local population. British efforts to stem the tide of violence had some effect by 2002, but untold damage had been done.

We should remember, however, that these conflicts are not representa-tive of vast and populous West Africa. Tens of millions of farmers and herders who manage to cope with fast-changing environments in this zone between desert and ocean live remote from the news-making conflicts along the monsoon coast. Time-honored systems continue to serve: for example, the local village markets that drive the traditional economy. Visit the countryside, and you will find that some village markets are not open every day but, rather, every three or four days. Such a system ensures that all villages get a share in the exchange ⏹18 network. These **periodic markets** represent one of the many traditions enduring here, even as the cities beckon the farmers and burst at the seams. This region's great challenges are economic survival and nation-building through political stability under some of the most difficult circumstances on Earth.

THE AFRICAN TRANSITION ZONE

As Figures 6-17 and 6-9 show, the African Transition Zone is unlike the four regions just discussed.

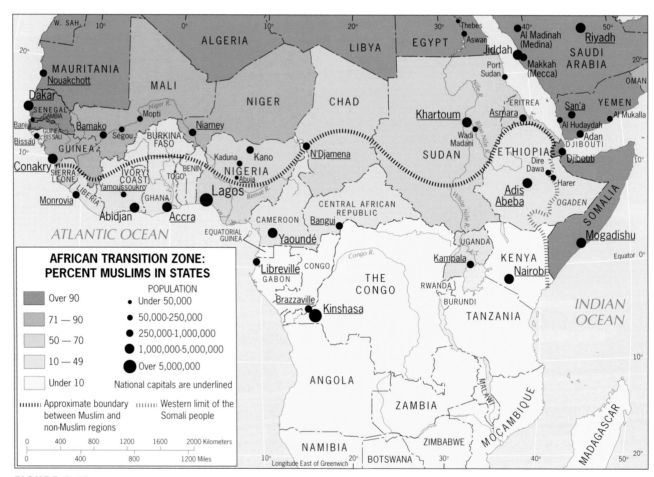

AFRICAN TRANSITION ZONE: PERCENT MUSLIMS IN STATES

		POPULATION
Over 90	●	Under 50,000
71 — 90	●	50,000-250,000
50 — 70	●	250,000-1,000,000
10 — 49	●	1,000,000-5,000,000
Under 10	●	Over 5,000,000

National capitals are underlined

⫶⫶⫶⫶ Approximate boundary between Muslim and non-Muslim regions

⫶⫶⫶⫶ Western limit of the Somali people

0 400 800 1200 1600 2000 Kilometers
0 400 800 1200 Miles

FIGURE 6-17

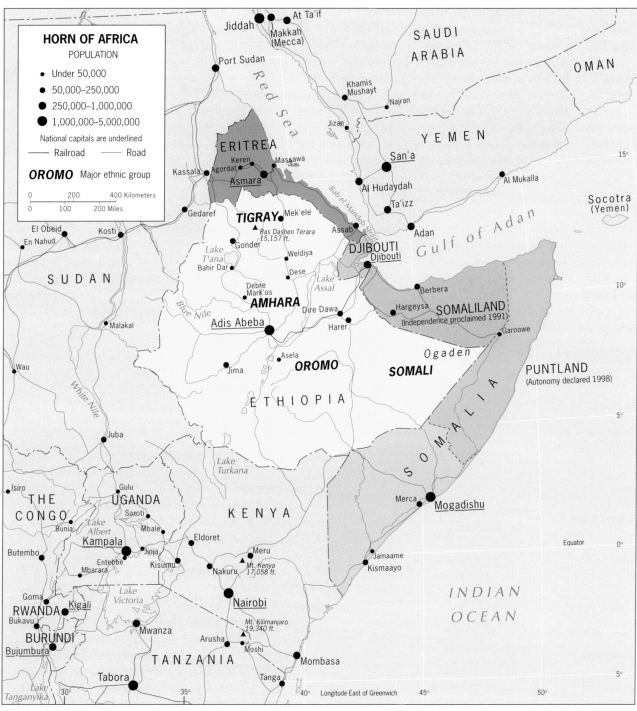

HORN OF AFRICA
POPULATION

- Under 50,000
- 50,000–250,000
- 250,000–1,000,000
- 1,000,000–5,000,000

National capitals are underlined
—— Railroad   —— Road

OROMO Major ethnic group

0   200   400 Kilometers
0   100   200 Miles

FIGURE 6-18

This is Subsaharan Africa's threshold to the world of Islam, and it is Islam's gateway into the Subsaharan African realm. It encompasses some entire countries that are also part of the formal regions on the map (such as Senegal and Burkina Faso), it divides others into Muslim and non-Muslim sectors (such as Chad and Sudan), and, in Africa's Horn, it comprises countries that do not form parts of other regions (Eritrea, Djibouti, Somalia). As elsewhere in the world where geographic realms meet or overlap, complications mark the African Transition Zone. In some areas, the transition from Muslim to non-Muslim society is gradual, as it remains in the northern parts of Ivory Coast and Nigeria despite the recent cultural problems erupting in these important states. Elsewhere the break is sharp, as it is in eastern Ethiopia where a traditional border between Christian/animist-African cultures and Muslim-African communities is sudden. The southern border of the African Transition Zone, the religious frontier sometimes referred to as **19** Africa's **Islamic Front**, is therefore neither static nor everywhere the same.

Conflict marks much of the Islamic Front in Africa: long-term in Sudan, where a 30-year war for independence by non-Islamic African rebels in the south against the Arabized, Muslim north has cost millions of lives; intermittent in Nigeria, where Islamic revivalism is currently clouding the country's prospects; and recent in Ivory Coast, where political rivalries spurred religious strife. Perhaps the most conflict-prone part of the African Transition Zone, however, lies in the east, involving not only Sudan but also the historic Christian state of Ethiopia, nearly encircled by dominantly Muslim societies.

The Horn of Africa (Fig. 6-18) is an especially volatile subregion of the African Transition Zone. In landlocked **Ethiopia**, nearly one-third of the population of more than 70 million is Muslim, virtually all of it concentrated in the east. A costly conflict with neighboring **Eritrea**, which is about 50 percent Muslim, has damaged the economies of both countries (that issue was not over religion but over boundary definition). The ministate of **Djibouti**, nearly 100 percent Muslim, has recently taken on added importance in the War on Terror since it lies directly across from Yemen and overlooks the narrow entry to the Red Sea, the Bab el Mandeb Strait. But the key component in the eastern sector of the African Transition Zone is **Somalia**, where more than 8 million people, virtually all Muslim, live at the mercy of a desert-dominated climate that enforces cross-border migration into Ethiopia's Ogaden area in pursuit of seasonal pastures. As many as three to five million Somalis live permanently on the Ethiopian side of the border, but this is not the only division the Somali "nation" faces. In reality, the Somali people is an assemblage of five major ethnic groups fragmented into hundreds of clans engaged in an endless contest for power as well as survival. In 2004, the nominal "state" of Somalia remained politically split into three segments: secession-minded Somaliland in the north; autonomy-proclaiming Puntland in the middle; and rump Somalia, with the capital of Mogadishu, in the south (Fig. 6-18). Dysfunctional and intractable, Somalia is of concern not only to its neighbors Ethiopia and Kenya, but also to other countries involved in the War on Terror.

As this chapter has underscored, Africa is a continent of infinite social diversity, and Subsaharan Africa is a realm of matchless cultural history. Comparatively recent environmental change widened the Sahara and separated north from south; the arrival of Islam turned the north to Mecca and left the south to Christian proselytism. Millions of Africans were carried away in bondage; the cruelties of colonialism subjugated those who were left behind in a political straitjacket forged in Berlin. Africa's equatorial heart is an incubator for debilitating diseases second to none on Earth, but Africa's health problems never were top priority in the medical world. When finally the richer countries abandoned their African empires, African peoples were left a task of reconstruction in which they not only got little help: they found their products uncompetitive on markets where rich-country producers basked in subsidies. We said at the beginning of this chapter that Africa's time and turn will come again, but it will take far more than is being done today to rectify half a millennium's malefactions.

DISCOVERY TOOLS
www.wiley.com/college/deblij

The *Concepts and Regions* Website, featuring *GeoDiscoveries*, offers many additional resources to enhance your understanding and experience of this chapter. Be sure to explore the following:

GeoDiscoveries Video Clips
Malaria

GeoDiscoveries Animations
Migration

GeoDiscoveries Interactivities
Naming the Region

Spatial Diffusion: Medical Geography

Photo Galleries
Dakar, Senegal
Niamey, Niger
Nairobi, Kenya
Meru, Kenya
Kariba Dam
Witbank, South Africa
Cape Town, South Africa
Mauritius

Virtual Field Trip
Timbuktu (Tombouctou): West Africa's Legendary City

More To Explore
Systematic Essay: Medical Geography

Issues in Geography: AIDS in Subsaharan Africa; A Green Revolution for Africa; The Berlin Conference; Distinctive Madagascar

Major Cities: Lagos, Nairobi, Johannesburg

Expanded Regional Coverage
Southern Africa, East Africa, Equatorial Africa, West Africa, and African Transition Zone

Chapter 7

North Africa/ Southwest Asia

CONCEPTS, IDEAS, AND TERMS

1. Cultural geography
2. Culture hearth
3. Cultural diffusion
4. Cultural environment
5. Cultural ecology
6. Hydraulic civilization theory
7. Climate change
8. Spatial diffusion
9. Expansion diffusion
10. Relocation diffusion
11. Contagious diffusion
12. Hierarchical diffusion
13. Islamization
14. Culture region
15. Religious revivalism
16. Wahhabism
17. Cultural revival
18. Stateless nation
19. Choke point
20. Nomadism
21. Buffer state

REGIONS

- EGYPT AND THE LOWER NILE BASIN
- MAGHREB AND ITS NEIGHBORS
- MIDDLE EAST
- ARABIAN PENINSULA
- EMPIRE STATES
- TURKESTAN

FROM MOROCCO ON the shores of the Atlantic to the mountains of Afghanistan, and from the Horn of Africa to the steppes of inner Asia, lies a vast geographic realm of enormous cultural complexity. It stands at the crossroads where Europe, Asia, and Africa meet, and it is part of all three (Fig. 7-1). Throughout history, its influences have radiated to these continents and to practically every other part of the world as well. This is one of humankind's primary source areas. On the Mesopotamian Plain between the Tigris and Euphrates rivers (in modern-day Iraq) and on the banks of the Egyptian Nile arose several of the world's earliest civilizations. In its soils, plants were domesticated that are now grown from the Americas to Australia. Along its paths walked prophets whose religious teachings are still followed by hundreds of millions. And at the opening of the twenty-first century, the heart of this realm is beset by some of the most bitter and dangerous conflicts on Earth.

STUDY TOOLS

3D GLOBE   FLASHCARDS   AUDIO PRONUNCIATION GUIDE   ANNOTATED WEBLINKS
AREA AND DEMOGRAPHIC DATA   MAP QUIZZES   CHAPTER QUIZ   LONELY PLANET WEBLINKS

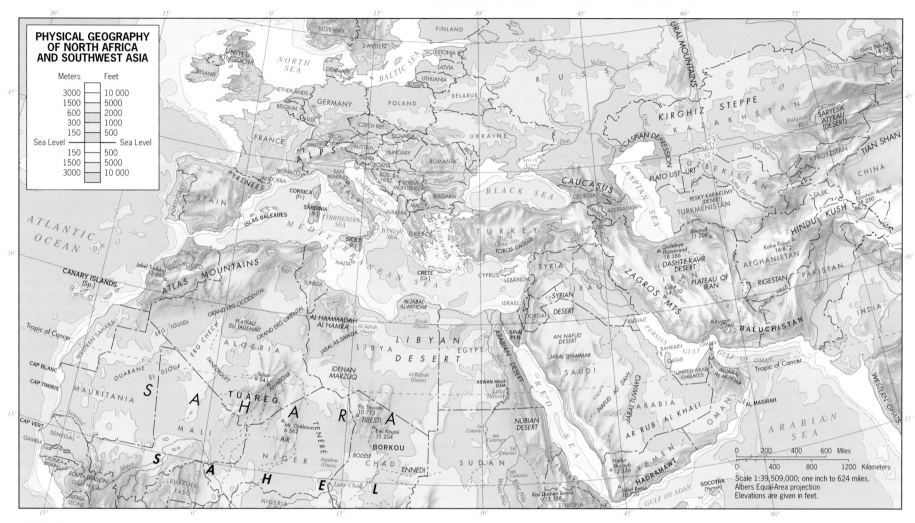

PHYSICAL GEOGRAPHY OF NORTH AFRICA AND SOUTHWEST ASIA

FIGURE 7-1

DEFINING THE REALM

It is tempting to characterize this geographic realm in a few words and to stress one or more of its dominant features. It is, for instance, often called the "Dry World," containing as it does the vast Sahara as well as the Arabian Desert. But most of the realm's people live where there is water—in the Nile Delta, along the hilly Mediterranean coastal strip (the *tell*, meaning "mound" in Arabic) of north-westernmost Africa, along the Asian eastern and northeastern shores of the Mediterranean Sea, in the Tigris-Euphrates Basin, in far-flung desert oases, and along the lower mountain slopes of Iran south of the Caspian Sea and of Turkestan to the northeast. We know this world region as one where water is al-most always at a premium, where peasants often struggle to make soil and moisture yield a small harvest, where nomadic peoples and their animals circulate across dust-blown flatlands, where oases are islands of sedentary farming and trade in a sea of aridity. But it also is the land of the Nile, the life-line of Egypt, the crop-covered *tell* of northwestern-

MAJOR GEOGRAPHIC QUALITIES OF NORTH AFRICA/SOUTHWEST ASIA

1. North Africa and Southwest Asia were the scene of several of the world's great ancient civilizations, based in its river valleys and basins.

2. From this realm's culture hearths diffused ideas, innovations, and technologies that changed the world.

3. The North Africa/Southwest Asia realm is the source of three world religions: Judaism, Christianity, and Islam.

4. Islam, the last of the major religions to arise in this realm, transformed, unified, and energized a vast domain extending from Europe to Southeast Asia and from Russia to East Africa.

5. Drought and unreliable precipitation dominate natural environments in this realm. Population clusters exist where water supply is adequate to marginal.

6. Certain countries of this realm have enormous reserves of oil and natural gas, creating great wealth for some but doing little to raise the living standards of the majority.

7. The boundaries of the North Africa/Southwest Asia realm consist of volatile transition zones in several places in Africa and Asia.

8. Conflict over water sources and supplies is a constant threat in this realm, where population growth rates are high by world standards.

9. The Middle East, as a region, lies at the heart of this realm; and Israel lies at the center of the Middle East conflict.

10. Religious, ethnic, and cultural discord frequently cause instability and strife in this realm.

most (coastal strip) Africa, the verdant shores of western Turkey, the meltwater-fed valleys of Central Asia. Compare Figure 7-1 to Figure G-8, and the dominance of *B* climates becomes evident. Also consult Figure G-9, and you will see how water-dependent this realm's clustered population is.

An "Arab World"?

North Africa/Southwest Asia is also often referred to as the Arab World. This term implies a uniformity that does not actually exist. First, the name *Arab* is applied loosely to the peoples of this area who speak Arabic and related languages, but ethnologists normally restrict it to certain occupants of the Arabian Peninsula—the Arab "source." In any case, the Turks are not Arabs, and neither are most Iranians or Israelis. Moreover, although the Arabic language prevails from Mauritania in the west across all of North Africa to the Arabian Peninsula,

Syria, and Iraq in the east, it is not spoken in other parts of this realm. In Turkey, for example, Turkish is the major language, and it has Ural-Altaic rather than Arabic's Semitic or Hamitic roots. The Iranian language belongs to the Indo-European linguistic family. Other "Arab World" languages that have separate ethnological identities are spoken by the Jews of Israel, the Tuareg people of the Sahara, the Berbers of northwestern Africa, and the peoples of the transition zone between North Africa and Subsaharan Africa to the south.

An "Islamic World"?

Yet another name given to this realm is the World of Islam. The prophet Muhammad (Mohammed) was born in Arabia in AD 571, and in the centuries after his death in 632, Islam spread into Africa, Asia, and Europe. This was the age of Arab conquest and expansion. Their armies penetrated Southern Europe,

their caravans crossed the deserts, and their ships plied the coasts of Asia and Africa. Along these routes they carried the Muslim (Islamic) faith, converting the ruling classes of the states of the West African savanna, threatening the Christian stronghold in the highlands of Ethiopia, penetrating the deserts of inner Asia, and pushing into India and even the island extremities of Southeast Asia. Today, the Islamic faith extends far beyond the limits of the realm under discussion (Fig. 7-2). Nor is the World of Islam entirely Muslim. Judaism, Christianity (notably in Egypt and Lebanon), and other faiths survive in the heartland of the Islamic World. So this connotation is not satisfactory either.

"Middle East"?

Finally, this realm is frequently called the Middle East. That must sound odd to someone in, say, India, who might think of a Middle West rather

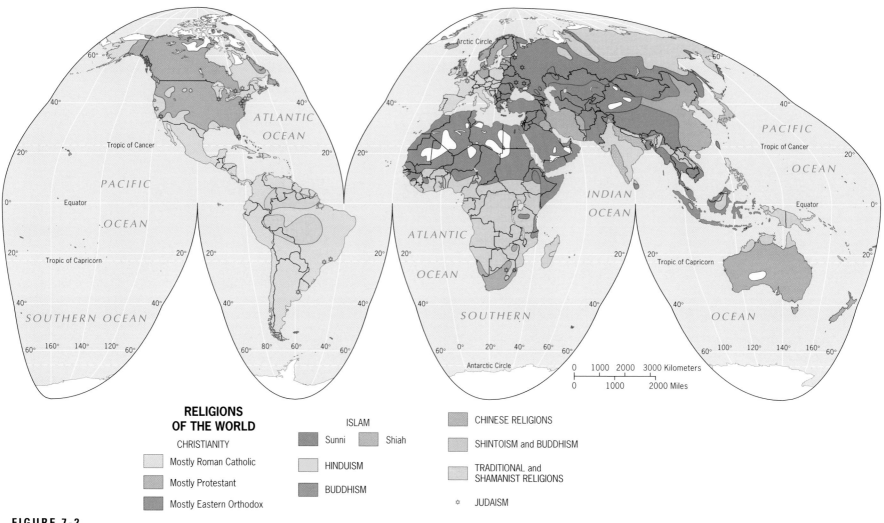

**RELIGIONS
OF THE WORLD**

CHRISTIANITY

Mostly Roman Catholic

Mostly Protestant

Mostly Eastern Orthodox

ISLAM

Sunni   Shiah

HINDUISM

BUDDHISM

CHINESE RELIGIONS

SHINTOISM and BUDDHISM

TRADITIONAL and
SHAMANIST RELIGIONS

✧   JUDAISM

FIGURE 7-2

than a Middle East! The name, of course, reflects the biases of its source: the "Western" world, which saw a "Near" East in Turkey, a "Middle" East in Egypt, Arabia, and Iraq, and a "Far" East in China and Japan. Still, the term has taken hold, and it can be seen and heard in everyday usage by scholars, journalists, and members of the United Nations. Even so, it should be applied only to one of the regions of this vast realm, not to the realm as a whole.

HEARTHS OF CULTURE

This geographic realm occupies a pivotal part of the world: here Eurasia, crucible of human cultures, meets Africa, source of humanity itself. A million years ago, the ancestors of our species walked from East Africa into North Africa and Arabia and spread from one end of Asia to the other. More than one hundred thousand years ago, *Homo sapiens* crossed these lands on the way to

Europe, Australia, and, eventually, the Americas. Ten thousand years ago, human communities in what we now call the Middle East began to domesticate plants and animals, learned to irrigate their fields, enlarged their settlements into towns, and formed the earliest states. One thousand years ago, the heart of the realm was stirred and mobilized by the teachings of Muhammad and the Quran (Koran), and Islam was on the march from North Africa to India. Today this realm is a cauldron of religious and political activity, weakened by conflict but empowered by oil, plagued by poverty but fired by a wave of religious fundamentalism (Muslims prefer the term religious *revivalism*).

In the Introduction (p. 13) we discussed the concept of culture and its regional expression in the **1** cultural landscape. **Cultural geography**, we noted, is a wide-ranging and comprehensive field that studies spatial aspects of human cultures, focusing not only on cultural landscapes but also on **2** **culture hearths**—the crucibles of civilization, the sources of ideas, innovations, and ideologies that changed regions and realms. Those ideas and innovations spread far and wide through a set of **3** processes that we study under the rubric of **cul-**

tural diffusion. Because we understand these processes better today, we can reconstruct ancient routes by which the knowledge and achievements of culture hearths spread (that is, diffused) to other areas. Another topic of cultural geography, also relevant in the context of the North Africa/Southwest **4** Asia realm, is the **cultural environment** that a dominant culture creates. Human cultures exist in long-term accommodation with (and adaptation to) their natural environments, exploiting opportunities that these environments present and coping with the extremes they can impose. The study of the relationship between human societies and natural environments has become a separate branch of **5** cultural geography called **cultural ecology**.

As we will see, the North Africa/Southwest Asia realm presents many opportunities to investigate cultural geography in regional settings.

Mesopotamia and the Nile

In the basins of the major rivers of this realm (the Tigris and Euphrates of modern-day Turkey, Syria, and Iraq, and the Nile of Egypt) lay two of the world's earliest culture hearths (Fig. 7-3). Mesopotamia, "land amidst the rivers," had fertile alluvial soils, abundant sunshine, ample water, and animals and plants that could be domesticated. Here, in the Tigris-Euphrates lowland between the Persian Gulf and the uplands of present-day Turkey, arose one of humanity's first culture hearths, a cluster of communities that grew into larger societies and, eventually, into the world's first states. (Early state development probably was going on simultaneously in East Asia's river basins as well.) Mesopotamians were innovative farmers who knew when to sow and harvest crops, water their fields, and store their surplus. Their knowledge diffused to villages near and far, and a *Fertile Crescent* evolved extending from Mesopotamia across southern Turkey into Syria and the Mediterranean coast beyond (Fig. 7-3).

Irrigation was the key to prosperity and power in Mesopotamia, and urbanization was its reward. Among many settlements in the Fertile Crescent, some thrived, grew, enlarged their hinterlands, and diversified socially and occupationally; others failed. What determined success? **6** One theory, the **hydraulic civilization theory**, holds that cities that could control irrigated farming over large hinterlands held power over others, used food as a weapon, and

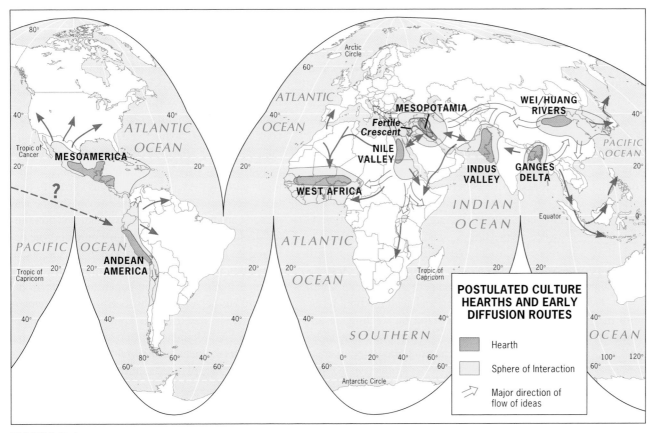

FIGURE 7-3

thrived. One such city, Babylon on the Euphrates River, endured for nearly 4000 years (from 4100 BC). A busy port, its walled and fortified center endowed with temples, towers, and palaces, Babylon for a time was the world's largest city.

Egypt's cultural evolution may have started even earlier than Mesopotamia's, and its focus lay upstream from (south of) the Nile Delta and downstream from (north of) the first of the Nile's series of rapids, or cataracts (Fig. 7-3). This part of the Nile Valley is surrounded by inhospitable desert, and unlike Mesopotamia (which lay open to all comers), the Nile provided a natural fortress here. The ancient Egyptians converted their security into progress. The Nile was their highway of trade and interaction; it also supported agriculture through irrigation. The Nile's cyclical ebb and flow was much more predictable than that of the Tigris-Euphrates river system. By the time Egypt finally fell victim to outside invaders (about 1700 BC), a full-scale urban civilization had emerged. Ancient Egypt's artist-engineers left a magnificent legacy of massive stone monuments, some of them containing treasure-filled crypts of god-kings called Pharaohs. These tombs have enabled archeologists to reconstruct the ancient history of this culture hearth.

To the east, separated from Mesopotamia by more than 1200 miles (1900 km) of mountain and desert, lies the Indus Valley (Fig. 7-3). By modern criteria, this eastern hearth lies outside the realm under discussion here; but in ancient times it had cultural and commercial ties with the Tigris-Euphrates region. Mesopotamian innovations reached the Indus region early, and eventually the cities of the Indus became power centers of a civilization that extended far into present-day northern India.

Today, the world continues to benefit from the accomplishments of the ancient Mesopotamians and Egyptians. They domesticated cereals (wheat, rye, barley), vegetables (peas, beans), fruits (grapes, apples, peaches), and many animals (horses, pigs, sheep). They also advanced the study of the calendar, mathematics, astronomy, government, engineering, metallurgy, and a host of other skills and technologies. In time, many of their innovations were adopted and then modified by other cultures in the Old World and eventually in the New World as well. Europe was the greatest beneficiary of these legacies of Mesopotamia and ancient Egypt, whose achievements constituted the foundations of "Western" civilization.

Decline and Decay

As Figure G-8 reminds us, many of the early cities of this culture realm lay in what is today desert territory. Assuming that there were no good reasons to build large settlements in the middle of deserts, we may ⑦ hypothesize that **climate change** sweeping over this region, not a monopoly over irrigation techniques, gave certain cities in the ancient Fertile Crescent an advantage over others. Climate change, associated with shifting environmental zones after the last Pleistocene glacial retreat, may have destroyed the last of the old civilizations. Perhaps overpopulation and human destruction of the natural vegetation contributed to the process. Indeed, some cultural geographers suggest that the momentous innovations in agricultural planning and irrigation technology were not "taught" by the seasonal flooding of the rivers but were forced on the inhabitants as they tried to survive changing environmental conditions.

The scenario is not difficult to imagine. As outlying areas began to fall dry and farmlands were destroyed, people congregated in the already crowded river valleys—and made every effort to increase the productivity of the land that could still be watered. Eventually overpopulation, destruction of the watershed, and perhaps reduced rainfall in the rivers' headwater areas dealt the final blow. Towns were abandoned to the encroaching desert; irrigation canals filled with drifting sand; croplands dried up. Those who could migrated to areas that were still reputed to be productive. Others stayed, their numbers dwindling, increasingly reduced to subsistence.

As old societies disintegrated, power emerged elsewhere. First the Persians, then the Greeks, and later the Romans imposed their imperial designs on the tenuous lands and disconnected peoples of North Africa/Southwest Asia. Roman technicians converted North Africa's farmlands into irrigated plantations whose products went by the boatload to Roman Mediterranean shores. Thousands of people were carried off as slaves to the cities of the new conquerors. Egypt was quickly colonized, as was the area we now call the Middle East. One region that lay distant, and therefore remote from these invasions, was the Arabian Peninsula, where no major culture hearth or large cities had emerged and where the turmoil had not affected Arab settlements and nomadic routes.

STAGE FOR ISLAM

In a remote place on the Arabian Peninsula, where the foreign invasions of the Middle East had had little effect on the Arab communities, an event occurred early in the seventh century that was to change history and affect the destinies of people in many parts of the world. In a town called Mecca (Makkah), about 45 miles (70 km) from the Red Sea coast in the Jabal Mountains, a man named Muhammad in the year AD 611 began to receive revelations from Allah (God). Muhammad (571–632) was then in his early forties and had barely 20 years to live. Convinced after some initial self-doubt that he was indeed chosen to be a prophet, Muhammad committed his life to fulfilling the divine commands he believed he had received. Arab society was in social and cultural disarray, but Muhammad forcefully taught Allah's lessons and began to transform his culture. His personal power soon attracted enemies, and in 622 he fled from Mecca to the safer haven of Medina (Al Madinah), where he continued his work. This moment, the *hejira* ("migration"), marks the starting date of the Muslim era, Year 1 on Islam's calendar. Mecca, of course, later became Islam's holiest place.

The precepts of Islam in many ways constituted a revision and embellishment of Judaic and Christian beliefs and traditions. All of these faiths have but one god, who occasionally communicates with human-kind through prophets; Islam acknowledges that Moses and Jesus were such prophets but considers Muhammad to be the final and greatest prophet. What is Earthly and worldly is profane; only Allah is pure. Allah's will is absolute; Allah is omnipotent and omniscient. All humans live in a world that Allah created for their use but only to await a final judgment day.

Islam brought to the Arab World not only the unifying religious faith it had lacked but also a new set of values, a new way of life, a new individual and collective dignity. Islam dictated observance of the Five Pillars: (1) repeated expressions of the basic creed, (2) the daily prayer, (3) a month each year of daytime fasting (Ramadan), (4) the giving of alms, and (5) at least one pilgrimage in each Muslim's lifetime to Mecca (the *hajj*). And Islam prescribed and proscribed in other spheres of life as well. It forbade alcohol, smoking, and gam-bling. It tolerated polygamy, although it acknowl-edged the virtues of monogamy. Mosques ap-peared in Arab settlements, not only for the (Friday) sabbath prayer, but also as social gather-ing places to knit communities closer together. Mecca became the spiritual center for a divided, widely dispersed people for whom a collective focus was something new.

The Arab-Islamic Empire

Muhammad provided such a powerful stimulus that Arab society was mobilized almost overnight. The prophet died in 632, but his faith and fame spread like wildfire. Arab armies carrying the banner of Islam formed, invaded, conquered, and converted wherever they went. As Figure 7-4 shows, by AD 700 Islam had reached far into North Africa, into Transcaucasia, and into most of Southwest Asia. In the cen-turies that followed, it penetrated Southern and Eastern Europe, Cen-tral Asia's Turkestan, West Africa, East Africa, and South and South-east Asia, even reaching China by AD 1000.

The spread of Islam provides a good illustration of a series of **8** processes called **spatial diffu-sion**, focusing on the way ideas, in-ventions, and cultural practices prop-agate through a population in space and time. In 1952, the Swedish geo-grapher Torsten Hägerstrand pub-lished a fundamental study of spatial diffusion entitled *The Propagation of Diffusion Waves*. He reported that diffusion takes place in two **9** forms: **expansion diffusion**, when propagation waves originate in a strong and durable source area and spread outward, affecting an ever larger region and population; and **10** **relocation diffusion**, in which an innovation, idea, or (for example) a virus is carried by migrants from the

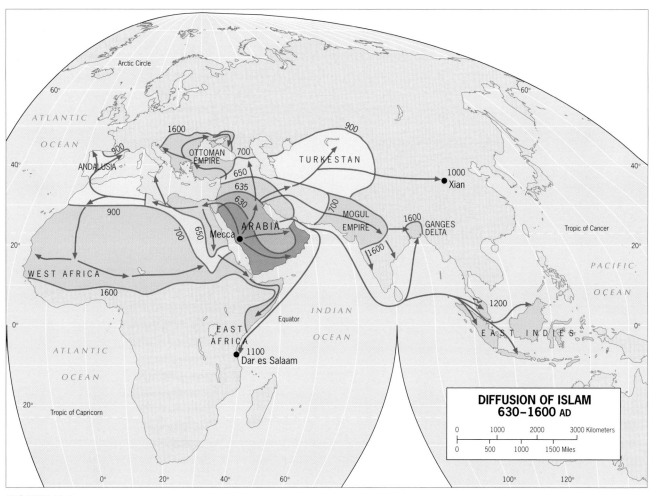

FIGURE 7-4

source to distant locations and diffuses from there. The global spread of AIDS is a case of relocation diffusion.

Both expansion diffusion and relocation diffusion include several types of processes. The spread of Islam, as Figure 7-4 shows, initially proceeded **11** by a form of expansion diffusion called **contagious diffusion** as the faith moved from village to village across the Arabian Peninsula and the Middle East. But Islam got powerful boosts when kings, chiefs, and other high officials were converted, who in turn propagated the faith downward through their bureaucracies to far-flung subjects. **12** This form of expansion diffusion, called **hierarchical diffusion**, served Islam well.

But the map leaves no doubt that Islam later spread by relocation diffusion as well, notably to the Ganges Delta in South Asia, to present-day Indonesia, and to East Africa. That process continues to this day, but the heart of Islam remains in Southwest Asia. There, Islam became the cornerstone of an Arab Empire with Medina as its first capital. As the empire grew by expansion diffusion, its headquarters were relocated from Medina to Damascus (in present-day Syria) and later to Baghdad on the Tigris River (Iraq). And it prospered. In architecture, mathematics, and science, the Arabs overshadowed their European contemporaries. The Arabs established institutions of higher learning in many cities including Baghdad, Cairo, and Toledo (Spain), and their distinctive cultural landscapes united their vast domain. Non-Arab societies in the path of the Muslim drive were not only Islamized, but also Arabized, adopting other Arab traditions as well. Islam had spawned a culture; it still lies at the heart of that culture today.

As we noted, Islam's expansion eventually was checked in Europe, Russia, and elsewhere. But a map showing the total area under Muslim sway in Eurasia and Africa reveals the enormous dimensions of the domain **13** affected by **Islamization** at one time or another (Fig. 7-5). Islam continues to expand, now mainly by relocation diffusion. There are Islamic communities in cities as widely scattered as Vienna, Singapore, and Cape Town, South Africa; Islam is also growing rapidly in the United States. With more than 1.3 billion adherents today, Islam is a vigorous and burgeoning cultural force around the world.

ISLAM DIVIDED

For all its vigor and success, Islam still fragmented into sects. The earliest and most consequential division arose after Muhammad's death. Who should be his legitimate successor? Some believed that only a blood relative should follow the prophet as leader of Islam. Others, a majority, felt that any devout follower of Muhammad was qualified. The first chosen successor was the fa-

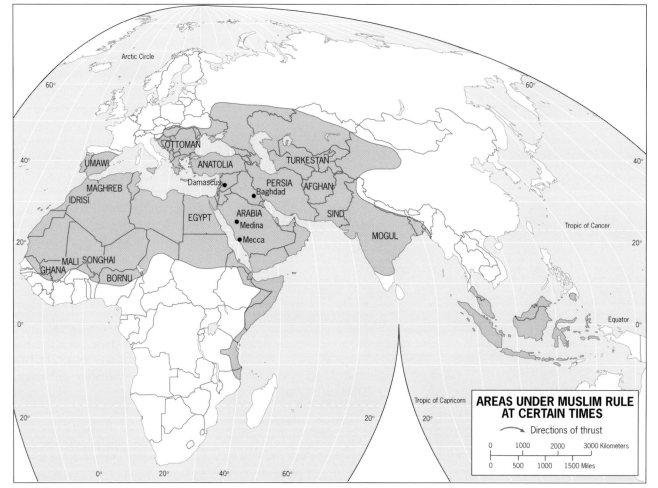

FIGURE 7-5

ther of Muhammad's wife (and thus not a blood relative). But this did not satisfy those who wanted to see a man named Ali, a cousin of Muhammad, made *caliph* (successor). When Ali's turn came, his followers, the Shi'ites, proclaimed that Muhammad finally had a legitimate successor. This offended the Sunnis, those who did not see a blood relationship as necessary for the succession. From the beginning of this disagreement, the numbers of Muslims who took the Sunni side far exceeded those who regarded themselves as Shiah (followers) of Ali. The great expansion of Islam was largely propelled by Sunnis; the Shi'ites survived as small minorities scattered throughout the realm. Today, about 85 percent of all Muslims are Sunnis.

But the Shi'ites vigorously promoted their version of the faith. In the early sixteenth century their work paid off: the royal house of Persia (modern-day Iran) made Shi'ism the only legal religion throughout its vast empire. That domain extended from Persia into lower Mesopotamia (modern Iraq), into Azerbaijan, and into western Afghanistan and Pakistan. As the map of religions (Fig. 7-2) shows, 14 this created for Shi'ism a large **culture region** and gave the faith unprecedented strength. Iran remains the bastion of Shi'ism in the realm today, and the appeal of Shi'ism continues to radiate into neighboring countries and even farther afield.

During the late twentieth century, Shi'ism gained unprecedented influence in the realm. In its heartland, Iran, a *shah* (king) tried to secularize the country and to limit the power of the *imams* (mosque officials); he provoked a revolution that cost him the throne and made Iran a Shi'ite Islamic republic. Before long, Iran was at war with neighboring, Sunni-dominated Iraq, and Shi'ite parties and communities elsewhere were invigorated by the newfound power of Shi'ism. From Arabia to Africa's northwestern corner, Sunni-ruled countries warily watched their Shi'ite minorities, newly imbued with religious fervor. Mecca, the holy place for both Sunnis and Shi'ites, became a battleground during the week of the annual pilgrimage, and for a time the (Sunni) Saudi Arabian government denied entry to Shi'ite pilgrims. Recently, that schism has healed somewhat, but intra-Islamic sectarian differences run deep.

Smaller Islamic sects further diversify this realm's religious landscape. Some of these sects play a disproportionately large role in the societies and countries of which they are a part, as we will see in our regional discussion.

Religious Revivalism in the Realm

Another cause of intra-Islamic conflict lies in the resurgence of religious fundamentalism, or as 15 Muslims refer to it, **religious revivalism**. In the 1970s, the imams in Shi'ite Iran wanted to reverse the shah's moves toward liberalization and secularization: they wanted to (and did) recast society in traditional, revivalist Islamic molds. An *ayatollah* (leader under Allah) replaced the shah in 1979; Islamic rules and punishments were instituted. Urban women, many of whom had been considerably liberated and educated during the shah's regime, resumed more traditional Islamic roles. Vestiges of Westernization, encouraged by the shah, disappeared. Under these new conditions even the war against Iraq (1980–1990), which began as a conflict over territory, became a holy war that cost more than a million lives.

Islamic revivalist fundamentalism did not rise in Iran alone, nor was it confined to Shi'ite communities. Many Muslims—Sunnis as well as Shi'ites—in all parts of the realm disapproved of the erosion of traditional Islamic values, the corruption of society by European colonialists and later by Western modernizers, and the declining power of the faith in the secular state. As long as economic times were good, such dissatisfaction remained submerged. But when jobs were lost and incomes declined, a return to fundamental Islamic ways became more appealing.

This set Muslim against Muslim in all the regions of the realm. Revivalists fired the faith with a new militancy, challenging the status quo from Afghanistan to Algeria. The militants forced their governments to ban "blasphemous" books, to re-segregate the sexes in schools, to enforce traditional dress codes, to legitimize religious-political parties, and to heed the wishes of the *mullahs* (teachers of Islamic ways). Militant Muslims proclaimed that democracy inherited from colonialists and adopted by Arab nationalists was incompatible with the rules of the Quran (Koran).

The rift between moderate Muslims and militant revivalists began to spill over into other parts of the world decades ago, from Pakistan to the African Transition Zone and from the Caucasus to the Philippines. Even while Islamic countries such as Iran and Algeria struggled to overcome the instability and damage arising from their internal conflicts, militant activists planned and carried out attacks against the allies of the moderates—oil-guzzling Western countries guilty of military intervention, propping up unrepresentative regimes, mistreatment of immigrant Muslims, and cultural corruption. Before the worst of these attacks (through mid-2004) destroyed the World Trade Center in New York and severely damaged the Pentagon outside Washington, D.C., the French had foiled a similar assault on the Eiffel Tower in Paris, the Russians had suffered hundreds of casualties in Moscow and Transcaucasia, and U.S. targets were hit in Lebanon, East Africa, Yemen, and elsewhere.

Although it has been argued that these attacks do not represent Islam as a religion in conflict with the West, they do have a religious context. During the 1990s, following the Cold War defeat of Soviet forces in Afghanistan and the subsequent abandonment of that country by the United States and its allies, an organization named *al-Qaeda* emerged, dedicated to punishing the perceived enemies of Islamic peoples by the most effective means at its disposal: terrorist attacks. However, the leader and chief financier of this movement, Usama bin Laden, had a bigger agenda. His ultimate goal was to overthrow the regime of rich princes ruling his home country, Saudi Arabia, accusing them of collusion with the infidel United States, of apostasy (faithlessness), and worse. Bin Laden wanted to make Saudi Arabia, birthplace of Muhammad and guarantor of holy Mecca, a true Islamic state.

Bin Laden was not the first Saudi to wish to return his country to its Islamic roots. During the eighteenth century an Islamic theologian named Muhammad ibn Abd al-Wahhab (1703–1792) founded a movement to bring Islamic society on the peninsula back to the most fundamental, traditional, and puritanical form of the faith. So strict were his teachings that he was expelled from his home town in 1744—a fateful exile because he moved to a place that happened to be the headquarters of Ibn Saud, ruler over a sizeable swath of the Arabian Peninsula. Ibn Saud formed an alliance with al-Wahhab, who had inherited a fortune when his wife died; between them, they set out on a campaign of conquest that made the Saudi dynasty the rulers of the region and **Wahhabism** the dominant form of Islam. Even the relatively liberal Ottoman sultans could not suppress either the Saudis or the Wahhabis, and in 1932, when the modern Kingdom of Saudi Arabia was established, Wahhabism became its official and dominant form of Islam.

Despite its reference to the name of the founder, Wahhabism today is a term used mostly by non-Muslims and outsiders. Followers of Wahhabi doctrines call themselves *Muwahhidun* or "Unitarians" to signify the strict and fundamental nature of their beliefs. These are the doctrines Usama bin Laden accuses the Saudi princes of having violated, and these are the harsh teachings so often cited in the Western media as emanating from the pulpits of Saudi Arabia's revivalist mosques.

Islam and Other Religions

Two other major faiths had their sources in the *Levant* (the area extending from Greece eastward around the Mediterranean coast to northern Egypt), and both were older than Islam. Islam's rise submerged many smaller Jewish communities, but the Christians, not the Jews, waged centuries of holy war against the Muslims, seeking, through the Crusades, not only to drive Islam back but to reestablish Christian communities where they had dominated before the Islamic expansion. The aftermath of that campaign still marks the realm's cultural landscape today. A substantial Christian minority (about one-fourth of the population) remains in Lebanon, and Christian minorities also survive in Israel, Syria, Egypt, and Jordan. Strained relations between the long-dominant Christian minority in Lebanon and the Muslim majority (itself divided into five sects) contributed to the disastrous armed struggle that engulfed this country in the 1970s and 1980s.

But the most intense conflict in modern times has pitted the Jewish state, Israel, against its Islamic neighbors near and far. Israel's United Nations-sponsored creation in 1948 precipitated more than a half-century of intermittent strife and attempts at mediation; it also caused friction among Islamic states in the region.

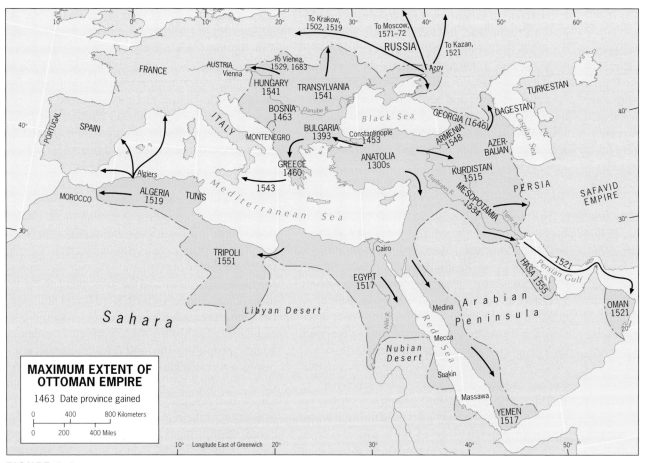

FIGURE 7-6

Jerusalem—holy city for Judaism, Christianity, and Islam—lies in the crucible of this confrontation.

The Ottoman Aftermath

It is a geographic twist of fate that Islam's last great advance into Europe resulted in the European occupation of Islam's very heartland. The Ottomans (named after their leader, Osman I), based in what is today Turkey, conquered Constantinople (now Istanbul) in 1453 and pushed into Eastern Europe. Soon Ottoman forces were on the doorstep of Vienna; they also invaded Persia, Mesopotamia,

and North Africa (Fig. 7-6). The Ottoman Empire under Suleyman the Magnificent, who ruled from 1522 to 1560, was the most powerful state in western Eurasia. As Figure 7-6 shows, its armies even advanced toward Moscow, Kazan, and Krakow from their base at Azov (near present-day Rostov in Russia), and the Turks also launched marine attacks on Sicily, Spain, and France.

The Ottoman Empire survived for more than four centuries, but it lost territory as time went on, first to the Hungarians, then to the Russians, and later to Greeks and Serbs until, after World War I, the European powers took over its provinces and

made them colonies—colonies we now know by the names of Syria, Iraq, Lebanon, and Yemen (Fig. 7-7). As the map shows, the French and the British took large possessions; even the Italians annexed part of the Ottoman domain.

The boundary framework that the colonial powers created to delimit their holdings was not satisfactory. As Figure G-9 reminds us, this realm's population of more than 530 million is clustered, fragmented, and strung out in river valleys, coastal zones, and crowded oases. The colonial powers laid out long stretches of boundary as ruler-straight lines across uninhabited territory; they saw no need to adjust these boundaries to cultural or physical features in the landscape. Other boundaries, even some in desert zones, were poorly defined and never marked on the ground. One product of this process was Iraq, delimited by (then) British Colonial Secretary Winston Churchill in 1921 as a kingdom with Baghdad as its capital; another was Jordan, centered on Amman. Later, when the colonies had become independent states, such boundaries led to quarrels, even armed conflicts, among neighboring Muslim states.

THE POWER AND PERIL OF OIL

Travel through the cities and towns of North Africa and Southwest Asia, talk to students, shopkeepers, taxi drivers, and migrants, and you will hear the same refrain: "Leave us in peace, let us do things our traditional way. Our problems—with each other, with the world—seem always to result from outside interference. The stronger countries of the world exploit our weaknesses and magnify our quarrels. We want to be left alone."

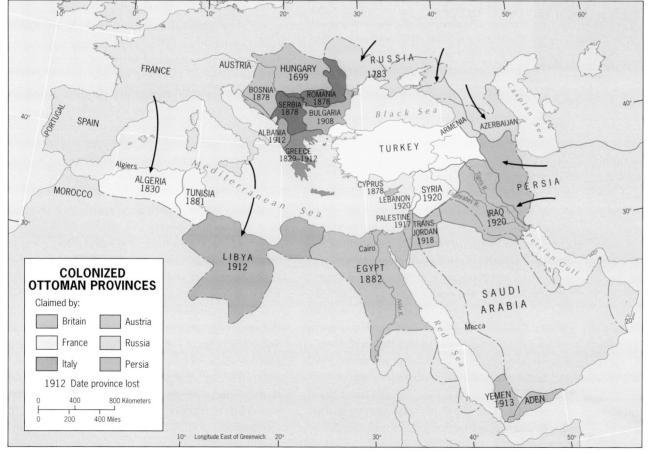

FIGURE 7-7

"The impact of oil wealth is nowhere better illustrated than it is along the waterfront of the famous Creek of Dubai (adjoining the city of the same name, largest in the United Arab Emirates). We stopped on the bridge that links the two sectors of Dubai to observe the old *dhows* (most of these wooden boats now with motors as well as triangular sail) still at their centuries-old moorings, overlooked not by traditional Arab buildings but by modern skyscrapers. Oil may have transformed the local economy, but don't count the role of the dhows out just yet. They still carry trade and contraband. We went to the dock and asked what was being transported. 'Jeans and cassettes to Iran,' we were told, 'and caviar and carpets from Iran. It's a good business.' So the traditional role of the dhows goes on, even in the age of oil."

That wish might be closer to fulfillment were it not for two relatively recent events: the creation of the state of Israel and the discovery of some of the world's largest oil reserves. We will discuss the founding of Israel in the regional section of this chapter. Here we focus on the realm's most valuable export product: oil.

Location and Dimensions of Known Reserves

About 30 of the world's countries have significant oil reserves. But five of these countries, all located in the North Africa/Southwest Asia realm, in combination possess larger reserves (ca. 77 percent of the world total) than the rest combined. The estimated reserves of this Big Five in 2003, averaged from several sources including the governments of the countries themselves and reported in billions of barrels, were as follows: (1) Saudi Arabia,

264; (2) Iraq, 117; (3) United Arab Emirates, 99; (4) Kuwait, 97; and (5) Iran, 90. The next-ranking country, Venezuela, had known reserves below 80 billion barrels; Russia below 60; the United States below 40; and Libya, Mexico, Nigeria, China, and Ecuador below 25. No other country's known reserves approached 20 billion barrels as of 2004.

In general terms, oil (and associated natural gas) exists in this realm in three discontinuous zones (Fig. 7-8). The most productive of these zones extends from the southern and southeastern part of the Arabian Peninsula northwestward around the rim of the Persian Gulf, reaching into Iran and continuing northward into Iraq, Syria, and southeastern Turkey, where it peters out. The second zone lies across North Africa and extends from north central Algeria eastward across northern Libya to Egypt's Sinai Peninsula, where it ends. The third zone begins on the margins of the realm in eastern Azerbaijan, continues eastward under the Caspian Sea into Turkmenistan

and Kazakhstan, and also reaches into Uzbekistan, Tajikistan, Kyrgyzstan, and Afghanistan.

Saudi Arabia is the world's largest oil exporter, but in recent years Russia has risen to second place. As we noted in Chapter 2, oil and natural gas are by far Russia's most valuable commodities, and the Russian state direly needs the revenues they produce. Thus Russia, with modest (though expanding) known reserves, is vigorously exporting to international markets and is now a leading factor in the global energy picture. In combination, however, production by the countries of the North Africa/Southwest Asia realm far exceeds that from all other sources. The United States, not surprisingly, is the world's leading importer even while it consumes almost all of its own production.

As Figure G-10 indicates, the production and export of oil and gas has elevated several of this realm's countries into the higher-income categories. But petroleum wealth also has enmeshed these Islamic societies and their governments in global strategic affairs. When regional conflicts create instability in producing countries that have the potential to disrupt supply lines, powerful consumers are tempted to intervene—and have done so.

When the colonial powers laid down the boundaries that partitioned this realm among themselves, no one knew about the riches that lay beneath the ground. A few wells had been drilled, and production in Iran had begun as early as 1908 and in Egypt's Sinai Peninsula in 1913. But the major discoveries came later, in some cases after the colonial powers had already withdrawn. Some of the newly independent countries, such as Libya, Iraq, and Kuwait, found themselves with wealth undreamed of when the Turkish Ottoman Empire collapsed. As Figure 7-8 shows, however, others were less fortunate. A few countries had (and still have) potential. The smaller, weaker emirates and sheikdoms on the Arabian Peninsula always feared that powerful neighbors would try to annex them (Kuwait faced this prospect in 1990 when Iraq invaded it). The unevenly distributed oil wealth,

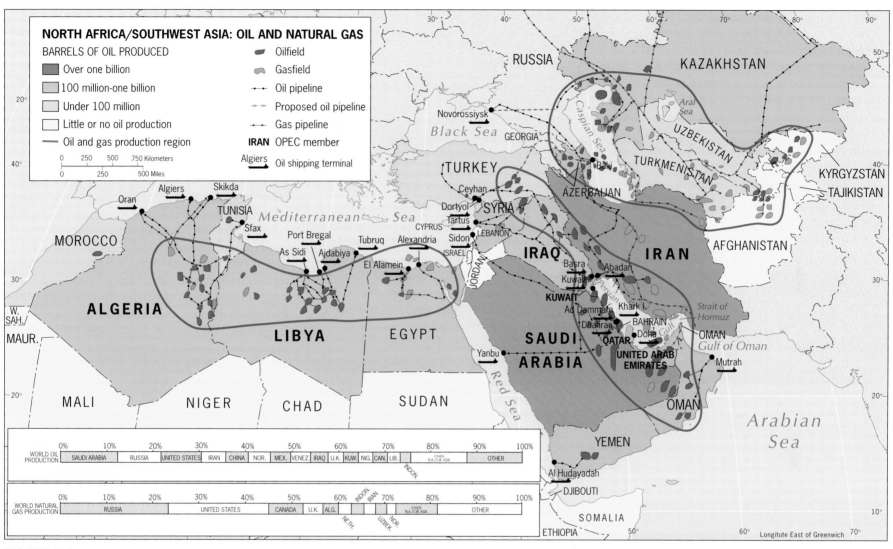

NORTH AFRICA/SOUTHWEST ASIA: OIL AND NATURAL GAS

BARRELS OF OIL PRODUCED

Over one billion
100 million-one billion
Under 100 million
Little or no oil production
Oil and gas production region

Oilfield
Gasfield
Oil pipeline
Proposed oil pipeline
Gas pipeline
IRAN OPEC member
Algiers Oil shipping terminal

FIGURE 7-8

therefore, created another source of division and distrust among Islamic neighbors.

A Foreign Invasion

The oil-rich countries of the realm found themselves with a coveted energy source, but without the skills, capital, or equipment to exploit it. These had to come from the Western world and entailed what many tradition-bound Muslims feared most: a strong foreign presence on Islamic soil, foreign intervention in political as well as economic affairs, and penetration of Islamic societies by the vulgarities of Western ways. Moreover, the advent of the oil industry created a veneer of modernity in the cultural landscape (where gleaming skyscrapers towered over ornate and historic minarets) as well as among the elites who benefited most directly from the energy riches. This set up tensions and even clashes between the modern and the traditional that continue to plague oil-rich Islamic countries today. It also created contrasts between countries that took different paths as their cultures changed. In the United Arab Emirates (UAE), it is no surprise to see a woman emerge from a supermarket, get into her car, and drive home with the groceries. That would be unthinkable in neighboring Saudi Arabia, where women are prohibited from driving cars.

In geographic terms, the impact of oil, its production and sale, in the exporting countries of this realm can be summarized as follows:

1. *High Incomes.* When oil prices on international markets were high, several countries in this realm ranked as the highest-income societies in the world. Even when oil prices declined, virtually all the petroleum-exporting states remained at least in the upper-middle-income category (Fig. G-10).

2. *Modernization.* Notably in Saudi Arabia but also in Kuwait and UAE, huge oil revenues have been used to modernize infrastructure. Futuristic city skyscrapers are only one manifestation of this; modern ports, airports, and superhighways are others.

3. *Industrialization.* Farsighted governments among those with oil wealth, realizing that petroleum reserves will not last forever, have invested some of their income in industrial plants that will outlast the era of oil. Petrochemical industries, plastics fabrication, and desalinization plants are among these facilities.

4. *Intra-Realm Migration.* The oil wealth has attracted millions of workers from less favored parts of the realm to work in the oilfields, in the ports, and in many other, mainly menial capacities. This has brought many Shi'ites to the countries of eastern Arabia; hundreds of thousands of Palestinians also work there as laborers. Saudi Arabia, with a population of 25.4 million, has more than 5 million foreign workers.

5. *Inter-Realm Migration.* The willingness of workers from such countries as Pakistan, India, and Sri Lanka to work for wages even lower than those the oil industry pays has attracted a substantial flow of temporary immigrants from outside the realm. These workers serve mostly as domestics, gardeners, refuse collectors, and the like.

6. *Regional Disparities.* Oil wealth and its manifestations in the cultural landscape create strong contrasts with areas not directly affected. The ultramodern east coast of Saudi Arabia is a world apart from large areas of its interior, where it becomes a land of desert, oasis, and camel, of vast distances, slow change, and isolated settlements. This phenomenon affects all oil-rich countries.

7. *Foreign Investment.* To some degree, governments and Arab businesspeople have invested oil-generated wealth in foreign countries. These investments have created a network of international involvement that links many of this realm's countries not only to the economies of foreign states, but also to growing Arab (and thus Islamic) communities in those states.

The map (Fig. 7-8) contains a warning that came home to the oil-rich states (not only in this realm but elsewhere as well) in the 1980s when, after a period of high oil prices and huge revenues, oil prices fell and incomes plummeted. Even the power of an 11-member cartel, OPEC (Organization of Petroleum Exporting Countries), could not recover the lost advantage. Figure 7-8 shows a system of oil and gas pipelines that strongly resembles the exploitative interior-to-coast railroad lines (such as we noted for much of Subsaharan Africa) in a mineral-rich colony of the past. Such a pattern spells disadvantage for the exporter, whether colony or independent country. Markets, not raw-material exporters, dominate international trade. Oil brought this realm into contact with the outside world in ways unforeseen just a century ago. Oil has strengthened and empowered some of its peoples; it has dislocated and imperiled others. It has truly been a double-edged sword.

REGIONS OF THE REALM

Identifying and delimiting regions in this vast geographic realm is quite a challenge. Population clusters are widely scattered in some areas, highly concentrated in others. Cultural transitions and landscapes—internal as well as peripheral—make it difficult to discern a regional framework. This, furthermore, is a highly changeable realm and always has been. Several centuries ago it extended into Eastern Europe; now it reaches into Central 17 Asia, where an Islamic **cultural revival** (the regeneration of a long-dormant culture through internal renewal and external infusion) is well under way.

The following are the regional components of this far-flung realm today (Fig. 7-9):

1. **Egypt and the Lower Nile Basin.** This region in many ways constitutes the heart of the realm as a whole. Egypt (together with Iran and Turkey) is one of the realm's three most populous countries. It is the historic focus of this part of the world and a major political and cultural force. It shares with its southern neighbor, Sudan, the waters of the lower Nile River.

2. **The Maghreb and Its Neighbors.** Western North Africa (the Maghreb) and the areas that border it also form a region, consisting of Algeria, Tunisia, and Morocco at the center, and Libya, Chad, Niger, Mali, and Mauritania along

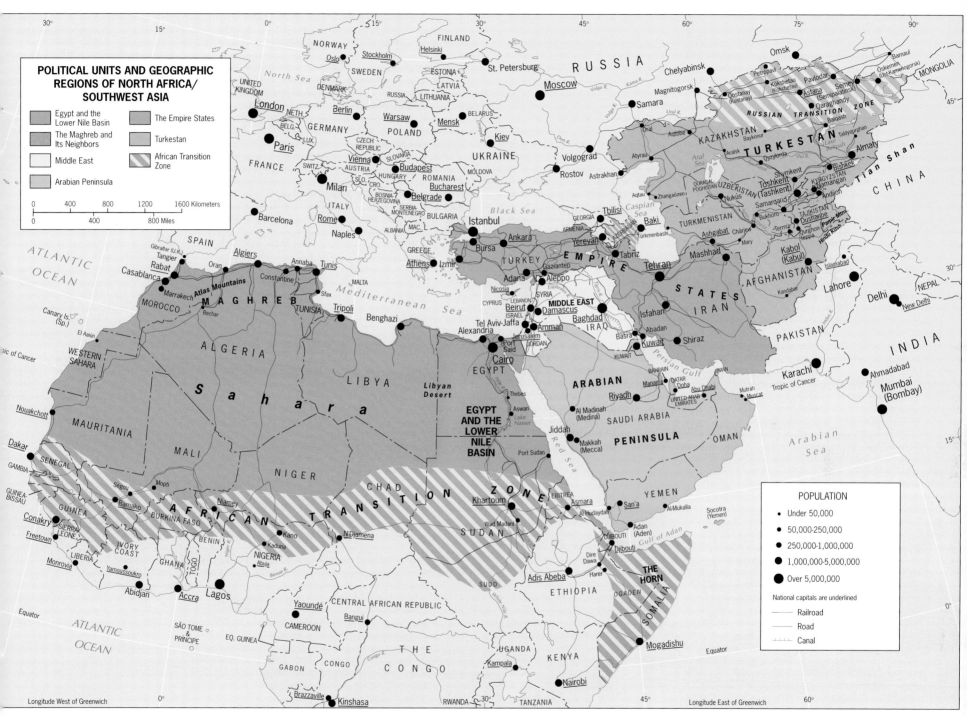

POLITICAL UNITS AND GEOGRAPHIC
REGIONS OF NORTH AFRICA/
SOUTHWEST ASIA

Egypt and the
Lower Nile Basin

The Maghreb and
Its Neighbors

Middle East

Arabian Peninsula

The Empire States

Turkestan

African Transition
Zone

| 0 | 400 | 800 | 1200 | 1600 Kilometers |
| 0 | | 400 | | 800 Miles |

POPULATION

• Under 50,000

• 50,000-250,000

● 250,000-1,000,000

● 1,000,000-5,000,000

● Over 5,000,000

National capitals are underlined

Railroad

Road

Canal

URE 7-9

the broad periphery. The last four of these countries also lie astride or adjacent to the broad transition zone where the Arab-Islamic realm of northern Africa merges into Subsaharan Africa.

3. **The Middle East.** This region includes Israel, Jordan, Lebanon, Syria, and Iraq. In effect, it is the crescent-like zone of countries that extends from the eastern Mediterranean coast to the head of the Persian Gulf.

4. **The Arabian Peninsula.** Dominated by the large territory of Saudi Arabia, the Arabian Peninsula also includes the United Arab Emirates (UAE), Kuwait, Bahrain, Qatar, Oman, and Yemen. Here lies the source and focus of Islam, the holy city of Mecca; here, too, lie many of the world's greatest oil deposits.

5. **The Empire States.** We refer to this region as the Empire States because two of the realm's giants, both the centers of historic empires, dominate its geography. Turkey, heart of the Ottoman Empire, now is the realm's most secular state and aspires to joining the European Union. Shi'ite Iran, the core of the former Persian Empire, has become an Islamic republic. To the north, Azerbaijan, once a part of the Persian Empire, lies in the turbulent, Muslim-infused Transcaucasian Transition Zone. To the south, the island of Cyprus is divided between Turkish and Greek spheres, the Turkish sector being another remnant of imperial times.

6. **Turkestan.** Turkish influence ranged far and wide in southwestern Asia, and following the Soviet collapse that influence proved to be durable and strong. In the five former Soviet Central Asian republics the strength and potency of Islam vary, and the new governments deal warily (sometimes forcefully) with Islamic revivalists. Russian influence continues, democratic institutions remain weak, and Western (chiefly American) involvement has increased since the War on Terror began and Afghanistan—which also forms part of this region—became a major target of that campaign.

EGYPT AND THE LOWER NILE BASIN

Egypt occupies a pivotal location in the heart of a realm that extends over 6000 miles (9600 km) longitudinally and some 4000 miles (6400 km) latitudinally. At the northern end of the Nile and of the Red Sea, at the eastern end of the Mediterranean Sea, in the northeastern corner of Africa across from Turkey to the north and Saudi Arabia to the east, adjacent to Israel, to Islamic Sudan, and to militant Libya, Egypt lies in the crucible of this realm. Because it owns the Sinai Peninsula (recently lost to and regained from Israel), Egypt, alone among states on the African continent, has a foothold in Asia, a foothold that gives it a coast overlooking the strategic Gulf of Aqaba (the northeasternmost arm of the Red Sea). Egypt also controls the Suez Canal, vital link between the Indian and Atlantic oceans and lifeline of Europe. The capital, Cairo (El Qahira), is the realm's largest city and a leading center of Islamic civilization. We hardly need to further justify Egypt's designation (together with northern Sudan) as a discrete region.

Egypt's Nile is the aggregate of two great branches upstream: the White Nile, which originates in the streams that feed Lake Victoria in East Africa, and the Blue Nile, whose source lies in Lake Tana in Ethiopia's highlands. The two Niles converge at Khartoum in modern-day Sudan.

About 95 percent of Egypt's 74.1 million people live within a dozen miles (20 km) of the great river's banks or in its delta (Fig. 7-10). It has always been this way: the Nile rises and falls seasonally, watering and replenishing soils and crops on its banks.

The ancient Egyptians used *basin irrigation*, building fields with earthen ridges and trapping the floodwaters with their fertile silt, to grow their crops. That practice continued for thousands of years until, during the nineteenth century, the construction of permanent dams made it possible to irrigate Egypt's farmlands year round. These dams, with locks for navigation, controlled floods, expanded the country's cultivable area, and allowed the farmers to harvest more than one crop per year on the same field. In a single century, all of Egypt's farmland was brought under *perennial irrigation*.

The greatest of all Nile projects, the Aswan High Dam (which began operating in 1968), creates Lake Nasser, one of the world's largest artificial lakes (Fig. 7-10). As the map shows, the lake extends into Sudan, where 50,000 people had to be resettled to

FROM THE FIELDNOTES

"On the eastern edge of central Cairo we saw what looked like a combination of miniature mosques and elaborate memorials. Here lie buried the rich and the prominent of times past in what locals call the 'City of the Dead.' But we found it to be anything but a dead part of the city. Many of the tombs here are so large and spacious that squatters have occupied them. Thus the City of the Dead is now an inhabited graveyard, home to at least one million people. The exact numbers are impossible to determine; indeed, whereas metropolitan Cairo in 2004 has an official population of just over 10 million, many knowledgeable observers believe that 16 million is closer to the mark."

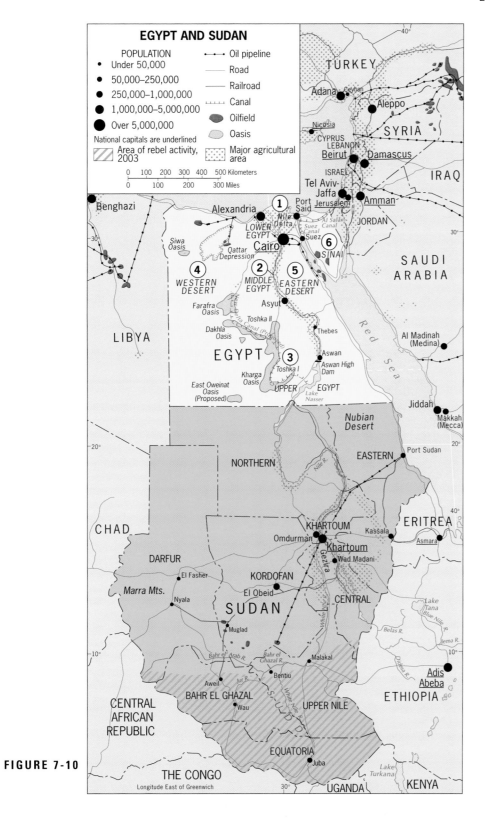

EGYPT AND SUDAN

POPULATION
- • Under 50,000
- • 50,000–250,000
- ● 250,000–1,000,000
- ● 1,000,000–5,000,000
- ● Over 5,000,000

National capitals are underlined
Area of rebel activity, 2003
Major agricultural area

Oil pipeline
Road
Railroad
Canal
Oilfield
Oasis

0 100 200 300 400 500 Kilometers
0 100 200 300 Miles

FIGURE 7-10

make way for it. The Aswan High Dam increased Egypt's irrigable land by nearly 50 percent and today provides the country with about 40 percent of its electricity. But, as is so often the case with megaprojects of this kind, the dam also produced serious problems. Snail-carried schistosomiasis and mosquito-transmitted malaria thrived in the dam's standing water, afflicting hundreds of thousands of people living nearby. By blocking most of the natural fertilizers in the Nile's annual floodwaters, the dam necessitated the widespread use of artificial fertilizers, proving very costly to small farmers and damaging to the natural environment. And the now fertilizer- and pesticide-laden Nile no longer supports the fish fauna offshore, reducing the catch and depriving coastal populations of badly needed proteins.

Egypt's elongated oasis along the Nile, just 3 to 15 miles (5 to 25 km) wide, broadens north of Cairo across a delta anchored in the west by the great city of Alexandria and in the east by Port Said, gateway to the Suez Canal. The delta contains extensive farmlands, but it is a troubled area today. The ever more intensive use of the Nile's water and silt upstream is depriving the delta of much needed replenishment. And the low-lying delta is geologically subsiding, raising fears of salt-water intrusion from the Mediterranean Sea that would damage soils here.

Egypt's millions of subsistence farmers, the *fellaheen*, still struggle to make their living off the land, as did the peasants of the Egypt of five millennia past. Rural landscapes seem barely to have changed; ancient tools are still used, and dwellings remain rudimentary. Poverty, disease, high infant mortality rates, and low incomes prevail. The Egyptian government is embarked on grandiose plans to expand its irrigated acreage, but without modernization and greater efficiency in the existing farmlands, the future is dim.

Egypt has six subregions, mapped in Figure 7-10. Most Egyptians live and work in Lower (i.e., northern) and Middle Egypt (regions ① and ②), the country's core area anchored by Cairo and flanked by the leading port and second industrial center, Alexandria. The economy has benefited from further oil discoveries in the Sinai (region ⑥) and in the

Western Desert (region ④), so that Egypt now is self-sufficient and even exports some petroleum. Cotton and textiles are the other major source of external income, but the important tourist industry has repeatedly been hurt by Islamic extremists. As the population mushrooms, the gap between food supply and demand widens, and Egypt must import grain. Since the late 1970s, Egypt has been a major recipient of U.S. foreign aid.

Egypt today is at a crossroads in more ways than one. Its planners know that reducing the high birth rate would improve the demographic situation, but revivalist Muslims object to any programs that promote family planning. Its accommodation with Israel helps ensure foreign aid but divides the people. Its government faces a fundamentalist challenge. Egypt's future, in this crucial corner of the realm, is uncertain.

Divided Sudan

As Figure 7-10 shows, Egypt is flanked by two countries that have posed challenges to its leadership: Sudan to the south and Libya to the west. Sudan, more than twice as large as Egypt and with nearly 35 million people, lies centered on the confluence of the White Nile (from Uganda) and the Blue Nile (from Ethiopia). Here the twin capital, Khartoum-Omdurman, anchors a large agricultural area where cotton was planted during colonial times. The British administration combined northern Sudan, which was Arabized and Islamized, with a large area to the south, which was African and where many villagers had been Christianized. After the British left, the regime in Khartoum wanted to impose its Islamic rule on this portion of the African Transition Zone that formed southern Sudan, and a bitter civil war ensued. The cost in human lives and dislocation over the better part of the past three decades is incalculable; in 2004 an estimated 4 million people remained refugees in their own land.

Sudan has a 300-mile (500-km) coastline on the Red Sea, where Port Sudan lies almost directly across from Jiddah and Makkah (Mecca) in Saudi Arabia. The country's economy for many years was typical of the energy-poor periphery, exchanging sheep, cotton, and sugar for oil, with Saudi Arabia the main trading partner. The pursuit of war in the African Transition Zone impoverished the Islamic regime in Khartoum, and per-capita income in Sudan was one of the world's lowest.

In the 1990s, the discovery of oil reserves in southern Kordofan Province changed the situation significantly (Fig. 7-10). As the map shows, these reserves lie close to the border with Bahr-el-Ghazal and Upper Nile Provinces, where rebels have the upper hand. Exploitation of these reserves, involving foreign companies, was accompanied by reports that the Sudanese army drove hundreds of thousands of villagers in the oil-rich areas from their land, killing those who resisted and burning their villages to the ground. Outraged protests in Canada persuaded a Canadian oil company to withdraw from the scene, but otherwise the world paid little heed.

Now oil that once flowed in the pipeline from Port Sudan to the capital flows the other way. By 2002 the Khartoum regime was earning U.S. $1 million per day from oil revenues, allowing it to double its military expenditures and giving it renewed advantage over the African rebels in the south. In mid-2004 negotiations were in progress that offered hope for a settlement of the issue, but even as this process wore on another conflict between Khartoum and its non-Muslim citizens broke out in western Darfur Province on the border with Chad. Hundreds of thousands of refugees crossed the border in another manifestation of strife along Africa's Islamic Front.

THE MAGHREB AND ITS NEIGHBORS

The countries of northwestern Africa are collectively called the *Maghreb*, but the Arab name for them is more elaborate than that: *Djezira-al-Maghreb*, or "Isle of the West," in recognition of

FROM THE FIELDNOTES

"Having seen the *Giralda* in Seville, Spain, I was interested to view the tower of the Koutoubia Mosque in Marrakech, Morocco, also built in the twelfth century and closely resembling its Andalusian counterpart. Some 220 feet (67 m) tall, the Koutoubia minaret was built by Spanish slaves and is a monument to the heyday of Islamic culture of the time. I soon discovered that it is not as easy as it looks to get a clear view of the tower, and a boy offered his help, guiding me around the block on which the mosque is situated. He was well informed on the building's history and on the varying fortunes of his home town. It was European mispronunciation of the name 'Marrakech' that led to Morocco being called Morocco, he reported. And Marrakech was Morocco's capital for a very long time, and still should be. 'But back to the mosque,' I said. 'What does the sign say?' He smiled, 'You'll just have to learn some Arabic, just as we know French,' he said. Obviously a teacher in the making, I thought."

the great Atlas Mountain range rising like a huge island from the Mediterranean Sea to the north and the sandy flatlands of the immense Sahara to the south.

The countries of the Maghreb (sometimes spelled *Maghrib*) are Morocco, last of the North African kingdoms; Algeria, a secular republic beset by the religious-political problems we noted earlier; and Tunisia, smallest and most Westernized of the three (Fig. 7-11). Libya, facing the Mediterranean between the Maghreb and Egypt, is unlike any other North African country: an oil-rich desert state whose population is almost entirely clustered in settlements along the coast.

Whereas Egypt is the gift of the Nile, the Atlas Mountains facilitate the settled Maghreb. These high ranges wrest from the rising air enough orographic rainfall to sustain life in the intervening valleys, where good soils support productive farming. From the vicinity of Algiers eastward along the coast into Tunisia, annual rainfall averages more than 30 inches (75 cm), a total more than three times as high as that recorded for Alexandria in Egypt's delta. Even 150 miles (240 km) inland,

the slopes of the Atlas still receive over 10 inches (25 cm) of rainfall. The effect of the topography can be read on the world map of precipitation (Fig. G-7): where the highlands of the Atlas terminate, desert conditions immediately begin.

The Atlas Mountains trend southwest-northeast and begin in Morocco as the High Atlas, with elevations close to 13,000 feet (4000 m). Eastward, two major ranges dominate the landscapes of Algeria proper: the Tell Atlas to the north, facing the Mediterranean, and the Saharan Atlas to the south, overlooking the great desert. Between these two mountain chains, each consisting of several parallel ranges and foothills, lies a series of intermontane basins (analogous to South America's Andean *altiplanos* but at lower elevations), markedly drier than the northward-facing slopes of the Tell Atlas. In these valleys, the rain shadow effect of the Tell Atlas is reflected not only in the steppe-like natural vegetation but also in land-use patterns: pastoralism replaces cultivation, and stands of short grass and bushes blanket the countryside.

During the colonial era, which began in Algeria in 1830 and lasted until the early 1960s, well over a

million Europeans came to settle in North Africa—most of them French, and a large majority bound for Algeria—and these immigrants soon dominated commercial life. They stimulated the renewed growth of the region's towns; Casablanca, Algiers, Oran, and Tunis became the urban foci of the colonized territories. Although the Europeans dominated trade and commerce and integrated the North African countries with France and the European Mediterranean world, they did not confine themselves to the cities and towns. They recognized the agricultural possibilities of the favored parts of the tell (the lower Tell Atlas slopes and narrow coastal plains that face the Mediterranean) and established thriving farms. Not surprisingly, agriculture here is Mediterranean. Algeria soon became known for its vineyards and wines, citrus groves, and dates; Tunisia has long been one of the world's leading exporters of olive oil; and Moroccan oranges went to many European markets.

Between desert and sea, the Maghreb states display considerable geographic diversity. **Morocco**, a conservative kingdom in a revolutionary region, is tradition-bound, and weak economically. Its core area lies in the north, anchored by four major cities; but the Moroccans' attention is focused on the south, where the government is seeking to absorb its neighbor, Western Sahara (a former Spanish dependency with about 300,000 inhabitants, many of them immigrants from Morocco). Even if this campaign is successful, it will do little to improve the lives of most of Morocco's 31 million people. Hundreds of thousands have emigrated to Europe, many of them via Spain's two tiny exclaves on the Moroccan coast, Ceuta and Melilla (Fig. 7-11).

Algeria, rich in oil and natural gas, fought a bitter war of liberation against the French colonizers and has been in turmoil virtually ever since, with devastating economic and social

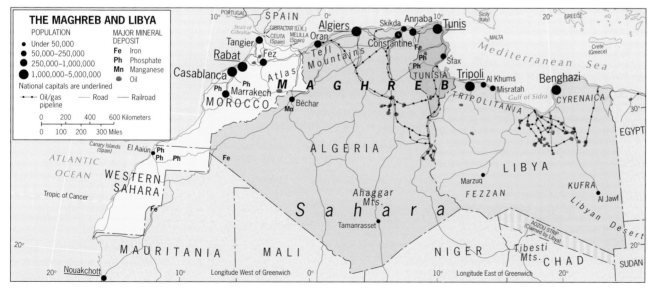

FIGURE 7-11

consequences. Algiers, Algeria's primate city, is centrally situated along its Mediterranean coast and is home to 3.3 million of the country's 32.5 million citizens—but there are more Algerians in France than in the capital today. Conflict between those wanting to make Algeria an Islamic republic and the military-backed regime has cost an estimated 100,000 lives; in recent years strife has threatened in the Kabylia region east of Algiers involving Algeria's Berber minority.

Almost rectangular in shape, **Libya**, (5.7 million) lies on the Mediterranean Sea between the Maghreb states and Egypt (Fig. 7-11). What limited agricultural possibilities exist lie in the district known as Tripolitania in the northwest, centered on the capital, Tripoli, and in the northeast in Cyrenaica, where Benghazi is the urban focus. But it is oil, not farming, that drives the economy of this highly urbanized country. The oilfields lie well inland from the Gulf of Sidra, linked by pipelines to coastal terminals. Libya's two interior corners, the desert Fezzan district in the southwest and the Kufra oasis in the southeast, are connected to the coast by two-lane roads subject to sandstorms.

Muammar Qaddafi, an army colonel, overthrew King Idris of Libya in 1969 and in the 1970s burnished his Islamic credentials by expelling Libya's Jewish community, supporting Palestinian causes, and promoting "Allah's language" throughout North Africa and the Middle East. In the 1980s Libya became involved in state-sponsored terrorism, and in the early 1990s the UN imposed sanctions because of Qaddafi's refusal to turn over information or suspects in the bombing of French and American civilian airliners.

Since the late 1990s, Libya's role in the world has taken some dramatic turns. First, Qaddafi turned his interest from Middle Eastern and Palestinian causes to Africa, using oil revenues to make large payments to African leaders and earning praise even from South Africa's Nelson Mandela. Next, Qaddafi cooperated with an international tribunal that charged two Libyans with the downing of a Pan American aircraft over Scotland, and agreed to make substantial payments to families of the victims. And in 2004 Qaddafi invited UN inspectors to Libya to study its weapons programs, including its nuclear installations. After 35 years in power, it appeared that Qaddafi hoped to end the isolation his policies had imposed on his country.

As Figure 7-11 shows, the Maghreb countries and Libya adjoin several desert-dominated states in the African Transition Zone. These are African countries with strong Islamic imprints, but as Figure 6-17 shows, they lie north of the Islamic Front with only one exception, Chad. Vast and sparsely populated **Mauritania** is the most Islamized of these countries, with about half its population of 2.7 million concentrated in and near the capital, Nouakchott, base of a small fishing fleet. Neighboring **Mali** depends on the waters of the upper Niger River, on which its capital of Bamako lies. A multiparty, democratic state, Mali is also multicultural, with a sizeable minority adhering to traditional beliefs and a small Christian minority. To the east of Mali lies **Niger**, which shares boundaries with Algeria and Libya; like Mali, this is one of the world's least urbanized countries (17 percent). Unlike Mali, however, Niger has only a short stretch of the middle course of the Niger River, where capital, Niamey, is situated. And **Chad**, directly south of Libya, has the lowest percentage of Muslim inhabitants but the strongest division between Islamized north and animist/Christian south. As Figure 6-17 reminds us, the capital, N'Djamena, lies on the southern edge of the African Transition Zone, directly on the Islamic Front. In 2004, even as Chad was experiencing an oil boom in the south, a cultural conflict in Sudan's neighboring Darfur Province sent more than 150,000 refugees across its border.

▶ THE MIDDLE EAST

The regional term *Middle East*, we noted earlier, is not satisfactory, but it is so common and generally used that avoiding it creates more problems than it solves. It originated when Europe was the world's dominant realm and when places were "near," "middle," and "far" from Europe: hence a Near East (Turkey), a Far East (China, Japan, Korea, and other countries of East Asia), and a Middle East (Egypt, Arabia, Iraq). If you check definitions used in the past, you will see that the terms were applied inconsistently: Syria, Lebanon, Palestine, even Jordan sometimes were included in the "Near" East, and Persia and Afghanistan in the "Middle" East.

Today, the geographic designation *Middle East* has a more specific meaning. And at least half of it has merit: this region, more than any other, lies at the middle of the vast Islamic realm (Fig. 7-9). To the north and east of it, respectively, lie Turkey and Iran, with Muslim Turkestan beyond the latter. To the south lies the Arabian Peninsula. And to the west lie the Mediterranean Sea and Egypt, and the rest of North Africa. This, then, is the pivotal region of the realm, its very heart.

Five countries form the Middle East (Fig. 7-12): Iraq, largest in population and territorial size, facing the Persian Gulf; Syria, next in both categories and fronting the Mediterranean; Jordan, linked by the narrow Gulf of Aqaba to the Red Sea; Lebanon, whose survival as a unified state has come into question; and Israel, Jewish nation in the crucible of the Muslim world. Because of the extraordinary importance of this region in world affairs, we focus in some detail on issues of cultural, economic, and political geography in the following discussion.

In the immediate aftermath of the 9/11 terrorist attacks in the United States in 2001, American forces overthrew the ruling revivalist Taliban regime in Afghanistan that had given refuge to Usama bin Laden, who plotted the assault. Soon, however, Afghanistan lost priority among U.S. concerns as **Iraq** was designated a security risk based on intelligence reports of its terrorist links, its alleged arsenal of weapons of mass destruction (WMD), its murderous regime, and the regional threat it posed. In March, 2003 the United States, helped by UK and Australian forces, attacked and invaded Iraq and following a brief struggle ended the rule of the Sunni-Muslim clique led by Saddam

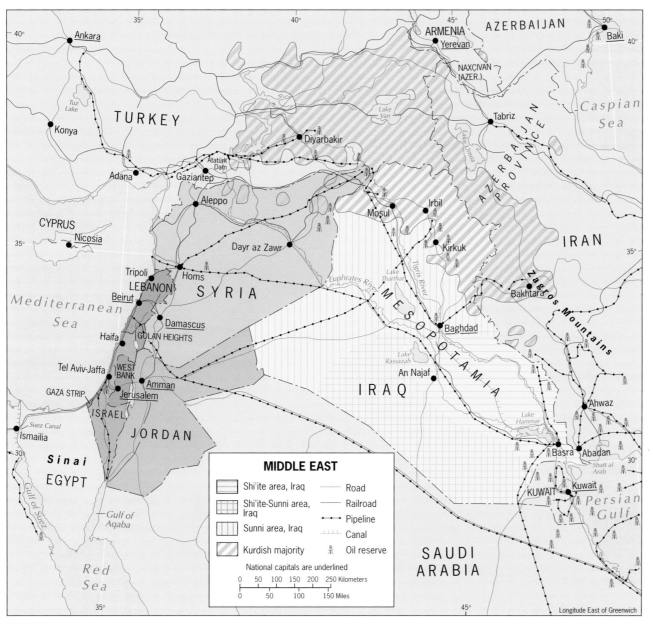

FIGURE 7-12

Hussein. While the search for WMD continued, the American government now faced the task of stabilizing Iraq, reconstructing its damaged infrastructure, and reconstituting its government.

It is therefore important to understand key aspects of the regional geography of California-sized Iraq and its official population of 24.8 million (which may in fact be two million less because of unregistered emigration). The country constitutes nearly 60 percent of the total area of the Middle East as we define it and has more than 40 percent of its population. With its world-class oilfields (revenues from which are to pay for much of Iraq's reconstruction), major gas reserves, and large areas of irrigable farmland, Iraq also is best endowed of all the region's countries with natural resources. Iraq is heir to the early Mesopotamian states and empires that emerged in the Tigris-Euphrates Basin, and the country is studded with matchless archeological sites and museum collections, to which disastrous damage was done during and after the 2003 war from combat and looting.

Today, Iraq is bounded by six neighbors and has in recent times had adversarial relations with most of them. To the north lies Turkey, source of both Iraq's vital rivers. To the east lies Iran, target of a destructive war during the 1980s. Kuwait, at the head of the Persian Gulf, was invaded by Iraq's armies in 1990. To the south, Saudi Arabia long was a friend of Iraq's enemies. And to the west, Iraq is adjoined by Jordan, which gave Iraq some help during the 1991 Gulf War but later turned its back on the Saddam

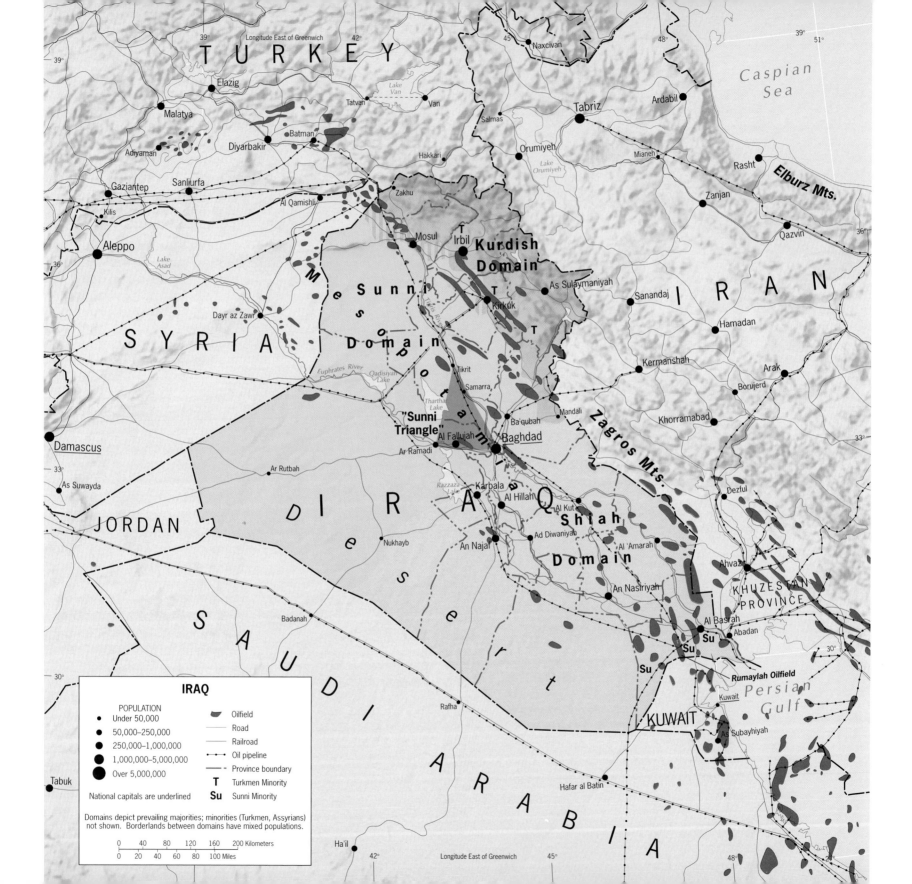

regime, and Syria, functionally most remote from Iraq but also a country ruled by a strongman and his minority clique. As (Fig. 7-13) shows, pipelines across Syria link Iraq's oilfields with Mediterranean terminals to the west.

This map also displays Iraq's regional components. The capital and core area, Baghdad and its environs, lie about midway between the Persian Gulf and the Turkish border astride the Tigris River. Three historic domains lie north and south of this heartland. To the north and northwest extends the Sunni stronghold (the "Sunni Triangle") from which Saddam's repressive regime was drawn. In the mountains along the Turkish-Iranian border lies the Kurdish zone that forms but a small part of a much larger ethnic Kurdish region. From the eastern suburbs of Baghdad southward stretches the third and largest domain in terms of both territory and population: the Shiah area extending all the way to Basra (Al Basrah) and the shores of the Persian Gulf. The vast, nearly empty desert bordering Saudi Arabia, Jordan, and Syria has few inhabitants and, as Figure 7-8 shows, no oil to attract settlers.

Population size, however, has not been matched by political power or influence in Iraq. By far the largest group is the Shi'ite sector, constituting between 60 and 65 percent of Iraq's population. Next comes the Sunni cluster, about 22 percent, historically the most powerful but fragmented along numerous clan and factional lines. The Kurds, victims of Saddam's chemical WMD in 1990, make up around 16 percent. There also are numerous smaller minorities, which include the ancient, mainly Christian, non-Arab Assyrian group living today scattered among the Sunnis and the Kurds in northern Iraq, and the Turkmen in Kirkuk and other locales along the Kurd-Sunni frontier.

The Shi'ites were ruthlessly suppressed during the rule by Saddam Hussein and his Sunni Ba'ath Party. When U.S. forces invaded Iraq from the south, they faced comparatively little resistance on

FIGURE 7-13

their way to Baghdad, and after the brief war ended, the Shiah south remained relatively calm. But in a truly democratic Iraq of the future, the Shi'ites would expect to play a dominant role, and over time they began to agitate for early elections. The same Shi'ite leader who had counseled patience during the early occupation, the cleric Ali al-Sistani, in early 2004 started to press for quick resolution, and mobilized large demonstrations to that end.

The situation in the Sunni domain (Fig. 7-13) was quite different. Sunnis known to have been part of Saddam's regime through membership in the Ba'ath Party or armed forces were arrested or dismissed from their positions, a practice that appeared to create more problems than it solved (even university professors who were party members lost their jobs although many joined as a matter of survival, not ideological commitment). While much of the Sunni area remained fairly calm, a major and costly insurrection arose in the so-called Sunni Triangle, within which Saddam Hussein's clan and that of others in his inner circle had their homes. This insurrection reached into Baghdad and other areas beyond the zone mapped in Figure 7-13, and to some extent destabilized the heart of the country.

The Kurds suffered severely in the old Iraq, but they benefited from the special protection they had during the period following the earlier Gulf War. They converted their security and their productive domain into stability and progress, and many feel that they would have a great deal to lose in a future Iraq dominated by the majority Shi'ites. Iraq's 4 million Kurds, we should note, are part of a greater Kurdish nation that extends into Turkey (with 14 million Kurds), Iran (5 million), and other neighboring countries. Together they form one of the **18** world's largest **stateless nations**, divided and often exploited by those who rule them.

American administrators hope to weld these disparate geographies into one democratic (possibly federal) Iraq, which would require enormous investment and much time—time that may not be available in the face of a specter of rising expectations.

Syria, too, is ruled by a minority. Although Syria's population of 18.1 million is about 75 percent Sunni Muslim, the ruling elite comes from a smaller Islamic sect based there, the Alawites. Leaders of this powerful minority have retained control over the country for decades, at times by ruthless suppression of dissent. In 2000, president-for-life Hafez al-Assad died and was succeeded by his son Bashar, signaling a continuation of the political status quo.

Like Lebanon and Israel, Syria has a Mediterranean coastline where crops can be raised without irrigation. Behind this densely populated coastal zone, Syria has a much larger interior than its neighbors, but its areas of productive capacity are widely dispersed. Damascus, in the southwest corner of the country, was built on an oasis and is considered to be the world's oldest continuously inhabited city. It is now the capital of Syria, with a population of 2.4 million.

The far northwest is anchored by Aleppo, the focus of cotton- and wheat-growing areas in the shadow of the Turkish border. Here the Orontes River is the chief source of irrigation water, but in the eastern part of the country the Euphrates Valley is the crucial lifeline. It is in Syria's interest to develop its eastern provinces, and recent discoveries of oil there will speed that process.

Jordan, Syria's southern neighbor, is a classic case (and victim) of changing relative location. This desert kingdom was a product of the Ottoman collapse, but when Israel was created it lost its window on the Mediterranean Sea, the (now Israeli) port of Haifa, previously in the British-administered Mandate of Palestine. Following its independence in 1946 with about 400,000 inhabitants, the creation of Israel bequeathed Jordan some 500,000 West Bank Palestinians and, later, a huge inflow of refugees. Today Palestinians outnumber original residents by more than two to one in the population of 5.5 million. It may be said that the 47-year rule by King Hussein, ending in 1999, was the key centripetal force that held the country together.

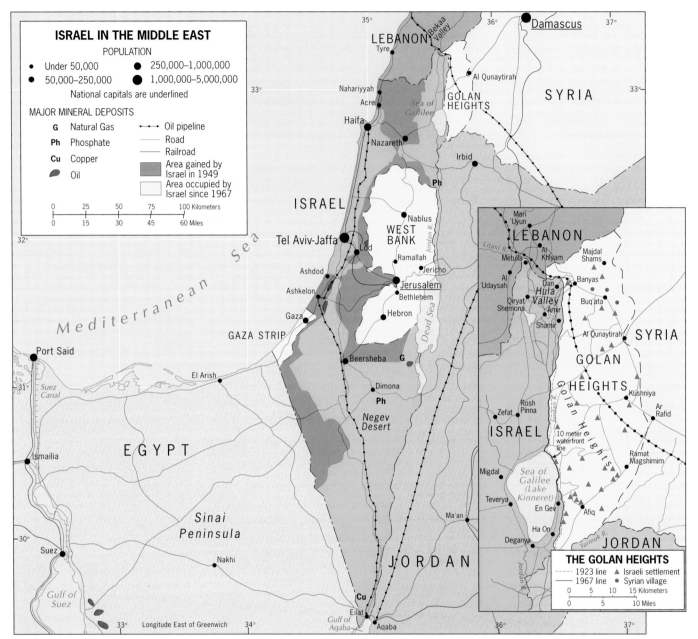

ISRAEL IN THE MIDDLE EAST

POPULATION

● Under 50,000   ● 250,000–1,000,000

● 50,000–250,000   ● 1,000,000–5,000,000

National capitals are underlined

MAJOR MINERAL DEPOSITS

G   Natural Gas

Ph   Phosphate

Cu   Copper

Oil

Oil pipeline

Road

Railroad

Area gained by Israel in 1949

Area occupied by Israel since 1967

THE GOLAN HEIGHTS

1923 line   ▲ Israeli settlement

1967 line   ● Syrian village

FIGURE 7-14

With an impoverished capital, Amman, without oil reserves, and possessing only a small and remote outlet to the Gulf of Aqaba, Jordan has survived with U.S., British, and other aid. It lost its West Bank territory in the 1967 war with Israel, including its sector of Jerusalem (then the kingdom's second-largest city). No third country has a greater stake in a settlement between Israel and the Palestinians than Jordan.

The map suggests that **Lebanon** has significant geographic advantages in this region: a lengthy coastline on the Mediterranean Sea; a world-class capital, Beirut, on its shoreline; oil terminals on its shores; and a major capital in its hinterland. The map at the scale of Figure 7-12 cannot reveal still another asset: the fertile, agriculturally productive Bekaa Valley in the eastern interior.

French colonialism in the post-Ottoman era created a territory which in 1930 was about equally Muslim and Christian. Beirut became known as the "Paris of the Middle East." After independence in 1946, Lebanon functioned as a democracy and did well economically as the region's leading banking and commercial center. But the Muslim sector of the population grew much faster than the Christian one, and in the late 1950s the Arabs launched their first rebellion against the established order. Following the first of several waves of influx by Palestinian refugees, Lebanon fell apart in 1975 in a civil war that wrecked Beirut, devastated the economy, and left the country at the mercy of Syrian forces which entered to stabilize the situation.

Today Lebanon has restabilized and Beirut is being rebuilt, but it will take generations to overcome the losses sustained during a quarter-century of dislocation.

Israel, the Jewish state, lies at the heart of the Arab World (Fig. 7-9). Since 1948, when Israel was created as a homeland for the Jewish people on the recommendation of a United Nations commission, the Arab-Israeli conflict has overshadowed all else in the region.

Figure 7-14 helps us understand the complex issues involved here. In 1946 the British, who

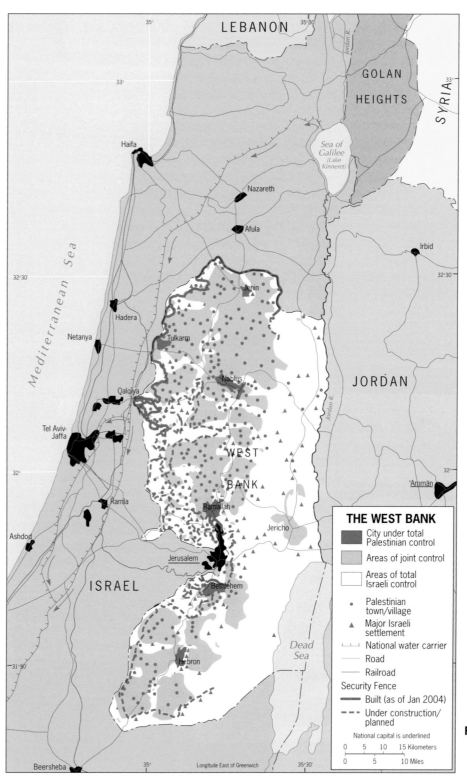

THE WEST BANK

- City under total Palestinian control
- Areas of joint control
- Areas of total Israeli control
- • Palestinian town/village
- ▲ Major Israeli settlement
- National water carrier
- Road
- Railroad

Security Fence
- Built (as of Jan 2004)
- Under construction/ planned

National capital is underlined

0   5   10   15 Kilometers
0   5   10 Miles

FIGURE 7-15

had administered this area in post-Ottoman times, granted independence to what was then called "Transjordan," the kingdom east of the Jordan River. In 1948, the orange-colored area became the U.N.-sponsored state of Israel—including, of course, land that had long belonged to Arabs in this territory called Palestine.

As soon as Israel proclaimed its independence, neighboring Arab states attacked it. Israel, however, not only held its own but pushed the Arab forces back beyond its borders, gaining the green areas shown in Figure 7-14. Meanwhile, Transjordanian armies crossed the Jordan River and annexed the yellow-colored area named the West Bank, including part of the city of Jerusalem. The king called his newly enlarged country Jordan.

More conflict followed. In 1967 a week-long war produced a major Israeli victory: Israel took the Golan Heights from Syria, the West Bank from Jordan, and the Gaza Strip from Egypt, and con-

quered the whole Sinai Peninsula all the way to the Suez Canal. In later peace agreements, Israel returned the Sinai but not the Gaza Strip.

All this strife produced a huge outflow of Palestinian Arab refugees and displaced persons. The Palestinians have been a stateless nation; about 1.25 million continue to live as Israeli citizens within the borders of Israel, but nearly 2.3 million are in the West Bank and more than 1.2 million in the Gaza Strip (the Golan Heights population is comparatively insignificant). An even larger number, however, live in neighboring and nearby countries including Jordan (2.5 million), Lebanon (500,000), Syria (450,000), and Saudi Arabia (340,000); another quarter of a million reside in Iraq, Egypt, Kuwait, and Libya. Many have been assimilated into the local societies, but tens of thousands of others continue to live in refugee camps. In 2004 the scattered Palestinian population was estimated to number nearly 9.5 million.

Israel is about the size of Massachusetts and has a population of 6.8 million (including 1.25 million Arabs), but because of its location amid adversaries and its strong international links, these data do not reflect Israel's importance. Israel has built a powerful military even as its regional Arab neighbors have grown stronger; its policies arouse Arab passions; and Palestinian aspirations for a territorial state have yet to bear fruit. As a democracy with strong Western ties, Israel has been a recipient of massive U.S. financial aid, and U.S. foreign policy has been to seek an accommodation between Jews and Palestinians as well as between Israel and its Arab neighbors.

The geographic obstacles to such an accommodation include the following:

1. *The West Bank.* Even after its capture by Israel in 1967, the West Bank might have become a Palestinian homeland (and possibly a state), but Jewish immigration to the area made such a future difficult. In 1977 only 5000 Jews lived in the West Bank; by 2004 there were over 200,000, making up more than 10 percent of the population and creating a seemingly inextricable jigsaw of Jewish and Arab settlements (Fig. 7-15).

2. *The Golan Heights.* The inset map in Figure 7-14 suggests how difficult the Golan Heights issue is. The "heights" overlook a large area of northern Israel, and they flank the Jordan River and crucial Lake Kinneret (the Sea of Galilee), the water reservoir for Israel. Relations with Syria are not likely to become normal again until the Golan Heights are returned, but in democratic Israel the political climate may make ceding this territory impossible.

3. *The Gaza Strip.* Small Jewish communities continue to live under armed protection in a poverty-stricken, barren strip of land inhabited by more than 1.2 million Palestinians, many of whom need jobs across the border in Israel to survive. Even now, a half-century later, refugee camps form a constant reminder of the consequences of Israel's creation.

FROM THE FIELDNOTES

"Cultural contrast pervades Jerusalem's urban landscape; social tension is palpable. From almost any high vantage point you can see the places of worship and the holy ground that means so much to Jew, Christian, and Muslim: synagogues, churches, and mosques, walled cemeteries, sacred shrines, historic sites, all juxtaposed as in this panorama. In the distance, in another direction, you can see the Jordan Valley; Jerusalem is where the West Bank meets Israel. When the Israeli government decided to make Jerusalem its political headquarters, it was a move designed to proclaim the nation's dominance here: Jerusalem would not become an international city. This makes Jerusalem another example of a forward capital, the vanguard of the state in potentially contested territory."

4. *Jerusalem.* The United Nations intended Tel Aviv to be Israel's capital, and Jerusalem an international city. But the Arab attack and the 1948–1949 war allowed Israel to drive toward Jerusalem (see Fig. 7-14, the green wedge into the West Bank). By the time a cease-fire was arranged, Israel held the western part of the city, and Arab forces the eastern sector. But in this eastern sector lay major Jewish historic sites, including the Western Wall. Still, in 1950 Israel declared the western sector of Jerusalem its capital, making this, in effect, a *forward capital.* Figure 7-16 shows the position of the armistice line, leaving most of the Old City in Jordanian hands. But then, in the 1967 war, Israel conquered all of the West Bank, including East Jerusalem, and in 1980 the Jewish state reaffirmed Jerusalem's status as capital, calling on all nations to move their embassies from Tel Aviv. Meanwhile, the government redrew the map of the ancient city, building Jewish settlements in a ring around East Jerusalem that would end the old distinction between Jewish west and Arab east. This enraged Palestinian leaders, who still view Jerusalem as the eventual headquarters of their hoped-for Palestinian state.

5. *Boundary Demarcation.* In 2003, the U.S. government announced its intention to support the creation of a Palestinian state. Simultaneously, the Israeli government began building an anti-terrorist "Security Fence" along the northern West Bank border, designed to keep terrorists out. The fence, however, did not conform to

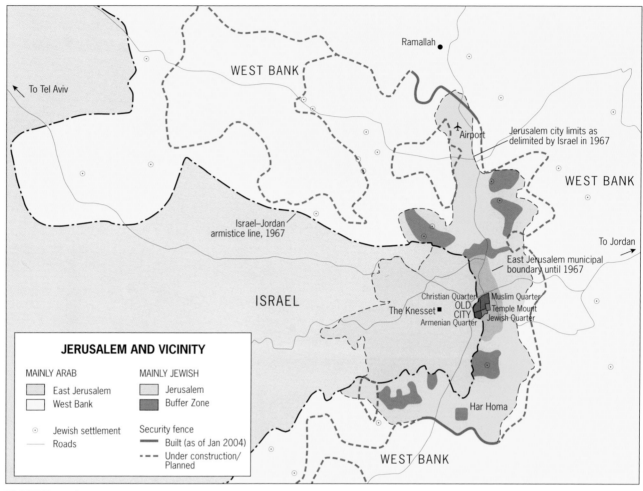

FIGURE 7-16

the *de facto* West Bank boundary, but cut about 10 percent of the West Bank's territory and, in effect, annexed it to Israel (see Fig. 7-15). This barrier and its heavily guarded gates symbolize the failure of compromise in the search for accommodation between Israelis and Arabs.

Israel lies at the center of a fast-moving geopolitical storm; the five issues raised above are only part of the overall problem (others include water rights, compensation for land expropriation, and the Palestinians' "right to return" to pre-refugee abodes). Now, in the age of nuclear, chemical, and biological weapons and longer-range missiles, Israel's search for an accommodation with its Arab neighbors is a race against time.

THE ARABIAN PENINSULA

The regional identity of the Arabian Peninsula is clear: south of Jordan and Iraq, the entire peninsula is encircled by water. This is a region of old-style emirates and sheikdoms made wealthy by oil; it also contains the site where Islam originated.

Saudi Arabia

Saudi Arabia itself has only 25.4 million inhabitants (including 5 million expatriate workers) in its vast territory, but we can see the kingdom's importance in Figure 7-8: the Arabian Peninsula contains the Earth's largest concentration of known petroleum reserves. Saudi Arabia occupies most of this area and by some estimates may possess as much as one-quarter of the world's oil deposits. As Figure 7-17 shows, these reserves lie in the eastern part of the country, particularly along the Persian Gulf coast and in the Rub al Khali (Empty Quarter) to the south.

As a region, the Arabian Peninsula is environmentally dominated by a desert habitat and politically dominated by the Kingdom of Saudi Arabia (Fig. 7-17). With 830,000 square miles (2,150,000 sq km), Saudi Arabia is the realm's fourth biggest state; only Kazakhstan, Algeria, and Sudan are larger. On the peninsula, Saudi Arabia's neighbors (moving clockwise from the head of the Persian Gulf) are Kuwait, Bahrain, Qatar, the United Arab Emirates, the Sultanate of Oman, and the Republic of Yemen (created in 1990 through the unification of former North Yemen and South Yemen). Together, these countries on the eastern fringes of the peninsula contain about 30 million inhabitants; the largest by far is Yemen, with 19.8 million.

Figure 7-17 reveals that most economic activities in Saudi Arabia are concentrated in a wide belt across the

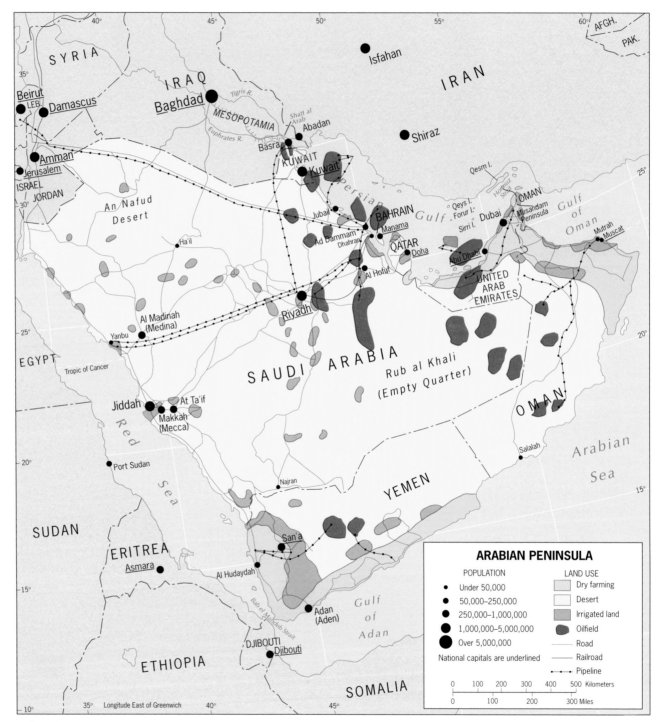

FIGURE 7-17

Map legend:

ARABIAN PENINSULA

POPULATION
- ● Under 50,000
- ● 50,000–250,000
- ● 250,000–1,000,000
- ● 1,000,000–5,000,000
- ● Over 5,000,000

National capitals are underlined

LAND USE
- Dry farming
- Desert
- Irrigated land
- Oilfield
- Road
- Railroad
- Pipeline

"waist" of the peninsula, from the oil boomtown of Dhahran on the Persian Gulf through the national capital of Riyadh in the interior to the Mecca–Medina area near the Red Sea. A modern transportation and communications network has recently been completed. But in the more remote zones of the interior, Bedouin nomads still ply their ancient caravan routes across the vast deserts. For decades, Saudi Arabia's royal families were virtually the sole beneficiaries of their country's wealth, and there was hardly any impact on the lives of villagers and nomads. When the oil boom arrived in the 1950s, foreign laborers were brought in (today there are about 5 million, many of them Shi'ite) to work in the oilfields, ports, factories, and as servants. The east boomed, but the rest of the country lagged behind.

Efforts to reduce Saudi Arabia's regional economic disparities have been impeded by the enormous cost of bringing water from deep-seated sources to the desert surface to stimulate agriculture, by the sheer size of the country, and by the high rate of population growth (2.9 percent, thereby doubling in just 24 years). Still, housing, health care, and education have seen major improvement,

and industrialization also has been stimulated in such places as Jubail on the Persian Gulf and Yanbu, near Medina, on the Red Sea.

Saudi Arabia's conservative monarchism, official friendship with the West, and social contrasts resulting from its economic growth have raised political opposition, for which no adequate channels exist. Relations with the United States were affected by the role of Saudi suicide terrorists in 9/11, Saudi financial support for Wahhabist extremists, anti-American sermons by revivalist clerics, discord over the war in Iraq, and other issues arising in the wake of the events of 2001, even as Saudi GDP declines dramatically and social benefits in the kingdom diminish. Long-term stability may be threatened in this, the region of Islam's birth.

On the Periphery

Five of Saudi Arabia's six neighbors on the Arabian Peninsula face the Persian and Oman gulfs (Fig. 7-17) and are monarchies in the Islamic tradition. All five also derive substantial revenues from oil. Their populations range from 0.6 to 3.6 million

in addition to hundreds of thousands of foreign workers; these are not strong or (OPEC aside) influential states. They do, however, display considerable geographic diversity. **Kuwait**, at the head of the Persian Gulf, almost cuts Iraq off from the open sea, an issue the Gulf War did not settle. **Bahrain** is an island-state, a tiny territory with dwindling oil reserves. Its approximately 680,000 people are 50 percent Shi'ite and only 35 percent Sunni; nearly two-thirds of its labor force is foreign. Neighboring **Qatar** consists of a peninsula jutting out into the Persian Gulf, a featureless, sandy wasteland made habitable by oil (now declining in importance) and natural gas (rising simultaneously). The **United Arab Emirates (UAE)**, a federation of seven emirates, faces the Persian Gulf between Qatar and Oman. The reigning sheik is an absolute monarch in each of the emirates, and the seven sheiks together form the Supreme Council of Rulers. In terms of oil revenues, however, there is no equality: two emirates—Abu Dhabi and Dubai—have most of the reserves.

The eastern corner of the Arabian Peninsula is occupied by the Sultanate of **Oman**, another absolute

FROM THE FIELDNOTES

"The port of Mutrah, Oman, like the capital of Muscat nearby, lies wedged between water and rock, the former encroaching by erosion, the latter crumbling as a result of tectonic plate movement. From across the bay one can see how limited Mutrah's living space is, and one of the dangers here is the frequent falling and downhill sliding of large pieces of rock. It took about five hours to walk from Mutrah to Muscat; it was extremely hot under the desert sun but the cultural landscape was fascinating. Oil also drives Oman's economy, but here you do not find the total transformation seen in Kuwait or Dubai. Townscapes (as in Mutrah) retain their Arab-Islamic qualities; modern highways, hotels, and residential areas have been built, but not at the cost of the older and the traditional. Oman's authoritarian government is slowly opening the country to the outside world after long-term isolation."

monarchy, centered on the capital, Muscat. Figure 7-17 shows that Oman consists of two parts: the large eastern corner of the peninsula and a small but critical cape to the north, the Musandam Peninsula, that protrudes into the Persian Gulf to form a **19** narrow **choke point**—the Hormuz Strait (Iran lies on the opposite shore). Tankers that leave the other Gulf states must negotiate this narrow channel at slow speed, and during politically tense times warships have had to protect them. Iran's claim to several small islands near the Strait that are owned by the UAE is a potential source of dispute.

This brings us to what is, in many ways, Saudi Arabia's most substantial peninsular neighbor: **Yemen.** The boundary between Saudi Arabia and Yemen has only recently been satisfactorily delimited in an area where there may be oil reserves. Another boundary, between former North Yemen and South Yemen, was erased in 1990, when the two countries joined to form the present state with a population that now totals just under 20 million. San'a, formerly the capital of North Yemen, retained that status; Adan (Aden), the only major port along a lengthy stretch of the peninsula, anchors the south. As Figure 7-17 shows, Yemen's southernmost tip forms one wall of the narrow, reef-studded entry to the Red Sea, another choke point Arabs call the Bab-el-Mandeb ("Gate of Grief"). Directly across lies the ministate of Djibouti, adjoined (officially) by Somalia. Since shortly after September 11, 2001, the United States has maintained a military presence in Djibouti.

▶ THE EMPIRE STATES

Two major states, both with imperial histories, dominate the region that lies immediately to the north of the Middle East (Fig. 7-9), where Arab ethnicity gives way but Islamic culture endures: Turkey and Iran. On its periphery lie two entities that are strongly entwined with these states: Azer-baijan to the north and the northern part of Cyprus to the south (Fig. 7-18).

Turkey

Earlier in this chapter we chronicled the historical geography of the Ottoman Empire, its expansion, cultural domination, and collapse. By the beginning of the twentieth century, the country we now know as Turkey lay at the center of this decaying and corrupt state, ripe for revolution and renewal. This occurred in the 1920s and thrust into prominence a leader who became known as the father of modern Turkey: Mustafa Kemal, known after 1933 as Atatürk, meaning "Father of the Turks."

The ancient capital of Turkey was Constantinople (now Istanbul), located on the Bosporus, part of the strategic straits connecting the Black and Mediterranean seas. But the struggle for Turkey's survival had been waged from the heart of the country, the Anatolian Plateau, and it was here that Atatürk decided to place his seat of government. Ankara, the new capital, possessed certain advantages: it would remind the Turks that they were (as Atatürk always said) Anatolians; it lay nearer the center of the country than Istanbul; and it could therefore act as a stronger unifier. Istanbul lies on the threshold of Europe, with the minarets and mosques of this largest and most varied Turkish city rising above a townscape that resembles one in Eastern Europe.

Although Atatürk moved the capital eastward and inward, his orientation was westward and outward. To implement his plans for Turkey's modernization, he initiated reforms in almost every sphere of life within the country. Islam, formerly the state religion, lost its official status, and Turkey became a secular state whose army ensured that the Islamists would not take over again. The state took over most of the religious schools that had controlled education. The Roman alphabet replaced the Arabic. A modified Western code supplemented Islamic law. Symbols of old—growing beards, wearing the fez—were prohibited. Monogamy was made law, and the emancipation of women was begun. The new government emphasized Turkey's separateness from the Arab World, and it has remained aloof from the affairs that engage other Islamic states.

From before Atatürk's time, Turkey has had a history of mistreating minorities. Soon after the outbreak of World War I, the (pre-Atatürk) regime decided to expel all the Armenians, concentrated in the country's northeast. Nearly 2 million Turkish Armenians were uprooted and brutally forced out; an estimated 600,000 died in a campaign that still arouses anti-Turkish emotions among Armenians today. In modern times, Turkey has been criticized for its treatment of its large and regionally concentrated Kurdish population. About one-fifth of Turkey's population of just under 70 million is Kurdish, and successive Turkish governments have mishandled relationships with this minority nation, even prohibiting the use of Kurdish speech and music in public places during one especially repressive period. The historic Kurdish homeland lies in the southeast of Turkey, centered on Diyarbakir, but millions of Kurds have moved to the shanty-towns around Istanbul—and to jobs in the countries of the European Union. The Kurdish issue has dimmed Turkey's aspirations to join the EU. Seeking to suppress a small but violent extremist group among the Kurds, the Turks violated the human rights of an entire minority.

Also problematic for Turkey is its role in Cyprus, the island 50 miles (80 km) to its south. Following a civil war between the majority Greeks and minority Turks in 1974, the Turkish government intervened with armed forces and the island was partitioned. In 1983, the northern 40 percent of the island, now an almost exclusively Turkish domain, declared itself the independent Turkish Republic of Northern Cyprus (TRNC). Against the wishes of the international community, Turkey alone recognized the TRNC as a state.

Turkey is a mountainous country of generally moderate relief and, as Figure G-8 indicates, considerable environmental diversity ranging from

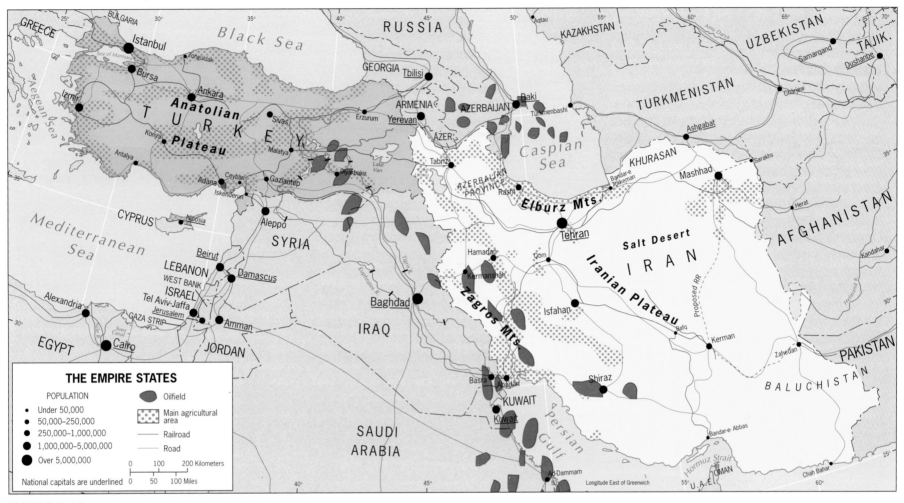

FIGURE 7-18

steppe to highland. On the dry Anatolian Plateau villages are small, and subsistence farmers grow cereals and raise livestock. Coastal plains are not large, but they are productive and densely populated. Textiles (from home-grown cotton) and farm products dominate the export economy, but Turkey also has substantial mineral reserves, some oil in the southeast, massive dam-building projects on the Tigris and Euphrates rivers, and a small steel industry based on domestic raw materials. Normally self-sufficient in staples, Turkey would

have even brighter prospects if its government and administration improved.

Iran

Iran, Turkey's neighbor to the east, also has a history of imperial conquest. In 1971, the then-reigning shah and his family celebrated the 2500th anniversary of Persia's first monarchy with unmatched royal splendor. But by 1979, revolution had engulfed Iran, and Shi'ite fundamentalists drove the

shah from power. The monarchy was replaced by an Islamic republic, and a frightful wave of retribution followed.

As Figure 7-18 shows, Iran also occupies a critical area in this turbulent realm. It controls the entire corridor between the Caspian Sea and the Persian Gulf. To the west it adjoins Turkey and Iraq, both historic enemies. To the north (west of the Caspian Sea) Iran borders Azerbaijan and Armenia, where once again Muslim confronts Christian. To the east Iran meets Pakistan and

Afghanistan, and east of the Caspian Sea lies volatile Turkmenistan.

Iran, as Figure 7-18 demonstrates, is a country of mountains and deserts. The heart of the country is an upland, the Iranian Plateau, that lies surrounded by even higher mountains, including the Zagros in the west, the Elburz in the north along the Caspian Sea coast, and the mountains of Khurasan to the northeast. This mountainous topography signals danger: here the Eurasian and Arabian tectonic plates converge, causing major and often devastating earthquakes (one recently struck the ancient eastern city of Bam, killing 28,000). The Iranian Plateau therefore is actually a huge highland basin marked by salt flats and wide expanses of sand and rock. The highlands wrest some moisture from the air, but elsewhere only oases break the arid monotony—oases that for countless centuries have been stops on the area's caravan routes.

In eastern Iran, neighboring Pakistan, and Afghanistan, some people still move with their camels, goats, and other livestock along routes that are almost as old as the human history of this realm. Usually they follow a seasonal and annual cycle, visiting the same pastures year after year, pitching their tents near the same stream. It is a lifestyle especially **20** associated with this realm: **nomadism**. In Iran as elsewhere, nomads are not aimless wanderers across boundless plains. They know their terrain intimately, and they carefully judge how long to linger and when to depart based on many years of experience along the route.

In ancient times, Persepolis in southern Iran (located near the modern city of Shiraz) was the focus of a powerful Persian kingdom, a city dependent on *qanats*, underground tunnels carrying water from moist mountain slopes to dry flatland sites many miles away. Today, Iran's population of 67.1 million is 66 percent urban, and the capital, Tehran, lies far to the north, on the southern slopes of the Elburz Mountains. This mushrooming metropolis of 7.3 million, lying at the heart of modern Iran's core area, still depends in part on the same kinds of qanats that sustained Persepolis more than 2000 years ago. As such, Tehran symbolizes the internal

contradictions of Iran: a country in which modernization has taken hold in the cities, but little has changed in the vast countryside, where the mullahs led their peasant followers in a revolution that overthrew a monarchy and installed a theocracy.

As Figure 7-8 shows, Iran has a large share of the oil riches in this part of the world. Petroleum and petroleum products provide about 90 percent of the country's income. The reserves lie in a zone along the southwestern periphery of Iran's territory, and Abadan became its "oil capital" near the head of the Persian Gulf. But Iran is a large and populous country, and the wealth oil generated could not transform it in the way the last shah intended, a transformation that had it occurred might have staved off the revolution. Modernization remained but a veneer: in the villages away from Tehran's polluted air, the holy men continued to dominate the lives of ordinary Iranians. As elsewhere in the Muslim world, urbanites, villagers, and nomads remained enmeshed in a web of production and profiteering, serfdom, and indebtedness that always characterized traditional society here. The revolution swept this system away, but it did not improve the lot of Iran's millions. A devastating war with Iraq (1980–1990), into which Iran ruthlessly poured hundreds of thousands of its young men, sapped both the coffers and energies of the state. When it was over, Iran was left poorer, weaker, and aimless, its revolution spent on unproductive pursuits.

In the early years of the twenty-first century, evidence abounded that the people of Iran remain divided between conservatives determined to protect the power of the mullahs and reformers intent on modernizing and liberalizing Iranian society. In 2002, a conservative court sentenced a university professor to death for publishing a proposal for an Islamic "enlightenment;" thousands of students took to the streets in protest. That such a sentence could be handed down at all is indicative of the gulf between the dogmatic and the rational in this historic and civilized society.

Revolutionary Iran disavowed its Persian imperial memories and ambitions, but that does not mean that its national interests now stop at its bor-

ders. Iran remains a major force in this region with cultural links to minorities in Iraq, Azerbaijan, and Afghanistan, and with economic interests in oil reserves and pipelines in the Caspian Sea area. Iran's revolution ended a monarchy, but it did not extinguish all ties to an imperial past.

TURKESTAN: THE SIX STATES OF CENTRAL ASIA

For centuries Turkish (Turkic) peoples held sway over a vast Central Asian domain that extended from Mongolia and Siberia to the Black Sea. Propelled by population growth and energized by Islam, they penetrated Iran, defeated the Byzantine Empire, and colonized much of Eastern Europe. Eventually, their power declined as Mongols, Chinese, and Russians invaded their strongholds. But these conquerors could not expunge them, as the names on the modern map prove (Fig. 7-19). The latest conquerors, the Russian czars and their communist successors, created Soviet Socialist Republics named after the majority peoples within their borders. Thus the Kazakhs, Turkmen, Kyrgyz, and other Turkic peoples retained some geographic identity in what was Soviet Central Asia.

Central Asia (Turkestan) is a still-changing region. In some areas, the cultural landscapes of neighboring realms extend into it, for example, in northern (Russian) Kazakhstan. In other areas, Turkestan extends into adjacent realms, as in Uyghur-influenced western China (Xinjiang). Some areas once penetrated by Turkic peoples are no longer dominated by them, for instance, Afghanistan. And certain peoples now living in Turkestan are not of Turkic ancestry, notably the Tajiks. This is a fractious region in sometimes turbulent transition.

The underlying cultural-geographic reason for this contentiousness—not only in Turkestan but in neighboring regions in Southwest and South Asia as well—is illustrated in Figure 7-20. This detailed map of ethnolinguistic groups actually is a generalization of an even more complex mosaic of peo-

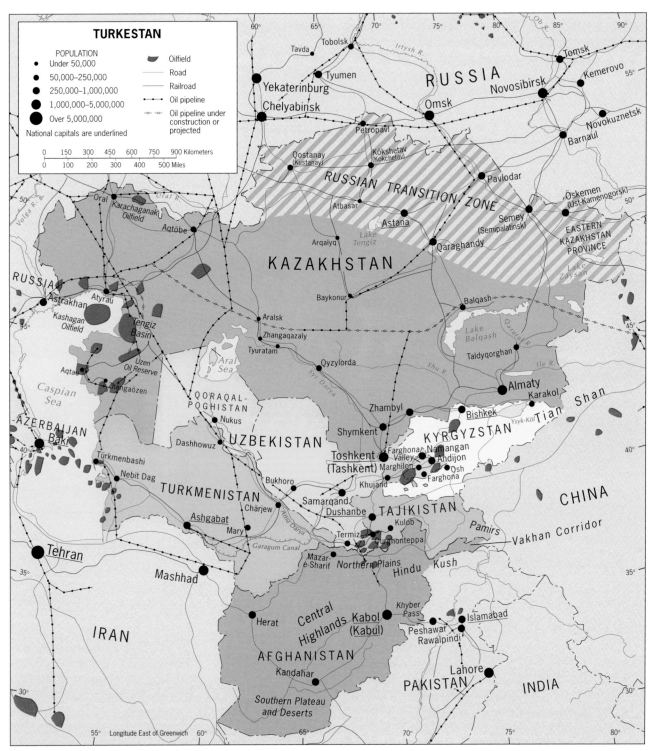

FIGURE 7-19

ples and cultures. Every cultural domain on the map also includes minorities that cannot be mapped at this scale, so that people of different faiths, languages, and ways of life rub shoulders everywhere. Often this results in friction, and at times such friction escalates into ethnic conflict.

As we define it, Turkestan includes most or all of six states: (1) Kazakhstan, territorially larger than the other five combined but situated astride an ethnic transition zone; (2) Turkmenistan, with important frontage on the Caspian Sea and bordering Iran and Afghanistan; (3) Uzbekistan, the most populous state and situated at the heart of the region; (4) Kyrgyzstan, wedged between powerful neighbors and chronically unstable; (5) Tajikistan, regionally and culturally divided as well as strife-torn; and (6) Afghanistan, engulfed in war almost continuously since it was invaded by Soviet forces in 1979.

During their hegemony over Central Asia, the Soviets tried to suppress Islam and install secular regimes (this was their objective when they invaded Afghanistan as well), but today Islam's revival is one of the defining qualities of this region. From Almaty to Samarqand, mosques are being repaired and revived, and Islamic dress again is part of the cultural landscape. National leaders have made high-profile visits to Mecca; they are sworn into office on the Quran. All of Central Asia's countries now observe Islamic holidays. In other ways, too, Turkestan reflects the norms of this realm: in its dry-world environments and the clustering of its population, its mountain-fed streams irrigating farms and fields, its sectarian conflicts, its

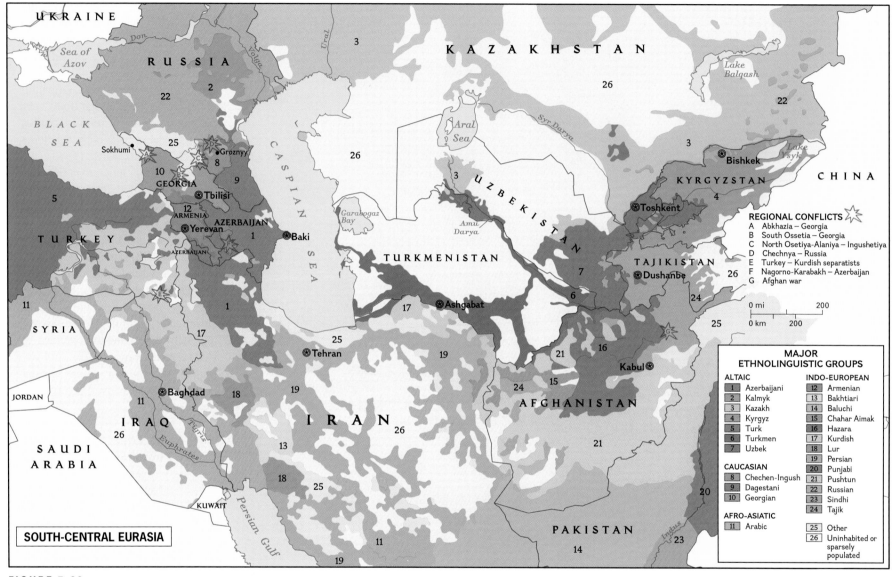

REGIONAL CONFLICTS
A Abkhazia – Georgia
B South Ossetia – Georgia
C North Osetiya-Alaniya – Ingushetiya
D Chechnya – Russia
E Turkey – Kurdish separatists
F Nagorno-Karabakh – Azerbaijan
G Afghan war

MAJOR ETHNOLINGUISTIC GROUPS

ALTAIC
1 Azerbaijani
2 Kalmyk
3 Kazakh
4 Kyrgyz
5 Turk
6 Turkmen
7 Uzbek

CAUCASIAN
8 Chechen-Ingush
9 Dagestani
10 Georgian

AFRO-ASIATIC
11 Arabic

INDO-EUROPEAN
12 Armenian
13 Bakhtiari
14 Baluchi
15 Chahar Aimak
16 Hazara
17 Kurdish
18 Lur
19 Persian
20 Punjabi
21 Pushtun
22 Russian
23 Sindhi
24 Tajik

25 Other
26 Uninhabited or sparsely populated

SOUTH-CENTRAL EURASIA

FIGURE 7-20

oil-based economies. It also is a region where democratic government remains an elusive goal, especially in Turkmenistan.

Kazakhstan is the territorial giant of the region and borders two greater giants: Russia and China (Fig. 7-19). During the Soviet period, northern Kazakhstan was heavily Russified and became,

in effect, part of Russia's Eastern Frontier (see Fig. 2-10). Rail and road links crossed the area mapped as the Russian Transition Zone, connecting the north to Russia. The Soviets made Almaty, in the heart of the Kazakh domain, the territory's capital. Today the Kazakhs are in control, and they in turn have moved the capital to Astana, right in the heart

of the Russian Transition Zone, where some 5 million Russians still live. Astana is clearly another *forward capital.*

Figure 7-19 reveals Kazakhstan's situation as a corridor between the Caspian Basin's oil reserves and China. Oil and gas pipelines across Kazakhstan could eliminate, or greatly reduce, China's dependence on oil carried by tankers along distant sea lanes.

Uzbekistan, the most populous state in this region (26.3 million), occupies the heart of Turkestan and borders every other state in it. Uzbeks not only make up 75 percent of this population, but also form substantial minorities in several neighboring states. The capital, Toshkent, lies in the far eastern core area of the country, where most of the people live in towns and farm villages, and the crowded Farghona Valley is the focus. In the west lies the shrinking Aral Sea, whose feeder streams were diverted into cotton fields and croplands during the Soviet occupation; heavy use of pesticides contaminated the groundwater, and countless thousands of local people suffered severe medical problems as a result.

In the post-Soviet period, Islamic (Sunni) revivalism has become a problem for Uzbekistan. Wahhabism, the particularly virulent form of Sunni fundamentalism, took root in eastern Uzbekistan; government efforts to suppress it have only increased its strength in the countryside.

Turkmenistan, the autocratic desert republic that extends all the way from the Caspian Sea to the borders of Afghanistan, has a population of just 5.8 million, of which about three-quarters are Turkmen. During Soviet times, the communist planners began work on a massive project: the Garagum (Kara Kum) Canal designed to bring water from Turkestan's eastern mountains into the heart of the desert. Today the canal is 700 miles (1100 km) long, and it has enabled the cultivation of some 3 million acres of cotton, vegetables, and fruits. The plan is to extend the canal all the way to the Caspian Sea, but meanwhile Turkestan has hopes of greatly expanding its oil and gas output from Caspian Basin reserves. But look again at Figure 7-19: Turkmenistan's relative location is not advantageous for export routes.

Kyrgyzstan's topography and political geography are reminiscent of the Caucasus. Shown in yellow in Figure 7-19, Kyrgyzstan lies intertwined with Uzbekistan and Tajikistan to the point of exclaves and enclaves along its borders. The Kyrgyz, for whom the Soviets established this "republic,"

constitute barely 60 percent of the population of 5.1 million. Uzbeks and other minorities form a complex cultural geography; mountains and valleys isolate communities and make nation-building difficult. The agricultural economy is weak, consisting of pastoralism in the mountains and farming in the valleys. About 70 percent of the people profess allegiance to Islam, and Wahhabism has gained a strong foothold here. The town of Osh is often referred to as the headquarters of the movement in Turkestan.

Tajikistan's mountainous scenery is even more spectacular than Kyrgyzstan's, and here, too, topography is a barrier to the integration of a multicultural society. The Tajiks, who constitute 62 percent of the population of 6.5 million, are of Persian (Iranian), not Turkic, origin and speak a Persian (and thus Indo-European) language. Most Tajiks, despite their Persian affinities, are Sunni Muslims, not Shi'ites. Small though Tajikistan is, regionalism plagues the state: the government in Dushanbe often is at odds with the barely connected northern area (see Fig. 7-19), a hotbed not only of Islamic revivalism but also of anti-Tajik, Uzbek activism. By 2003, what had amounted to a civil war had abated.

Afghanistan, the southernmost country in this region, exists because the British and Russians, competing for hegemony in this area during the nineteenth century, agreed to tolerate it as a cushion, **21** or **buffer state**, between them. This is how Afghanistan acquired the narrow extension leading from the main territory eastward to the Chinese border—the Vakhan Corridor (Fig. 7-21). As the colonialists delimited it, Afghanistan adjoined the domains of the Turkmen, Uzbeks, and Tajiks to the north, Persia (now Iran) to the west, and the western flank of British India (now Pakistan) to the east.

Geography and history seem to have conspired to divide Afghanistan. As Figure 7-21 shows, the towering Hindu Kush range dominates the center of the country, creating three broad environmental zones: the relatively well-watered and fertile northern plains and basins; the rugged, earthquake-prone central highlands; and the desert-dominated southern plateaus. Kabol (Kabul), the capital, lies on the southeastern slope of the Hindu Kush, linked by

narrow passes to the northern plains and by the Khyber Pass to Pakistan.

Across this variegated landscape moved countless peoples: Greeks, Turks, Arabs, Mongols, and others. Some settled here, their descendants today speaking Persian, Turkic, and other languages. Others left archeological remains or no trace at all. The present population of Afghanistan (29.2 million) has no ethnic majority. This is a country of minorities in which the Pushtuns (or Pathans) of the east are the most numerous but make up less than 40 percent of the total (substantially less if those across the borders are subtracted). The second-largest minority are the Tajiks, a world away across the Hindu Kush, concentrated in the zone near Afghanistan's border with Tajikistan. The Hazaras of the central highlands and the south, the Uzbeks and Turkmen in the northern border areas, the Baluchi of the southern deserts, and other, smaller groups scattered across this riven country create one of the world's most complex cultural mosaics (Fig. 7-20). Two major languages, Pushtun and Dari (the local variant of Persian), plus several others create a veritable Tower of Babel here.

Episodes of conflict have marked the history of Afghanistan, but none was as costly as its involvement in the Cold War. Following the Soviet intervention of 1979, the United States supported the Muslim opposition, the *Mujahideen* ("strugglers"), with modern weapons and money, and the Soviets were forced to withdraw. Soon the factions that had been united during the anti-Soviet campaign were in conflict, delaying the return of some 4 million refugees who had fled to Pakistan and Iran. The situation resembled the pre-Soviet past: a feudal country with a weak and ineffectual government in Kabol.

In 1994, what at first seemed to be just another warring faction appeared on the scene: the so-called *Taliban* ("students of religion") from religious schools in Pakistan. Their avowed aim was to end Afghanistan's chronic factionalism by instituting strict Islamic law. Popular support in the war-weary country, especially among the Pushtuns, led to a series of successes, and by 1996 the Taliban had taken Kabol.

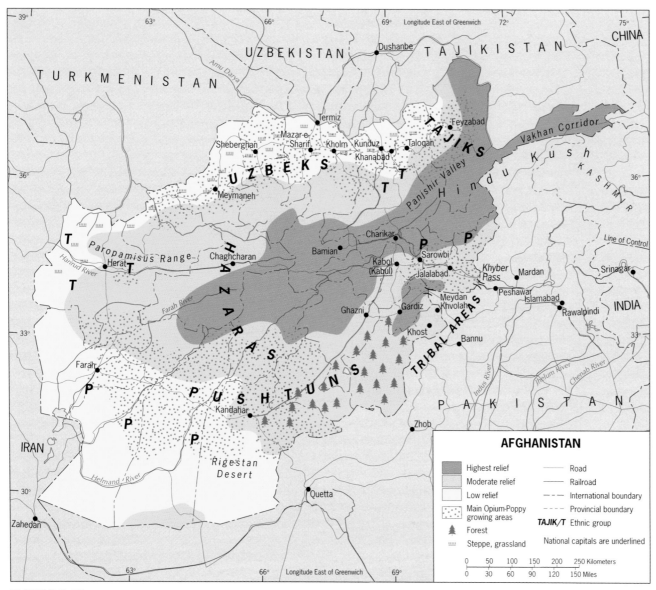

FIGURE 7-21

from the Cold War, they also benefited from Afghanistan's huge, illicit opium trade. In 2000, according to United Nations data, Afghanistan produced between 70 and 79 percent of the world's opium. Much of the revenue found its way into the coffers of the conspirators.

With such resources, the militants were able to organize and launch attacks against several Western targets in the realm and elsewhere, but events took a fateful turn in 1996 that would empower them as never before. During the conflict with the Soviets, the Mujahideen cause had been supported not only by the United States but also by a devout and revivalist Saudi Muslim named Usama bin ("son of") Laden. The child of a construction billionaire who had more than 50 children including 22 sons, Usama graduated from a university in Jiddah in 1979 and headed for Afghanistan with an inherited fortune estimated at about $300 million to help the anti-Soviet campaign. Well-connected in Saudi Arabia and now a pivotal figure in Afghanistan, bin Laden saw his fame as well as his war chest grow as the Soviet intervention collapsed. Following the withdrawal of the communist forces, he returned to Saudi Arabia, where he denounced his government for allowing U.S. troops on Saudi soil during the Gulf War. The Saudi regime responded by stripping him of his citizenship and expelling him, and bin Laden fled to Sudan. There he set up several legitimate businesses to facilitate his now-global financial transactions, but he also established terrorist training camps. Under international pressure, the Khartoum regime ousted him in 1996, and bin Laden returned to a country he knew would welcome him again: Afghanistan.

The Taliban's imposition of Islamic law was so strict and severe that Islamic as well as non-Islamic countries objected. Restrictions on the activities of women ended their professional education, employment, and freedom of movement, and had a devastating impact on children as well. Public amputations and stonings enforced the Taliban's code. In the process, Afghanistan became a haven

for groups of revolutionaries whose goals went far beyond those of the Taliban: they plotted attacks on Western interests throughout the realm and threatened Arab regimes they deemed compliant with Western priorities. Taliban-ruled, cave-riddled, remote and isolated Afghanistan was an ideal locale for these outlaws. Already in possession of arms and ammunition (Soviet as well as American) left over

Bin Laden's fateful return to Afghanistan coincided with the Taliban's conquest of Kabol, and now he helped its forces push northward into the fertile and productive northern plains. Meanwhile, a terrorist organization named al-Qaeda took root in the country, a global network that would further the aims of the revolutionaries once loosely allied. Afghanistan became al-Qaeda's headquarters, and bin Laden its director; its exploits were funded by numerous Muslim sources and ranged from terrorist-training in local camps to lethal attacks on American targets, including a warship in Yemen's port of Adan and two U.S. embassies in East Africa.

On February 26, 1993, terrorists exploded a massive car bomb in the basement garage of the World Trade Center in New York, but their objective—to topple the 110-story tower—failed. Even as those responsible went on trial, al-Qaeda's leaders were planning the suicide attacks of September 11, 2001 that destroyed the buildings, killed thousands, and caused billions of dollars in damage. Several weeks later, United States and British forces, with the acquiescence of Pakistan, attacked both the Taliban regime and the al-Qaeda infrastructure in Afghanistan, the first stage of a global "War on Terror" that would leave no country unaffected.

Proof of bin Laden's and al-Qaeda's complicity in the September 11 assault was found in videotape and documentary form.

The Taliban and al-Qaeda leaderships may not have counted on Pakistan's compliance with Western demands, but they surely knew what the consequences of 9/11 would be for Afghanistan and its people. With a weak postwar government, historic ethnic divisions, and fanatical warlords still controlling remote provinces, Afghanistan risks reverting to its fractured past. But to the zealots of al-Qaeda, it is a price to be paid in order to show the Muslim world the consequences of dealing with, and hosting, infidels.

DISCOVERY TOOLS   www.wiley.com/college/deblij

The *Concepts and Regions* Website, featuring *GeoDiscoveries*, offers many additional resources to enhance your understanding and experience of this chapter. Be sure to explore the following:

GeoDiscoveries Video Clips

Diversifying Its Portfolio

Omanization

Something's Fishy Here

A Future Without Oil

Dubai

The Strategic Significance of Oman

GeoDiscoveries Animations

Diffusion

Expansion Diffusion

Hierarchical Diffusion

Relocation Diffusion

An Example of Relocation Diffusion

GeoDiscoveries Interactivities

Refining the Region: Oil

In the Pipeline

Photo Galleries

Creek of Dubai

Cairo, Egypt

Jerusalem, Israel

Mutrah, Oman

Morocco

Marrakech, Morocco

Western Wall, Israel

Bahai Temple, Israel

Virtual Field Trip

Jerusalem: Sacred Place–Contested Space

More To Explore

Systematic Essay: Cultural Geography

Issues in Geography: Diffusion Processes, The Flowering of Islamic Culture, A Future Kurdistan?; Choke Points: Danger on the Sea Lanes

Major Cities: Cairo, Istanbul

Expanded Regional Coverage

Egypt and the Lower Nile Basin; Maghreb and Its Neighbors, Middle East; Arabian Peninsula, Empire States, Turkestan

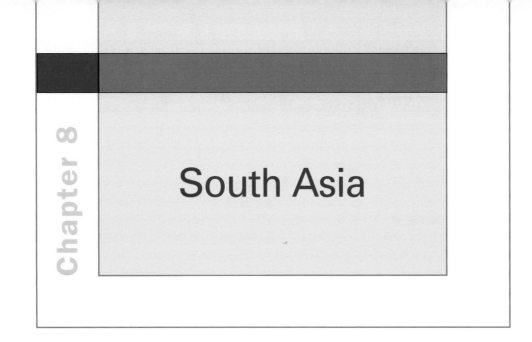

Chapter 8

South Asia

CONCEPTS, IDEAS, AND TERMS

1. Wet monsoon
2. Social stratification
3. Refugees
4. Population geography
5. Population distribution
6. Population density
7. Physiologic density
8. Rate of natural population increase
9. Doubling time
10. Demographic transition
11. Population explosion
12. Forward capital
13. Irredentism
14. Caste system
15. Intervening opportunity
16. Natural hazards
17. Insurgent state

REGIONS

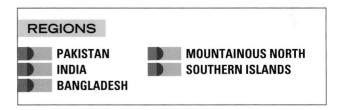

- PAKISTAN
- INDIA
- BANGLADESH
- MOUNTAINOUS NORTH
- SOUTHERN ISLANDS

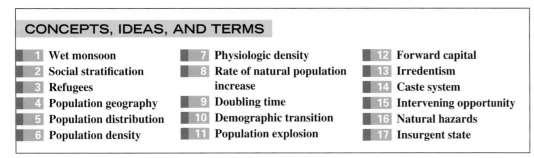

FROM IBERIA TO Arabia and from Malaysia to Korea, Eurasia is a landmass fringed by peninsulas. The largest of all is the great triangle of India that divides the northern Indian Ocean into two seas: the Arabian Sea to the west and the Bay of Bengal to the east (Fig. 8-1). The peninsula of India forms the heart of South Asia, a vast, varied, and volatile geographic realm.

DEFINING THE REALM

Mountains, deserts, and coastlines combine to make South Asia one of the world's most vividly defined physiographic realms. To the north, the Himalaya Mountains create a natural wall between South Asia and China. To the east, mountain ranges and dense forests mark the boundary between South and Southeast Asia. To the west, rugged highlands and expansive deserts separate South Asia from its neighbors.

Within these confines lies a geographic realm that is more densely populated than any other. If current population trends continue, it will soon be the most populous realm on Earth as well.

STUDY TOOLS

3D GLOBE	FLASHCARDS	AUDIO PRONUNCIATION GUIDE	ANNOTATED WEBLINKS
AREA AND DEMOGRAPHIC DATA	MAP QUIZZES	CHAPTER QUIZ	LONELY PLANET WEBLINKS

MAJOR GEOGRAPHIC QUALITIES OF SOUTH ASIA

1. South Asia is clearly defined physiographically and is bounded by mountains, deserts, and ocean; the Indian peninsula is Eurasia's largest.

2. South Asia is a poverty-afflicted realm, with low average incomes, low levels of education, poorly balanced diets, and poor overall health.

3. With only 3 percent of the world's land area but 22 percent of its population, more than half of it engaged in subsistence farming, South Asia's economic prospects are bleak.

4. Population growth rates in South Asian countries are among the highest in the world; India's population surpassed the 1 billion mark in 1999.

5. The North Indian Plain, the lower basin of the Ganges River, contains the heart of the world's second largest population cluster.

6. Despite encircling mountain barriers, invaders from ancient Greeks to later Muslims penetrated South Asia and complicated its cultural mosaic.

7. British colonialism unified South Asia under a single flag, but the British South Asian Empire fragmented into several countries along cultural lines after Britain's withdrawal in 1947.

8. Pakistan, South Asia's western region, lies on the flanks of two realms: largely Muslim North Africa/Southwest Asia and dominantly Hindu South Asia.

9. India is the world's largest federation and most populous democracy, but its political achievements have not been matched by enlightened economic policies.

10. Religion remains a powerful force in South Asia. Hinduism in India, Islam in Pakistan, and Buddhism in Sri Lanka all show tendencies toward fundamentalism and nationalism.

11. Active and potential boundary problems involve internal areas (notably between India and Pakistan in Kashmir) as well as external locales (between India and China in the northern mountains).

South Asia consists of five regions (Fig. 8-2). Its keystone is India, whose population passed the 1 billion mark in 1999. In the west lies Pakistan. South Asia's eastern flank is centered on Bangladesh. The northern region consists of the mountainous lands of Kashmir, Nepal, and Bhutan. And the southern region includes the islands of Sri Lanka and the Maldives. As the map shows, India divides into several subregions.

South Asia's physiographic boundaries are formidable barriers, but they have not prevented conquerors or proselytizers from penetrating it. As a result, today this realm is a patchwork of religions, languages, traditions, and cultural landscapes. So complex is this mosaic that, remarkably, its political geography numbers just seven states (and only five on the mainland).

Among the cultural infusions was Islam. Today, Pakistan is an Islamic republic, and Islam provides the cement for the state. Pakistan's eastern border with India is a cultural divide in more ways than one: dominantly Hindu India is a secular, not a theocratic, state. Why, then, do we include Pakistan in the South Asia rather than the Southwest Asia/North Africa realm? One criterion is ethnic continuity, which links Pakistan to India rather than to Afghanistan or Iran. Another is historical geography. Pakistan was part of Britain's South Asian Empire, and it originated from the partition of that domain between Muslim and Hindu majorities. Although Urdu is the official national language of Pakistan, English is the *lingua franca*, as it is in India. Furthermore, the border between India and Pakistan does not signify the eastern frontier of Islam in Asia. About 150 million of India's more than 1 billion citizens are Muslims, and in South Asia's eastern region, Bangladesh (population: 140 million) is more than 85 percent Muslim. Finally, Pakistan and India are locked in a struggle to control a vital mountainous area in the far north, where the British withdrawal left the boundary between them unresolved. Even as Indian and Pakistani cricket teams play each other on sun-baked pitches, their armies face off in a deadly conflict—a conflict that has the potential to unleash a nuclear war.

PHYSIOGRAPHIC REGIONS OF SOUTH ASIA

Before we look into South Asia's complex and fascinating cultural geography, we need to discuss the physical stage of this populous realm (Fig. 8-1). South Asia is a subcontinent of immense physiographic variety, of snowcapped peaks and forest-clad slopes, of vast deserts and broad river basins, of high plateaus and spectacular shores. The collision of two of the Earth's great tectonic plates created the world's highest mountain ranges, their icy crests yielding meltwaters for great rivers below. The workings of the Earth's atmosphere put South Asia in the path of tropical cyclones and produce reversing seasonal windflows known as *monsoons*. This is a realm of almost infinite variety, a world unto itself.

In general terms, we can recognize three clearly defined physiographic zones in South Asia: the northern mountains, the southern peninsular plateaus, and between them a belt of river lowlands. Superimposed on this configuration is an east-west precipitation gradient from wet (Bangladesh) to dry (western Pakistan) that is clearly visible in Figures G-7 and G-8,

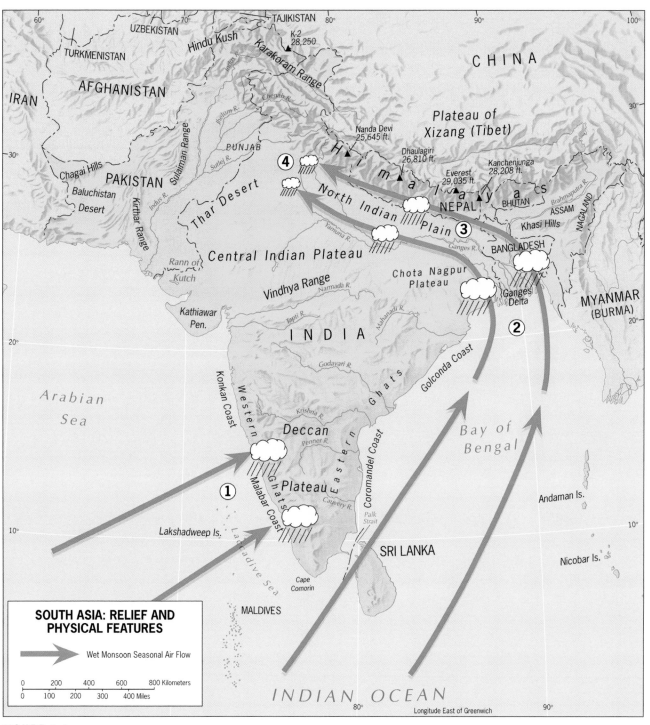

FIGURE 8-1

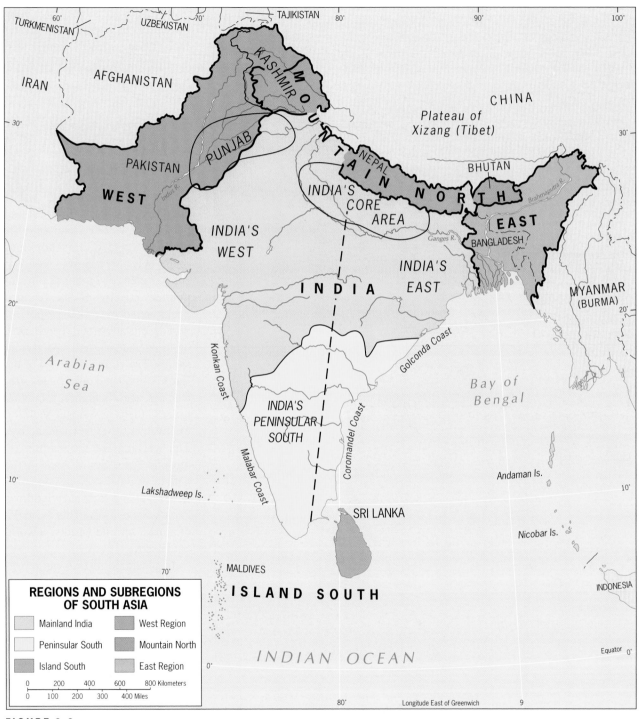

FIGURE 8-2

broken only by the strip of high moisture along India's southwestern Malabar Coast.

Northern Mountains

The northern mountains extend from the Hindu Kush and Karakoram ranges in the northwest through the Himalayas in the center (Mount Everest, the world's tallest peak, lies in Nepal) to the ranges of Bhutan and the Indian State of Arunachal Pradesh in the east (Fig. 8-1). Dry and barren in the west on the Afghanistan border, the ranges become green and tree-studded in Kashmir, forested in the lower sections of Nepal, and even more densely vegetated in Arunachal Pradesh. Transitional foothills, with many deeply eroded valleys cut by rushing meltwaters, lead to the river basins below.

River Lowlands

The belt of river lowlands extends eastward from Pakistan's lower Indus Valley (the area known as Sind) through the wide plain of the Ganges Valley of India and on across the great double delta of the Ganges and Brahmaputra in Bangladesh (Fig. 8-1). In the east, this physiographic region often is called the North Indian Plain. To the west lies the lowland of the Indus River, which rises in Tibet, crosses Kashmir, and then bends southward to receive its major tributaries from the Punjab ("Land of Five Rivers").

Southern Plateaus

Peninsular India is mostly plateau country, dominated by the massive

Deccan, a tableland built of basalt that poured out when India separated from Africa during the breakup of Gondwana (see Fig. 6-2). The Deccan (meaning "South") tilts to the east, so that its highest areas are in the west and the major rivers flow into the Bay of Bengal. North of the Deccan lie two other plateaus, the Central Indian Plateau to the west and the Chota-Nagpur Plateau to the east (Fig. 8-1). On the map, note the Eastern and Western Ghats ("hills") that descend from Deccan plateau elevations to the narrow coastal plains below.

The Annual Wet Monsoon

India and Bangladesh form what climatologists call the "type area" for a major environmental **1** phenomenon: the annual **wet monsoon**. By June of each year, a strong low-pressure system develops over northern India, drawing moist, warm oceanic air onto the peninsula (Fig. 8-1). As this air is pulled over the Western Ghats ①, it is cooled, its moisture condenses, and rains begin, which may last 60 days or more. Other streams of air come from the Bay of Bengal ② and get caught up in the convection over northeastern India and Bangladesh. Seemingly endless summer rains now inundate the whole North Indian Plain; the Himalayas to the north create a corridor that sends the rain westward ③ until eventually it dries out as it reaches Pakistan ④. By the end of summer the system breaks down, the wet monsoon gives way to periodic rains and, eventually, another dry season—and the anxious wait begins again for next year's life-giving monsoon.

THE HUMAN SEQUENCE

Great river basins mark the physiography of South Asia; in one of these basins, that of the Indus River in present-day Pakistan, lies evidence of the realm's oldest major civilization. It existed at the same time, and interacted with, ancient Mesopotamia, and it was centered on large, well-organized cities (Fig. 8-3). From here, influences and innovations diffused into India. In fact, India's very name is believed to derive from the ancient Sanskrit word *sindhu*.

Eventually, such cities as Harappa and Mohenjo-Daro seem to have experienced the same fate as those of Mesopotamia, perhaps also because of environmental change. About 3500 years ago, Aryan (Indo-European) speaking peoples migrated into the region, penetrating India and starting the process of welding the Ganges Basin's isolated tribes and villages into an organized system. Having absorbed much of the culture of the Indus Valley, they brought a new order to this region. A new belief system, *Hinduism*, arose and with it a new way of life.

2 A complex **social stratification** developed in which Brahmans, powerful priests, stood at the head of a *caste system* in which soldiers, merchants, artists, peasants, and all others had their place.

Hinduism was restrictive, especially for those of the lower castes, and in the sixth century BC (more than 2500 years ago) a prince born into one of northern India's kingdoms sought a better way.

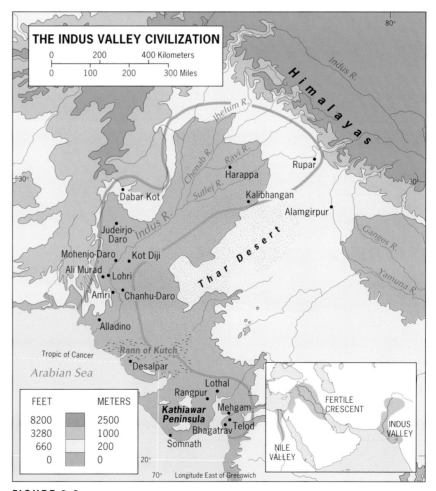

FIGURE 8-3

Prince Siddhartha, better known as Buddha, gave up his royal position to teach religious salvation through meditation, the rejection of Earthly desires, and a reverence for all forms of life. His teachings did not have a major impact on the Hindu-dominated society of his time, but what he taught was not forgotten. Centuries later, the ruler of a powerful Indian state decided to make *Buddhism* the state religion, following which the faith spread far and wide.

But first South Asia was penetrated by other foreign influences: the Persians were followed by the Greeks under Alexander the Great in the late fourth century BC. This is why, on a world map of languages, Hindi (the major domestic language of India) lies at the eastern end of the region of Indo-European languages, in the same family as Persian, Italian, and English. But the map of India's languages (Fig. 8-4) shows that southern India's languages are *not* Indo-European. Indeed, while northern India was a subregion of cultural infusion and turmoil, the south lay remote and isolated. This part of the peninsula had apparently been settled long before the Indus and Ganges civilizations arose, and southern India developed into a distinctive subregion with its own cultures. The *Dravidian* languages spoken here—Telugu, Tamil, Kanarese (Kannada), and Malayalam—have long literary histories.

Aśoka's Mauryan Empire

When the Greeks withdrew from the Ganges Basin and the Hindu heartland was once again free, a powerful empire arose there—the first true empire in the realm. This, the Mauryan Empire, extended its influence over India as far west as the Indus Valley (thus incorporating the populous Punjab) and as far east as Bengal (the double delta of the Ganges and Brahmaputra); it reached as far south as the modern city of Bangalore.

This Mauryan Empire was led by a series of capable rulers who achieved stability over a vast domain. Undoubtedly the greatest of these leaders was Aśoka, who reigned for nearly 40 years during the middle of the third century BC. Aśoka was a believer in Buddhism, and it was he who elevated this religion from obscurity to regional and ultimately global importance.

In accordance with Buddha's teachings, Aśoka reordered his government's priorities from conquest and expansion to a Buddhist-inspired search for stability and peace. He sent missionaries to the outside world to carry Buddha's teachings to distant peoples, thereby also contributing to the diffusion of Indian culture. As a result, Buddhism became permanently established as the dominant religion in Sri Lanka (formerly Ceylon), and it established footholds as far afield as Southeast Asia and Mediterranean Europe. Ironically, Buddhism thrived in these remote places even as it declined in India itself. With Aśoka's death, the faith lost its strongest supporter.

The Mauryan Empire represented India's greatest political and cultural achievements in its day, and when the empire collapsed, late in the second century AD, India fragmented into a patchwork of states. Once again, India lay open to infusions from the west and northwest, and across present-

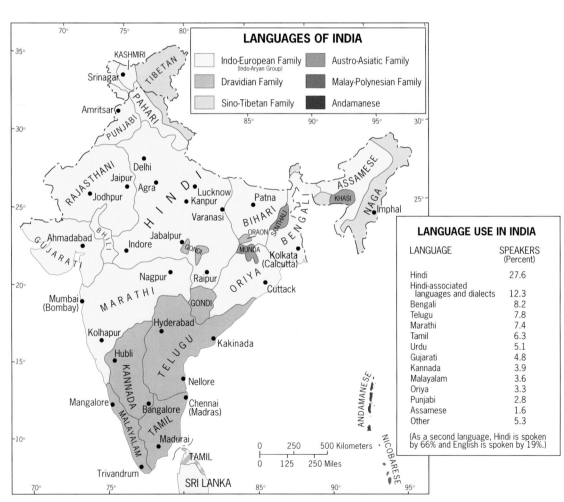

LANGUAGE USE IN INDIA	
LANGUAGE	SPEAKERS (Percent)
Hindi	27.6
Hindi-associated languages and dialects	12.3
Bengali	8.2
Telugu	7.8
Marathi	7.4
Tamil	6.3
Urdu	5.1
Gujarati	4.8
Kannada	3.9
Malayalam	3.6
Oriya	3.3
Punjabi	2.8
Assamese	1.6
Other	5.3

(As a second language, Hindi is spoken by 66% and English is spoken by 19%.)

FIGURE 8-4

day Pakistan they came: Persians, Afghans, Turks, and others driven from their homelands or attracted by the lands of the Ganges.

The Power of Islam

In the late tenth century, Islam came rolling like a giant tide across the subcontinent, spreading from Persia in the west and Afghanistan in the northwest. Of course, the Indus Valley lay directly in the path of this Islamic advance, and virtually everyone was converted. Next the Muslims penetrated the Punjab, the subregion that lies astride the present Pakistan-India border, and there perhaps as many as two-thirds of the inhabitants became converts. Then Islam crossed the bottleneck where Delhi is situated and diffused east- and southeastward into the Gangetic Plain and the subregion known as Hindustan—India's evolving core area. Here Islam's proselytizers had less success, persuading perhaps one in eight Indians to become Muslims. In the meantime, Islam arrived at the Ganges Delta by boat, and present-day Bangladesh became overwhelmingly Islamic. (To the south of the Ganges heartland, however, Islam's diffusion wave lost its energy: Dravidian India never came under Muslim influence.)

Islam's vigorous, often violent, onslaught changed Indian society. As in West Africa, Islam frequently was superimposed through political control: when the rulers were converted, their subjects followed. By the early fourteenth century, a sultanate centered at Delhi controlled more of the subcontinent than even the Mauryan Empire had earlier. Later, the Islamic Mogul Empire (the similarity to the word "Mongol" is by no means a coincidence) constituted the largest political entity ever to unify the realm in precolonial times. To many Hindus of lower caste, Islam represented a welcome alternative to the rigid socioreligious hierarchy in which they were trapped at the bottom. Thus Islam was the faith of the ruling elites and of the disadvantaged, a powerful cultural force in the heartland of Hinduism.

Just as Islam weakened in Southern Europe, so its force ultimately became spent in vast and populous India. For all the Muslims' power, they never managed to convert a majority of South Asians. They dominated the northwest corner of the realm (present-day Pakistan), where Lahore became one of Islam's greatest cities. But in all of what is today India, less than 15 percent of the population became and remained Muslim. And throughout the period of Islamic intervention, the struggle for cultural supremacy continued. Placid Hinduism and aggressive Islam did not easily coexist.

The European Intrusion

Into this turbulent complexity of religious, political, and linguistic disunity yet another element began to intrude after 1500: European powers in search of raw materials, markets, and political influence. Because the Europeans profited from the Hindu-Muslim contest, they exploited local rivalries, jealousies, and animosities. British merchants gained control over the trade with Europe in spices, cotton, and silk goods, ousting the French, Dutch, and Portuguese. The British East India Company's ships also took over the intra-Asian sea trade between India and Southeast Asia, which had long been in the hands of Arab, Indonesian, Chinese, and Indian merchants. In effect, the East India Company (EIC) became India's colonial administration.

In time, however, the EIC faced problems that made it increasingly difficult to combine commerce with administration. Eventually, a mutiny among Indian troops in the service of the EIC led to the abolition of the company. The British government took over in 1857 and maintained its rule (*raj*) until 1947.

Colonial Transformation

When the British took power over South Asia, they controlled a realm with considerable industrial development (notably in metal goods and textiles) and an active trade with both Southwest and South-

east Asia. The colonialists saw this as competition, and soon India was exporting raw materials and importing manufactured goods—from Europe, of course. Local industries declined and Indian merchants lost their markets.

Unifying their realm was a tougher task for the British. In 1857, about 750,000 square miles (2 million sq km) of South Asian territory still was beyond British control, including hundreds of entities that had been guaranteed autonomy by the EIC during its administration. These "Native States," ranging in size from a few acres to Hyderabad's 80,000 square miles (200,000 sq km), were assigned British advisors, but in fact India was a near-chaotic amalgam of modern colonial and traditional feudal systems.

Colonialism did produce assets for India. The country was bequeathed one of the best transport networks of the colonial domain, especially the railroad system (although the network focused on interior-seaport linkages rather than fully interconnecting the various parts of the country). British engineers laid out irrigation canals through which millions of acres of land were brought into cultivation. Settlements that had been founded by Britain developed into major cities and bustling ports, led by Bombay (now Mumbai), Calcutta (now Kolkata), and Madras (now Chennai). These three cities are still three of India's largest urban centers, and their cityscapes bear the unmistakable imprint of colonialism. Modern industrialization, too, was brought to India by the British on a limited scale. In education, an effort was made to combine English and Indian traditions; the Westernization of India's elite was supported through the education of numerous Indians in Britain. Modern practices of medicine were also introduced. Moreover, the British administration tried to eliminate features of Indian culture that were deemed undesirable by any standards—such as the burning alive of widows on their husbands' funeral pyres, female infanticide, child marriage, and the caste system. Obviously, the task was far too great to be achieved in barely three generations of colonial rule, but

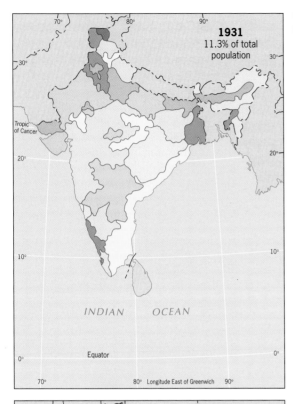

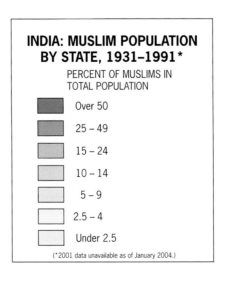

**INDIA: MUSLIM POPULATION
BY STATE, 1931–1991***

PERCENT OF MUSLIMS IN
TOTAL POPULATION

▓	Over 50
▓	25 – 49
▓	15 – 24
░	10 – 14
░	5 – 9
░	2.5 – 4
□	Under 2.5

(*2001 data unavailable as of January 2004.)

FIGURE 8-5

independent India itself has continued these efforts where necessary.

Partition

Even before the British government decided to yield to Indian demands for independence, it was clear that British India would not survive the coming of self-rule as a single political entity. As early as the 1930s, the idea of a separate Pakistan was being promoted by Muslim activists, who circulated pamphlets arguing that British India's Muslims were a nation distinct from the Hindus and that a separate state consisting of Sind, Punjab, Baluchistan, Kashmir, and a portion of Afghanistan should be created from the British South Asian Empire in this area. The first formal demand for such partitioning was made in 1940, and, as later elections proved, the idea had almost universal support among the realm's Muslims.

As the colony moved toward independence, a political crisis developed: India's majority Congress Party would not even consider partition, and the minority Muslims refused to participate in any future unitary government. But partition would not be a simple matter. True, Muslims were in the majority in the western and eastern sectors of British India, but Islamic clusters were scattered throughout the realm (Fig. 8-5). Any new boundaries between Hindus and Muslims to create an Islamic Pakistan and a Hindu India would have to be drawn right through areas where both sides coexisted. People by the millions would be displaced.

Nor were Hindus and Muslims the only people affected by partition. The Punjab area, for example, was home to millions of Sikhs, whose leaders were fiercely anti-Muslim. But a Hindu-Muslim border based only on those two groups would leave the Sikhs in Pakistan. Even before independence day, August 15, 1947, Sikh leaders talked of revolt, and there were some riots. But no one could have foreseen the dreadful killings and mass migrations that followed the creation of the boundary and the formation of independent Pakistan and India. Just how many people felt compelled to participate in the ensuing migrations will never be known; 15 million is the most common estimate. It was human suffering on an incomprehensible scale.

3 Even so large a flow of cross-border **refugees**, however, hardly began to "purify" India of Muslims. After the initial mass exchanges, there still were tens of millions of Muslims in India (Fig. 8-5). Today, the Muslim minority in Hindu-dominated India is about as large as the whole Islamic population of Pakistan, having more than tripled since the late 1940s. It is the world's largest minority, far more than a mere remnant of the days when Islam ruled the realm. This force will play a growing role in the India of the future.

SOUTH ASIA'S POPULATION DILEMMA

South Asia is a poverty-plagued realm, with low average incomes, poorly balanced diets, and poor overall health. Poverty's persistence also is related to *demographic issues*, that is, the size, distribution, and growth of population in a realm, region, country, or province and factors associated **4** with these conditions. The field of **population geography**, the spatial expressions of demography, focuses on such matters. It is useful to refer back to Figure G-9, which displays the continuing concentration of population in South Asia's major river valleys, underscoring the enduring dependence of the majority of the people on these ribbons of life.

Figure G-9 reveals a crucial facet of population **5** geography: global and regional **population distribution**. Using various symbols (in this case, dots representing quantity), geographers represent the demographic situation spatially so that we can observe from a glance at the map the overall location of densely and sparsely peopled areas. But the general population distribution does not tell **6** us much about **population density**, a more specific measure that reports the number of people who reside in a unit area (e.g., a square mile or square kilometer). Column 4 of Table G-1 lists population density by country, per square mile. That average is often cited in official statistics, but without additional data it is not very meaningful. When millions of people live in a river valley and almost nobody lives in deserts or mountains, what does Pakistan's "arithmetic" average of 487 really mean? Even parts of India, populous as it is, are very sparsely populated. A more insightful measure of density is based on the amount of agriculturally productive land. How many people are there per unit area of land suitable for farming or **7** pastoralism? This **physiologic density** measure, listed in column 5 of Table G-1, shows that Sri Lanka and Bangladesh, not India, are the most densely populated countries in the realm. (The very high figure for the tiny islands of the Maldives reflects the virtual absence of agricultural potential there.)

Critical to any assessment of South Asia's social **8** and economic geography is the realm's **rate of natural population increase**, the number of births minus the number of deaths, usually given as a percentage or per thousand in the population. By world standards these rates, for the realm's individual countries, remain high at around 1.8 percent per year. When populations grow this rapidly on already-large bases, economic advancement is impeded. From the rate of annual population increase, we can calculate a useful measure of a country's **9** demographic condition: its **doubling time**, the number of years it would take, at current rates, for a population to grow twice as large. Some Muslim countries have high growth rates and thus low doubling times, reflecting the youthfulness (and restiveness) of their populations. Check column 9 of Table G-1 and compare South Asia's collective and national doubling times with those of other areas of the world.

Dynamics of Population Growth

South Asia is a realm of massive population clusters. Its five mainland countries today contain about 1.4 billion people—more than one-fifth of all humankind. India alone has 1.086 billion inhabitants, second only to China among countries of the world, and is on course to overtake it. Pakistan, with a population of 150 million, and Bangladesh, with 140 million, also rank among the dozen most populous countries of the world.

Not only is South Asia populous, but its population also is growing rapidly at 1.8 percent per year, yielding a doubling time of only 39 years. Pakistan is growing even faster at 2.1 percent, doubling in just 33 years. As a result, economic gains are being overtaken by growing numbers, and although the Green Revolution narrowed the food deficit, hundreds of millions of children do not get balanced meals or adequate calories.

Three-quarters of South Asia's population live in India, so we should consider this global giant in the context of its population geography and its impact on the realm as a whole. Comparing the map of the world economies (Fig. G-10) with the list of population growth rates in Table G-1, we note a clear pattern: the bulk of population growth is occurring in the lower-income economies. In many of the high-income economies, population growth is small, has leveled off, or is even negative. These higher-income economies have gone through the **10** so-called **demographic transition**, a four-stage sequence that took them from high birth rates and high death rates in preindustrial times to very low birth rates and very low death rates today (Fig. 8-6). Stages 2 and 3 in this model constitute the **11** **population explosion**, a hallmark of the twentieth century: death rates in the industrializing and urbanizing countries dropped, but birth rates took longer to decline. In 1900, the world's population was about 1.5 billion; by 2000, it had surpassed 6 billion.

When the British ruled India during the nineteenth century, the country still was in the first stage, with high birth rates and high death rates; the high death rates were caused not only by a high incidence of infant and child mortality but also by famines and epidemics. As Figure 8-6 indicates, the population during Stage 1 does not grow or decline much, but it is not stable. Famines and disease outbreaks kept erasing the gains made during better times. But then India entered the second stage. Birth rates remained high, but death rates declined because medical services improved (soap came into widespread use), food distribution networks became more effective, farm production expanded, and urbanization developed. In the 1920s, India's population still was growing at a rate of only 1.04 percent, but by the 1970s, that rate had shot up to 2.22 percent per year (Fig. 8-7). Note that India gained 28 million people during the 1920s but a staggering 135 million during the 1970s.

Has India entered the third stage, when the death rate begins to level off and birth rates decline substantially, narrowing the gap and slowing the annual increase? The rate of increase suggests it has: from 2.22 percent during the 1970s, it dropped to 2.11 percent in the 1980s, 1.88 percent in the 1990s, and a projected 1.61 percent during our current decade that began in 2000 (Fig. 8-7). But India has another problem. During its population explosion, its numbers grew so large that even a declining rate of natural increase continues to add ever greater numbers to its total. In Figure 8-7, we see that while the decadal rate of increase dropped from 2.22 to 1.88 between 1970 and 2000, the millions added grew from 135 in the 1970s to 161 in the 1980s to 175 during the 1990s—taking the total past 1 billion in 1999. If India has indeed entered the third stage of the demographic transition, it will not feel its effects for some time.

Some population geographers theorize that all countries' populations will eventually stabilize at some level, just as Europe's did. Certain governments, notably China's, have instituted regulations to limit family size, but this policy is more easily implemented by dictatorships than democracies. And even if such stabilization were just one doubling time away in India, the country still would

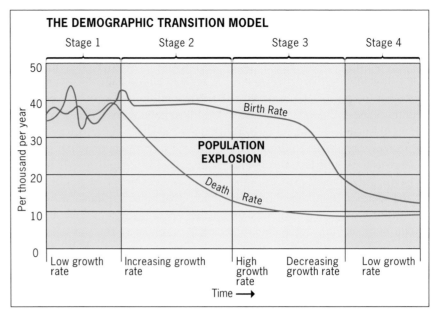

FIGURE 8-6

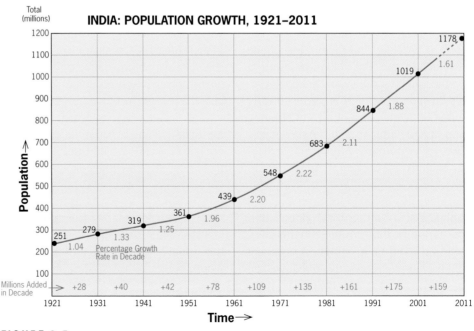

FIGURE 8-7

have an astronomical total of more than 2 billion residents.

Geography of Demography

Statistics for a country as large as India tend to lose their usefulness unless they are put in geographic context. In its demographic as in so many of its other aspects, there is not just one India but several regionally different and distinct Indias. Figure 8-8 takes population growth down to the State level and provides a comparison between the census periods of 1981–1991 and 1991–2001. (The names of India's States are shown in Fig. 8-11.)

During the period from 1981 to 1991, the highest growth rates were recorded in the States of the northeast, and only eight States had growth rates below 2 percent. But between 1991 and 2001, no fewer than ten States, had growth rates below 2 percent. Comparing these two maps tells us that little has changed in India's heartland, where the populous States of West Bengal and Madhya Pradesh show slight decreases but Uttar Pradesh and Bihar display similarly slight increases. The most important reductions in rates of natural increase are recorded in parts of the northeast, the east, and the south. In the tip of the peninsula, Kerala and neighboring Tamil Nadu have growth rates lower than that of the region's leading country, Sri Lanka.

One reason for these contrasts lies in India's federal system: individual States pursued population-control policies to reduce population growth, including mass sterilizations. But another, as we note in more detail in the regional section, relates to spatial differences in India's development. As in the world at large, India's economically better-off States have lower rates of population growth.

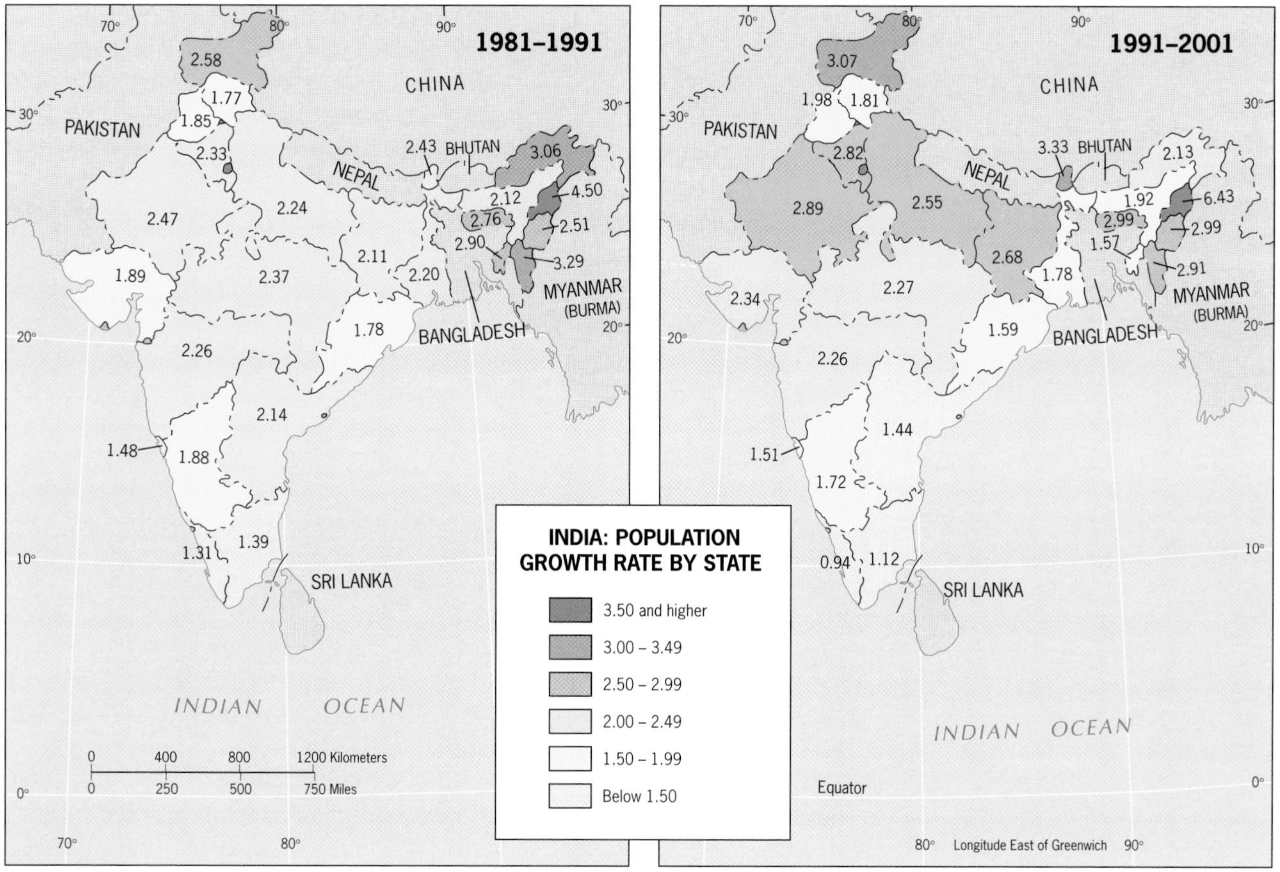

India is a Hindu-dominated, officially secular country in which population policies can be debated and implemented. Pakistan, on the other hand, is a strictly Islamic state in which no such options exist. Population-control policies are regarded as incompatible with Islamic tenets, and Pakistan remains one of the world's fast-growing nations with an annual rate of natural increase of 2.1 percent.

FIGURE 8-8

A REALM OF POVERTY

As we have noted, South Asia at the opening of the twenty-first century accounts for more than one-fifth of the world's population. This single realm, however, also contains two-thirds of the poorest inhabitants on Earth. Literacy rates in South Asia are among the lowest in the world. Nearly half of the people in this realm earn less than the equivalent of one U.S. dollar per day. It is estimated that half the children in South Asia are malnourished and underweight, most of them girls. South Asia is often called the most deprived realm in the world.

A combination of geographic factors underlies this tragic picture. With 22 percent of the world's population but just 3 percent of its land area, South Asia cannot raise living standards for its hundreds of millions of subsistence farmers. Governmental policies contribute to the problem: while East and Southeast Asia forged ahead by looking outward, encouraging exports and foreign investment, and spending heavily on literacy and technical education, health care, and land reform, South Asian governments tended to adopt bureaucratic controls and state planning. Cultural traditions also play their role. Resistance to change and reluctance by the privileged to open doors of opportunity to the less advantaged inhibit economic advancement for all.

Here is a grim reality: we will note repeatedly in this chapter that South Asia is self-sufficient in food and that India produces enough grain to not only be able to feed its own population but to fill the needs of its neighbors. In early 2003, official statistics showed that India had a stored wheat surplus of more than 50 million metric tons and a rice surplus of 5 million tons, enough to feed the entire realm's population for more than a year, with another good harvest in sight. But this apparent triumph of the Green Revolution did not end malnourishment in India. Despite growing annual surpluses, India's children remain among the most malnourished in the world: at the turn of the millennium, 47 percent were reported to be either severely malnourished or moderately malnourished. While food rots in government warehouses in Punjab, children die in villages in the neighboring State of Rajasthan.

How can a government hoard food in huge quantities while letting the poor die of starvation? The answer is economics—and politics. India's government responds to the pressures of powerful farm lobbies, raising the price of grain and then buying it from commercial farmers at higher prices for storage and redistribution thereby in-creasing the market cost for poor villagers who cannot afford to buy it. While millions of villagers go hungry, the government is also under pressure from international organizations, which lend India money, to limit food subsidies to "consumers." In addition, India's nationwide subsidized food distribution system, set up four decades ago, suffers from inefficiency, corruption, and massive theft. As a result, poverty and government mismanagement combine to condemn tens of millions of India's rural poor to malnutrition and even starvation at a time when official statistics proclaim that the country produces enough food not only to feed itself but even to export grain to hungry neighbors. We will revisit this issue later in the chapter.

A REALM OF FLASHPOINTS

As Table G-1 reminds us, this is a realm of few countries. In the two realms we discussed in Chapters 6 and 7, there are 75 countries: 45 in Subsaharan Africa and 30 in North Africa/Southwest Asia. In South Asia, the number is just seven—but South Asia's population is much larger than that of all those 75 countries combined. Parts of South Asia are among the most densely peopled rural areas on the planet, and the cultural mosaic here is complex and intricate. Moreover, the realm is bordered by some fractious neighbors, from Afghanistan to China. Thus, while South Asia may be internationally less fragmented than Africa or other realms of Asia, like other parts of the world it suffers from persistent geopolitical trouble spots. Some of these flashpoints have more than regional significance because they involve countries with nuclear weapons. In 2004, concern focused on the following issues:

1. **Kashmir.** As we will note in the regional section, this mountainous and remote territory was not divided between India and Pakistan when their boundary was delimited under British auspices in 1947. Repeated wars (all *before* the two countries acquired nuclear capabilities) have led only to an armistice line, not a settlement.

2. **Arunachal Pradesh.** China claims virtually all of this northeastern State of India and publishes maps showing this area as part of the People's Republic. In 2003, the heads of state of India and China met on matters of mutual interest, but the Chinese leadership did not yield on this potentially explosive issue.

3. **The Tribal Areas.** Along the border between Pakistan and Afghanistan, on the Pakistani side, lies a comparatively small area marked "Tribal Areas" that is not formally a part of any of the country's provinces. This area, traditionally under the control of mullahs and tribal chiefs, is a hilly warren of villages and tracks where government authority fails. Long a hideout for rebels and refugees, it has taken on added significance during the War on Terror when it was believed to be the hiding place of Usama bin Laden, Mullah Omar, and others in the al-Qaeda–Taliban alliance.

4. **Nepal.** A Maoist-communist insurgency in the western hills of the mountain kingdom has caused political instability, stalled economic growth, and undermined social development programs.

5. **Sri Lanka.** Civil war has raged almost continuously in this island republic since 1983 at a cost of nearly 70,000 lives and immeasurable social and economic damage. The secessionist Liberation Tigers of Tamil Eelam (LTTE) fought to establish an independent state for Hindu Tamils in this Sinhalese Buddhist country; negotiations begun under Norwegian auspices in 2002 continued in 2004.

Each of these issues has implications beyond South Asia itself, ranging from the risk of a first nuclear exchange between India and Pakistan to the disputed designation of the LTTE as a terrorist organization in the current climate marking the War on Terror. Here again is evidence that local problems have global consequences.

REGIONS OF THE REALM

PAKISTAN: ON SOUTH ASIA'S WESTERN FLANK

If India is the dominant entity in South Asia, why focus first on Pakistan? There are several reasons, both historic and geographic. Here lay South Asia's earliest urban civilizations, whose innovations radiated into the great peninsula. Here, too, lies South Asia's Muslim frontier, contiguous to the great Islamic realm to the west and irrevocably linked to the enormous Muslim minority to its east. Pakistan's cultural landscapes bear witness to its transitional location. Teeming, disorderly Karachi is the typical South Asian city; as in India, the largest urban center lies on the coast. Historic, architecturally Islamic Lahore is reminiscent of the scholarly centers of Muslim Southwest Asia. In Pakistan's eastern borderland, the postcolonial boundary divides a Punjab that stretches beyond the horizon on both sides of the line, a contiguous cultural landscape of villages, wheatfields, and irrigation ditches. In the northwest, Pakistan resembles Afghanistan in its huge migrant populations and its mountainous frontier. And in the far north, Pakistan and India are locked in a deadly conflict over Jammu and Kashmir. The western flank is South Asia's most critical region.

If, as is so often said, Egypt is the gift of the Nile, then Pakistan is the gift of the Indus. The Indus River and its principal tributary, the Sutlej, nourish the ribbons of life that form the heart of this populous country (Fig. 8-9). Territorially, Pakistan is not large by Asian standards; its area is about the same as that of Texas plus Louisiana. But Pakistan's population of 149.6 million makes it one of the world's ten most populous states. Among Muslim countries (officially, it is known as the Islamic Republic of Pakistan) only Southeast Asia's Indonesia is larger, but Indonesia's Islam is much less pervasive than Pakistan's.

Upon independence in 1947, Pakistan consisted not only of the territory we know as Pakistan today, which was known as West Pakistan, but also of "East Pakistan"—present-day Bangladesh. That union was based on their shared adherence to Islam, but it did not last long. A political crisis in 1971 led to conflict and East Pakistan's declaration of independence as the People's Republic of Bangladesh.

When Pakistan became a sovereign state following the partition of British India in 1947, its capital was Karachi on the south coast, near the western end of the Indus Delta. As the map shows, the present capital is Islamabad, near the larger city of Rawalpindi in the north, not far from Kashmir. By moving the capital from the "safe" coast to the embattled interior and by placing it on the doorstep of contested territory, Pakistan announced its intent to stake a claim to its northern frontiers. And by naming the city Islamabad, Pakistan proclaimed its Muslim foundation, here in the face of the Hindu challenge. This politico-geographical use of a national capital can be assertive, and Islamabad exemplifies **12** the principle of the **forward capital**.

The Kashmir Issue

The contested territory, of course, was (and is) the northern area of Jammu and Kashmir (Kashmir for short). Mountainous, remote, and unassigned when independence came in 1947, Kashmir was 75 percent Muslim but ruled by a Hindu elite. The Maharajah wanted an autonomous state that would remain outside both India and Pakistan, but a Muslim uprising put an end to that notion. Indian and Pakistani armies entered the fray, but their stalemate only postponed renewed wars in 1965 and 1971. The result was the armistice *Line of Control* mapped in Figure 8-10.

Pakistan is firmly in control of its sector of Kashmir, and the question now centers on the future of Indian-held Kashmir, part of which has been taken by China (Fig. 8-10). Even here, Muslims are in the majority, heavily concentrated in the Vale of Kashmir; the area named Ladakh on the map is strongly Buddhist; and the center and southeast are mostly minority Hindu. A referendum would undoubtedly go in Pakistan's favor, but secular India cannot abandon its minorities to the kind of Islamic dominance prevailing there. Skirmishes and acts of terrorism continue in the shadow of Islamabad and under a cloud of potential nuclear catastrophe.

Forging Centripetal Forces

At independence, Pakistan had a bounded national territory, a capital, a cultural core, and a population—but it had few centripetal forces to bind state and nation, especially while East Pakistan was still a part of it. The disparate regions of Pakistan shared the Islamic faith and an aversion for Hindu India, but little else. Karachi and the coastal south, the desert of Baluchistan, the city of Lahore and the Punjab, the rugged northwest along Afghanistan's border, and the mountainous far north are worlds apart, and a Pakistani nationalism to match that of India at independence did not exist. Successive Pakistani governments, civilian as well as military, turned to Islam to provide the common bond that history and geography had denied the nation. In the process, Pakistan became one of the world's most theocratic states; its common law, based on the English model, was gradually transformed into a Quranic (Koranic) system with Islamic Sharia courts and associated punishments.

But even Islam itself is not unified in restive Pakistan. About 80 percent of the people are Sunni Muslims, and the Shiah minority numbers about 16 percent. Sunni fanatics intermittently attack Shi'ites, leading to retaliation and creating grounds for subsequent revenge.

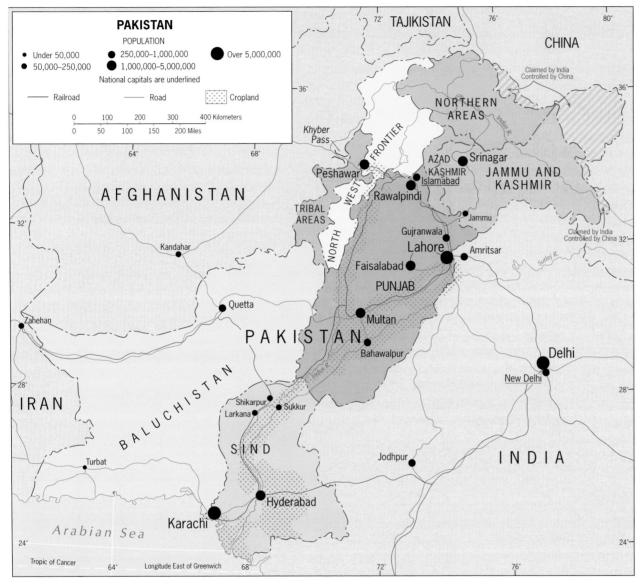

FIGURE 8-9

Despite the Islamization of Pakistan's plural society, it remains a strongly regionalized country in which Urdu is the official language and English is still the *lingua franca* of the elite. Yet several other major languages prevail in diverse parts, and lifeways vary from nomadism in Baluchistan to irrigated farming in the Punjab to pastoralism in the northern highlands.

The Provinces

As Figure 8-9 shows, Pakistan is administratively divided into four provinces: Punjab, Sind, North West Frontier, and Baluchistan. Pakistan was founded as a federal state (although it has been ruled by authoritarian and military governments plagued by corruption and inefficiency), and the relations between these provinces are important to the cohesion of the country. These relations have been difficult.

In large part this is because the *Punjab* is disproportionately dominant in Pakistan: it is the country's core area, home to about 60 percent of the entire population, contains the capital (Islamabad), the primate city (Lahore, 2000 years old, a great Muslim center with magnificent architecture), and leads the nation in almost every economic category. To the south, *Sind* lies centered on the chaotic port city of Karachi, and here the rice and wheat fields in the lower Indus Basin form the breadbasket of Pakistan. Commercially, cotton is king, supporting major textile industries in the province's cities and towns. To the west of Sind lies desert *Baluchistan*, land of ancient caravan routes still operating and home to a sizeable Shi'ite minority, reflecting its proximity to Iran. Recent oil and gas discoveries may transform this remote province in the near future. The *North West Frontier*, the aptly-named, mountainous province closest to the core area of Afghanistan anchored by Peshawar and adjoined by the unruly Tribal Areas, has repeatedly taken the brunt of events occurring across the border. When Soviet armed forces tried to install a secular regime in Kabol during the 1980s, several million Pushtun refugees streamed out of Afghanistan, settled in refugee camps in the North West Frontier, and became a major political (and Islamic-cultural) force here. Many of these refugees wanted political rights in Pakistan and campaigned to bring strict Islamic

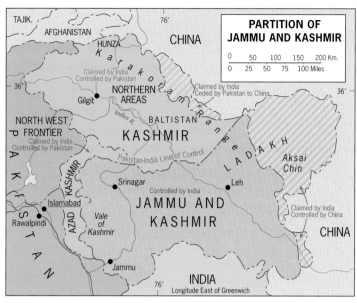

PARTITION OF JAMMU AND KASHMIR

0   50   100   150   200 Km.

0   25   50   75   100 Miles

FIGURE 8-10

government to the North West Frontier Province, receiving cross-border support from their kinspeople **13** still in Afghanistan. This example of **irredentism** posed serious problems for Pakistan, and it worsened when the repressive Taliban regime took power during the mid-1990s and many more refugees crossed the border. All this turmoil made the North West Frontier Province a haven for radicals, and today the provincial government is revivalist Islamic, challenging the authority of the federal government in Islamabad.

Pakistan's Prospects

Until September 2001, Pakistan remained a typical low-income country with a troubled economy, huge debts, worrisome social indicators (see Table G-1), and unstable government, an army general having recently ousted his civilian adversary and taken control of the state. As the only country with day-to-day relations with the Taliban regime in neighboring Afghanistan, Pakistan faced another round of disputes over military versus civilian rule,

Islamic versus federal courts, education policies, and other issues. But in the aftermath of the 9/11 attacks and the formal start of the War on Terror, Pakistan's ruler, General Pervez Musharraf, had to make a choice between support for the United States and the West or neutrality. It was a difficult decision: support for the United States would energize Islamic militants in his country, while neutrality would bring adversarial relations with a long-term Pakistani ally.

When Pakistan joined the West in the War on Terror and U.S. troops, investigators, journalists, and others arrived, the benefits were financial and military and the costs were social and political. Pakistan moved to close the revivalist *madrassas* (Islamic religious schools) that had produced many of the Taliban extremists, but, as we noted, militants took over the regional government of the North West Frontier Province. Acts of defiance and terrorism occurred from Karachi to Peshawar, and it is virtually certain that the al-Qaeda and Taliban leadership escaped to, and were given refuge in, the Tribal Areas. Political tensions in Pakistan

escalated, but the quick collapse of the Taliban regime helped defuse the risk.

Still, the further radicalization of Pakistani society is likely. United by Sunni Islam but by little else, facing domestic and regional strife from Karachi to Kashmir, allied with a Western power not known for long-term commitments, located on the threshold of Turkestan and on the border of turbulent Afghanistan, Pakistan is a state that direly lacks the cohesion it needs.

▶ INDIA: FEDERAL AND FREE

Nearly three-quarters of the great land triangle of South Asia is occupied by a single country—India, the world's most populous democracy and, in terms of human numbers, the world's largest federation.

That India has endured as a unified country is a politico-geographical miracle. India is a cultural mosaic of immense ethnic, religious, linguistic, and economic diversity and contrast; it is a state of many nations. The period of British colonialism gave India the underpinnings of unity: a single capital, an interregional transport network, a *lingua franca*, a civil service. Upon independence in 1947, India adopted a federal system of government, giving regions and peoples some autonomy and identity, and allowing others to aspire to such status. Unlike Africa, where federal systems failed and where military dictatorships replaced them, India remained essentially democratic and retained a federal framework in which States have considerable local authority.

This political, democratic success has been achieved despite the presence of powerful centrifugal forces in this vast, culturally diverse country. Relations between the Hindu majority and the enormous Muslim minority, better in some States than in others, have at times threatened to destabilize the entire federation. Local rebellions, demands by some minorities for their own States, frontier wars, even involvement in a foreign but nearby civil war (in Sri Lanka) have buffeted the system—which has bent but not collapsed. India has succeeded where others in the postcolonial world have failed.

This success has not been matched in the field of economics, however. After more than 50 years of independence India remains a very poor country, and not all of this can be blamed on the colonial period or on population growth, although overpopulation remains a strong impediment to improvement of living standards. Much of it results from poor and inconsistent economic planning, too much state ownership of inefficient industries, excessive government control over economic activities, bureaucratic suppression of initiative, corruption, and restraints against foreign investment. As we shall presently see, a few bright spots in some of India's States contrast sharply to the overwhelming poverty of hundreds of millions.

States and Peoples

The map of India's political geography shows a federation of 28 States, 6 Union Territories (UTs), and 1 National Capital Territory (NCT) (Fig. 8-11). The federal government retains direct authority over the UTs, all of which are small in both territory and population. The NCT, however, includes Delhi and the capital, New Delhi, and has over 15 million inhabitants.

The political spatial organization shown in Figure 8-11 is mainly the product of India's restructuring following independence from Britain. Its State boundaries reflect the broad outlines of the country's cultural mosaic: overall the system recognizes languages, religions, and cultural traditions. Indians speak 14 major and numerous minor languages, and while Hindi is the official language (and English is the *lingua franca*), it is by no means universal. The map is the product of endless compromise—endless because demands for modifications of it continue to this day; as recently as late 2000, the federal government authorized the creation of three new States. In the northeast lie very small States established to protect the local tradi-

tions of small populations; minority groups in the larger States ask why they should not receive similar recognition.

With only 28 States for a national population of 1.086 billion, several of India's States contain more people than many countries of the world. As Figure 8-11 shows, the (territorially) largest States lie in the heart of the country and on the great southward-pointing peninsula. Uttar Pradesh (just over 166 million according to the 2001 census) and Bihar (about 83 million) constitute much of the Ganges River Basin and are the core area of modern India. Maharashtra (almost 97 million), anchored by the great coastal megacity of Bombay (renamed Mumbai in 1996), also has a population larger than that of most countries. West Bengal, the State that adjoins Bangladesh, has more than 80 million residents, 14.3 million of whom live in its urban focus, Calcutta (renamed Kolkata in 2000).

These are staggering numbers, and they do not decline much toward the south. Southern India consists of four States linked by a discrete history and by their distinct Dravidian languages. Facing the Bay of Bengal are Andhra Pradesh (76 million) and Tamil Nadu (62 million), both part of the hinterland of the city of Madras (renamed Chennai in 1997) and located on the coast near their joint border. Facing the Arabian Sea are Karnataka (53 million) and Kerala (32 million). Kerala, often at odds with the federal government in New Delhi, has long had the highest literacy rate in India and one of the lowest rates of population growth owing to strong local government and strictly enforced policies. "It's a matter of geography," explained a teacher in the Kerala city of Cochin. "We are here about as far away as you can get from the capital, and we make our own rules."

As Figure 8-11 shows, India's smaller States lie mainly in the northeast, on the far side of Bangladesh, and in the northwest, toward Jammu and Kashmir. North of Delhi, India is flanked by China and Pakistan, and physical as well as cultural landscapes change from the flatlands of the Ganges to

the hills and mountains of spurs of the Himalayas. In the State of Himachal Pradesh, forests cover the hillslopes and relief reduces living space; only 6 million people live here, many in small, comparatively isolated clusters. Before independence and political consolidation, the colonial government called this area the "Hill States."

But the map becomes even more complex in the distant northeast, beyond the narrow corridor between Bhutan and Bangladesh. The dominant State here is Assam (27 million), famed for its tea plantations and important because its oil and gas production amounts to more than 40 percent of India's total.

In the Brahmaputra Valley, Assam resembles the India of the Ganges. But in almost all directions from Assam, things change. To the north, in sparsely populated Arunachal Pradesh (1.1 million), we are in the Himalayan offshoots again. To the east, in Nagaland (2 million), Manipur (2.4 million), and Mizoram (0.9 million), lie the forested and terraced hillslopes that separate India from Myanmar (Burma). This is an area of numerous ethnic groups (more than a dozen in Nagaland alone) and of frequent rebellion against Delhi's government. And to the south, the States of Meghalaya (2.3 million) and Tripura (3.2 million), hilly and still wooded, border the teeming floodplains of Bangladesh. Here in the country's northeast, where peoples are always restive and where population growth is still soaring, India faces one of its strongest regional challenges.

Here, too, India faces a challenge from its powerful neighbor China. As Figure 8-11 shows, China's version of its southern border with India lies deep within the Northeast, even coinciding with the Assam boundary part of the way. This dispute has its origins in the Simla Conference of 1913–1914, when the British negotiated a treaty with Tibet that defined their boundary in accordance with the terms laid down on the map by their chief negotiator, Sir Henry McMahon. Essentially, this *McMahon Line*, as it came to be known, ran along the most prominent crestline of the Himalayas. However, China's

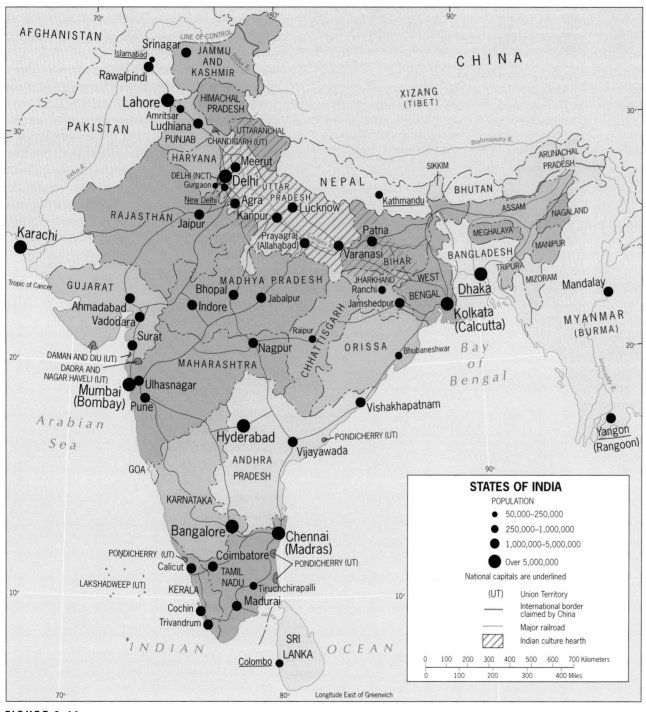

FIGURE 8-11

representatives at the conference refused to sign the agreement, arguing that Tibet was part of China and had no power to enter into treaties with foreign governments. In 1962, Chinese armed forces crossed the border and occupied Indian territory, but then withdrew. Today the matter remains unresolved.

India's Changing Map

After independence, the Indian government began phasing out the privileged "Princely States" which the British had protected during the colonial period. Next, the government reorganized the country on the basis of its major regional languages (see Fig. 8-4). Hindi, spoken by more than one-third of the population, was designated the country's official language, but 13 other major languages also were given national status by the Indian constitution, including the four Dravidian languages of the south. English, it was anticipated, would become India's common language, its *lingua franca* at government, administrative, and business levels. Indeed, English not only remained the language of national administration but also became the chief medium of commerce in growing urban India. English was the key to better jobs, financial success, and personal advancement, and the language constituted a common ground in higher education.

The newly devised framework based on the major regional languages, however, proved to be unsatisfactory to many communities in India. In the first place, many more

"More than a half-century after the end of British rule, the centers of India's great cities continue to be dominated by the Victorian-Gothic buildings the colonists constructed here. Here is evidence of a previous era of globalization, when European imprints transformed urban landscapes. Walking the streets of Mumbai (the British called it Bombay) you can turn a corner and be forgiven for mistaking the scene for London, double-deckered buses and all. One of the British planners' major achievements was the construction of a nationwide railroad system, and railway stations were given prominence in the urban architecture. I had walked up Naoroji Road, having learned to dodge the wild traffic around the circles in the Fort area, and watched the throngs passing through Victoria Station. Inside, the facility is badly worn, but the trains continue to run, bulging with passengers hanging out of doors and windows."

languages are in use than the 14 that had been officially recognized. Demands for additional States soon arose. As early as 1960, the State of Bombay was divided into two language-based States, Gujarat and Maharashtra.

This devolutionary pressure has continued throughout India's existence as an independent country. In 2000, three new States were recognized: Jharkhand, carved from southern Bihar State on behalf of 18 poverty-stricken districts there; Chhattisgarh, where tribal peoples had been agitating since the 1930s for separation from the State of Madhya Pradesh; and Uttaranchal, which split from India's most populous, Ganges Basin, core-area State of Uttar Pradesh on the basis of its highland character and lifeways (Fig. 8-11).

For many years India has faced quite a different set of cultural-geographic problems in its northeast, where numerous ethnic groups occupy their own niches in a varied, forest-clad topography. The Naga, a cluster of peoples whose domain had been incorporated into Assam State, rebelled soon after India's independence. A protracted war brought federal troops into the area; after a truce and lengthy negotiations, Nagaland was proclaimed a State in 1961. This led the way for other politico-geographical changes in India's problematic northeastern wing.

The Sikhs

A further dilemma involves India's Sikh population. The Sikhs (the word means "disciples") adhere to a religion that was created about five centuries ago to unite warring Hindus and Muslims into a single faith. This faith's principles rejected negative aspects of Hinduism and Islam, and it gained millions of followers in the Punjab and adjacent areas. During the colonial period, many Sikhs supported British administration of India, and by doing so they

won the respect and trust of the British, who employed tens of thousands of Sikhs as soldiers and policemen. By 1947, there was a large Sikh middle class in the Punjab. When independence came, many left their rural homes and moved to the cities to enter urban professions. Today, they still exert a strong influence over Indian affairs, far in excess of the less than 2 percent of the population (about 19 million) they constitute.

After independence, the Sikhs demanded that the original Indian State of Panjab (Punjab) be divided into a Sikh-dominated northwest and a Hindu-majority southeast. The government agreed, so that Punjab as now constituted (Fig. 8-11) is India's Sikh stronghold, whereas neighboring Haryana State is mainly Hindu.

The Muslims

These ethnic, cultural, and regional problems are but a sample of the stresses on India's federal framework. There is no Muslim State in India, but India, according to its 2001 census, has about 150 million Muslims within its borders—the largest cultural minority in the world. As Figure 8-5 shows, the percentage of Muslims is highest in remote Jammu and Kashmir, but it also is substantial in such widely dispersed States as Kerala, Assam, and Uttar Pradesh. Moreover, the Muslim population today (approximately 14 percent) constitutes a larger percentage than it did after partition (9.9 percent). This Islamic minority also ranks among the most rapidly growing sectors of India's population, and is strongly urbanized as well—nearly one-third of the population of India's largest city, Mumbai (Bombay), is Muslim.

Relations between the Hindu majority and Muslim minority are complex. What makes the news is conflict—for example, in 2002 when Muslims attacked a train carrying Hindus to a contested holy site in Ayodhya, killing dozens, which was followed by retaliation in several towns (but not others) in Gujarat. What does not make the news is that a Muslim population surpassing 150 million lives and participates in the kind of democracy that is all but unknown in the Muslim world itself.

Centrifugal Forces: From India to Hindustan?

In Chapter 1 we introduced the concept of centrifugal and centripetal forces affecting the fabric of the state. No country in the world exhibits greater cultural diversity than India, and variety in India comes on a scale unmatched anywhere else on Earth. Such diversity spells strong centrifugal forces, although, as we will see, India also has powerful consolidating bonds.

Among the centrifugal forces, Hinduism's stratification of society into castes remains pervasive. Under Hindu dogma, *castes* are fixed layers in society whose ranks are based on ancestries, family ties, **14** and occupations. The **caste system** may have its origins in the early social divisions into priests and warriors, merchants and farmers, craftspeople and servants; it may also have a racial basis, for the Sanskrit term for caste is color. Over the centuries, its complexity grew until India had thousands of castes, some with a few hundred members, others containing millions. Thus, in city as well as in village, communities were segregated according to caste, ranging from the highest (priests, princes) to the lowest (the untouchables). The term *untouchable* has such negative connotations that some scholars object to its use. Alternatives include *dalits* (oppressed), the common term in Maharashtra State but coming into general use; *harijans* (children of God), which was Gandhi's designation, still widely used in the State of Bihar; and *Scheduled Castes*, the official government label.

A person was born into a caste based on his or her actions in a previous existence. Hence, it would not be appropriate to counter such ordained caste assignments by permitting movement (or even contact) from a lower caste to a higher one. Persons of a particular caste could perform only

FROM THE FIELDNOTES

"The streets of India's cities often seem to be one continuous market, with people doing business in open storefronts, against building walls, or simply on the sidewalk. Here the formal and informal sectors of India's economy intermingle. I walked this way in Delhi every morning, and the store selling mattresses and pillows was always open. But the women in the foreground, selling handkerchiefs and other small items from a portable iron rack, were sometimes here, and sometimes not; one time I ran into them about a half-mile down the road. As I learned one tumultuous day, every time the government tries to exercise some control over the street hawkers and sidewalk sellers, massive opposition results and chaos can ensue. A few days later, everything is as it was. Change comes very slowly here."

certain jobs, wear only certain clothes, worship only in prescribed ways at particular places. They or their children could not eat, play, or even walk with people of a higher social status. The untouchables occupying the lowest tier were the most debased, wretched members of this rigidly structured social system. Although the British ended the worst excesses of the caste system, and postcolonial Indian leaders—including Mohandas (Mahatma) Gandhi (the great spiritual leader who sparked the independence movement) and Jawa-

harlal Nehru (the first prime minister)—worked to modify it, a few decades cannot erase centuries of class consciousness. In traditional India, caste provided stability and continuity; in modernizing India, it constitutes an often painful and difficult legacy.

Today we can discern a geography of caste—a degree of spatial variation in its severity. Cultural geographers estimate that about 15 percent of all Indians are of lower caste, about 40 percent of backward caste (one important rank above the lower caste), and some 18 percent of upper caste, at the top of which are the Brahmans, men in the priesthood. (The caste system does not extend to the Muslims, Sikhs, and other non-Hindus in India, which is why these percentages do not total 100.) The colonial government and successive Indian governments have tried to help the lowest castes. This effort has had more effect in the urban than in the rural areas of India. In the isolated villages of the countryside, the untouchables often are made to sit on the floor of their classroom (if they go to school at all); they are not allowed to draw water from the village well because they might pollute it; and they must take off their shoes, if they wear any, when they pass higher-caste houses. But in the cities, untouchables have reserved for them places in the schools, a fixed percentage of State and federal government jobs, and a quota of seats in national and State legislatures. Gandhi, who took a special interest in the fate of the untouchables in Indian society, accomplished much of this reform.

The caste system remains a powerful centrifugal force, not only because it fragments society but also because efforts to weaken it often result in further division. Gandhi himself was killed, only a few months after independence, by a Hindu fanatic who opposed his work for the least fortunate in Indian society. But progress is being made, and while efforts to help the poorest are not always popular among the better-off, the future of India depends on it.

Hinduvta

Another growing centrifugal force in India has to do with a concept known as *Hindutva* or Hinduness—a desire to remake India as a society in which Hindu principles prevail. This concept has become the guiding agenda for a political party that became a powerful component of the federal government, and it is variously expressed as Hindu nationalism, Hindu patriotism, and Hindu heritage. This naturally worries Muslims and other minorities, but it also concerns those who understand that India's secular-

ism, its separation of religion and state, are indispensable to the survival of its democracy. *Hindutva* enthusiasts want to impose a Hindu curriculum on schools, change the flexible family law in ways that would make it unacceptable to Muslims, inhibit the activities of non-Hindu religious proselytizers, and forge an India in which non-Hindus are essentially outsiders. Moderate Hindus and non-Hindus in India oppose such notions, which are as divisive as any India has faced. They nevertheless acknowledged the appeal of the Bharatiya Janata Party (BJP), which swept to power on a Hindu-nationalist platform, and realized that only the constraints of a coalition government kept the party from pursuing its more extreme goals.

The radicalization of Hinduism and the infusion of Hindu nationalism into federal politics loom today as twin threats to India's unity. The BJP and other Hindu-nationalist parties are polarizing electorates at State as well as federal levels; the recent spate of name changes on India's map is one manifestation of this polarization. If Indian politics were to fragment along religious lines, the miracle of Indian unity would be at risk.

Centripetal Forces

In the face of all these divisive forces, what bonds have kept India unified for so long? Without question, the dominant binding force in India is the cultural strength of Hinduism, its sacred writings, holy rivers, and influence over Indian life. For most Indians, Hinduism is a way of life as much as it is a faith, and its diffusion over virtually the entire country (regardless of the Muslim, Sikh, and Christian minorities) brings with it a national coherence that constitutes a powerful antidote to regional divisiveness. Over the long term, however, the key ingredients of this Hinduism have been its gentility and introspection, radical outbursts notwithstanding.

Another centripetal force lies in India's democratic institutions. In a country as culturally diverse and as populous as India, reliance on democratic institutions has been a birthright ever since indepen-

dence, and democracy's survival—raucous, often corrupt, always free—has been a crucial unifier.

Furthermore, communications are better in much of India than in many other countries in the global periphery, and the continuous circulation of people, ideas, and goods helps bind the disparate state together. Before independence, opposition to British rule was a shared philosophy, a strong centripetal force. After independence, the preservation of the union was a common objective, and national planning made this possible.

India's capacity for accommodating major changes and its flexibility in the face of regional and local demands are also a centripetal force. Boundaries have been shifted; internal political entities have been created, relocated, or otherwise modified; and secessionist demands have been handled with a mixture of federal power and cooperative negotiation. Indians in South Asia have accomplished what Europeans in former Yugoslavia could not, and India's history of success is itself a centripetal force.

Still another centripetal force in India is education. Although literacy rates remain comparatively low and reflect women's disadvantage (the 2001 census reported the national rate is 65.4 percent while the rate for females is 54.3 percent), these numbers are far higher than those of neighboring countries and reflect Indians' determination to avail themselves of every educational opportunity, another colonial legacy. Private institutes teaching English abound in the cities, and when service-outsourcing opportunities arose on the international scene, India had the educated class to seize them.

Finally, no discussion of India's binding forces would be complete without mentioning the country's strong leadership. Gandhi, Nehru, and their successors did much to unify India by the strength of their compelling personalities. For many years, leadership was a family affair: Nehru's daughter, Indira Gandhi, twice took decisive control (in 1966 and 1980) after weak governments, and her son, Rajiv Gandhi (who, in 1991, like his mother seven years earlier, also was assassinated), was prime minister in the late 1980s. Since then, political leadership of India has been less dynastic—and also less cohesive.

Urbanization

India is famous for its great and teeming cities, but India is not yet an urbanized society. Only 28 percent of the population lived in cities and towns in 2004—but in terms of sheer numbers, that 28 percent amounts to over 300 million people, more than the entire population of the United States.

And India's rate of urbanization is on the upswing. People by the hundreds of thousands are arriving in the already-overcrowded cities, swelling urban India by about 5 percent annually, almost three times as fast as the overall population growth. Not only do the cities attract as they do everywhere; many villagers are driven off the land by the desperate conditions in the countryside. As villagers manage to establish themselves in Mumbai or Kolkata or Chennai, they help their relatives and friends to join them in squatter settlements that often are populated by newcomers from the same area, bringing their language and

customs with them and cushioning the stress of the move.

As a result, India's cities display staggering social contrasts. Squatter shacks without any amenities at all crowd against the walls of modern highrise apartments and condominiums. Hundreds of thousands of homeless roam the streets and sleep in parks, under bridges, on sidewalks. As crowding intensifies, social stresses multiply. Disorder never seems far from the surface; sporadic rioting, often attributable to rootless urban youths unable to find employment, has become commonplace in India's cities.

India's modern urbanization has its roots in the colonial period, when the British selected Calcutta (Kolkata), Bombay (Mumbai), and Madras (Chennai) as regional trading centers and fortified ports. Madras was fortified as early as 1640; Bombay (1664) had the situational advantage of being the closest of all Indian ports to Britain; and Calcutta (1690) lay on the margin of India's largest population cluster and had the most productive hinterland, to which the Ganges Delta's countless channels connected it. This natural transport network made Calcutta an ideal colonial headquarters, but the population of Bengal was often rebellious. In 1912 the British moved their colonial government from Calcutta to the safer interior city of New Delhi, built adjacent to the old Mogul capital of Delhi.

Figure 8-11 displays the distribution of major urban centers in India. Except for Delhi-New Delhi, the largest cities have coastal locations: Kolkata dominates the east, Mumbai the west, and Chennai the south. But urbanization also has expanded in the interior, notably in the core area. The surface interconnections among India's cities remain inadequate (notably the road network), but an Indian urban system is emerging.

FROM THE FIELDNOTES

"Searing social contrasts abound in India's overcrowded cities. Even in Mumbai (Bombay), India's most prosperous large city, hundreds of thousands of people live like this, in the shadow of modern apartment buildings. Within seconds we were surrounded by a crowd of people asking for help of any kind, their ages ranging from the very young to the very old. Somehow this scene was more troubling here in well-off Mumbai than in Kolkata (Calcutta) or Chennai (Madras), but typified India's urban problems everywhere."

Economic Geography

If India has faced problems in its great effort to achieve political stability and national cohesion,

the past decade, an estimated 26 percent of the population still lives below it.

India's economy in the early 2000s is on the move, but its heavy dependence on agriculture continues. Farming still accounts for nearly 70 percent of employment and yields 25 percent of India's GDP, which makes the entire economy vulnerable to the whim of the annual wet monsoon.

Agriculture

Growing contradictions mark agriculture in India. In Punjab, Haryana, and western Uttar Pradesh, farming remains relatively large-scale and comparatively efficient; from these areas come much of the huge grain surplus India has been accruing over the last several years. In Gujarat, Chhattisgarh, and Bihar, some children still starve and tens of millions are malnourished as staples do not reach them or are unaffordable when they do.

Traditional farming methods persist as they do in Subsaharan Africa, and yields per acre are low by world standards for virtually every crop grown this way. Transport systems are inadequate, hampering the movement of farm produce. In 2005, only about half of India's 600,000 villages were accessible by motorable road, and in this era of modern transportation animal-drawn carts still outnumber vehicles nationwide.

As the total population grows, the amount of cultivated land per person declines. Today, this physiologic density is 1688 per square mile (652 per sq km). However, this is nowhere near as high as the physiologic density in neighboring Bangladesh, where the figure is more than twice as great (3800 and 1467, respectively). But India's farming is so inefficient that this comparison is deceptive. Fully two-thirds of India's huge working population depends directly on the land for its livelihood, but the great majority of Indian farmers are poor and cannot improve their soils, equipment, or yields. Those areas in which India has substantially modernized its agriculture (as in Punjab's wheat zone) remain islands in a sea of agrarian stagnation.

these problems are more than matched by the difficulties that lie in the way of economic growth and development. The large-scale factories and power-driven machinery of the colonial rulers wiped out a good part of India's indigenous industrial base. Indian trade routes were taken over. European innovations in health and medicine sent the rate of population growth soaring, without introducing solutions for the many problems this spawned. Surface communications improved and food distribution systems became more efficient, but it would be a mis-

take to assume that this adequately protected Indian peasants from starvation. Local and regional food shortages occurred as El Niño-generated droughts struck India repeatedly, grain prices rose beyond what villagers could afford, and the colonial (and later sovereign) governments failed to intervene effectively. Today, more than 70 percent of Indians still live on the land, where, as we noted earlier, their social, economic, and political disadvantages sometimes spell disaster. Although tens of millions of Indians have been lifted above the poverty line over

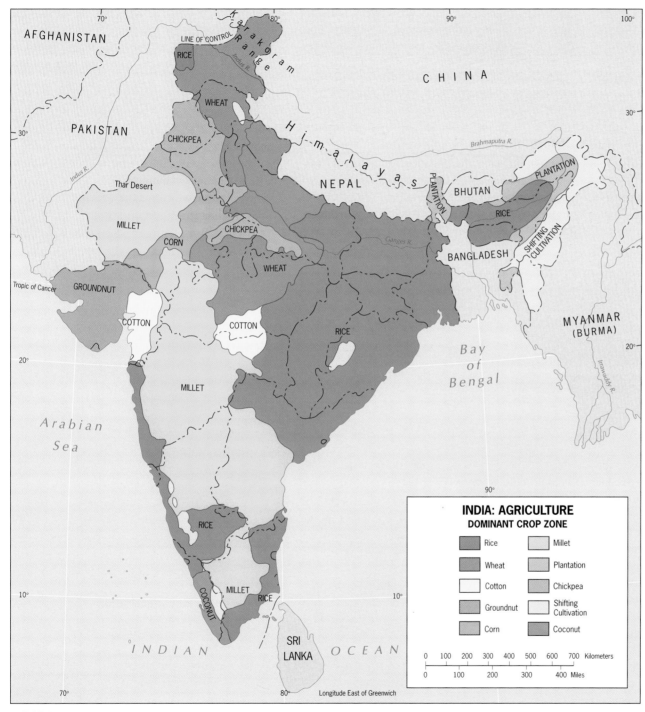

FIGURE 8-12

This stagnation has persisted in large measure because India, after independence, failed to implement a much-needed nationwide land reform program. Roughly one-quarter of India's entire cultivated area is still owned by less than 5 percent of the country's farming families, and little land redistribution was taking place. Perhaps half of all rural families own either as little as an acre or no land at all. Independent India inherited inequities from the British colonial period, but the individual States of the federation would have had to cooperate in any national land reform program. As always, the large landowners retained considerable political influence, so the program never got off the ground.

To make matters worse, much of India's farmland is badly fragmented as a result of local rules of inheritance, thereby inhibiting cooperative farming, mechanization, shared irrigation, and other opportunities for progress. Not surprisingly, land consolidation efforts have had only limited success except in the States of Punjab, Haryana, and parts of Uttar Pradesh, where modernization has gone farthest. Official agricultural development policy, at the federal and State levels, has also contributed to India's agricultural malaise and the uneven distribution of progress. Unclear priorities, poor coordination, inadequate information dissemination, and other failures have been reflected in the country's disappointing output.

It is instructive to compare Figure 8-12, showing the distribution of crop regions in India, with Figure G-7, which shows mean annual precipitation in India and the world. In the comparatively dry northwest, notably in the Punjab and neighboring areas of the upper Ganges, wheat is the leading cereal crop. Here, India has made major gains in annual production through the introduction of high-yielding grain varieties developed under the banner of the Green Revolution, the international research program that played a key role in overcoming the food crises of the 1960s. These "miracle crops" led to the expansion of cultivated areas, the construction of new irrigation systems, and the more intensive use of fertilizer (a mixed blessing, for fertilizers tend to be expensive and the "miracle" crops are more heavily dependent on them).

Toward the moister east, and especially in the wet-monsoon-drenched areas (Fig. 8-12), rice becomes the dominant staple. About one-fourth of India's total farmland lies under rice cultivation, most of it in the States of Assam, West Bengal, Bihar, Orissa, and eastern Uttar Pradesh and along the Malabar-Konkan coastal strip facing the Arabian Sea. These areas receive over 40 inches (100 cm) of rainfall annually, and irrigation supplements precipitation where necessary.

India devotes more land to rice cultivation than any other country, but yields per acre remain among the world's lowest—despite the introduction of "miracle rice." Nevertheless, the gap between demand and supply has narrowed, and in the late 1980s India actually exported some grain to Africa as part of a worldwide effort to help refugees there. But the situation remains precarious. As the population map (Fig. G-9) shows, there is a considerable degree of geographic covariation between India's rice-producing zones and its most densely populated areas. India is just one poor-harvest year away from another food crisis.

Subsistence remains the fate of tens of millions of Indian villagers who cannot afford fertilizers, cannot cultivate the new and more productive strains of rice or wheat, and cannot escape the cycle of poverty. Perhaps as many as 175 million of these people do not even own a plot of land and must live as tenants, always uncertain of their fate. This is the enduring reality against which optimistic predictions of im-

proved nutrition in India must be weighed. True, rice and wheat yields have increased at slightly more than the rate of population growth since the Green Revolution. But food security remains elusive, and India continues to face the risks inherent in the ever-growing needs of its burgeoning population.

Industrialization

In 1947, India inherited the mere rudiments of an industrial framework. After more than a century of British control over the economy, only 2 percent of India's workers were engaged in industry, and manufacturing and mining combined produced only about 6 percent of the national income. Textile and food processing were the dominant industries. Although India's first iron-making plant opened in 1911 and the first steel mill began operating in 1921, the initial major stimulus for heavy industrialization did not come until after the outbreak of World War II. Manufacturing was concentrated in the largest cities: Kolkata led, Mumbai was next, and Chennai ranked third.

The geography of manufacturing still reflects those beginnings, and industrialization in India has proceeded slowly, even after independence (Fig. 8-13). Kolkata now anchors India's eastern industrial region—the Bihar-Bengal District—where

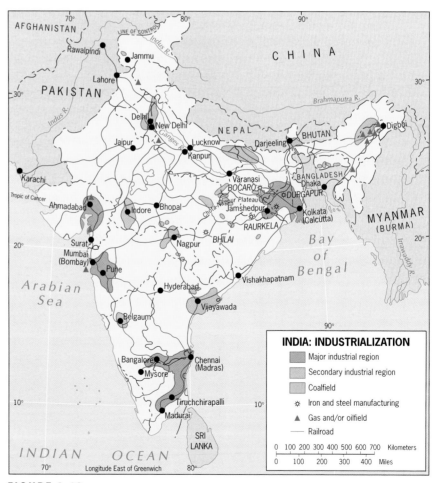

FIGURE 8-13

FROM THE FIELDNOTES

"Negotiating the traffic in India's chaotic cities is always a challenge as creaking buses, vintage cars, taxis large and small, scooters, and bicycles mingle in a mass of movement that sometimes makes streets look like rivers of humanity. I watched the scene from a vantage point on Nungambakkan Road in Chennai (formerly Madras), impressed that so much high-speed congestion produced no accidents. 'You have to have a sense of humor to be part of all that,' said a man who stopped to chat. 'Just down the street, make a left and then a right, and you'll see that even the authorities do.' He was referring to the sign (at the right) appealing to drivers to obey the seemingly nonexistent rules. India is in dire need of highway and road improvement; in cities like Chennai, the road system is pretty much the way the British left it in the 1940s when India became independent."

jute manufactures dominate, but cotton, engineering, and chemical industries also operate. On the nearby Chota-Nagpur Plateau to the west, coal-mining and iron and steel manufacturing have developed.

On the opposite side of the subcontinent, two industrial areas dominate the western manufacturing region: one is centered on Mumbai and the other on Ahmadabad. This dual region, lying in Maharashtra and Gujarat States, specializes in cotton and chemicals, with some engineering and food processing. Cotton textiles have long been an industrial mainstay in India, and this was one of the few industries to benefit from the nineteenth-century economic order the British imposed. With the local cotton harvest, the availability of cheap yarn, abundant and inexpensive labor, and the power supply from the Western Ghats' hydroelectric stations, the industry thrived.

The southern industrial region consists chiefly of a set of linear, city-linking corridors focused on Chennai, specializing in textile production and light engineering activities. Today, all of India's manufacturing regions are increasing their output of ready-to-wear garments—another legacy of the early development of cotton textiles. Clothing has become India's second-leading export by value; the production of gems and jewelry, another growing specialization, ranks first.

An important development in the south is occurring in and around Bangalore in the State of Karnataka, India's "Silicon Plateau." Several hundred software companies are based here, one-third of them foreign with names such as IBM, Texas Instruments, and Motorola. What attracts them, and makes them profitable, are the modest salaries of

India's software engineers, who earn about onefifth of what their foreign colleagues earn. The emigration of technicians is becoming a problem, but replacements are still plentiful. Bangalore is proof of India's potential in the modern world.

Despite some imbalances and inefficiencies, India's industrial resource base is well endowed. Limited high-quality coal deposits are exploited in the Chota-Nagpur area. In combination with large lower-grade coalfields elsewhere, the country's total output is high enough to rank it among the world's ten leading coal producers. With no known major petroleum reserves (some oil comes from Assam, Gujarat, Punjab, and offshore from Mumbai), India must spend heavily on fuel imports every year. Major investments have been made in hydroelectric plants, especially multipurpose dams that provide electricity, enhance irrigation, and facilitate flood control. India's iron ore deposits in Bihar (northwest of Kolkata) and Karnataka (in the heart of the Deccan) may rank among the largest in the world. Jamshedpur, located west of Kolkata in the eastern industrial region, has become India's steel-making and metals-fabrication center. Yet India still exports iron ore as a raw material to the higher-income industrialized countries, mainly Japan. For low-income, revenue-needy countries, entrenched practices are difficult to break.

Improving Prospects

Nevertheless, India's economy is showing signs of breaking out of its long-term mold. Despite India's low current GNP of U.S. $2340 per person, the sheer size of its economically active population makes this one of the world's largest economies, ranking sixth according to the latest data. A middle class of nearly 300 million is fueling demand for goods ranging from mobile phones (recent sales were about 1 million per month!) to motor bikes (10,000 a day according to press reports in late 2003). India's enormous market is attracting more foreign investment each year, and government efforts to constrain corruption and reduce red tape are paying off. High-tech industries exporting software, and service-industry outsourcing, are transforming places like Bangalore and Gurgaon into ultramodern, globally-connected centers. Until 1999, India did not have a single modern shopping mall. In 2004 it had more than 100, with dozens of others under construction. India's economy is on the move (recording a soaring growth rate of 10.4 percent in the final quarter of 2003), with huge implications for the global economic picture.

India East and West

The most commonly cited, and most clearly evident, regional division of India is between north and south. The north is India's heartland, the south its Dravidian appendage; the north speaks Hindi as its *lingua franca*, the south prefers English over Hindi; the north is bustling and testy, the south seems slower and less agitated.

But there is another, as yet less obvious, but potentially more significant divide across India. In Figure 8-13, draw a line from Lucknow, on the Ganges River, south to Madurai, near the southern tip of the peninsula (also review Figure 8-2). To the west of this line, India is showing signs of economic progress, the kind of economic activity that has brought Pacific Rim countries such as Thailand and Indonesia a new life. To the east, India has more in common with less promising countries also facing the Bay of Bengal: Bangladesh and Myanmar (Burma).

As with other regional divides, there are exceptions to our east-west delineation. Indeed, our map seems to suggest that much of India's industrial strength lies in the east. But what the map cannot reveal is the profitability of those industries. True, the east is rich in iron and coal, but the heavy industries built by the state in the 1950s are now outdated, uncompetitive, and in decline. The hinterland of Kolkata now contains India's Rustbelt. The government keeps many industries going but at a high cost. Old industries, such as carpetmaking and cottonweaving, continue to use child labor to remain viable. The State of Bihar represents the stagnation that afflicts much of India east of our line: by several measures it ranks among the poorest of the 28 States.

Compare this to western India. The State of Maharashtra, the hinterland of Mumbai, leads India in many categories, and Mumbai leads Maharashtra. Many smaller, private industries have emerged here, manufacturing goods ranging from umbrellas to satellite dishes and from toys to textiles. Across the Arabian Sea lie the oil-rich economies of the Arabian Peninsula. Hundreds of thousands of workers from western India have found jobs there, sending money back to families from Punjab to Kerala. More importantly, many have used their foreign incomes to establish service industries back home. Outward-looking western India, in contrast to the inward-looking east, has begun to establish other ties to the outside world. Satellite links have enabled Bangalore to become the center of a growing software-producing complex reaching world markets. The beaches of Goa, the small State immediately to the south of Maharashtra, appeal to the tourist markets of Europe. This is, in fact, a classic case of **intervening opportunity** because resorts have sprung up along Goa's coast, and European tourists who once went to the more distant Maldives and Seychelles are coming to Goa. Maharashtra's economic success also has spilled over into Gujarat to the north, and even landlocked Rajasthan (the next State to the north) is experiencing the beginnings of what, by Indian standards, is a boom.

The boom has created political problems, however. Not only is Maharashtra State a rising economic power; it also is the base of a strong Hindu nationalist political movement whose leaders object to foreign intrusions and have blocked major development projects and other enterprises. They halted a huge industrial scheme about halfway through and closed a fast-food operation that they deemed incompatible with local culture. Such clashes between foreign interests and domestic traditions continue even as India's economy forges ahead.

Nevertheless, India's east-west divide shows a growing contrast that puts the west far ahead. The hope is that Maharashtra's success will spread northward and southward along the Arabian Sea coast and will ultimately diffuse eastward as well. But for this to happen, India will have to bring its population spiral under control.

▶ BANGLADESH: CHALLENGES OLD AND NEW

On the map of South Asia, Bangladesh looks like another State of India: the country occupies the area of the double delta of India's great Ganges and Brahmaputra rivers, and India almost completely surrounds it on its landward side (Fig. 8-14). But Bangladesh is an independent country, born in 1971 after its brief war for independence against Pakistan, with a territory about the size of Wisconsin. Today it remains one of the poorest and least developed countries on Earth, with a population of 139.5 million that is growing at an annual rate of 2.2 percent.

Not only is Bangladesh a poor country; it also is **16** highly susceptible to damage from **natural hazards**. During the twentieth century, eight of the ten deadliest natural disasters in the entire world struck this single country. In 1991, a cyclone (as hurricanes are called in this part of the world) killed over 150,000 people.

The reasons for Bangladesh's vulnerability can be deduced from Figures 8-14 and 8-1. Southern Bangladesh is the deltaic plain of the Ganges-Brahmaputra river system, combining fertile alluvial soils that attract farmers with the low elevations that endanger them when water rises. The shape of the Bay of Bengal forms a funnel that sends cyclones and their storm surges of wind-whipped water barreling into the delta coast. Without money to build seawalls, floodgates, elevated shelters in sufficient numbers, or adequate escape routes, hundreds of thousands of people are at continuous risk, with deadly consequences. And as if this is not enough, millions of people have been found to be exposed to excessive (natural) arsenic in the drinking water from their wells.

Bangladesh remains a nation of subsistence farmers; urbanization is at only 23 percent, and Dhaka, the megacity capital, and the southeastern port of Chittagong are the only urban centers of consequence. Moreover, Bangladesh has one of the highest physiologic densities in the world (3800 people per square mile/1467 per sq km), and only higher-yielding varieties of rice and the introduction of wheat in the crop rotation (where climate allows) have improved diets and food security. But diets remain poorly balanced, and, overall, the people's nutrition is unsatisfactory. The textile industry provides most of Bangladesh's foreign revenues, but the once-thriving jute industry continues its decline. The discovery of a natural gas reserve is now the subject of a national debate: home consumption or money-making export?

Bangladesh is a dominantly Muslim society, but not one dominated by revivalists. For example, 30 seats in the national legislature are reserved for women. Its relations with neighboring India have at times been strained over water resources (India's control over the Ganges that is Bangladesh's lifeline), cross-border migration (10 percent of the population is Hindu), and transit between parts of India across Bangladesh's north (refer to Figure 8-13 to see the reason). All the disadvantages of the global periphery afflict this populous, powerless

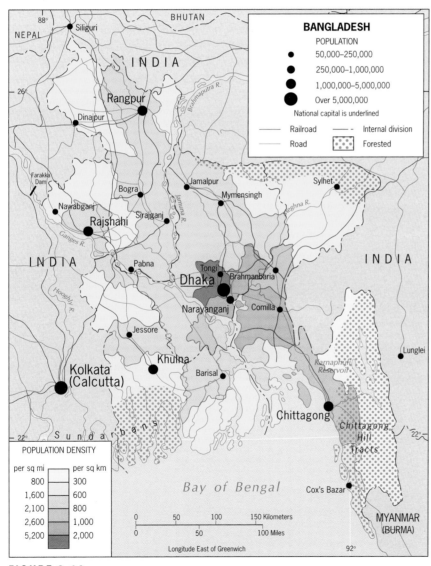

FIGURE 8-14

country where survival is the leading industry and all else is luxury.

THE MOUNTAINOUS NORTH

As Figures 8-1 and 8-2 show, a tier of landlocked countries and territories lies across the mountainous zone that walls India off from China. One of them, Kashmir, is in a state of near-war. Another, Sikkim, was absorbed by India in 1975 and made into one of its federal States. But Nepal and Bhutan retain their independence.

Nepal, northeast of India's Hindu core, has a population of 25 million and is the size of Illinois. It has three geographic zones (Fig. 8-15): a southern, subtropical, fertile lowland called the Terai; a central belt of Himalayan foothills with swiftly flowing streams and deep valleys; and the spectacular high Himalayas themselves (topped by Mount Everest) in the north. The capital, Kathmandu, lies in the east-central part of the country in an open valley of the central hill zone.

Nepal is materially poor but culturally rich. The Nepalese are a people of many sources, including India, Tibet, and interior Asia. About 90 percent are Hindu, and Hinduism is the country's official religion; but Nepal's Hinduism is a unique blend of Hindu and Buddhist ideals. Thousands of temples and pagodas ranging from the simple to the ornate grace the cultural landscape, especially in the valley of Kathmandu, the country's core area. Although over a dozen languages are spoken, 90 percent of the people also speak Nepali, a language related to Indian Hindi.

As the data in Table G-1 suggest, Nepal is a troubled country suffering from severe underdevelopment—and its GNP is the lowest in the entire realm. It also faces strong centrifugal social and political forces. Environmental degradation, crowded farmlands and soil erosion, and deforestation scar the countryside. The Himalayan peaks form a world-renowned tourist attraction, but tourist spending in Nepal, always relatively modest, has been cut back because of the current revival of Maoist-communist terrorism, not only in the western hills where it has long been active, but also in the capital itself.

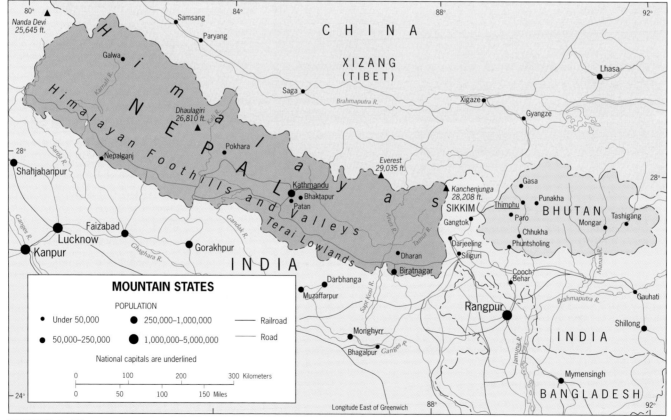

FIGURE 8-15

In fact, Nepal's political geography has long been troubled. The end of absolute monarchy in 1991 and the advent of democracy did not end the country's regional divisions: the southern Terai with its tropical lowlands is a world apart from the hills of central Nepal, and the peoples of the west have origins and traditions different from those in the east. Moreover, in 2001 the assassination of Nepal's king threatened the disintegration of the state, narrowly averted by the leadership of the Nepal Congress Party. Then in 2002 the country's stability was imperiled again as the Maoist insurgency spread. The tourist industry collapsed, and the economy withered in this scenic, divided, endangered state.

Mountainous **Bhutan**, wedged between India and China's Tibet, is the only other buffer between Asia's giants. In landlocked, fortress-like Bhutan, time seems to have stood still. Bhutan is officially a constitutional monarchy, but its king rules the country with virtually absolute power; economic subsistence and political allegiance are the norms of life for most of the population of just under one million. Thimphu, the capital, has about 50,000 inhabitants. The symbols of Buddhism, the state religion, dominate its cultural landscape. Social tensions arise from the large but diminishing Nepalese minority, most of whom are Hindus and some of whom have been persecuted by the dominant Bhutia.

Forestry, hydroelectric power, and tourism all have potential here, and Bhutan has considerable mineral resources. But isolation and inaccessibility preserve traditional ways of life in this mountainous buffer state.

◢ THE SOUTHERN ISLANDS

As Figure 8-2 shows, South Asia's continental landmass is flanked by several sets of islands: Sri Lanka off the southern tip of India, the Maldives in the Indian Ocean to the southwest, and the Andaman Islands (belonging to India) marking the eastern edge of the Bay of Bengal.

The **Maldives** consists of more than a thousand tiny islands whose combined area is just 115 square miles (less than 300 sq km) and whose highest elevation is barely over 6 feet (2 m) above sea level. Its population of under 300,000 from Dravidian and Sri Lankan sources is now 100 percent Muslim, one-quarter of which is concentrated on the capital island named Maale. The Maldives might be unremarkable, except that, as Table G-1 shows, this country has by far the realm's highest GNP per capita. The locals have translated their palm-studded, beach-fringed islands into a tourist mecca that attracts tens of thousands of mainly European visitors annually.

Sri Lanka: South Asian Tragedy

Sri Lanka (known as Ceylon before 1972), the compact, pear-shaped island located just 22 miles (35 km) across the Palk Strait from India, became independent from Britain in 1948 (Fig. 8-16). There were good reasons to create a separate sovereignty for Sri Lanka. This is neither a Hindu nor a Muslim country: the majority of its nearly 19 million people—about 70 percent—are Buddhists. Furthermore, unlike India or Pakistan, Sri Lanka is a plantation country, and commercial farming still is the mainstay of the agricultural economy.

The great majority of Sri Lanka's people are descended from migrants who came to this island from northwest India beginning about 2500 years ago. Those migrants introduced the advanced culture of their source area, building towns and irrigation systems and bringing Buddhism. Today, their descendants, known as the Sinhalese, speak a language (Sinhala) that belongs to the Indo-European language family of northern India.

The Dravidians who lived on the mainland, just across the Palk Strait, came later and in far smaller numbers—until the British colonialists intervened.

FROM THE FIELDNOTES

"Sri Lanka is a dominantly Buddhist country, the religion of the majority Sinhalese. Most areas of the capital, Colombo, have numerous reminders of this form of architecture and statuary: shrines to Buddha, large and small, abound. But walk into the Tamil parts of town, and the cultural landscape changes drastically. This might as well be a street in Chennai or Madurai: elaborate Hindu shrines vie for space with storefronts and Buddhist symbols are absent. The people here seemed to be less than enthused about the Tamil Tigers' campaign for an independent state. 'This would never be a part of it anyway,' said the fellow walking toward me as I took this photograph. 'We're here for better or worse, and for us the situation up north makes it worse.' But he added, Sri Lankan governments of the past had helped create the situation by discriminating against Tamils."

FIGURE 8-16

During the nineteenth century the British brought hundreds of thousands of Tamils to work on their tea plantations, and soon a small minority became a substantial segment of Ceylonese society. The Tamils brought their Dravidian tongue to the island and introduced their Hindu faith. At the time of independence, they constituted more than 15 percent of the population; today they total about 18 percent.

When Ceylon became independent, it was one of the great hopes of the postcolonial world. The country had a sound economy and a democratic government, and it was renowned for its tropical beauty. Its reputation soared when a massive campaign succeeded in eradicating malaria and when family-planning campaigns reduced population growth while the rest of the realm was experiencing a population explosion. Rivers from the cool, forested interior highlands fed the paddies that provided ample rice; crops from the moist southwest paid the bills, and the capital, Colombo, grew to reflect the optimism that prevailed.

In the midst of this glowing scenario, the seeds of disaster were already being sown. Sri Lanka's Tamil minority soon began proclaiming its sense of exclusion, demanding better treatment from the Sinhalese majority. Although the government recognized Tamil as a "national language" in 1978, sporadic violence marked the Tamil campaign, and in 1983 full-scale civil war began. Now many in the Tamil community demanded a separate Tamil state to encompass the north and east of the country (see Fig. 8-16), and a rebel army called the Tamil Tigers confronted Sri Lanka's national forces.

The sequence of events fits the model of the evolution of the **insurgent state** discussed in Chapter 5. In Sri Lanka, the *equilibrium* stage was reached in the early 1990s, when the Tamil Tigers claimed the Jaffna Peninsula and set up their headquarters there. The *counteroffensive* stage is now in progress, but the Tamil forces strike at will (they destroyed Colombo's international airport in July 2001) and a negotiated settlement, now being pursued with the help of Norwegian mediators, appears inevitable. Even if the Tamils do not succeed in securing the independent state they call Eelam, they will likely force the government to make some territorial concessions.

Once-promising Sri Lanka is paying dearly for its politicians' failures. The cost of the civil war is enormous. The economic consequences are incalculable, with the tourist industry having been reduced to a fraction of what it could be. And South Asia has lost a beacon of opportunity and progress.

DISCOVERY TOOLS

www.wiley.com/college/deblij

The *Concepts and Regions* Website, featuring *GeoDiscoveries*, offers many additional resources to enhance your understanding and experience of this chapter. Be sure to explore the following:

GeoDiscoveries Video Clips

The Monsoon and Daily Life

Drought in Indian Villages

GeoDiscoveries Animations

A-B-C's of Climate

How Do Monsoons Work?

GeoDiscoveries Interactivities

Painting India by Numbers

Photo Galleries

Khyber Pass, Pakistan-Afghanistan

Chennai, India

Udaipur, India

Goa, India

Mumbai, India,

Rural India

Delhi, India

Marmagao, India

Colombo, Sri Lanka

Maldives

Virtual Field Trip

Sri Lanka's "Teascapes"

More To Explore

Systematic Essay: Population Geography

Issues in Geography: South Asia's Life-Giving Monsoons, The Problem of Kashmir, Solace and Sickness from the Holy Ganges

Major Cities: Mumbai, Kolkata, Delhi New and Old

Expanded Regional Coverage

Pakistan, India, Bangladesh, Mountainous North, Southern Islands

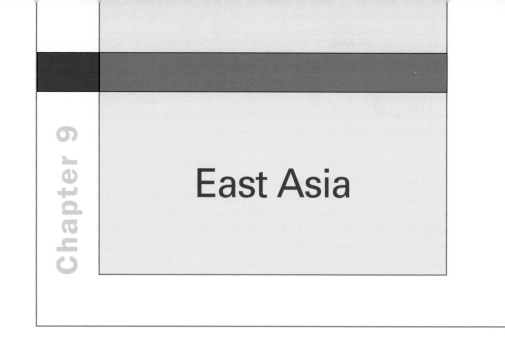

Chapter 9

East Asia

CONCEPTS, IDEAS, AND TERMS

1. Pacific Rim
2. Confucianism
3. Sinicization
4. Extraterritoriality
5. Special Administrative Region (SAR)
6. Economic restructuring
7. Core area
8. Geography of development
9. Overseas Chinese
10. Special Economic Zone (SEZ)
11. Regional state
12. Economic tiger
13. Buffer state
14. Jakota Triangle
15. Modernization
16. Relative location
17. Areal functional organization
18. Regional complementarity
19. State capitalism

REGIONS

- CHINA PROPER
- XIZANG (TIBET)
- XINJIANG
- MONGOLIA
- JAKOTA TRIANGLE (JAPAN-KOREA-TAIWAN)

$\mathcal{E}$AST ASIA IS a geographic realm like no other. At its heart lies the world's most populous country. On its periphery lies one of the globe's most powerful national economies. Along its coast-line, on its peninsulas, and on its islands an economic boom has transformed cities and country-sides. Its interior contains the world's highest moun-tains and vast deserts. It is a storehouse of raw mate-rials. The basins of its great rivers produce food that can sustain more than a billion people.

DEFINING THE REALM

The East Asian geographic realm consists of six political entities: China, Mongolia, North Korea, South Korea, Japan, and Taiwan. Note that we refer here to "political entities" rather than "states." In changing East Asia, the distinction is significant. Taiwan, which its government offi-cially calls the Republic of China, functions as a state but is regarded by mainland China (the Peo-ple's Republic of China) as a temporarily wayward

STUDY TOOLS

| 3D GLOBE | FLASHCARDS | AUDIO PRONUNCIATION GUIDE | ANNOTATED WEBLINKS |
| AREA AND DEMOGRAPHIC DATA | MAP QUIZZES | CHAPTER QUIZ | LONELY PLANET WEBLINKS |

MAJOR GEOGRAPHIC QUALITIES OF EAST ASIA

1. East Asia is encircled by snowcapped mountains, vast deserts, cold climates, and Pacific waters.

2. East Asia was one of the world's earliest culture hearths, and China is one of the world's oldest continuous civilizations.

3. East Asia is the world's most populous geographic realm, but its population remains strongly concentrated in its eastern regions.

4. China, the world's largest nation-state demographically, is the current rendition of an empire that has expanded and contracted, fragmented and unified many times during its long existence.

5. China today remains a mainly rural society, and its vast eastern river basins feed hundreds of millions in a historic pattern that still continues.

6. China's sparsely peopled western regions are strategically important to the state, but they lie exposed to minority pressures and Islamic influences.

7. Along China's Pacific frontage an economic transformation is taking place, affecting all the coastal provinces and creating an emerging Pacific Rim region.

8. Increasing regional disparities and fast-changing cultural landscapes are straining East Asian societies.

9. Japan, the economic giant of the East Asian realm, has a history of colonial expansion and wartime conduct that still affects international relations here.

10. East Asia may witness the rise of the world's next superpower as China's economic and military strength and influence grow—and if China avoids the devolutionary forces that fractured the Soviet Union.

11. The political geography of East Asia contains a number of flashpoints that can generate conflict, including Taiwan, North Korea, and several island groups in the realm's seas.

---

province. North Korea is not a full member of the United Nations, and the division of the Korean Peninsula may be temporary.

As defined here, East Asia lies between the vast expanses of Russia to the north and the populous countries of South and Southeast Asia to the south. This geographic realm extends from the deserts of Central Asia to the Pacific islands of Japan and Taiwan. Environmental diversity is one of its hallmarks.

East Asia also is the hub of the evolving regional 1 phenomenon called the **Pacific Rim**. From Japan to Taiwan and from South Korea to Hong Kong (Xianggang), the Pacific frontage of East Asia is being transformed. Skyscrapers tower over old cities whose traditional housing is being swept away. Enormous industrial complexes disgorge products that flood world markets. Millions of people are on the move, abandoning farms and villages for assembly lines and sweatshops. The process started in Japan, and soon encompassed Taiwan, South Korea, Hong Kong, and Singapore. And when China's political climate changed, almost all of coastal East Asia was swept up in one of the greatest regional transformations in history.

NATURAL ENVIRONMENTS

Figure 9-1 illustrates the complex physical geography of the East Asian realm. In the southwest lie ice-covered mountains and plateaus, the Earth's crust in this region crumpled up like the folds of an accordion. A gigantic collision of tectonic plates is creating this landscape as the Indian Plate pushes north-ward into the underbelly of the Eurasian Plate (Fig. G-4). The result is some of the world's most spectacular scenery, but snow, ice, and cold are not the only dangers to human life here. Earthquakes and tremors occur almost continuously, causing landslides and avalanches. As the map shows, the high mountains and plateaus widen from a relatively narrow belt in the Karakoram to form Tibet's vast Qinghai-Xizang Plateau, flanked by the Himalayas to the south. Then, east of Tibet, the mountain ranges converge again and bend southward into Southeast Asia, where they lose their high relief.

As Figure G-9 shows, this Asian interior is one of the world's most sparsely populated areas, but it is nevertheless critical to the lives of hundreds of millions of people. In these high mountains, fed by the melting ice and snow, rise the great rivers that flow

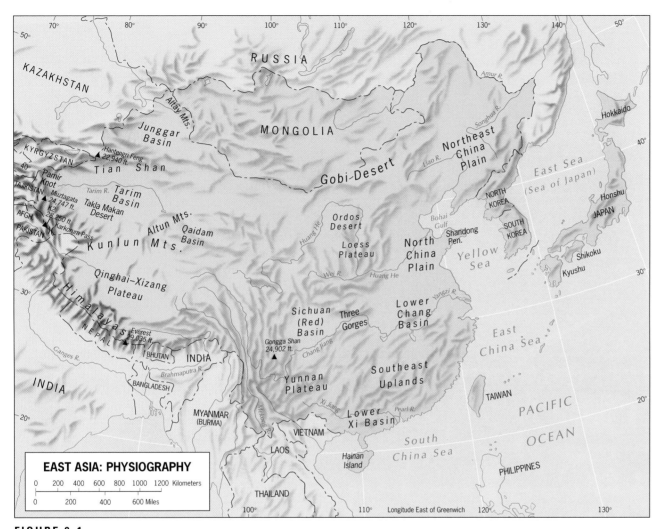

EAST ASIA: PHYSIOGRAPHY

0 200 400 600 800 1000 1200 Kilometers

0 200 400 600 Miles

FIGURE 9-1

erate conditions, comparable to those of the U.S. Southeast; North Korea has a harsh continental climate like that of North Dakota.

From the high-relief interior come three major river systems that have played crucial roles in the human drama. In the north, the Huang He (Yellow River) arises deep in the high mountains, crosses the Ordos Desert and the Loess Plateau, and deposits its fertile sediments in the vast North China Plain, where East Asia's earliest states emerged. In the center, the Chang Jiang (Long River), called the Yangzi downstream, crosses the Sichuan Basin and the Three Gorges, where a huge dam project is under way, and waters extensive ricefields in the Lower Chang Basin. And in the south, the Xi Jiang (West River) originates on the Yunnan Plateau and becomes the Pearl River in its lowest course. Its estuary, flanked by several of China's largest urban-industrial complexes, has become one of the hubs of the evolving Pacific Rim.

Further scrutiny of Figure 9-1 indicates that a fourth river system plays a role in China: the Liao River in the northeast and its basin, the Northeast China Plain. As the map suggests, however, the Liao is not comparable to the great rivers to its south, its course being shorter and its basin, in this higher-latitude zone, much smaller.

In the interior, note the Loess Plateau located south of the Ordos Desert, where the Huang He (Yellow River) makes its giant loop. Glacial action pulverized rocks and produced the wind-blown, fertile deposit known as *loess*, and where water is available it can sustain a dense agricultural population. To the south, deep in the interior, lies the Sichuan (Red) Basin, crossed by the

eastward across China and southward across Southeast and South Asia. Throughout the Holocene, these rivers have been eroding the uplands and depositing their sediments in the lowlands, in effect creating the alluvium-filled basins that now sustain huge populations. Fertile alluvial soils and adequate growing seasons, combined with ample water and millions of hands to sow the wheat and plant the rice, have allowed the emergence of one of the great population concentrations on Earth.

Physiography, therefore, has much to do with East Asia's population distribution, but even the more habitable and agriculturally productive east has its limitations. The northeast suffers from severe continentality, with long and bitterly cold winters. High relief encircles the river basins north of the Yellow Sea and dominates much of the northern part of the Korean Peninsula, creating strong local environmental contrasts. South Korea, as Figure G-8 shows, experiences relatively mod-

Chang Jiang. This basin has supported human communities for a long time, and you can actually see its current population cluster in Figure G-9, the map of world population distribution. The Sichuan Basin, encircled as it is by mountains, is one of the world's most clearly defined physiographic regions, and the concentration of its approximately 120 million inhabitants reflects that definition.

Still farther to the south lies the Yunnan Plateau, source of the tributaries that feed the Xi River. Much of southeastern China has comparatively high relief; it is hilly and in places mountainous. This high relief has helped limit contacts between China and Southeast Asia.

East Asia's Pacific margin is a jumble of peninsulas and islands. The Korean Peninsula looks like a near-bridge from Asia to Japan, and indeed it has served as such in the past. The Liaodong and Shandong peninsulas protrude into the Yellow Sea, which continues to silt up from the sediments of the Huang and Liao rivers. Off the mainland lie the islands that have played such a crucial role in the modern human geography of Asia and, indeed, the world: Japan, Taiwan, and Hainan. Japan's environmental range is expressed by cold northern Hokkaido and warm southern Kyushu, but Japan's core area lies on its main island, Honshu. As Figure 9-1 shows, myriad smaller islands flank the mainland and dot the East and South China Seas. As we will discover, some of these smaller islands have major significance in the human geography of this realm.

HISTORICAL GEOGRAPHY

Hominid and human histories in East Asia are lengthy and complex. Many archeological sites in this realm have yielded evidence of *Homo erectus*, including perhaps the most famous of all: Peking Man, found in a cave not far from Beijing in the 1920s. Current anthropological theory holds that *Homo sapiens* arrived in East Asia between 40,000 and 60,000 years ago and eliminated the hominids in short order.

Early Cultural Geography

Humans have inhabited the plains and river basins, foothills, and islands of this realm for a very long time. Hunting sustained both the hominids and the early human communities; fishing drew them to the coasts and onto the islands. The first crossing into Japan may have occurred as long as 10,000 to 12,000 years ago, possibly much earlier, when the Jomon people, a Caucasoid population of uncertain geographic origins, entered the islands; their modern descendants, the Ainu, spread throughout the archipelago. Today, only about 20,000 persons living in northernmost Hokkaido trace their ancestry to Ainu sources.

About 2300 years ago, the Yayoi people, rice farmers who had settled in Korea, appear to have crossed by boat to Kyushu, Japan's southernmost island, from where they advanced northward. The Ainu, who subsisted by fishing, trapping, and hunting, were driven back, but gene-pool studies show that much mixing of the groups took place; they also show that the Yayoi invasion was followed by other incursions from the Asian mainland. By then, powerful dynastic states had already arisen in what is today China, and early Chinese culture traits thus found their way into Japan through the process we know as *relocation diffusion*.

On the Asian mainland, plant and animal domestication had begun as early as anywhere on Earth. We in the Western world take it for granted that these momentous processes began in what we now call the Middle East and diffused from the Fertile Crescent to other parts of Eurasia and the rest of the world. But the taming of animals and the selective farming of plants may have begun as early, or earlier, here in East Asia. As in Southwest Asia, the fertile alluvial soils of the great river basins and the ebb and flow of stream water cre-

ated an environment of opportunity, and millet and rice were being harvested between 7000 and 8000 years ago.

Even during this Neolithic period of increasingly sophisticated stone tools, East Asia was a mosaic of regional cultures. Their differences are revealed by the tools they made and the decorations on their bowls, pots, and other utensils. An especially important discovery of two 8000-year-old pots in the form of a silkworm cocoon, from China's Hebei Province, suggests a very ancient origin for one of the region's leading historic industries.

As noted earlier, plant and animal domestication produced surpluses and food storage, enabling population growth and requiring wider regional organization. Here as elsewhere during the Neolithic, settlements expanded, human communities grew more complex, and power became concentrated in a small group, an *elite*.

This process of *state formation* is known to have occurred in only a half dozen regions of the world, and China was one of these. But evidence about China's earliest states has long been scarce. Today, however, archeologists are focusing on the lower Yi-Luo River Valley in the western part of Henan Province, where the first documented Chinese dynasty, the *Xia* Dynasty (2200–1770 BC) existed. The capital of this ancient state, Erlitou, has been found, and archeologists now refer to the Xia Dynasty as the Erlitou culture. Secondary centers are being discovered, and Erlitou tools and implements in a wider area prove that the Xia Dynasty represents a substantial state.

All early states were ruled by elites, but China's political history is chronicled in *dynasties* because here the succession of rulers came from the same line of descent, sometimes enduring for centuries. In the transfer of power, family ties counted for more than anything else. Dynasties were overthrown, but the victors did not change this system. Dynastic rule lasted into the twentieth century.

The Xia Dynasty may have been the earliest Chinese state, but it lay in the area where, later, more powerful dynastic states arose: the North China Plain. Here the tenets of what was to become Chinese society were implanted early and proved to be extremely durable. From this culture hearth ideas, innovations, and practices diffused far and wide. From agriculture to architecture, poetry to porcelain making, influences radiated southward into Southeast Asia, westward into interior Asia, and eastward into Korea and Japan. In the North China Plain lay the origins of what was to become the Middle Kingdom, which its citizens considered the center of the world.

Dynastic China

If the Xia period was indeed China's first dynasty, dynastic rule in this part of East Asia lasted 4000 years, ending only in 1911 when the last emperor of the Qing (Manchu) Dynasty, a six-year-old boy, was forced to abdicate the Chinese throne. As Figure 9-2 shows, the Chinese sphere expanded quite rapidly, reaching its greatest extent during that last dynasty (there also were times when China experienced competing dynasties, internal division, and temporary losses). The lower Huang He (Yellow River) and the Wei basins were the original heartland of China during the *Shang* Dynasty (ca. 1766–ca. 1080 BC), but the following dynasty, the *Zhou* Dynasty (ca. 1027 BC–221 BC) was crucial. Taoism arose, Buddhism arrived, and Confucius (Kongfuzi in the modern Pinyin spelling) produced what was to become China's guiding philosophy for more than 2000 years.

The *Han* Dynasty (206 BC–AD 220), however, was in many ways China's formative period: enormous territorial expansion accompanied a flowering of Chinese culture. The Han Empire had about the same territorial dimensions as the Roman Empire, which existed about the same time; Xian, the "Rome of China," was one of the world's greatest cities. This was the time of the Silk Route, of penetrations into Central and Southeast Asia, of authoritarian government and disciplined armies. To this

day, the people in China refer to themselves as the *People of Han*.

After the Han Dynasty, China went into one of its periods of division and decline, but revival began with the brief *Sui* Dynasty (581–618) and reached a golden age during the *Tang* Dynasty (618–907), when Xian was China's cultural capital and the largest city in the world, the Silk Route was loaded with trade, Arab and Persian seafarers were arriving in Chinese ports, and Chinese influences penetrated Korea and Japan. But the glorious Tang Dynasty, too, was followed by a period of political instability until the Northern and Southern *Song* Dynasties (960–1279) brought a momentous development: the invasion and takeover of China by the Mongols and Kublai Khan. This was the prelude to the Mongol-dominated *Yuan* Dynasty (1264–1368), when the northern city, Beijing, grew into a major metropolis. The Mongols ruled, but instead of converting the Chinese to Mongol ways it was the Mongol rulers who adopted Chinese culture.

Eventually, Mongol power waned, and now another indigenous Chinese dynasty assumed control: the *Ming* Dynasty (1368–1644). During this era, China made major advances in science and technology; great oceangoing ships sailed in huge fleets into the Indian Ocean and reached East Africa long before the Europeans in their much smaller vessels did. Farming expanded, and with it population—until climate change caused famines and havoc, the fleets were burned, and the Ming rulers turned isolationist and ineffective.

This created an opportunity for another group of outsiders to establish what turned out to be China's final dynasty: the Manchu, a people with Tatar links living in present-day Northeast China. The *Manchu (Qing)* Dynasty (1644–1911) began when this minority of about 1 million seized control of the nation of several hundred million by taking power in Beijing. They retained the Ming system of administration and kept Ming officials in office, and began a campaign of territorial expansion that created the

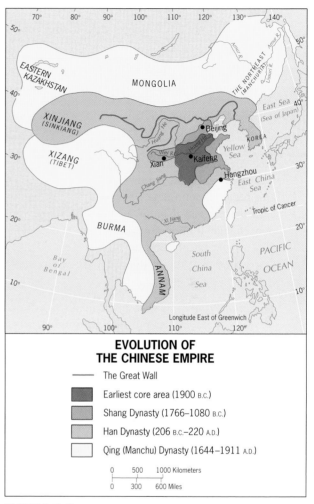

EVOLUTION OF THE CHINESE EMPIRE

— The Great Wall

Earliest core area (1900 B.C.)

Shang Dynasty (1766–1080 B.C.)

Han Dynasty (206 B.C.–220 A.D.)

Qing (Manchu) Dynasty (1644–1911 A.D.)

0   500   1000 Kilometers
0   300   600 Miles

FIGURE 9-2

largest China-centered empire ever (Fig. 9-2), including Mongolia, much of Turkestan, Xizang (Tibet), Myanmar (Burma), Indochina, Korea, and Taiwan. But the Qing rulers had the misfortune of being in charge when the European powers and Japan arrived in force; revolution and collapse of the dynasty followed.

As Figure 9-2 suggests, the growth and expansion of dynastic China was the dominant and formative process in this realm. Many of the territorial acquisitions made during the Qing Dynasty form part of the modern Chinese state today; others are historic justifications for potential claims on "lost" areas such as the Russian Far East and Mongolia. Large Chinese minorities now reside in Southeast Asian countries, an emigration that began during the Qing Dynasty and changed the social and political landscape of colonial as well as modern Southeast Asia. Only Japan, while influenced by Chinese tenets, escaped incorporation into China's East Asian sphere throughout its history.

REGIONS OF THE REALM

East Asia presents us with an opportunity to illustrate the changeable nature of regional geography. Our regional delimitation is based on current circumstances, and it predicts ways the framework may change. It is anything but static. At the beginning of the twenty-first century, we can identify five geographic regions in the East Asian realm (Fig. 9-3). These are:

1. *China Proper*. Almost any map of China's human geography—population distribution, urban centers, surface communications, agriculture, industry—emphasizes the strong concentration of Chinese activity in the country's eastern sector. This is the "real" China, where its great cities, populous farmlands, and historic sources are located. Long ago, scholars called this *China Proper*, and it is a good regional designation. But China is a large and complex country, and a number of subregions are nested within China Proper. Some of these, such as the North China Plain and the Sichuan Basin, are old and well-established geographic units. One in particular is new: China's Pacific Rim, still growing, yet poorly defined, and shown as "formative" in Figure 9-3.

2. *Xizang (Tibet)*. The high mountains and plateaus of Xizang, ruled by China but still widely known by its older name of Tibet, form a stark contrast to teeming China Proper. Here, next to one of the world's largest and most populous regions, lies one of the emptiest and, in terms of inhabited space, smallest regions.

3. *Xinjiang*. The vast desert basins and encircling mountains of Xinjiang form a third East Asian region. Again, physical as well as human geographic criteria come into play: here China meets Islamic Central Asia.

4. *Mongolia*. The desert state of Mongolia forms East Asia's fourth region. Like Tibet, landlocked Mongolia, vast but sparsely peopled, stands in stark contrast to populous China Proper.

5. *Jakota Triangle*. East Asia's fifth region is defined by its economic geography. *Ja*pan, South *Ko*rea, and *Ta*iwan (the name "Jakota" derives from the first two letters of each) were transformed by Pacific Rim economic developments during the second half of the twentieth century. Its recent emergence foreshadows further changes in the decades ahead as Korean unification becomes a possibility and contrasts between the Jakota Triangle and Pacific Rim China diminish.

CHINA PROPER

When we in the Western world chronicle the rise of civilization, we tend to focus on the historical geography of Southwest Asia, the Mediterranean, and Western Europe. Ancient Greece and Rome were the crucibles of culture; Mediterranean and Atlantic waters were the avenues of its diffusion. China lay remote, so we believe, barely connected to this Western realm of achievement and progress. When an Italian adventurer named Marco Polo visited China during the thirteenth century and described the marvels he saw there, his work did little to change European minds. Europe was and would always be the center of civilization.

The Chinese, naturally, take a different view. Events on the western edge of the great Eurasian landmass were deemed irrelevant to theirs, the most advanced and refined culture on Earth. Roman emperors were rumored to be powerful, and Rome was a great city, but nothing could match the omnipotence of China's rulers. Certainly the Chinese city of Xian far eclipsed Rome as a center of sophistication. Chinese civilization existed long before ancient Greece and Rome emerged, and it was still there long after they collapsed. China, the Chinese teach themselves, is eternal. It was, and always will be, the center of the civilized world.

We should remember this notion when we study China's regional geography because 4000 years of Chinese culture and perception will not change overnight—not even in a generation. Time and again, China overcame the invasions and depredations of foreign intruders, and afterward the Chinese would close off their vast country against the outside world. A mere 35 years ago, in the early 1970s, there were just a few *dozen* foreigners in the entire country with its (then) nearly 1 billion inhabitants. The institutionalization of communism required this insularity, and even the Soviet advisors had been thrown out. But by the early 1970s, China's rulers decided that an opening to the Western world would be advantageous, and so U.S. President Richard Nixon was invited to visit Beijing. That historic occasion, in 1972, ended this latest period of isolation—as always, on China's terms. Since then, China has been open to tourists and businesspeople, teachers and investors. Tens of thousands of Chinese students have been sent to study at American and other Western institutions.

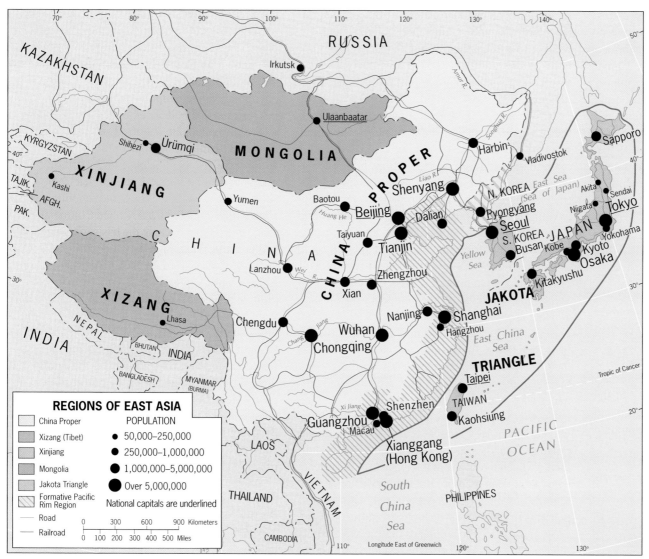

FIGURE 9-3

phy. In other words, China's "splendid isolation" was made possible by geography.

Earlier we noted the role of relief and desert in encircling the culture hearth of East Asia, but equally telling is the factor of distance. Until recently, China lay far from the modern source areas of innovation and change. True, China—as the Chinese emphasize—was itself such a hearth, but China's contributions to the outside world remained limited, essentially, to finely made arts and crafts. China did interact with Korea, Japan, Taiwan, and parts of Southeast Asia, and eventually millions of Chinese emigrated to neighboring countries. But compare these regional links to those of the Arabs, who ranged worldwide and who brought their knowledge, religion, and political influence to areas from Mediterranean Europe to Bangladesh and from West Africa to Indonesia. Later, when Europe became the center of intellectual and material innovation, China found itself farther removed, by land or sea, than almost any other part of the world.

Today, modern communications notwithstanding, China still is distant from almost anywhere else on Earth. Going by rail from Beijing to Moscow, the capital of China's Eurasian neighbor, involves a tedious journey that takes the better part of a week. Direct surface connections with India are practically nonexistent. Overland linkages with Southeast Asian countries, though improving, remain tenuous.

But for the first time in its history, China now lies near a world-class hearth of technological innovation and financial power: Japan. This proximity to an industrial and financial giant is critical for

Long-suppressed ideas flowed into China, and a pro-democracy movement arose and climaxed in 1989. China's rulers knew that their violent repression of this movement would anger the world, but that did not matter because they deemed foreign condemnation irrelevant. Foreigners in China had done much worse. Moreover, Westerners had no business interfering in China's domestic affairs.

RELATIVE LOCATION

Throughout their nation's history, the Chinese have at times decided to close their country to any and all foreign influences; the most recent episode of exclusion occurred just a few decades ago. Exclusion is one of China's recurrent traditions, made possible by China's relative location and Asia's physiogra-

the momentous economic developments taking place in China's coastal provinces. Japanese investments and business partnerships have transformed the economic landscape of Pacific-coast China. American and European trade links also are important, but Japan's role was crucial. Japan's economic success set the Pacific Rim engine in motion, and Japan's best financial years happened to coincide with China's reopening to foreigners in the 1970s. Geographic and economic circumstances combined to transform the map and made "Pacific Rim" a household word around the world.

EXTENT AND ENVIRONMENT

China's total area is slightly smaller than that of the United States including Alaska: each country has about 3.7 million square miles (9.6 million sq km). As Figure 9-4 reveals, the longitudinal extent of China and the 48 contiguous U.S. States also is similar. Latitudinally, however, China is considerably wider. Miami, near the southern limit of the United States, lies halfway between Shanghai and Guangzhou. Thus China's lower-latitude southern

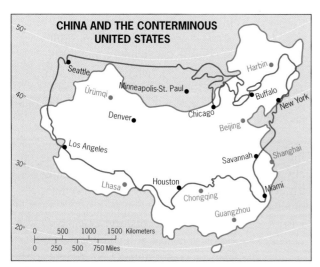

CHINA AND THE CONTERMINOUS UNITED STATES

FIGURE 9-4

region takes on characteristics of tropical Asia. In the Northeast, too, China incorporates much of what in North America would be Quebec and Ontario. Westward, China's land area becomes narrower and physiographic similarities increase. But, of course, China has no west coast.

Now compare the climate maps of China and the United States in Figure 9-5 (which are enlargements of the appropriate portions of the world climate map in Figure G-8). Note that both have a large southeastern climatic region marked *Cfa* (that is, humid, temperate, warm-summer), flanked in China by a zone of *Cwa* (where winters become drier). Westward in both countries, the *C* climates yield to colder, drier climes. In the United States, moderate *C* climates develop again along the Pacific coast. China, however, stays dry and cold as well as high in elevation at equivalent longitudes.

Note especially the comparative location of the U.S. and Chinese *Cfa* areas in Figure 9-5. China's lies much farther to the south. In the United States, the *Cfa* climate extends beyond 40° North latitude, but in China, cold and generally winter-dry *D* climates take over at the latitude of Virginia. Beijing has a warm summer but a bitterly cold and long winter. Northeast China, in the general latitudinal range of Canada's lower Quebec and Newfoundland, is much more severe than its North American equivalent. Harsh environments prevail over vast regions of China, but, as we will see, nature compensates in spectacular fashion. From the climatic zone marked *H* (for highlands) in the west come the great life-giving rivers whose wide basins contain enormous expanses of fertile soils. Without these waters, China would not have a population more than four times that of the United States.

EVOLVING CHINA

Even if China is not the world's longest continuous civilization (Egypt may claim this distinction), no other state on Earth can trace its cultural heritage as far back as China can. China's fortunes rose and fell,

but over more than 40 centuries its people created a society with strong traditions, values, and philosophies. The teachings of Kongfuzi (551–479 BC), still **2** known as **Confucius**, dominated Chinese life and thought for over 20 centuries. Kongfuzi not only taught and championed the poor and indigent: his revolutionary ideas extended to the rulers as well as the ruled. He abhorred supernatural mysticism and dismissed notions of divine ancestries of dynastic rulers. Competence and merit, not some godly legacy, he insisted, should determine a person's place in society. He wrote lengthy treatises from which emerged the so-called Confucian Classics, 13 texts that became the basis for education in China until Western influences during the Qing Dynasty began to erode their relevance. Ranging from government to morality and from law to religion, the Classics, especially the revised *Analects*, were Chinese civilization's guide. But the communists, who took power in 1949, attacked Kongfuzi thought on all fronts, substituting indoctrination for education and relegating even the family, one of Kongfuzi's predilections, to the "dustbin of history." Here the communists miscalculated. They were unable to eradicate two millennia of cultural conditioning in a few decades, and even today the spirit of Kongfuzi still haunts China's physical and mental landscapes.

As we try to gain a better understanding of China's present political, economic, and social geography, we should keep its continuities in mind. Yes, China is changing radically today, but it has undergone convulsive change in the past, and has endured and regained its coherence. Larger and more populous than Europe, China, too, had its divisive feudal periods, but unlike Europe, China always came together again under a single flag. In the process, often-dictatorial China incorporated minorities ranging from Koreans and Mongols to Uyghurs and Tibetans; in the far south, it now includes several peoples with Southeast Asian affinities. Like the Chinese citizenry itself, these minorities have experienced both benevolent government and brutal subjugation. But it has always been

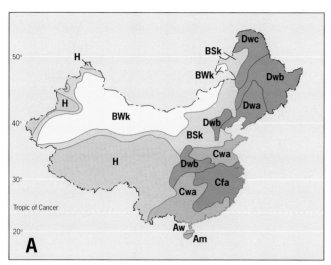

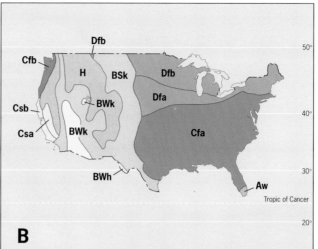

CLIMATES OF CHINA AND THE CONTERMINOUS UNITED STATES
After Köppen-Geiger

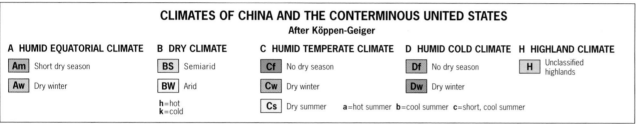

A HUMID EQUATORIAL CLIMATE	B DRY CLIMATE	C HUMID TEMPERATE CLIMATE	D HUMID COLD CLIMATE	H HIGHLAND CLIMATE
Am Short dry season	**BS** Semiarid	**Cf** No dry season	**Df** No dry season	**H** Unclassified highlands
Aw Dry winter	**BW** Arid	**Cw** Dry winter	**Dw** Dry winter	
	h=hot k=cold	**Cs** Dry summer a=hot summer b=cool summer c=short, cool summer		

FIGURE 9-5

3 China's wish to **Sinicize** them, to endow them with the elements of Chinese culture.

A Century of Convulsion

When the European colonialists appeared in East Asia, China long withstood them with a self-assured superiority based on the strength of its culture and the reassuring continuity of the state. There was no market for the British East India Company's rough textiles in a country long used to finely fabricated silks and cottons. There was little interest in the toys and trinkets the Europeans produced in the hope of barter for Chinese tea and porcelain. Even key European inventions, such as the mechanical clock, though considered amusing and entertaining, were ignored and even deprecated as irrelevant to Chinese culture.

The Ming emperors were particularly dismissive of European manufactures, but the Ming rulers' confidence in their home-made products was sometimes misplaced. When the Manchu forces invaded in 1644, their bows and arrows proved superior to Chinese-manufactured muskets, which were so heavy and difficult to load that they were almost useless. As we know, the Manchu emperors kept Ming administrators in office, and the prevailing attitude continued. Even when Europe's sailing ships made way for steam-driven vessels and newer and better European products (including weapons) were offered in trade for China's tea and silk, China continued to reject European imports and resisted commerce in general. The Chinese kept the Europeans confined to small peninsular outposts, such as Macau, and minimized interaction with them. Long after India had succumbed to mer-

cantilism and economic imperialism, China maintained its established order. This was no surprise to the Chinese. After all, they had held a position of undisputed superiority in their Celestial Kingdom as long as could be remembered, and they had dealt with foreign invaders before.

A (Lost) War on Drugs and Its Aftermath

The Manchus, however, had the misfortune of ruling China when the balance of power shifted decisively in favor of the colonialists. On two fronts in particular, the economic and the political, the European powers destroyed China's invincibility. Economically, they succeeded in lowering the cost and improving the quality of manufactured goods, especially textiles, and the handicraft industries of China began to collapse in the face of unbeatable competition. Politically, the demands of the British merchants and the growing English presence in China led to conflicts. In the early part of the nineteenth century, the central issue was the importation into China from British India of opium, a dangerous and addictive intoxicant. Opium was destroying the very fabric of Chinese culture, weakening the society, and rendering China easy prey for colonial profiteers. As the Qing (Manchu) government moved to stamp out the opium trade in 1839, armed hostilities broke out, and soon the Chinese found themselves losing a war on their own territory. The First Opium War (1839–1842) ended in disaster: China's rulers were forced to yield to British demands, and the breakdown of Chinese sovereignty was under way.

British forces penetrated up the Chang Jiang and controlled several areas south of it (Fig. 9-6); Beijing hurriedly sought a peace treaty by which it granted leases and concessions to foreign merchants. In

CHINA:
COLONIAL SPHERES, TERRITORIAL LOSSES
Manchu Dynasty at its greatest extent
The Long March, 1934–1935
Administered by Japan, 1937–1945
Territorial losses 19th and 20th Century
People's Republic of China (1949–)
19TH CENTURY COLONIAL INFLUENCE
Russian
British
French
German
0 300 600 900 Kilometers
0 100 200 300 400 500 Miles
RUSSIAN
TO RUSSIA (KAZAKHSTAN)
MONGOLIA (INDEPENDENT)
(TO RUSSIA)
RUSSIAN
RUSSIAN
XINJIANG (SINKIANG)
XIZANG (TIBET)
TO BRITAIN (INDIA)
BRITISH
GERMAN
FRENCH
TO BRITAIN (MYANMAR)
TO FRANCE (VIETNAM, LAOS, CAMBODIA)
TO JAPAN (N. & S. KOREA)
FORMOSA (TAIWAN) (TO JAPAN)
SAKHALIN
KARAFUTO
JAPAN
East Sea (Sea of Japan)
Yellow Sea
East China Sea
South China Sea
PACIFIC OCEAN
Ryukyu Islands
Tropic of Cancer
Beijing
Liaodong Peninsula
Shandong Peninsula
Qingdao
Yanan
Nanjing
Shanghai
Chongqing
Ruijin
Guangzhou
Macau
Hong Kong (Xianggang)
Zhanjiang
Amur R.
Songhua R.
Ussuri R.
Liao R.
Huang He
Wei R.
Chang Jiang
Xi Jiang
Longitude East of Greenwich

FIGURE 9-6

addition, China ceded Hong Kong Island to the British and opened five ports, including Guangzhou (Canton) and Shanghai, to foreign commerce. No longer did the British have to accept a status that was inferior to the Chinese in order to do business; henceforth, negotiations would be pursued on equal terms. Opium now flooded into China, and its impact on Chinese society became even more devastating. Fifteen years after the First Opium War, the Chinese again tried to stem the disastrous narcotic tide, and the foreigners who had attached themselves to their country again defeated them. Now the government legalized cultivation of the opium poppy in China itself. Chinese society was disintegrating; the scourge of this drug abuse was not defeated until after the revival of Chinese power in the twentieth century.

But before China regained control of its destiny, it lost heavily to the colonial powers. The Germans in 1898 obtained a "lease" on the city of Qingdao on the Shandong Peninsula. The French acquired a sphere of influence in the far south at Zhanjiang (Fig. 9-6). The Portuguese confirmed their hold over Macau. The Russians took control over Liaodong in the Northeast. Even Japan got into the act by annexing the Ryukyu Islands (1879) and, more importantly, Formosa (Taiwan) in 1895.

One of the most humiliating practices with a geographic dimension, which the Europeans forced upon the Chinese, was the doctrine known as 4 **extraterritoriality**. Under this doctrine, foreign states and their representatives are immune from the jurisdiction of the country in which they are based, for example, in the case of embassies and diplomatic personnel. But in Qing (Manchu) China, it went far beyond that. The European and Japanese invaders established as many as 90 so-called *treaty ports* which were, in effect, extraterritorial enclaves where traders as well as diplomats were exempt from Chinese law. The best residential areas in major cities were declared "extraterritorial" and made inaccessible to Chinese citizens, including Sha Mian Island on the Pearl River waterfront in the city of Canton (now Guangzhou). Christian missionaries fanned out into China, their residences and churches fortified with extraterritorial security. In many places, Chinese citizens found themselves unable to enter parks and buildings without permission from foreigners. This involved a loss of face and built a resentment that exploded in the Boxer Rebellion of 1900.

Figure 9-6 shows the fate of the territory acquired in the early period of Manchu rule. Much of the Northeast was taken by Russia; the Japanese colonized Korea; Mongolia became independent under foreign auspices; large parts of Turkestan were lost, also to Russia; the French took over in Indochina and the British in Burma. But in the early twentieth century, the Chinese had had enough. Bands of revolutionaries roamed city and countryside, attacking the hated foreigners as well as Chinese who had adopted Western ways. The Boxer Rebellion (after a loose translation of the Chinese name for these revolutionaries) was put down with much bloodshed by what today would be called a "multinational force:" British, French, German, Italian, Russian, Japanese, and American soldiers participated. But at the same time a domestic Chinese revolutionary movement was gaining support, aimed against the Qing Dynasty itself. In 1911, the emperor's garrisons were attacked all over China, and in a few months the 267-year-old dynasty was overthrown.

The so-called Nationalist forces that ousted the last emperor proclaimed a republican China and negotiated an end to the extraterritorial treaties, but China remained a badly divided country. Even while the famous Chinese revolutionary leader

Sun Yat-sen tried to unify the country from his base in Canton (Guangzhou) in the south, another government tried to rule from Beijing, the old imperial headquarters in the north. Meanwhile, a group of intellectuals in Shanghai founded the Chinese Communist Party. A prominent member of this group was a man named Mao Zedong.

Nationalists and Communists

During these chaotic 1920s, the Nationalists and the Communist Party at first cooperated, with the remaining foreign presence their joint target. After Sun Yat-sen's death in 1925, Chiang Kai-shek became the Nationalists' leader, and by 1927 the foreigners were on the run, escaping by boat and train or falling victim to rampaging Nationalist forces. But soon the Nationalists began purging communists even as they pursued foreigners, and in 1928, when Chiang established his Nationalist capital in the city of Nanjing, it appeared that the Nationalists would emerge victorious from their campaigns. They had driven the communists ever deeper into the interior, and by 1933 the Nationalist armies were on the verge of encircling the last communist stronghold in the area of Ruijin in Jiangxi Province.

This led to a momentous event in Chinese history: the *Long March*. Nearly 100,000 people— soldiers, peasants, leaders—marched westward from Ruijin in 1934, a communist column that included Mao Zedong and Zhou Enlai. The Nationalist forces rained attack after attack on the marchers, and of the original 100,000, about three-quarters were killed. But new sympathizers joined along the way (see the route marked on Fig. 9-6), and the 20,000 survivors found a refuge in the mountainous interior of Shaanxi Province, 6000 miles (10,000 km) away. There, they prepared for a renewed campaign that would bring them to power.

Japan in China

Although many foreigners fled China during the 1920s and 1930s, others seized the opportunity

presented by the contest between the Nationalists and communists. The Japanese took control over the Northeast, and when the Nationalists proved unable to dislodge them, they set up a puppet state there, appointed a Manchu ruler to represent them, and called their possession Manchukuo.

The inevitable full-scale war between the Chinese and the Japanese broke out in 1937, with the Nationalists bearing the brunt of it (which gave the communists an opportunity to regroup). The gray boundary on Figure 9-6 shows how much of China the Japanese conquered. The Nationalists moved their capital to Chongqing, and the communists controlled the area centered on Yanan. China had been broken into three pieces.

The Japanese committed unspeakable atrocities in their campaign in China. Millions of Chinese citizens were shot, burned, drowned, subjected to gruesome chemical and biological experiments, and otherwise wantonly victimized. Years later, when China's economic reforms of the 1980s and 1990s led to a renewed Japanese presence in China, the Chinese public and its leaders called for Japan to acknowledge and apologize for these wartime abuses. In Japan, this pitted apologists against strident nationalists, causing a political crisis. In 1992, Emperor Akihito visited China and referred to the war but stopped short of a formal apology. The book is not yet closed on this most sensitive issue.

Communist China Arises

After the U.S.-led Western powers defeated Japan in 1945, the civil war in China quickly resumed. The United States, hoping for a stable and friendly government in China, sought to mediate the conflict but at the same time recognized the Nationalists as the legitimate government. The United States also aided the Nationalists militarily, destroying any chance of genuine and impartial mediation. By 1948, it was clear that Mao Zedong's well-organized militias would defeat Chiang Kai-shek. Chiang kept moving his capital—back to

"To pass through the gate from Tiananmen Square and enter the Manchu Emperors' Forbidden City was a riveting experience. Look back now toward the gate, and you see throngs of Chinese visiting what less than a century ago was the exclusive domain of the Qing Dynasty's absolute rulers—so absolute that unauthorized entry was punished by instant execution. The vast, walled complex at the heart of Beijing not only served as the imperial palace; it was a repository of an immense collection of Chinese technological and artistic achievements. Some of this is still here, but most of all you are awed by the lingering atmosphere, the sense of what happened here in these exquisitely constructed buildings that epitomized what high Chinese culture could accomplish. Inevitably you think of the misery of millions whose taxes and tribute paid for what stands here."

Guangzhou, seat of Sun Yat-sen's first Nationalist government, then to Chongqing. Late in 1949, after a series of disastrous defeats in which hundreds of thousands of Nationalist forces were killed, the remnants of Chiang's faction gathered Chinese treasures and valuables and fled to the island of Taiwan. There, they took control of the government and proclaimed their own Republic of China.

Meanwhile, on October 1, 1949, standing in front of the assembled masses at the Gate of Heavenly Peace on Beijing's Tiananmen Square, Mao Zedong proclaimed the birth of the People's Republic of China.

CHINA'S HUMAN GEOGRAPHY

After more than a half-century of communist rule, China is a society transformed. It has been said that the year 1949 actually marked the beginning of a new dynasty not so different from the old, an autocratic system that dictated from the top. In that view, Mao Zedong simply bore the mantle of his dynastic predecessors. Only the family lineage had fallen away; now communist "comrades" would succeed each other.

And certainly some of China's old traditions continued during the communist era, but in many other ways Chinese society was totally overhauled. Benevolent or otherwise, the dynastic rulers of old China headed a country in which—for all its splendor, strength, and cultural richness—the fate of landless people and of serfs often was undescribably miserable; in which floods, famines, and diseases could decimate the populations of entire regions without any help from the state; in which local lords could (and often did) repress the people with impunity; in which children were sold and brides were bought. The European intrusion made things even worse, bringing slums, starvation, and deprivation to millions who had moved to the cities.

The communist regime, dictatorial though it was, attacked China's weaknesses on many fronts, mobilizing virtually every able-bodied citizen in the process. Land was taken from the wealthy; farms were collectivized; dams and levees were built with the hands of thousands; the threat of hunger for millions receded; health conditions improved; child labor was reduced. But China's communist planners also made terrible mistakes. The *Great Leap Forward*, requiring the reorganization of the peasantry into communal brigade teams to speed industrialization and make farming more productive, had the opposite effect and so disrupted agriculture that between 20 and 30 million people died of starvation between 1958, when the program was implemented, and 1962, when it was abandoned.

Mao ruled China from 1949 to 1976, long enough to leave lasting marks on the state. Another of his communist dictums had to do with population. Like the Soviets (and influenced by a horde of Soviet advisors and planners), Mao refused to impose or even recommend any population policy, arguing that such a policy would represent a capitalist plot to constrain China's human resources. As a result, China's population grew explosively during his rule.

Yet another costly episode of Mao's rule was the so-called *Great Proletarian Cultural Revolution*, launched by Mao Zedong during his last decade in power (1966–1976). Fearful that Maoist communism was being contaminated by Soviet "deviationism" and worried about his own stature as its revolutionary architect, Mao unleashed a campaign against what he viewed as emerging elitism in society. He mobilized young people living in cities and towns into cadres known as Red Guards and ordered them to attack "bourgeois" elements throughout China, criticize Communist Party officials, and root out "opponents" of the system. He shut down all of China's schools, persecuted untrustworthy intellectuals, and encouraged the Red Guards to engage in what he called a renewed "revolutionary experience." The results were disastrous: Red Guard factions took to fighting among themselves, and anarchy, terror, and economic paralysis followed. Thousands of China's leading intellectuals died, moderate leaders were purged, and teachers, elderly citizens, and older revolutionaries were tortured to make them confess to "crimes" they did not commit. As the economy suffered, food and industrial production declined. Violence and famine killed as many as 30 million people as the Cultural Revolution spun out of control. One of those who survived was a Communist Party leader who had himself been purged and then been reinstated—Deng Xiaoping. Deng was destined to lead the country in the post-Mao period of economic transformation.

Political and Administrative Divisions

Before we investigate the emerging human geography of contemporary China, we should acquaint ourselves with the country's political and administrative framework (Fig. 9-7). For administrative purposes, China is divided into the following units:

4 Central-government-controlled Municipalities (*Shi's*)

5 Autonomous Regions

22 Provinces

2 Special Administrative Regions

The four central-government-controlled municipalities are the capital, Beijing; its nearby port city, Tianjin; China's largest metropolis, Shanghai; and the Chang River port of Chongqing, in the interior. These *Shi's* form the cores of China's most populous and important subregions, and direct control over them from the capital entrenches the central government's power.

We should note that the administrative map of China continues to change—and to pose problems for geographers. The city of Chongqing was made a *shi* in 1996, and its "municipal" area was enlarged to incorporate not only the central urban area but a huge hinterland covering all of eastern Sichuan Province. As a result, the "urban" population of Chongqing is officially 30 million, making this the world's largest metropolis—but in truth, the central urban area has no more than about 6 million inhabitants. And because Chongqing's population is officially *not* part of the province that borders it to the west (Fig. 9-7), the official population of Sichuan declined by 30 million when the *Chongqing Shi* was created.

The five Autonomous Regions were established to recognize the non-Han minorities living there. Some laws that apply to Han Chinese do not apply to certain minorities. As we saw in the case of the former Soviet Union,

POLITICAL DIVISIONS OF CHINA

— - — International boundary
— - - — Province boundary
National capital is underlined

FIGURE 9-7

however, demographic changes and population movements affect such regions, and the policies of the 1940s may not work in the twenty-first century. Han Chinese immigrants now outnumber several minorities in their own regions. The five Autonomous Regions (A.R.'s) are: (1) Nei Mongol A.R. (Inner Mongolia); (2) Ningxia Hui A.R. (adjacent to Inner Mongolia); (3) Xinjiang Uyghur A.R. (China's northwest corner); (4) Guangxi Zhuang A.R. (far south, bordering Vietnam); and (5) Xizang A.R. (Tibet).

China's 22 Provinces, like U.S. States, tend to be smallest in the east and largest toward the west. The territorially smallest are the three easternmost provinces on China's coastal bulge: Zhejiang, Jiangsu, and Fujian. The two largest are Qinghai, flanked by Tibet, and Sichuan, China's Midwest.

As with all large countries, some provinces are more important than others. The Province of Hebei nearly surrounds Beijing and occupies much of the core of the country. The Province of Shaanxi is centered on the great ancient city of Xian. In the southeast, momentous economic developments are occurring in the Province of Guangdong, whose urban focus is Guangzhou. When, in the pages that follow, we refer to a particular province or region, Figure 9-7 is a useful locational guide.

In 1997, the British dependency of Hong Kong (Xianggang) was taken over by China and became the **5** country's first **Special Administrative Region (SAR)**. In 1999, Portugal similarly transferred Macau, opposite Hong Kong on the Pearl River estuary, to Chinese control, creating the second SAR under Beijing's administration.

Population Issues

Many Chinese provinces, like many of India's States, have populations larger than most of the world's countries. With Chongqing Shi, Sichuan Province has almost 120 million inhabitants. Approaching 100 million are Henan and Shandong provinces (see Fig. 9-7). There are almost as many people in Guangdong Province as in Germany. Jiangsu Province has 14 million more inhabitants than France.

With a population of over 1.3 billion in 2004, China is the largest na-

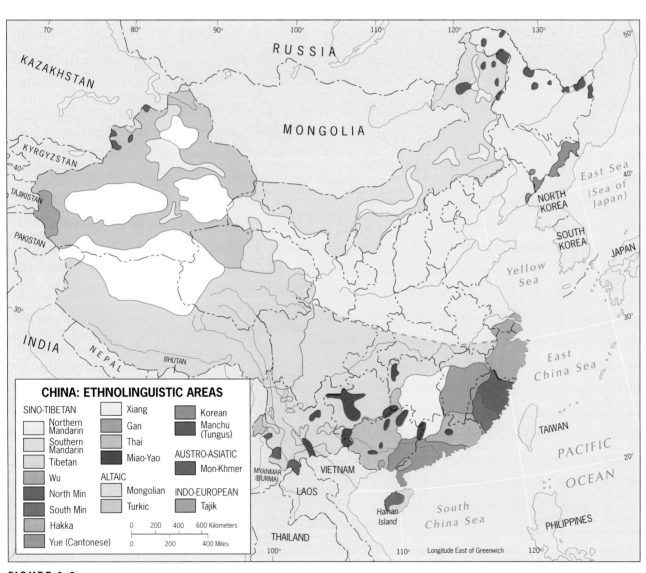

CHINA: ETHNOLINGUISTIC AREAS

SINO-TIBETAN
- Northern Mandarin
- Southern Mandarin
- Tibetan
- Wu
- North Min
- South Min
- Hakka
- Yue (Cantonese)
- Xiang
- Gan
- Thai
- Miao-Yao

ALTAIC
- Mongolian
- Turkic

- Korean
- Manchu (Tungus)

AUSTRO-ASIATIC
- Mon-Khmer

INDO-EUROPEAN
- Tajik

0   200   400   600 Kilometers

0   200   400 Miles

FIGURE 9-8

tion on Earth, and it inherited from the communist period a high rate of natural increase. Aware of the economic costs of rapid population growth, China after Mao's death embarked on a vigorous population-control program. In the early 1970s, the annual rate of natural increase was about 3 percent; by the mid-1980s, it was down to 1.2 percent. Families were ordered to have one child only, and those who violated the policy were penalized by losing tax advantages, educational opportunities, and even housing privileges. Today, China officially reports a moderate growth rate of 0.7 percent.

The Minorities

When the Soviet Union collapsed, some observers described that event as the "end of empire"—the breakdown of the last empire on Earth. But those observers forgot about China. Diminished from its Qing Dynasty dimensions as China may be, the Beijing government controls territories and peoples that are in effect colonized. The ethnolinguistic map (Fig. 9-8) apparently confirms this proposition. This map should be seen in context, however. It does not show local-area majority populations but instead reveals where minorities are concentrated. For example, the Mongolian population in China is shown to be clustered along the southeastern border of Mongolia in the Autonomous Region called Nei Mongol (see Fig. 9-7). But even in that A.R., the Han Chinese, not the Mongols, are now in the majority.

Nevertheless, the map gives definition to the term *China Proper* as the home of the People of Han, the ethnic (Mandarin-speaking) Chinese depicted in tan and light orange in Figure 9-8. From the upper Northeast to the border with Vietnam and from the Pacific coast to the margins of Xinjiang, this Chinese majority dominates. When you compare this map to that of population distribution (to be discussed next), it will be clear that the minorities constitute only a small percentage of the country's total. The Han Chinese form the largest and densest clusters.

In any case, China controls non-Chinese areas that are vast, if not populous. The Buddhist Tibetan group numbers under 3 million, but it extends over all of settled Xizang. Muslim Turkic peoples inhabit large areas of Xinjiang. Thai, Vietnamese, and Korean minorities also occupy areas on the margins of Han China. As we will see later, the Southeast Asian minorities in China have participated strongly in the Pacific Rim developments on their doorsteps. Hundreds of thousands have migrated from their Autonomous Regions to the economic opportunities along the coast.

Numerically, Chinese dominate in China to a far greater degree than Russians dominated their Soviet Empire. But territorially, China's minorities extend over a proportionately larger area. The Ming and Manchu rulers bequeathed the People of Han an empire.

PEOPLE AND PLACES OF CHINA PROPER

A map of China's population distribution (Fig. 9-9) reveals the continuing relationship between the physical stage and its human occupants. In technologically advanced countries, we have noted, people shake off their dependence on what the land can provide; they cluster in cities and in other areas of economic opportunity. This depopulates rural areas that may once have been densely inhabited. In China, that stage has not yet been reached. While China has large cities, as in India a sizeable majority of the people (62 percent) still live on—and from—the land. Thus the map of population distribution reflects the livability and productivity of China's basins, lowlands, and plains. Compare Figures 9-5A and 9-9, and China's continuing dependence on soil, water, and warmth will be evident.

The population map also suggests that in certain areas environmental limitations are being overcome. Industrialization in the Northeast, irrigation in the Inner Mongolia Autonomous Region, and

oil-well drilling in Xinjiang enabled millions of Chinese to migrate from China Proper into these frontier zones, where they now collectively outnumber the indigenous minorities.

Nevertheless, physiography and demography remain closely linked in China. To grasp this relationship, it is useful to compare Figures 9-1 (physiography) and 9-9 (population) as our discussion proceeds. On the population map, the darker the color, the denser the population: in places we can follow the courses of major rivers by this means. Look, for example, at China's Northeast. A ribbon of population follows the lower Liao and Songhua rivers. Also note the huge, nearly circular population concentration on the western edge of the red zone; this is the Sichuan Basin, in the upper course of the Chang Jiang. Four major river basins contain more than three-quarters of China's 1.3 billion people:

1. The Liao-Songhua Basin or the Northeast China Plain
2. The Lower Huang He (Yellow River) Basin, known as the North China Plain
3. The Upper and Lower Basins of the Chang Jiang (Yangzi River)
4. The Basins of the Xi (West) and Pearl River

Not all the people living in these river plains are farmers, of course. China's great cities also have developed in these populous areas, from the aging industrial centers of the Northeast to the Pacific Rim upstarts of the South. Both Beijing and Shanghai lie in major river basins. And, as the map shows, the hilly areas between the river basins are not exactly sparsely populated either. Note that the hill country south of the Chang Basin (opposite Taiwan) still has a minimal density of 130 to 260 people per square mile (50 to 100 per sq km).

Northeast China Plain

The Northeast China Plain is the heartland of China's Northeast, ancestral home of the Manchus

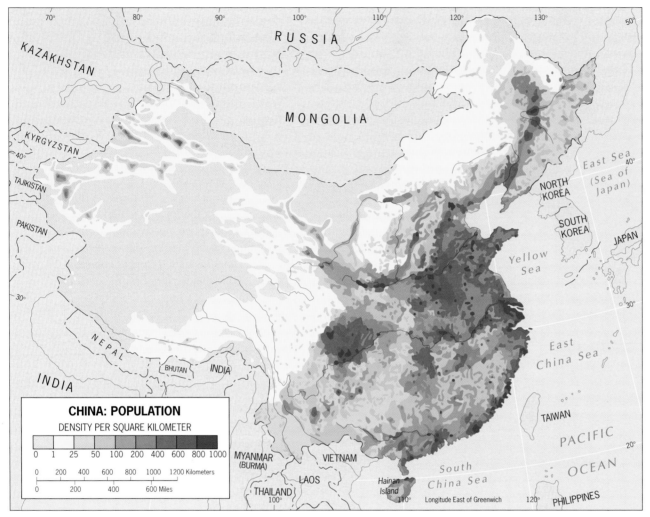

FIGURE 9-9

power and made the industrial development of the Northeast a priority. From the 1950s until the 1970s, the Northeast led the nation in manufacturing growth. Its population, just a few million in the 1940s, mushroomed to more than 100 million. Towns and cities grew exponentially.

All this growth was based on the Northeast's considerable mineral wealth (see the symbols in Fig. 9-10), and for a time, it worked. During the 1970s, the Northeast contributed fully one-quarter of the entire country's industrial output.

6 But then came the **economic restructuring** of the post-Mao era, resulting from a new, market-driven economic order. The huge, inefficient, state-supported industries could not adapt, and in short order the Northeast became a Rustbelt, its contribution to China's industrial output plummeting below 10 percent. As factories closed, hundreds of thousands of workers found themselves destitute, their factory jobs and pensions eliminated and their families impoverished. In the frigid winters of the Northeast, they staged protests, blocking roads, railroads and even airport runways to demand government help, a rarity in tightly-controlled China.

who founded China's last dynasty, battleground between Japanese and Russian invaders, Japanese colony, heartland of communist industrial development, and now losing ground to the industries of the Pacific Rim. This used to be called Manchuria. As the administrative map (Fig. 9-7) shows, there are three provinces here: Liaoning in the south, facing the Yellow Sea; Jilin in the center; and Heilongjiang, by far the largest, in the north.

The Northeast has experienced many ups and downs. Although Japanese colonialism was ruthless and exploitive, Japan did build railroads, roads, bridges, factories, and other components of the regional infrastructure. (At one time, half of the entire railroad mileage in China was in the Northeast.) After the Japanese were ousted, the Soviets looted the area of machinery, equipment, and other assets. During the late 1940s, the Northeast was a ravaged frontier. But then the communists took

Recently, the government has begun to address the Northeast's problems, in part by awarding laid-off workers a small stipend but also by seeking ways to revive the region's fortunes. One initiative involves Russia, whose sliver of coastal territory landlocks the Northeast's hard-hit Jilin Province (Fig. 9-10). The Chinese want Russia to lease them a coastal enclave where China would build a small Special Economic Zone (see p. 322) complete with port facilities, linking Jilin and thus the Northeast

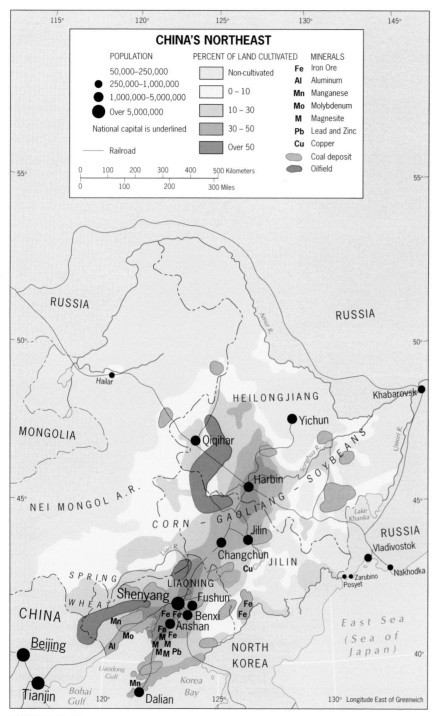

FIGURE 9-10

to the outside world. The Russians, fearing a Chinese takeover, have resisted, arguing that their own small ports, Zarubino and Posyet, can handle any transit trade China's Northeast will generate. China proposed building a facility where oil and natural gas can be imported from Russia; the Russians responded by suggesting a joint venture with China instead of a Chinese lease on any part of their coast. There the matter stood in mid-2004, an obstacle to Chinese hopes of bringing the troubled Northeast into the Pacific Rim sphere.

North China Plain

The North China Plain is one of the world's most heavily populated agricultural areas. Figure 9-9 shows that most of it has a density of more than 1000 people per square mile (400 per sq km), and in some parts the density is twice as high. Here, the ultimate hope of the Beijing government lay less in land redistribution than in raising yields through improved fertilization, expanded irrigation facilities, and the more intensive use of labor. A series of dams on the Huang River, including the Xiaolangdi Dam upstream in Henan Province, now reduce the flood danger, but outside the irrigated areas the ever-present problem of rainfall variability and drought persists. The North China Plain has not produced any substantial food surplus even under normal circumstances; thus when the weather turns unfavorable the situation soon becomes precarious. The specter of famine may have receded, but the food situation is still uncertain in this very critical part of China Proper.

The North China Plain is an excellent example **7** of the concept of a national **core area**. Not only is it a densely populated, highly productive agricultural zone, but it also is the site of the capital and other major cities, a substantial industrial complex, and several ports, among which Tianjin ranks as one of China's largest (Fig. 9-11). Tianjin, on the Bohai Gulf, is linked by rail and highway (less than a two-hour drive) to Beijing. Like many of China's harbors, that of Tianjin's river port is not

FROM THE FIELDNOTES

"We flew over the North China Plain for an hour, and one could not miss the remarkably regular spacing of the villages on this seemingly table-flat surface. Many of these villages became communes during the communist reorganization of China's social order. The elongated buildings (many added since the 1950s) each were designed to contain one or more extended families. After the demise of Mao and his clique the collectivization effort was reversed and the emphasis returned to the small nuclear family. In this village and others we overflew, the long sides of the buildings almost always lay east-west, providing maximum sun exposure to windows and doors, and enhancing summer ventilation from longitudinal breezes. The more distant view looks hazy, because the North China Plain during the spring is wafted by loess-carrying breezes from the northwest. On the ground, a yellow-gray hue shrouds the atmosphere until a passing weather front temporarily clears the air."

tropolis and the center of one of its leading industrial complexes.

Beijing, unlike Tianjin, is China's political, cultural, and educational center. Its industrial development has not matched Tianjin's. The communist administration did, however, greatly expand the municipal area of Beijing, which (as we noted), is not controlled by the Province of Hebei but is directly under the central government's authority. In one direction, Beijing was enlarged all the way to the Great Wall—30 miles (50 km) to the north—so that the "urban" area includes hundreds of thousands of farmers. In central Beijing, old, traditional neighborhoods called *hutongs* are being razed to make way for 8-lane highways and 40-story buildings. Beijing, with its 10.9 million inhabitants, ranks among the world's most dynamic megacities.

Inner Mongolia

Northwest of the core area as defined by the North China Plain, along the border with the state of Mongolia, lies Inner Mongolia, administratively defined as the Nei Mongol Autonomous Region (see Fig. 9-7). Originally established to protect the rights of the approximately 5 million Mongols who live outside the Mongolian state, Inner Mongolia has been the scene of massive immigration by Han Chinese. Today, Chinese outnumber Mongols here by nearly four to one. Irrigation and industry have created an essentially Chinese landscape, so different from independent Mongolia that you can observe the location of the border from an aircraft. Fences and eroded land mark the Chinese side; grassy steppe, the Mongolian. The A.R.'s capital, Hohhot, has been eclipsed by Baotou on the Huang He, which supports a corridor of farm settlements as it crosses the dry land here on the margins of the Gobi and Ordos deserts. With about 24 million inhabitants, Inner Mongolia still cannot compare to the huge numbers that crowd the North China Plain, but recent mineral discoveries have boosted industry in Baotou and livestock herding is ex-

particularly good; but Tianjin is well situated to serve not only the northern sector of the North China Plain and the capital, but also the Upper Huang Basin and Inner Mongolia beyond. Tianjin, again like several other Chinese ports, had its modern start as a treaty port, but the city's major growth awaited communist rule. For decades it was a center for light industry and a flood-prone

harbor, but after 1949 the communists constructed a new artificial port and flood canals. They also chose Tianjin as a site for major industrial development and made large investments in the chemical industry (in which Tianjin still leads China), iron and steel production, heavy machine manufacturing, and textiles. Today, with a population of 9.3 million, Tianjin is China's fourth largest me-

FROM THE FIELDNOTES

"Beijing seems to be one gigantic construction project, with six-lane ringroads and arteries connecting clusters of high-rises still sprouting from the ground. But there is another side to this. Old, historic Beijing's neighborhoods are being swept away, often without any consideration either for the heritage being lost or for the families being displaced. Here, near the old city wall in the southwest, part of a neighborhood already has been bulldozed, and locals salvage what they can (including reusable bricks) from the rubble. Laundry still hangs from lines in front of the still-occupied houses in the background, but any day they, too, will be gone."

panding. Nei Mongol may retain its special administrative status, but it functions as part of Han China in all but name.

Basins of the Chang/Yangzi

In contrast to the contiguous, flat agricultural-urban-industrial North China Plain defined by the sediment-laden Huang River, the basins and valleys of the Chang Jiang (Long River) display variation in elevation and relief. The Chang River, whose lower-course name becomes the Yangzi, is

an artery like no other in China. Near its mouth lies the country's largest city, Shanghai. Part of its middle course is now being transformed by a gigantic engineering project. Farther upstream, the Chang crosses the populous, productive Sichuan Basin. And unlike the Huang, the Chang Jiang is navigable to oceangoing ships for over 600 miles (1000 km) from the coast all the way to Wuhan (Fig. 9-11). Smaller ships can reach Chongqing, even after dam construction. Several of the Chang River's tributaries also are navigable, so that 18,500 miles (30,000 km) of water transport

routes serve its drainage basin. Thus the Chang Jiang constitutes one of China's leading transit corridors. With its tributaries it handles the trade of a vast area, including nearly all of middle China and sizeable parts of the north and south. The North China Plain may be the core area of China, but in many ways the Lower Chang Basin is its heart.

As Figure 9-12 shows, the Lower Chang Basin is an area of both rice and wheat farming, offering further proof of its pivotal situation between south and north in the heart of China Proper. Shanghai (population: 12.7 million) lies at the coastal gateway to this productive region on a small tributary of the Chang (now Yangzi) River, the Huangpu. The city has an immediate hinterland of some 20,000 square miles (50,000 sq km)—half the size of Ohio—containing more than 50 million people. About two-thirds of this population are farmers who produce food, silk filaments, and cotton for the city's industries.

Travel upriver along the Yangzi/Chang Jiang, and you meet an unending stream of vessels large and small, including numerous "barge trains"—as many as six or more barges pulled by a single tug in an effort to save fuel (they are slow, but time is not the primary concern). The traffic between Shanghai and Wuhan—Wuhan is short for Wuchang, Hanyang, and Hankou, a coalescing conurbation of three cities—makes the Yangzi one of the world's busiest waterways. Smoke-belching boat engines create a persistent plume of pollution here, worsening the regional smog created by the factories of Nanjing and others perched on the waterfront. Here China's Industrial Revolution is in full gear, with all its environmental consequences.

Above Wuhan the river traffic dwindles because the depth of the Chang reduces the size of vessels that can reach Yichang. River boats carry coal, rice, building materials, barrels of fuel, and many other items of trade. But the middle course of the Chang River is becoming more than a trade route. In the northward bend of the great river between Yichang and Chongqing, where the Chang

FIGURE 9-11

has cut deep troughs on its way to the coast, one of the world's largest engineering projects is in progress (with a completion date of 2009). It has several names: the Sanxia Project, the Chang Jiang Water Transfer Project, the Three Gorges Dam, and the New China Dam. It is most commonly called the Three Gorges Dam because it is here that the great river flows through a 150-mile (240-km) series of steep-walled valleys less than 360 feet (110 m) wide. Near the lower end of this natural trough, a concrete dam has risen to a height of over 600 feet (180 m) above the valley floor to a width of 1.3 miles (2.1 km), creating a reservoir that will inundate the Three Gorges and extend more than 385 miles (over 600 km) upstream. The rising water required the relocation of more than 1.2 million people, but China's project engineers say that the dam will end the river's rampaging flood cycle, enhance navigation, stimulate development along the new lake perimeter, and provide at least one-tenth, and perhaps as much as one-eighth, of China's electrical power supply. As such it will transform the heart of China, which is why many Chinese like to compare this gigantic project to the Great Wall and the Grand Canal, and call it the New China Dam.

The Three Gorges Dam project will strongly affect the fortunes of the city of Chongqing, upstream from the reservoir. A diversion channel around the dam will enhance Chongqing's river-port functions, allowing bigger vessels to reach it. Although Chengdu is Sichuan's capital, Chongqing, whose status was recently ele-

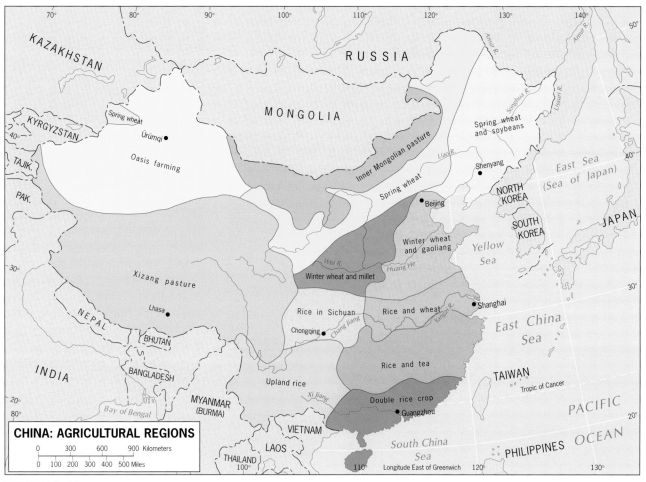

FIGURE 9-12

CHINA: AGRICULTURAL REGIONS
0 300 600 900 Kilometers
0 100 200 300 400 500 Miles

Lower Chang Basin. And, as Figure 9-9 shows, the population in this southern part of China is smaller than that in the more northerly heartlands.

Again, unlike the Huang and Chang rivers, which rise among snow-covered interior mountains, the Xi is a shorter stream whose source is on the Yunnan Plateau (see Fig. 9-1). Just up the coast from its delta, however, lies one of the most important areas of modern China. As we will note later, the mouth of the Pearl River is flanked by some of China's fastest-growing economic complexes, including Guangzhou, capital of Guangdong Province, Hong Kong, the former British dependency, and adjoining Shenzhen, one of the world's fastest-growing cities.

This is all the more remarkable because, as the maps emphasize, South China is not especially well endowed with geographic advantages, locational or otherwise. The Pearl River is navigable for larger ships only to Guangzhou, and the Xi for small river boats no farther than Wuzhou. Guangdong Province, the key province of southern China, lies far from China's heartlands. Nor is South China blessed with the resources that propelled the industrial development of the Northeast and other parts of China. Indeed, one of the principal objectives of the Three Gorges Dam project is to provide South China with ample electricity. As Figure 9-13 shows, northern and western China have most of the coal, oil, and natural gas resources, not the South.

China Proper, with its populous and powerful subregions, is in many ways the dominant region in the East Asian geographic realm. But China Proper is only one part of China. To the west lie two vast regions with relatively sparse populations: Xizang

vated to *shi*, is the province's key outlet. With 120 million people, energy resources, and farms that produce crops ranging from grains to tea, sugarcane, fruits, and vegetables, Sichuan is a Chinese breadbasket, and Chongqing may become the country's number-one growth pole.

Basins of the Xi (West) and Pearl River

As Figure 9-11 shows, the Xi River and its basins are no match for the Chang or Huang, not even for the Liao. This southernmost river even seems to have the wrong name: Xi means West! It reaches the coast in a complex area where it forms a delta immediately adjacent to the estuary of the Pearl River. In this subtropical part of China, local relief is higher than in the lowlands and basins of the center and north, so that farmlands are more confined. Especially in the interior areas, water supply is a recurrent problem. On the other hand, its warm climate permits the double-cropping of rice. Regional food production, however, has never approached that of the North China Plain or the

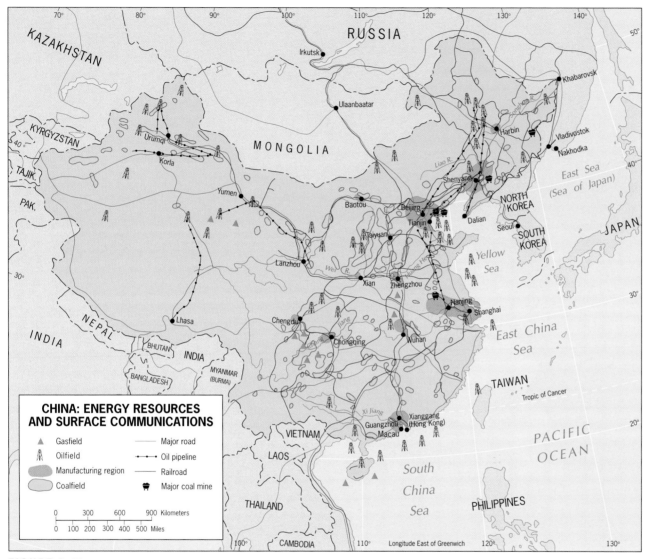

FIGURE 9-13

CHINA: ENERGY RESOURCES
AND SURFACE COMMUNICATIONS

▲	Gasfield	——	Major road
⛏	Oilfield	•–•–•	Oil pipeline
(region)	Manufacturing region	——	Railroad
(region)	Coalfield	🛒	Major coal mine

the climate is comparatively mild, and some cultivation is possible. Here lies Tibet's main population cluster, including the crossroads capital of Lhasa. The Chinese government has made investments to develop these valleys, which contain excellent sites for hydroelectric power projects (some have been used for a few light industries) and promising mineral deposits. The second area of interest is the Qaidam Basin to the north (Fig. 9-14). This basin lies thousands of feet below the surrounding Kunlun and Altun Mountains and has always contained a concentration of nomadic pastoralists. Recently, however, exploration has revealed the presence of oilfields and coal reserves below the surface of the Qaidam Basin, and these resources are now being developed.

As Figure 9-2 reminds us, Tibet came under Chinese domination during the Manchu (Qing) Dynasty in 1720, but it regained its separate status in the late nineteenth century. China's communist regime took control after the invasion of 1950; in 1959, China crushed an uprising after Tibetan villagers tried to resist the Chinese presence. Tibetan society had been organized around the fortress-like monasteries of Buddhist monks who paid allegiance to their supreme leader, the Dalai Lama. The Chinese wanted to modernize this feudal system, but the Tibetans clung to their traditions. In 1959, they proved no match for the Chinese armed forces; the Dalai Lama was ousted, and the monasteries were emptied. The Chinese destroyed much of Tibet's cultural heritage, looting its religious treasures and works of art. Their harsh rule devastated Tibetan society, but after Mao's death in 1976 the Chinese relaxed their tight control. Amends were made, religious treasures were

(Tibet) and Xinjiang. And here, in this remote western periphery adjoining Turkestan, China meets Muslim Central Asia.

XIZANG (TIBET)

As the climate and physiographic maps demonstrate, harsh physical environments dominate this colony of China. It is a vast, sparsely peopled region designated as an Autonomous Region, but in fact Xizang (Tibet) is an occupied society. In terms of its human geography, we should take note of two subregions.

The first is the core area of Tibetan culture, located on the southeastern edge of the Qinghai-Xizang Plateau just north of the Himalayas. In this area, some valleys lie below 7000 feet (2130 m);

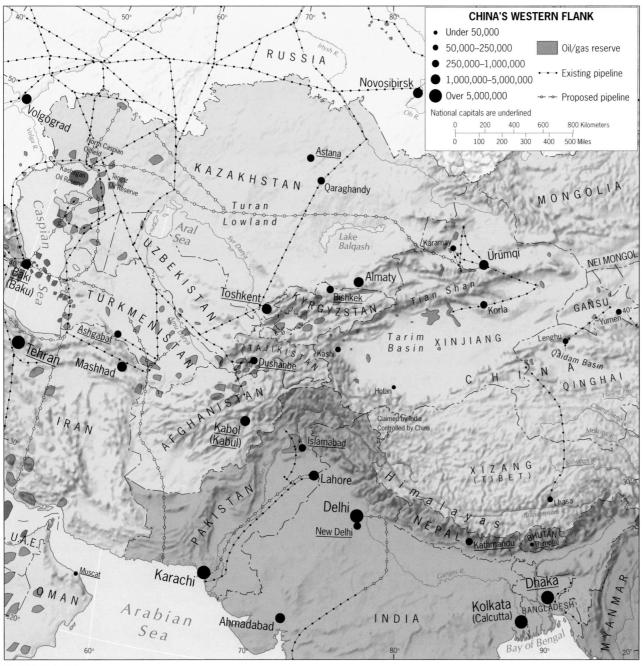

CHINA'S WESTERN FLANK

- Under 50,000
- 50,000–250,000
- 250,000–1,000,000
- 1,000,000–5,000,000
- Over 5,000,000

Oil/gas reserve

•—•—• Existing pipeline

•—○—• Proposed pipeline

National capitals are underlined

| 0 | 200 | 400 | 600 | 800 Kilometers |
| 0 | 100 | 200 | 300 | 400 | 500 Miles |

FIGURE 9-14

returned to Xizang, monastery reconstruction was permitted, and Buddhist religious life resumed. Since its formal annexation in 1965, Xizang has been administered as an Autonomous Region. Although it is large in area, its population is only 2.6 million.

XINJIANG

China's northwestern corner consists of the giant Xinjiang Uyghur Autonomous Region, constituting over one-sixth of the country's land area and situated in a strategic part of Central Asia (Fig. 9-14). Linked to China Proper via the vital Hexi Corridor in Gansu Province to the east, Xinjiang ("New Territory") is China's largest single administrative area.

But even today, after more than a half-century of vigorous Sinicization, Chinese remain a minority here. About 40 percent of the population of 19 million is Chinese (up from 5 percent in 1949), and the Chinese control virtually all aspects of life, including the clock: although Xinjiang lies 1500 miles (2500 km) west of the capital, it must keep Beijing time! The majority are Muslim Uyghurs, who make up approximately half the total, and Kazakhs, Kyrgyz, and others with cultural affinities across the border to all five republics of former Soviet Central Asia. The Uyghurs like to call this area "East Turkestan," and some defiantly set their watches two hours behind Beijing's.

The physical geography of Xinjiang is dominated by high mountain ranges and vast basins (Fig. 9-1). The southern margin is defined by the Kunlun Shan (Mountains) beyond which lies Xizang (Tibet). Across the north-center of the region lies the mountain wall of the Tian Shan. Between the Kunlun and Tian Shan extends the vast, dry Tarim Basin within which lies the Takla Makan Desert, and north of the Tian Shan lies another depression, the Junggar Basin (Fig. 9-1). Rivers rising on the mountain slopes disappear beneath the sediments of the basins, sustaining a ring of oases where their waters come near the surface or where wells and *qanats* make irrigation possible (Fig. 9-15).

This would not seem to be the most favorable geography, but Xinjiang has other assets that China exploits. The Qing (Manchu) armies that fought the nomadic Muslim warlords here in the 1870s saw the agricultural possibilities and planted the first crops. The communist rulers after 1949 exiled political opponents as well as criminals to notorious prison camps in this remote frontier. Then Xinjiang turned out to have extensive reserves of oil and natural gas (Figs. 9-14, 9-15). Here, too, China could experiment with space technology, rocketry, and atomic weaponry far from densely populated areas (and shielded from foreign spies). And now, as Figure 9-14 reminds us, China's westernmost borders lie relatively close to the great energy reservoirs of the Caspian Sea Basin—and only one neighbor lies in the way. Pipeline construction across Kazakhstan to Xinjiang is in progress.

Xinjiang's evolving human geography is revealed by Figure 9-15. Most of the Han Chinese are concentrated in the north-central cluster of cities and towns between the

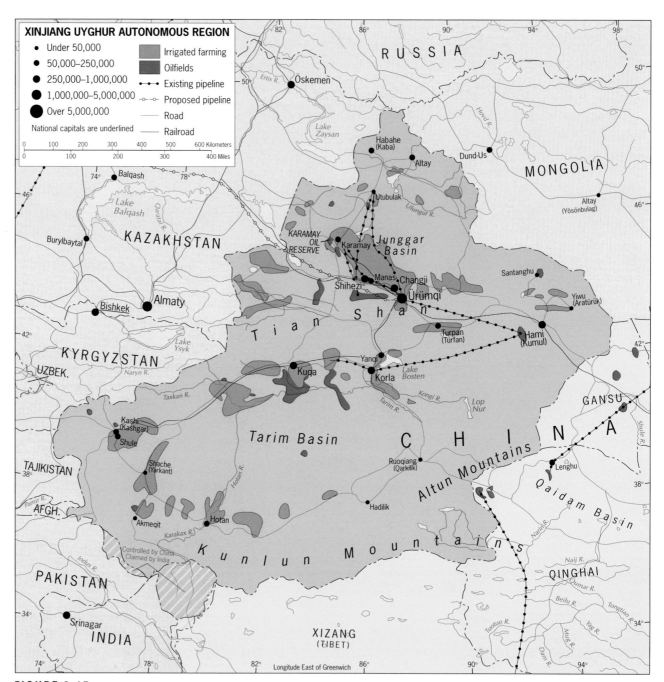

FIGURE 9-15

capital, Ürümqi, and the oil center of Karamay. Also centered on the Ürümqi-Karamay corridor is Xinjiang's largest agricultural zone, where cotton and fruits yield export revenues. As the map shows, a ring of oases and clusters of irrigated agriculture encircles both the Tarim and Junggar Basins.

Beijing has made extensive efforts to integrate Xinjiang into the Chinese state. A railroad and highway link Ürümqi and Lanzhou via the Hexi Corridor. Westward from Ürümqi, a railroad crosses the border with Kazakhstan and connects to the Öskemen-Almaty line; another railroad, opened in 2000, connects the capital to Kashi (Kashgar) in the distant southwest (Fig. 9-15). But vast spaces in Xinjiang remain distant and inaccessible. As the map shows, just one road, recently upgraded, loops around the vast southern and eastern flank of the Tarim Basin from Kashi via Hotan to Korla.

China's western frontier regions in some ways resemble the Soviet Union's former Central Asian "republics." Here Chinese modernizers meet traditional Buddhism and Islam; here locals argue that they are colonized by Beijing and at times demand independence. But the map leaves no doubt as to these regions' importance and value to China. Soviet-style devolution is unlikely to happen here.

CHINA'S PACIFIC RIM: EMERGING REGION

From Dalian on the doorstep of the Northeast to Hainan Island off the country's southernmost point, China is booming. In the cities, old neighborhoods are being bulldozed to make way for modern skyscrapers. In the countryside of the coastal provinces, thousands of factories employing millions of workers who used to farm the land are changing life in the towns and villages. New roads, airports, power plants, and dams are being built. It is an uneven, sometimes chaotic process that is affecting certain stretches of China's coast more than others. It

is penetrating parts of China's interior, but meanwhile social gaps are widening: between advantaged and disadvantaged, well-off and poor, urbanite and villager. It is creating regional economic disparities that China has never seen before. And it is producing what may become a separate geographic region along China's Pacific littoral, a region that has more in common with Japan, South Korea, and Taiwan than with Yunnan or Gansu.

In the Introduction, we discussed issues relating to the geography of development, emphasizing that, in this complex, globalizing world, the concept of developed versus underdeveloped countries (DCs and UDCs) is no longer tenable. By almost any measure, China remains what used to be called a UDC. As Figure G-10 shows, its per-capita GNP still classifies it as a lower-income economy, although the most recent World Bank analysis raises it to "lower-middle income" from "low-income." But China now has its own core-periphery duality: in its Pacific coast provinces, per-capita incomes today are far above the national average, while hundreds of millions of villagers in China's interior earn much less. In its burgeoning Pacific Rim, China resembles a modernizing economy with "developed" characteristics. In much of its interior, it does not. This contrast is one reason a discrete Pacific Rim region appears to be forming in China Proper.

Geography of Development

Many yardsticks are used to gauge a country's development. These include: GNP per capita (as we noted); percentage of workers in farms, factories, or other kinds of employment; amount of energy consumed; efficiency of transport and communications; use of manufactured metals like aluminum and steel in the economy; productivity of the labor force; and social measures such as literacy, nutrition, medical services, and savings rates. All these data are calculated per person, yielding an overall measure that ranks the country among all those providing statistics. These data

do not, however, tell us *why* some countries, or parts of countries, exhibit the level of development they do.

8 The **geography of development** leads us into the study of raw-material distributions, environmental conditions, cultural traditions, historic factors (such as the lingering effects of colonialism), and forces of location. In the emergence of China's Pacific Rim, relative location played a major role, as we shall see.

Geographers tend to view the development process spatially, whereas other scholars focus on structural aspects. In the 1960s, economist Walt Rostow formulated a global model of the development process that is still much discussed today. This model suggested that all developing countries follow an essentially similar path through five interrelated growth stages. In the earliest of these stages, a *traditional society* engages mainly in subsistence farming, is locked in a rigid social structure, and resists technological change. When it reaches the second stage, *preconditions for takeoff*, progressive leaders move the country toward greater flexibility, openness, and diversification. Old ways are abandoned, workers move from farming into manufacturing, and transport improves. This leads to the third stage, *takeoff*, when the country experiences a type of industrial revolution. Sustained growth takes hold, industrial urbanization proceeds, and technological and mass-production breakthroughs occur. If the economy continues to expand, the fourth stage, *drive to maturity*, brings sophisticated industrial specialization and increasing international trade. Some countries reach a still more advanced fifth stage of *high mass consumption* marked by high incomes, widespread production of consumer goods and services, and most workers employed in the tertiary and quaternary economic sectors.

This model takes little account of core-periphery contrasts within individual countries. In China, takeoff conditions (stage 3) exist in much of the Pacific Rim, but other major areas of this vast country remain in stage 1. China has passed through its transition from tradition-bound to progressive leadership, but the effect of that leadership's reformist policies has not been the same in all parts of the country. The most fertile ground lies along the Pacific, where China has a history of contact with foreign enterprises, where the country is most open to the world, and from where many Chinese departed for Southeast Asian (and other) countries. These **9** **Overseas Chinese** were positioned to play a crucial role in China's Pacific Rim development: many had left kin and community there, and those links facilitated the investment of hundreds of millions of dollars when the opportunity arose. Many of the Pacific Rim's factories were built by means of this repatriated money.

Transforming the Economic Map

China's communist era has witnessed the metamorphosis of the world's most populous nation. Despite the dislocations of collectivization, communization, the Cultural Revolution, power struggles, policy reversals, and civil conflicts, China as a communist state in the aftermath of Mao managed to stave off famine and hunger, control its population growth rate, strengthen its military capacity, improve its infrastructure, and enhance its world position. It took over Hong Kong from the British and Macau from the Portuguese, closing the chapter on colonialism. When economic turbulence struck its Pacific Rim and Southeast Asian neighbors, China withstood the onslaught. Its trade with neighbors near and far is growing. Foreign investment in China has mushroomed. And still its government espouses communist dogma.

How could the government manipulate the coexistence of communist politics and market economics? That was the key question confronting Deng Xiaoping and his comrades when they took power in 1979. In terms of ideology, this objective would seem unattainable; an economic "open-door" policy would surely lead to rising pressures for political democracy.

But Deng thought otherwise. If China's economic experiments could be spatially separated

from the bulk of the country, the political impact would be kept at bay. At the outset, the new economic policies would apply mainly to China's bridgehead on the Pacific Rim, leaving most of the vast country comparatively unaffected. Accordingly, the government introduced a complicated **10** system of **Special Economic Zones (SEZs)**, so-called *Open Cities*, and *Open Coastal Areas*, which would attract technologies and investments

from abroad and transform the economic geography of eastern China (Fig. 9-16).

In these economic zones, investors are offered many incentives. Taxes are low. Import and export regulations are eased. Land leases are simplified. The hiring of labor under contract is allowed. Products made in the economic zones may be sold on foreign markets and, under some restrictions, in China as well. Even Taiwanese enterprises may

FIGURE 9-16

operate here. And profits made may be sent back to the investors' home countries.

When Deng's government made the decisions that would reorient China's economic geography, location was a prime consideration. Beijing wanted China to participate in the global market economy, but it also wanted to cause as little impact on interior China as possible—at least in the early stages. The obvious answer was to position the Special Economic Zones along the coast. Initially, the government established four SEZs, all with particular locational properties (Fig. 9-16):

1. *Shenzhen*, adjacent to then-booming British Hong Kong on the Pearl River Estuary in Guangdong Province
2. *Zhuhai*, across from then still-Portuguese Macau, also on the Pearl River Estuary in Guangdong Province
3. *Shantou*, opposite southern Taiwan, a colonial treaty port, also in Guangdong Province, source of many Chinese now living in Thailand
4. *Xiamen*, on the Taiwan Strait, also a colonial treaty port (then known as Amoy in the local dialect), in Fujian Province, source of many Chinese now based in Singapore, Indonesia, and Malaysia

In 1988 and 1990, respectively, two additional SEZs were proclaimed:

5. *Hainan Island*, declared an SEZ in its entirety, its potential success linked to its location closest to Southeast Asia
6. *Pudong*, across the river from Shanghai, China's largest city, different from other SEZs because it was a giant state-financed project designed to attract large multinational companies

China also opened 14 other coastal cities for preferential treatment of foreign investors (Fig. 9-16). Again, most of these cities had been treaty ports in the colonial extraterritoriality period. They were chosen for their size, overseas trading histories, links to emigrated Chinese, level of industrial-ization, and pool of local talent and labor. These 14 Open Cities are, from north to south:

Dalian (Liaoning Province)	Ningbo (Zhejiang)
Qinhuangdao (Hebei)	Wenzhou (Zhejiang)
Tianjin Shi	Fuzhou (Fujian)
Yantai (Shandong)	Guangzhou (Guangdong)
Qingdao (Shandong)	Zhanjiang (Guangdong)
Lianyungang (Jiangsu)	Beihai (Guangxi Zhuang A.R.)
Nantong (Jiangsu)	
Shanghai Shi	

When all these Pacific Rim initiatives were implemented, China's market-conscious communist leaders had reason to expect an economic boom in the coastal provinces. (On the map, note that the new economic policy affected every coastal province, as well as the single coastal A.R.) Japan already was a highly developed economy (at the final stage of the Rostow model). South Korea and Taiwan were well beyond takeoff. Hong Kong proved what could be done on this side of the Pacific. Singapore, Southeast Asia's most successful state, is 77 percent Chinese. In truth, China's pragmatic planners were unsure of just what forces they might unleash, which is why foreign investment rules were stricter in the Open Cities and Open Coastal Areas than in the freewheeling SEZs.

CHINA'S PACIFIC RIM TODAY

When China's planners laid out the SEZs, they held the highest hopes for Shenzhen, immediately adjacent to the thriving British dependency of Hong Kong. They knew that the kinds of attractions the SEZs offered to foreign investors would entice many factory owners paying Hong Kong wages to move across the border to Shenzhen. Successful Chinese living in Southeast Asia could use Hong Kong as a base for investment in Shenzhen; Taiwanese, even during the political standoff be-tween Beijing and Taipei, could channel investments into this favored SEZ.

Those hopes were not disappointed. In the late 1970s, Shenzhen was just a sleepy fishing and duck-farming village of about 20,000. By 2004, its population was just under 4 million—the fastest growth of any urban center on Earth in human history. Thousands of factories, large and small, moved into the SEZ from Hong Kong. Workers by the hundreds of thousands came from Guangdong Province, Guangxi Zhuang A.R., and beyond to look for jobs. High-rise buildings housing banks, corporate offices, hotels, and other service facilities made Shenzhen look like a new Hong Kong.

Shenzhen's economic impact was far-reaching. Persuading Hong Kong industries to cross the border precipitated a sharp modification of Hong Kong's own economy; attracting investments from Overseas Chinese, mainly in Southeast Asia, opened financial coffers long closed to China; and enabling Taiwanese investors to participate in China's Pacific Rim boom created links with the "wayward province" that have economic as well as political benefit. By its very success (in its 1990s heyday it produced 15 percent of all of China's exports by value), Shenzhen signaled the way for the other SEZs and Open Cities.

Guangdong Province and the Pearl River Estuary

As Figure 9-16 shows, eight provinces (including Hainan Island) and one Autonomous Region face the Pacific and form the basis for recognizing an emerging Pacific Rim region in eastern China. But among these provinces, one stands out by making the largest contribution by far to the national economy: Guangdong Province. Although Hong Kong and Macau, both SARs, are not administratively a part of Guangdong Province, they are functionally crucial parts of the vast economic complex that fringes the estuary of the Pearl River (Fig. 9-17). Indeed, without Hong Kong, there would not have been a Shenzhen of the kind described above.

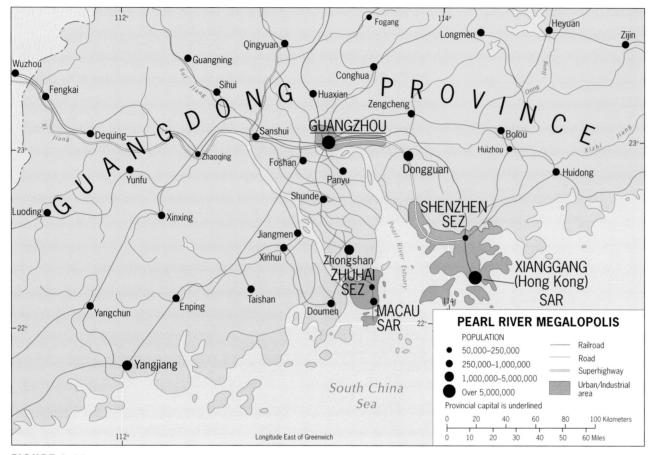

FIGURE 9-17

The economic core of Guangdong consists of Guangzhou (the capital), Shenzhen SEZ, Zhuhai SEZ, as well as Hong Kong (Xianggang) and Macau. In a few years this Pearl River hub will have three cities with more than 5 million inhabitants, five with more than 1 million, and six with more than a half-million. Per-capita incomes here are estimated to be ten times the national average.

11 The Pearl River hub fits the model of the **regional state** as defined by the Japanese scholar Kenichi Ohmae. These "natural economic zones," as he called them, defy old political borders and are shaped by the global economy of which they are parts; their representatives deal directly with foreign partners and negotiate the best terms they can with the national governments under which they operate (Europe, we noted, has several examples of this kind). The Pearl River Megalopolis has moved so far ahead of China as a whole that it is a political as well as an economic force in the country.

Role of Hong Kong

The process started with the economic success of Hong Kong while it was still a British dependency, empowered by a unique set of geographic circumstances. Here, on the left bank of the Pearl River Estuary, more than 7 million people (97 percent of Chinese descent) crowded onto 400 square miles (1000 sq km) of fragmented, hilly territory under the tropical sun and created an economy larger than that of a hundred countries.

Hong Kong (the name means "Fragrant Harbor") was ceded by the Chinese to Britain during the colonial period in three parts: the islands in 1841, the Kowloon Peninsula in 1860, and the New Territories (a lease) in 1898 (Fig. 9-18). Until the 1950s, Hong Kong was just another trading colony, but then the Korean War and the United Nations embargo on communist China cut Hong Kong off from its hinterland. With a huge labor force and no domestic resources, the colony needed an alternative—and it came in the form of a textile industry based on imported raw materials. Soon Hong Kong textiles were selling on worldwide markets; capital was invested in light industries; and the industrial base diversified into electrical equipment, appliances, and countless other consumer goods. Meanwhile, Hong Kong quietly served as a back door to China, which rewarded the colony by providing fresh water and staple foods. The fast-developing colony **12** had become an **economic tiger**.

Hong Kong also became one of the world's leading financial centers. When China changed course and embarked on its "Open Door" policy in the 1980s, Hong Kong was ready. Hong Kong factory owners moved their plants to Shenzhen, where wages were even lower; Hong Kong banks financed Shenzhen industries; and Hong Kong's economy adjusted once again to a new era. Hong Kong's port, however, continued to transfer Shen-

FIGURE 9-18

China made enormous infrastructure investments to enhance the attractiveness and efficiency of these growth poles. In Shanghai, a massive development project transformed the right bank of the Huangpu River (across from the old downtown) into an ultramodern complex whose skyline rivals that of Hong Kong. *Pudong*, as this area is called, is designed to offer competition to the Pearl River Megalopolis as part of the Beijing government's effort to reduce the imbalance between north and south along its Pacific Rim.

None of the other SEZs has come close to what had happened in Shenzhen or Pudong. Xiamen was established especially to attract Taiwanese investment; Shantou has benefited greatly from investment by Overseas Chinese in Southeast Asia (notably Thailand); Zhuhai is benefiting from the fact that its wages are even lower than Shenzhen's, causing the transfer of many industries from Shenzhen; and Hainan lags because of relative location and poor administration. The SEZs are no workers' paradise: thousands of factories producing goods ranging from toys to textiles create a pollution-choked landscape, working conditions often are dreadful, and wages are marginal. Unfortunately, China's new Industrial Revolution has some of the features of the old one.

Toward the Interior

Economic progress on the Pacific Rim has created growing regional disparities in the new China. This is of great concern to China's leaders because such disparities can lead to political unrest and instability. Thus the government has begun to grant SEZ-like privileges to a part of Yunnan Province, which borders Southeast Asian countries; the notion of a Pacific Rim-linked corridor extending westward from Shanghai along the Yangzi/Chang River to the Sichuan Basin also is gaining ground. China's government is well aware of the need not only to diffuse economic growth into Yunnan, Sichuan, and the Northeast, but also to improve the living

zhen's products. During the 1990s, about 25 percent of all of China's trade passed through Hong Kong harbor.

When the British lease on Hong Kong expired in 1997 and the British agreed to yield authority to China, the Chinese promised to allow Hong Kong's freewheeling economy to continue, in effect creating "one country, two systems." At midnight on June 30, Hong Kong became the Xianggang Special Administrative Region (SAR). True to the letter (if not always the spirit) of this commitment, the name "Hong Kong" has continued in general use. And on December 20, 1999 China took control over the last remaining European colony—Portuguese Macau—and made it a SAR as well, guaranteeing (as in Hong Kong) the continuation of its existing social and economic system for 50 years.

The Other SEZs

Not only did China grant economic advantages to corporations doing business in the coastal SEZs;

FROM THE FIELDNOTES

"China's new consumer culture is perhaps nowhere more evident than in burgeoning, modernizing, globalizing Shanghai. I walked today down Nanjing Road, exactly 20 years after I first saw it in 1982, but only four years after my last excursion here. The rate of transformation is increasing. Thousands of locals have taken up the fast-food habit and are eating at Kentucky Fried Chicken (the colonel's face is everywhere) or McDonald's (I counted five of the latter along my route). A section of Nanjing Road has been made into a pedestrian mall, and here too is evidence of a new habit: private automobile transportation. Car sales lined the street, and potential buyers thronged the displays. Given the existing congestion on local roads, where will these cars fit?"

and working conditions of farmers and others in the interior provinces generally. Restricting migration to the cities of the burgeoning east can only be a temporary remedy for a major challenge facing the communist government.

CHINA: GLOBAL SUPERPOWER?

China in the early twenty-first century appears likely to become more than an economic force of world proportions: China also appears on course to achieve global superpower stature. As long as it retains its autocratic form of government (which allowed it to impose draconian population policies and comprehensive economic experiments without having to consult the electorate), China will be able to practice the kind of state capitalism that—in another guise—made South Korea an economic power. And unlike Japan or South Korea, China has no constraints on its military power. Its People's Liberation Army served as security forces during the pro-democracy turbulence of 1989; China's military is the largest standing army in the world, with nearly 3 million soldiers and some 1.2 million reserves. Its military equipment is being updated, and the government made a major investment in its military by raising its armed forces budget enormously in 2001. Already, China is a nuclear power and has a growing arsenal of medium-range and intercontinental ballistic missiles. China also has orbited an astronaut and declared its intention to reach the Moon.

During the twentieth century, the United States and the Soviet Union were locked in a 45-year Cold War that repeatedly risked nuclear conflict. That fatal exchange never happened, in part because it was a struggle between superpowers who understood each other comparatively well. While the politicians and military strategists were plotting, the cultural doors never closed: American audiences listened to Prokofiev and Shostakovich, watched Russian ballet, and read Tolstoy and Pasternak even as the Soviets cheered Van Cliburn, read Hemingway, and lionized American dissidents. In short, it was an *intracultural* Cold War, which reduced the threat of mutual destruction.

The twenty-first century may witness a far more dangerous geopolitical struggle in which the adversaries may well be the United States and China. U.S. power and influence still prevail in the western Pacific, but it is easy to discern areas where Chinese and American interests will diverge (Taiwan is only one example). American bases in Japan, thousands of American troops in South Korea, and American warships in the East and South China Seas are potential grounds for dispute. All this might generate the world's first *intercultural* Cold War, in which the risk of fatal misunderstanding is incalculably greater than it was during the last.

How can such a Cold War be averted? Trade, scientific and educational links, and cultural exchanges are obvious remedies: the stronger our interconnections, the less likely is a deepening conflict. We Americans should learn as much about China as we can, to understand it better, to appreciate its cultural characteristics, and to recognize the historico-geographical factors underlying China's views of the West.

For more than 40 centuries, China has known authoritarianism of both the brutal and the benevolent kind, has been fractured by regionalism only to unify again, and has depended on communalism to survive environmental and despotic depredations. For nearly 25 centuries Confucianism has guided it. Time and again, China has been opened to, and ruled by, foreigners, and time and again it has retreated into isolationism when things went wrong. Such a reversal may no longer be possible, given what we have seen in this chapter. But now a renewed force is rising in modernizing China: nationalism. A spate of recent books published in China, including one by Song Qiang et al. titled *China Can Say No*, reflects the frustrations of many Chinese over what they view as American arrogance and insensitivity to China's traditions and interests. The "inadvertent" bombing of the Chinese embassy in Belgrade by American aircraft operating under NATO command in 1999, and the mid-air collision between what the Chinese re-

garded as a U.S. "spy plane" and one of its aircraft near the island of Hainan in 2001, are the kinds of incidents that contribute to this view and that energize Chinese nationalism. American scorn for China's human-rights failings is another source of irritation.

In 2003, China failed two tests of its new image. In the spring, an outbreak of a disease called Severe Acute Respiratory Syndrome (SARS) occurred in Guangdong Province and spread rapidly, but the Chinese government concealed crucial information about it even when it reached the capital. This prevarication outraged Chinese citizens as well as others affected by the deadly epidemic and its economic impact. In the summer, Hong Kong's Beijing-approved leadership tried to institute new laws that would limit freedoms supposedly guaranteed to Hong Kong's residents under the "one country, two systems" principle. This brought a half-million protesters into the streets, and the new laws were "reconsidered." But the SARS episode and the Hong Kong initiative cast doubt on China's much-heralded new governmental accountability, and on Beijing's dependability to keep its word. As we will see, both incidents had consequences in Taiwan.

China today is on the world stage, globally engaged and internationally involved. Constraining the forces that tend to lead to world-power competition between China and the United States will be the geopolitical challenge of the twenty-first century.

MONGOLIA

Between China's Inner Mongolia and Xinjiang to the south and Russia's Eastern Frontier region to the north lies a vast, landlocked, isolated country called Mongolia that constitutes a discrete region of East Asia (Fig. 9-3). With only 2.5 million inhabitants in an area larger than Alaska, Mongolia is a steppe- and desert-dominated vacuum between two of the world's most powerful countries.

From the late 1600s until the revolutionary days of 1911, Mongolia was part of the Chinese Empire. With Soviet support, the Mongols held off Chinese attempts to regain control, and in the 1920s the country became a People's Republic on the Soviet model. Free elections in 1990 ushered in a new political era, but Mongolia's economy, still largely based on animal products (chiefly cashmere wool), is troubled by mismanagement and environmental problems.

The map reflects Mongolia's difficulties. Despite its historic associations, ethnic affinities, and cultural involvement with China, Mongolia's incipient core area, including the capital of Ulaanbaatar, adjoins Russia's Eastern Frontier. The country's 800,000 herders and their 17 million livestock, mostly sheep, follow nomadic tracks along the fenceless fringes of the vast Gobi Desert. When Siberian cold descends on the steppes, as happened in 2000 and 2001, human and livestock losses are severe and the economy breaks down.

This is a weak country in a vulnerable part of
13 inner Asia, a **buffer state** wedged, Tibet-like, between populous and powerful neighbors. Will it escape the fate of other Asian buffers?

THE JAKOTA TRIANGLE REGION

We turn now to a region that exemplifies the future of East Asia. Along the Pacific Rim, from Japan (through South Korea, Taiwan, China's Guangdong) to Singapore, rapid economic development has transformed community and society. In China, as we saw, a Pacific Rim region is in the making, still discontinuous today but likely to extend all along the coast in the future. In Japan, South Korea, and Taiwan we can observe that future—today. That is why we recognize an East Asian region we
14 call the **Jakota Triangle**, consisting of Japan, Korea (South Korea at present but probably a reunited Korea later), and Taiwan (Fig. 9-19). This is

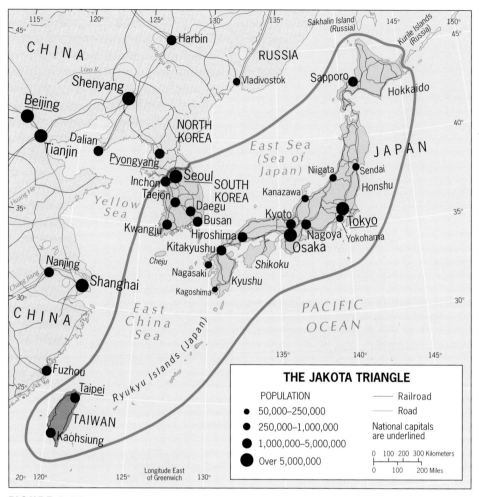

FIGURE 9-19

the Ryukyu Islands, and Taiwan, as well as the southern half of Sakhalin Island (called Karafuto). Not even a disastrous earthquake, which destroyed much of Tokyo in 1923 and killed 143,000 people, could slow the Japanese drive.

World War II saw Japan expand its domain farther than the architects of the 1868 "modernization" could have anticipated. By early December 1941, Japan had conquered large parts of China Proper, all of French Indochina to the south, and most of the small islands in the western Pacific. Then, on December 7, 1941, Japanese-built aircraft carriers moved Tokyo's warplanes within striking range of Hawai'i, and the surprise attack on Pearl Harbor underscored Japan's confidence in its war machine. Soon the Japanese overran the Philippines, the (then) Netherlands East Indies, Thailand, and British Burma and Malaya, and drove a wide corridor through the heart of China to the border with Vietnam.

A few years later, Japan's expansionist era was over. Its armies had been driven from virtually all its possessions, and when American nuclear bombs devastated two Japanese cities in 1945, the country lay in ruins. But once again, Japan, aided this time by an enlightened U.S. postwar administration, surmounted disaster.

Japan's economic recovery and its rise to the status of world economic superpower was the success story of the second half of the twentieth century. Japan lost the war and its empire, but it scored many economic victories in a new global arena. Japan became an industrial giant, a technological pacesetter, a fully urbanized society, a political power, and an affluent nation. No city in the world today is without Japanese cars in its streets; few photography stores lack Japanese cameras and film; laboratories the world over use Japanese optical equipment. From microwave ovens to DVDs, from oceangoing ships to digital cameras, Japanese goods flood the world's markets.

Japan's brief colonial adventure helped lay the groundwork for other economic successes along the western Pacific Rim. The Japanese ruthlessly

a region of great cities, huge consumption of raw materials from all over the world, voluminous exports, and global financial linkages. It also is a region of social problems, political uncertainties, and economic vulnerabilities.

JAPAN

When we assess China's prospects of becoming a superpower, we should remember what happened in Japan in the nineteenth century. In 1868, a group of reform-minded modernizers seized power from an old guard, and by the end of the century Japan was a military and economic force. From the factories in and around Tokyo and from urban-industrial complexes elsewhere poured forth a stream of weapons and equipment the Japanese used to embark on colonial expansion. By the mid-1930s, Japan lay at the center of an empire that included all of the Korean Peninsula, the whole of China's Northeast (which the Japanese called Manchukuo),

FROM THE FIELDNOTES

"Visiting the Peace Memorial Park in Hiroshima is a difficult experience. Over this site on August 6, 1945 began the era of nuclear weapons use, and the horror arising from that moment in history, displayed searingly in the museum, is an object lesson in this time of nuclear proliferation and enhanced risk. In the museum is a model of the city immediately after the explosion (the red ball marks where the detonation occurred), showing the total annihilation of the entire area at an immediate loss of more than 80,000 people and the death from radiation of many more subsequently. In the park outside, the Atomic Bomb Memorial Dome, the only building to partially survive the blast, has become the symbol of Hiroshima's devastation and of the dread of nuclear war."

exploited Korean and Formosan (Taiwanese) natural and human resources, but they also installed a new economic order there. After World War II, this infrastructure facilitated an economic transition—and soon made both Taiwan and South Korea competitors on world markets.

Spatial Limitations

After the modernizers took control of Japan in 1868—an event known as the *Meiji Restoration* (the return of "enlightened rule" centered on the Emperor Meiji)—they turned to Britain for guidance in reforming their nation and its economy. In the decades that followed, the British advised the Japanese on the layout of cities and the construction of a railroad network, on the location of industrial plants, and on the organization of education. The British influence still is visible in the Japanese cultural landscape today: the Japanese, like the British, drive on the left side of the road. Consider how this affects the effort to open the Japanese market to U.S. automobiles!

The Japanese reformers of the late nineteenth century undoubtedly saw many geographic similarities between Britain and Japan. At that time, most of what mattered in Japan was concentrated on the country's largest island, Honshu (literally, "mainland"). The ancient capital, Kyoto, lay in the interior, but the modernizers wanted a coastal, outward-looking headquarters. So they chose the town of Edo, on a large bay where Honshu's eastern coastline turns sharply (Fig. 9-20). They renamed the place *Tokyo* ("eastern capital"), and little more than a century later it was the largest urban agglomeration in the world. Honshu's coasts were near mainland Asia, where raw materials and potential markets for Japanese products lay. The notion of a greater Japanese empire followed naturally from the British example.

But in other ways, the British and Japanese archipelagoes, at opposite ends of the Eurasian landmass, differed considerably. In total area, Japan is larger. In addition to Honshu, Japan has three other large islands—Hokkaido to the north and Shikoku and

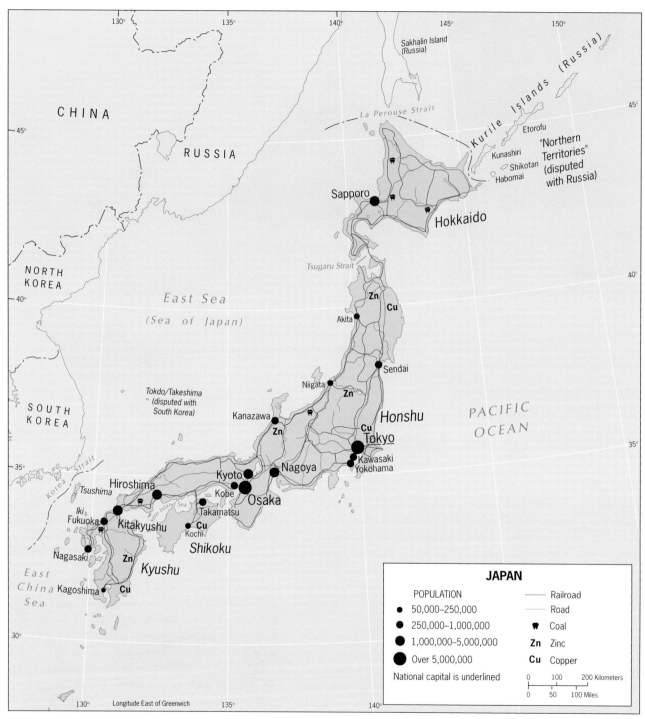

FIGURE 9-20

Kyushu to the south—as well as numerous small islands and islets, for a total land area of about 146,000 square miles (377,000 sq km). Much of this area is mountainous and steep-sloped, geologically young, earthquake-prone, and studded with volcanoes. Britain has lower relief, is older geologically, does not suffer from severe earthquakes, and has no active volcanoes. And in terms of the raw materials for industry, Britain was much better endowed than Japan. Self-sufficiency in iron ore and high-quality coal gave Britain the head start that lasted a century.

Japan's high-relief topography has been an ever-present challenge. All of Japan's major cities, except the ancient capital of Kyoto, are perched along the coast, and virtually all lie partly on artificial land claimed from the sea. Sail into Kobe harbor, and you will pass artificial islands designed for high-volume shipping and connected to the mainland by automatic space-age trains. Enter Tokyo Bay, and the refineries and factories to your east and west stand on huge expanses of landfill that have pushed the bay's shoreline outward. With 127.4 million people, the vast majority (78 percent) living in towns and cities, Japan uses its habitable living space intensively—and expands it wherever possible.

As Figure 9-21 shows, farmland in Japan is both limited and regionally fragmented. Urban sprawl has invaded much cultivable land. In the hinterland of Tokyo lies the Kanto Plain; around Nagoya, the Nobi Plain; and surrounding Osaka, the Kansai District—each a major farming zone under relentless urban pressure. All three of these plains lie within Japan's fragmented but well-defined core area (delimited by the red line on the map), the heart of Japan's prodigious manufacturing complex.

Modernization

The reformers who set Japan on a new course in 1868 probably did not anticipate that three generations later their country would lie at the heart of a major empire sustained by massive military might. They set into motion **15** a process of **modernization**, but they managed to build on, not replace, Japanese cultural traditions. We in the Western world tend to equate modernization with Westernization: urbanization, the spread of transport and communications facilities, the establishment of a market (money) economy, the breakdown of local traditional communities, the proliferation of formal schooling, the acceptance and adoption of foreign innovations. In the non-Western world, the process often is viewed differently. There, "modernization" is seen as an outgrowth of colonialism, the perpetuation of a system of wealth accumulation introduced by foreigners driven by greed. In this view, the local elites who replaced the colonizers in the newly independent states only continue the disruption of traditional societies, not their true modernization. Traditional societies, they argue, can be modernized without being Westernized.

In this context, Japan's modernization is unique. Having long resisted foreign intrusion, the Japanese

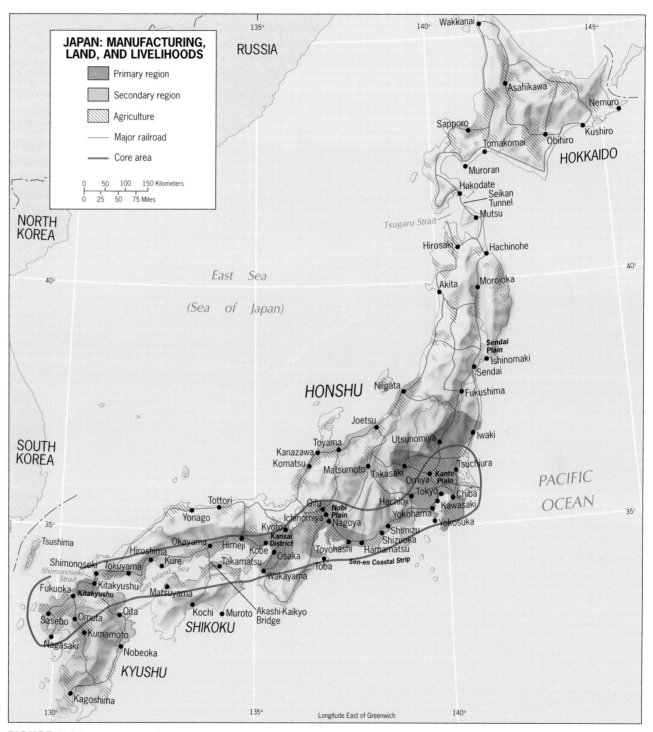

FIGURE 9-21

did not achieve the transformation of their society by importing a Trojan horse; it was done by Japanese planners, building on the existing Japanese infrastructure, to fulfill Japanese objectives. Certainly Japan imported foreign technologies and adopted innovations from the British and others, but the Japan that was built, a unique combination of modern and traditional elements, was basically an indigenous achievement.

Relative Location

Japan's changing fortunes over the past century **16** reveal the influence of **relative location** in the country's development. When the Meiji Restoration took place, Britain, on the other side of the Eurasian landmass, lay at the center of a global empire. The colonization and Europeanization of the world were in full swing. The United States was still a developing country, and the Pacific Ocean was an avenue for European imperial competition. Japan, even while it was conquering and consolidating its first East Asian colonies (the Ryukyus, Taiwan, Korea), lay remote from the mainstream of global change.

Then Japan became embroiled in World War II and dealt severe blows to the European colonial armies in Asia. The Europeans never recovered: the French lost Indochina, and the Dutch were forced to abandon their East Indies (now Indonesia). When the war ended, Japan was defeated and devastated, but at the same time the Japanese had done much to diminish the European presence in the Pacific Basin. Moreover, the global situation had changed dramatically. The United States, Japan's trans-Pacific neighbor, had become the world's most powerful and wealthiest country, whereas Britain and its global empire were fading. Suddenly Japan was no longer remote from the mainstream of global action: now the Pacific was becoming the avenue to the world's richest markets. Japan's relative location—its situation relative to the economic and political foci of the

world—had changed. Therein lay much of the opportunity the Japanese seized after the postwar rebuilding of their country.

Japan's Spatial Organization

Imagine this: more than 127 million people crowded into a territory the size of Montana (population: 960,000), most of it mountainous, subject to frequent earthquakes and volcanism, with no domestic oilfields, little coal, few raw materials for industry, and not much level land for farming. If Japan today were an underdeveloped country in need of food relief and foreign aid, explanations would abound: overpopulation, inefficient farming, energy shortages.

True, only an estimated 18 percent of Japan's national territory is designated as habitable. And Japan's large population is crowded into some very big cities. Moreover, Japan's agriculture is not especially efficient. But Japan defeated the odds by calling on old Japanese virtues: organizational efficacy, massive productivity, dedication to quality, and adherence to common goals. Even before the Meiji Restoration, Japan was a tightly organized country of some 30 million citizens.

Areal Functional Organization

All this proved invaluable to the modernizers when they set Japan on its new course. The new industrial growth of the country could be based on the urban and manufacturing development that was already taking place. As we noted, Japan does not possess major domestic raw-material sources, so no substantial internal reorganization was necessary. However, some cities were better sited and enjoyed better situations relative to those limited local resources and, more importantly, to external sources of raw materials than others. As Japan's regional organization took shape, a hierarchy of cities developed; Tokyo took and kept the lead, but other cities grew rapidly into industrial centers.

This process was governed by a geographic **17** principle that Allen Philbrick called **areal functional organization**, a set of five interrelated tenets that help explain the evolution of regional organization, not only in Japan but throughout the world. Human activity, Philbrick reasoned, has spatial *focus*. It is concentrated in some locale, whether a farm or factory or store. Every one of these establishments occupies a particular *location*; no two of them can occupy exactly the same spot on the Earth's surface (even in high-rises, there is a vertical form of absolute location). Nor can any human activity proceed in total isolation, so that *interconnections* develop among these various establishments. This system of interconnections grows more complex as human capacities and demands expand. Each system (for example, farmers sending crops to market and buying equipment at service centers) forms a unit of areal functional *organization*. In the Introduction we referred to functional regions as systems of spatial organization; we can map Philbrick's units of areal functional organization as regions. These regions evolve because of what he called "creative imagination" as people apply their total cultural experience and their technological know-how when they organize and rearrange their living space. Finally, Philbrick suggests, we can recognize levels of *development* in areal functional organization, a ranking of places and regions based on the type, extent, and intensity of exchange.

Coastal Development

Japan's level of development is reflected by its organizational map (Fig. 9-21): it is the highest of Philbrick's three categories (subsistence, transitional, and exchange). And the map tells us much about the nature of Japan's exchange economy, its external orientation and dependence on foreign trade. All primary and secondary regions lie on the coast.

Dominant among these regions is the *Kanto Plain* (Fig. 9-21), which is the heart of Japan's core area, lies focused on the Tokyo urban area, and contains about one-third of the country's population. Among its advantages are an unusually extensive area of low relief, a fine natural harbor at Yokohama, a relatively mild and moist climate, and a central location with respect to the country as a whole. Its major disadvantage lies in its vulnerability to earthquakes. The Kanto Plain and its Tokyo-centered megalopolis (population: 26.7 million) lie at the convergence of three tectonic plates, and Toyko has a centuries-long history of devastating earthquakes that have struck about every 70 years since 1633 (Fig. G-4).

The second-ranking economic region in Japan is the *Kansai District*, comprised by the Osaka-Kobe-Kyoto triangle and located at the eastern end of the Seto Inland Sea. Osaka and Kobe are major industrial centers and busy ports, and the Kansai District also continues to yield large harvests of rice, Japan's staple. Between the Kanto Plain and the Kansai District lies the *Nobi Plain* (Fig. 9-21), where Nagoya is the key city. And, as the map shows, the Japanese core area is anchored in the west by the conurbation centered on *Kitakyushu*, the only part of it not situated on Honshu Island but, rather, on northern Kyushu. This five-city conurbation is growing rapidly, favored by its location relative to Japan's Pacific Rim partners.

Japan's Pacific Rim Prospects

During the late 1990s and the first years of the new century, Japan's economy failed to sustain the growth of previous decades. A large part of the problem involved mismanagement and government inefficiency, and some of it had to do with a downturn in Southeast Asian economies with which Japan's economy was closely linked. But Japan is also facing tougher competition from other economies on the Pacific Rim, including South Korea (which has cut into Japan's lead in automobile sales) and Taiwan (which dominates the computer field).

Japan also has unfinished World War II business that hurts its status. Tokyo never signed a peace treaty with the Soviets in the aftermath of the Second World War because the Russians occupied and kept four sets of small islands northeast of Hokkaido (Fig. 9-20). Negotiations for the return of these "Northern Territories" failed despite Japan's offer of a massive aid-and-development program to develop the Russian Far East, including Vladivostok. This has cost Japan a role in the economic development of a hinterland from which it might have gained vast energy as well as mineral supplies. Relations with South Korea also have been eroded by Korean memories of Japanese misbehavior during World War II and Japan's refusal to acknowledge its misconduct. Similar issues damage Japan's carefully nurtured links with China.

All this should be seen against the backdrop of Japan's changing society. Its population of 127 million is aging rapidly; growth is projected to stop in 2007, after which it will decline to about 100 million in 2050 and only 67 million in 2100. Over the long term, Japan will need millions of immigrants to keep its economy going; for culturally homogeneous, ethnically conscious Nippon (Japan's name for itself), which has historically resisted immigration, this will be a wrenching reality.

Thus Japan's future in the geopolitical and economic frameworks of the western Pacific is uncertain. More than a half-century after the end of World War II, American forces are still based on Japanese territory, guaranteeing Japanese security. China and Korea accept this arrangement, which constrains Japan's rearming, in view of what happened during World War II. But in Japan, where nationalism is rising, and concern over a nuclear-armed North Korea is deepening, domestic military preparedness is a growing political issue. The situation is fraught with potential problems.

Japan is therefore a pivotal country in the western Pacific Rim that stands at economic, social, and geopolitical crossroads. Its future will profoundly influence not only East Asia and the Pacific, but the world as a whole.

KOREA

On the Asian mainland, directly across the Sea of Japan (East Sea), lies the peninsula of Korea (Fig. 9-22), a territory about the size of the State of Idaho, much of it mountainous and rugged, and containing a population of 73 million. Unlike Japan, however, Korea has long been a divided country, and the Koreans a divided nation. For uncounted centuries Korea has been a pawn in the struggles of more powerful neighbors. It has been a dependency of China and a colony of Japan. When it was freed from Japan's oppressive rule at the end of World War II (1945), the victorious Allied powers divided Korea for administrative purposes. That division gave North Korea (north of the 38th parallel) to the forces of the Soviet Union and South Korea to those of the United States. In effect, Korea traded one master for two new ones. The country was not reunited for the rest of the century because North Korea immediately fell under the communist ideological sphere and became a dictatorship in the familiar (but in this case extreme) pattern. South Korea, with massive American aid, became part of East Asia's capitalist perimeter. Once again, it was the will of external powers that prevailed over the desires of the Korean people.

The Korean War (1950–1953), which began when communist forces from North Korea invaded the south in a forced-unification drive, devastated much of the peninsula before a cease-fire line became the *de facto* boundary between South and North Korea (Fig. 9-22). Tens of thousands of U.S. troops continue to guard this border between ideological adversaries; in the decades since the war ended, South Korea, with 45 percent of the peninsula's land but two-thirds of its population, has become an economic powerhouse. Although

the two Koreas are in a potential situation of
18 **regional complementarity**, in which North
Korea has raw materials South Korea needs and
South Korea produces food that North Korea needs,
the political status of the two sides stops them from
taking advantage of it.

South Korea, as shown on Figure 9-22, is a key
component of the Jakota Triangle, whereas North
Korea is not even part of it. From the ravages of war
it emerged as the world's largest shipbuilding nation, a major automobile manufacturer, and a producer of iron and steel as well as chemicals. Despite
corrupt dictatorial rule, political instability, and social unrest, successive South Korean regimes managed to keep the country's economy growing, its
19 **state capitalism** propelled by powerful industrial conglomerates in cahoots with the politicians.
Eventually, democratic government did take hold in
South Korea, and the transition, coupled with the
economic downturn in other parts of Asia in the
1990s, slowed the country's growth somewhat.

By global standards, however, South Korea is a
prosperous country, and its economic prowess is evident from the map (Fig. 9-22). The capital, Seoul,
with some 10 million inhabitants, ranks among the
world's megacities and is the anchor of a huge industrial complex facing the Yellow Sea at the waist
of the Korean Peninsula. Hundreds of thousands of
farm families migrated to the Seoul area after the
end of the Korean War (today only about 20 percent
of South Koreans remain on the land). They also
moved to Busan, the nucleus of the area called
Kyongsang, which constitutes the country's second
largest manufacturing zone, located on the Korea
Strait opposite the western tip of Honshu. And the
government-supported, urban-industrial drive here
continues. Just 30 years ago, Ulsan City, 40 miles
(60 km) north of Pusan along the coast, was a fishing
center with perhaps 50,000 inhabitants; today its
population exceeds 1.6 million, nearly half of them
the families of workers in the Hyundai automobile
factories and the local shipyards and docks.

The third industrial area shown in Figure 9-22,
anchored by the city of Kwangju, has advantages

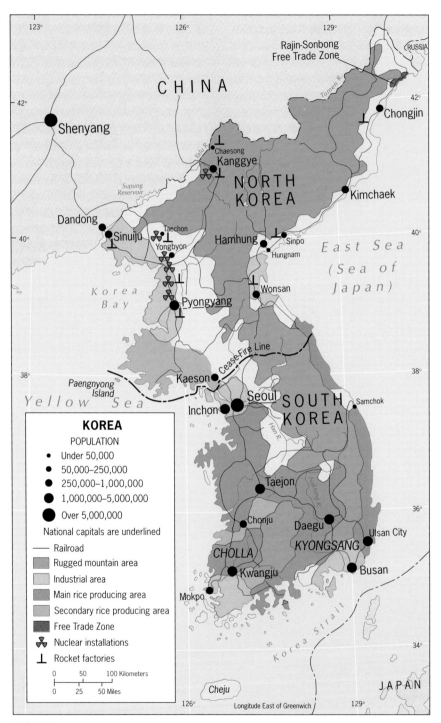

FIGURE 9-22

FROM THE FIELDNOTES

"South Korea was one of the four 'economic tigers' marking the upsurge of the western Pacific Rim in the 1980s and 1990s, its products ranging from computers to cars selling on world markets and its GDP rising rapidly. Walking through the prosperous central business district of globalizing Seoul you see every luxury name brand on display amid plentiful evidence of wealth and well-being. But you also see evidence of the survival of traditions long lost elsewhere, in the huge number of bookstores (South Koreans still read voraciously), in the stores selling traditional as well as modern musical instruments, in the school uniforms. And in other ways: when I got on a bus with all seats taken, a boy got up and offered me his place."

of East Asia's most productive and prosperous societies. But in the foreseeable future the prospect for reunification is dim, and the map of Korea continues to illustrate the high cost of the Cold War.

TAIWAN

The third component of the region we call the Jakota Triangle is another economic success story on the Asian Pacific Rim: Taiwan (Fig. 9-23). Along with a small archipelago in the Taiwan Strait and two small islands on the very doorstep of mainland China, Taiwan in 1949 became the last refuge of China's Nationalists under Chiang Kai-shek, who fled there with their weapons and treasure when the communists triumphed. Arriving in the capital, Taipei, they took control and proclaimed the Republic of China (ROC), the "legitimate" government of the country now ruled from Beijing by Mao Zedong and his comrades.

Taiwan's earliest inhabitants were Malay–Polynesians, and Taiwan became a formal part of China only during the last (Qing) Dynasty, following a hunger-driven Chinese emigration from Fujian Province. As on the mainland, colonial powers established "treaty ports" here, and in 1895 the Manchus were forced to cede all of Taiwan to the Japanese. During their half-century of rule, the Japanese exploited Taiwan but also endowed it with major infrastructure including railways, roads, hydroelectric schemes, irrigation projects, factories, and mines. Although damaged during the war, this infrastructure proved crucial to the Nationalists when they took control, for it enabled them to speed a massive reconstruction project, with American help, that quickly revived Taiwan's economy.

In the process, Taiwan became one of the "economic tigers" on the Pacific Rim. As Figure 9-23 shows, this is not a large island. In fact, it is smaller than Switzerland but has a much larger population (23 million), most of which is concentrated in an arc lining the western and northern coasts. The Chungyang Mountains, an area of high elevations

of relative location that will spur its development. This area is called Cholla, and it has traditionally faced a bias in Seoul that has worked to its disadvantage, an historic prejudice that is only now diminishing. Geography will help Cholla overcome this history.

North Korea, after six decades of communist rule, has become one of the poorest, hungriest, and most regimented nations on Earth. The capital, Pyongyang, has a population perhaps one-fifth that of Seoul; in the industrial zone facing Korea Bay, industries from mining (coal) to manufacturing (textiles) are outdated and inefficient. Collectivized farming produces far less than North Korea's 23.6

million people need, but Pyongyang issues no information on such matters. Refugees' and escapees' stories tell of poverty and misery. But North Korea did progress on one significant front: nuclear capability and associated weaponry (Fig. 9-22). The country's nuclear capability became an international concern in 2003, when North Korea itself proclaimed its aspirations and achievements. In 2004, a multinational effort by the United States, Russia, China, Japan, and South Korea was under way to defuse this dangerous case of nuclear proliferation.

Many people on both sides of the ideological divide hope for the eventual reunification of the two Koreas, which over the long term could create one

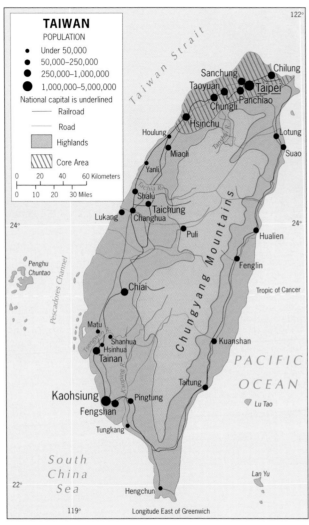

FIGURE 9-23

outport, to export nearby coal, but now the raw materials flow the other way. Taiwan imports raw cotton for its textile industry, bauxite (for aluminum) from Indonesia, oil from Brunei, and iron ore from Africa. Taiwan has a developing iron and steel industry, nuclear power plants, shipyards, a large chemical industry, and modern transport networks. Increasingly, however, Taiwan is exporting products of its budding high-technology industries: personal computers, telecommunications equipment, and precision electronic instruments.

Taiwan's economy was boosted by the creation of China's SEZs, where the rules permitted Taiwanese companies to set up factories, and by the SAR status of Hong Kong, which allows Taiwanese businesspeople to enter China via this back door (direct travel is prohibited). The Taiwanese have skillfully exploited their opportunities, resulting in one of the world's strongest economies. Today, per-capita GNP in Taiwan exceeds U.S. $15,000, which is on a par with that of many European countries.

But Taiwan faces serious political difficulties. The communist regime in Beijing regards Taiwan as a "wayward province" to be reunited with the motherland, possibly by force. In 1971, the ROC was ousted from the United Nations, its seat taken by the PRC; later, the United States, Taiwan's only ally, was forced to publicly subscribe to a "One China" policy. Taiwan's ROC leaders never proclaimed an independent state on their island, and now the Taiwan question is a complicated, uneasy standoff. Unlike communist mainland China, Taiwan has progressed from authoritarianism to democracy; it is the ROC, not the PRC, that has managed to combine economic success with democratization.

There are signs of hope. A majority of Taiwan's voters have recently signaled that they are not in favor of risky moves toward sovereignty; Beijing's rulers have shown signs of flexibility by agreeing to talk directly with Taiwanese representatives. All sides seem to be aware that an

(some over 10,000 feet [3000 m]), steep slopes, and dense forests, dominate the eastern half of the island. Westward, these mountains yield to a zone of hilly topography and, facing the Taiwan Strait, a substantial coastal plain. Streams from the mountains irrigate the paddyfields, and farm production has more than doubled since 1950 even as hundreds of thousands of farmers left the fields for work in Taiwan's expanding industries.

Today, the lowland urban-industrial corridor of western Taiwan is anchored by the capital, Taipei (Taibei), at the island's northern end and rapidly growing Kaohsiung (Gaoxiong) in the far south.* The Japanese developed Chilung (Jilong), Taipei's

*The Taiwanese have retained the old Wade-Giles spelling of place names: China now uses the *pinyin* system. Names in parentheses are written according to the pinyin system.

armed conflict over Taiwan would produce no winners, although some Taiwanese politicians do like to test the waters. In September 2003, for example, Taiwanese passport holders found the name *Taiwan* beneath the (routine) Republic of China on the cover, an identification not previ- ously used. And some prominent politicians raise the prospect of referendums to test the people's views on matters that might be camouflaged as a vote on independence. This enrages leaders in mainland China, who do not wish such views to be tested.

The East Asian realm is as fraught with risk as it is filled with promise. As China's power rises and Japan's wanes, the U.S. role in this realm will change, a transition that will substantially define the route to the New World Order to which we al- luded in our introduction to this book.

DISCOVERY TOOLS   www.wiley.com/college/deblij

The *Concepts and Regions* Website, featuring *GeoDiscoveries*, offers many additional resources to enhance your understanding and experience of this chapter. Be sure to explore the following:

GeoDiscoveries Video Clips
Decorating the Interior
Wasting Away in Northeast China
Shanghai
Guangdong Becomes a Worldly Crossroads

GeoDiscoveries Interactivities
Capitalism and Communism
A Picture is Worth a Thousand Words

Photo Galleries
Gobi Desert
Tiananmen Square, Beijing, China
Pudong, Shanghai, China
Guilin
Guangxi-Zhuang
Kyoto, Japan
Tokyo, Japan
Busan, South Korea
Anyang, China
Shenzhen, China
Loess Plateau, China
Beijing, China
North China Plain
Shanghai, China
Hiroshima, Japan
Seoul, South Korea

Virtual Field Trip
Korea: Along the Demilitarized Zone (DMZ)

More To Explore
Systematic Essay: Geography of Development
Issues in Geography: Names and Places; Kongfuzi (Confucius); Extraterritoriality; When the Big One Strikes; The Wages of War; Obsolescent North Korea
Major Cities: Xian, Beijing, Shanghai, Tokyo, Seoul

Expanded Regional Coverage
China Proper, Xizang (Tibet), Xinjiang, Mongolia, Jakota Triangle

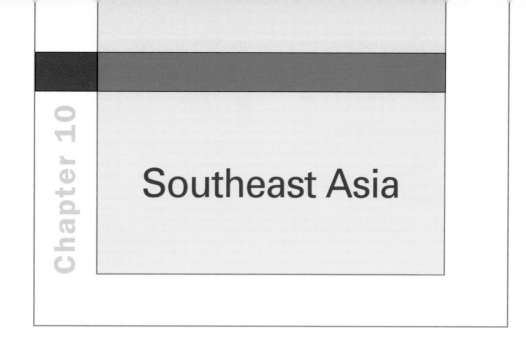

Southeast Asia

CONCEPTS, IDEAS, AND TERMS

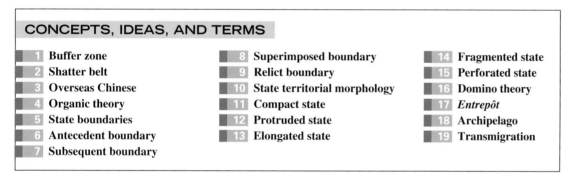

1. Buffer zone
2. Shatter belt
3. Overseas Chinese
4. Organic theory
5. State boundaries
6. Antecedent boundary
7. Subsequent boundary
8. Superimposed boundary
9. Relict boundary
10. State territorial morphology
11. Compact state
12. Protruded state
13. Elongated state
14. Fragmented state
15. Perforated state
16. Domino theory
17. *Entrepôt*
18. Archipelago
19. Transmigration

REGIONS

MAINLAND SOUTHEAST ASIA

INSULAR SOUTHEAST ASIA

SOUTHEAST ASIA IS a realm of peninsulas and islands, a corner of Asia bounded by India on the northwest and China on the northeast (Fig. 10-1). Its western coasts are washed by the Indian Ocean, and to the east stretches the vast Pacific. From all these directions, Southeast Asia has been penetrated by outside forces. From India came traders; from China, settlers; from across the Indian Ocean, Arabs to engage in commerce and Europeans to build empires; and from across the Pacific, the Americans. Southeast Asia has been the scene of countless contests for power and primacy—the competitors have come from near and far.

Southeast Asia's geography in some ways resembles that of Eastern Europe. It is a mosaic of smaller countries on the periphery of one of the **1** world's largest states. It has been a **buffer zone 2** between powerful adversaries. It is a **shatter belt** in which stresses and pressures from without and within have fractured the political geography. Like Eastern Europe, Southeast Asia exhibits great cultural diversity. It is a realm of hundreds of cultures, numerous languages and dialects, and several major religions.

339

STUDY
TOOLS

3D GLOBE
AREA AND DEMOGRAPHIC DATA

FLASHCARDS
MAP QUIZZES

AUDIO PRONUNCIATION GUIDE
CHAPTER QUIZ

ANNOTATED WEBLINKS
LONELY PLANET WEBLINKS

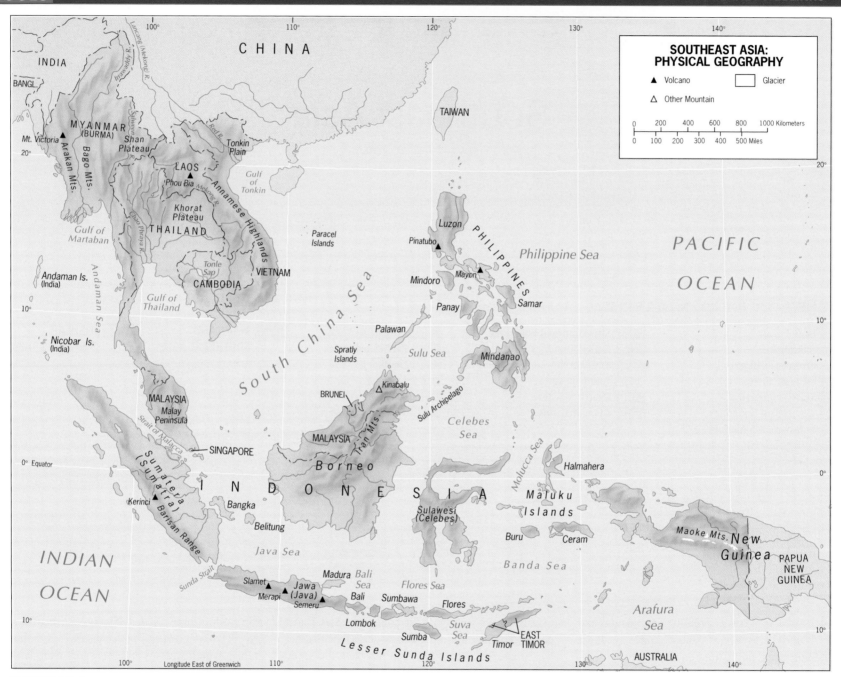

SOUTHEAST ASIA:
PHYSICAL GEOGRAPHY

▲ Volcano   ☐ Glacier

△ Other Mountain

FIGURE 10-1

DEFINING THE REALM

Figure 10-2 shows the dimensions of the Southeast Asian geographic realm, but note the disconformity between the eastern boundary of the realm and the eastern limits of its most populous state, Indonesia. The easternmost part of Indonesia is the western half of the island of New Guinea, where indigenous cultures are not Southeast Asian but Pacific. Today Indonesia rules what is in effect a Pacific island colony, although Papua is officially one of its provinces. While we refer in this chapter to Papua because of its association with Indonesia, we discuss all of New Guinea under the Pacific Realm in Chapter 12.

Because the politico-geographical map (Fig. 10-2) is so complicated, it should be studied atten-tively. One good way to strengthen your mental map of this realm is to follow the mainland coast-line from west to east. The westernmost state in the realm is Myanmar (called Burma before 1989 and still referred to by that name), the only country in Southeast Asia that borders both India and China. Myanmar shares the "neck" of the Malay Penin-sula with Thailand, heart of the *mainland* region. The south of the peninsula is part of Malaysia—except for Singapore, at the very tip of it. Facing the Gulf of Thailand is Cambodia. Still moving generally eastward, we reach Vietnam, a strip of land that extends all the way to the Chinese border. And surrounded by its neighbors is landlocked Laos, remote and isolated. This leaves the islands that constitute *insular* Southeast Asia: the Philip-pines in the north and Indonesia in the south, and between them the offshore portion of Malaysia, situated on the largely Indonesian island of Bor-neo. Also on Borneo lies the ministate of Brunei, small but, as we will see, important in the regional picture. And finally, a new state has just appeared on the map in the realm's southeastern corner: East Timor, a former Portuguese colony annexed by In-donesia in 1976, released to United Nations super-vision in 1999, and independent in 2002.

These are countries of a geographic realm that has no dominant state—no China, no India, no Brazil—although one country, Indonesia, contains 40 per-cent of its total population and has the potential to

MAJOR GEOGRAPHIC QUALITIES OF SOUTHEAST ASIA

1. Southeast Asia extends from the peninsular mainland to the archipelagos offshore. Because Indonesia controls part of New Guinea, its functional region reaches into the neighboring Pacific geographic realm.

2. Southeast Asia, like Eastern Europe, has been a shatter belt between powerful adversaries and has a fractured cultural and political geography shaped by foreign intervention.

3. Southeast Asia's physiography is dominated by high relief, crustal instability marked by volcanic activity and earth-quakes, and tropical climates.

4. A majority of Southeast Asia's more than half-billion people live on the islands of just two countries: Indonesia, with the world's fourth-largest population, and the Philippines. The rate of population increase in the insular region of Southeast Asia exceeds that of the mainland.

5. Although the overwhelming majority of Southeast Asians have the same ancestry, cultural divisions and local traditions abound, which the realm's divisive physiography sustains.

6. The legacies of powerful foreign influences, Asian as well as non-Asian, continue to affect the cultural landscapes of Southeast Asia.

7. Southeast Asia's political geography exhibits a variety of boundary types and several categories of state territorial morphology.

8. The Mekong River, Southeast Asia's Danube, has its source in China and borders or crosses five Southeast Asian coun-tries, sustaining tens of millions of farmers, fishing people, and boat owners.

9. The realm's giant in terms of territory as well as population, Indonesia, has not asserted itself as the dominant state be-cause of mismanagement and corruption; but Indonesia has enormous potential.

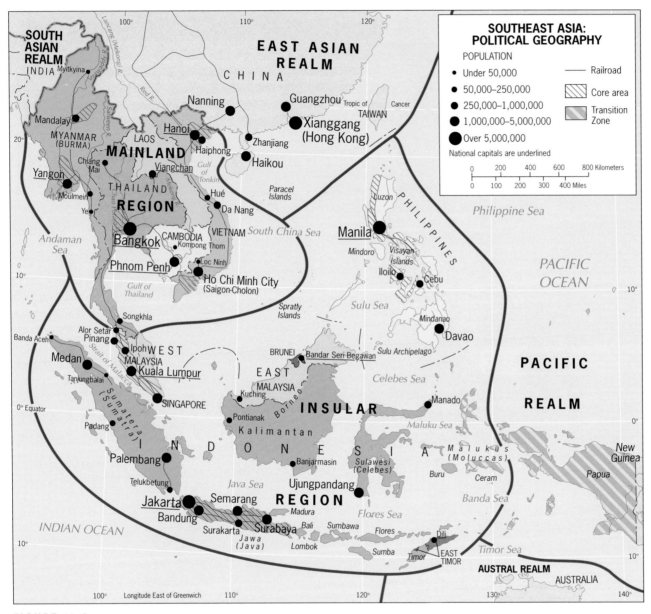

FIGURE 10-2

emerge as a commanding force. Neither did any single, dominant core of indigenous culture develop here as it did in East Asia. In the river basins and on the plains of the mainland, as well as on the islands offshore, a flowering of cultures produced a diversity of societies whose languages, religions, arts, music, foods, and other achievements formed an almost infinitely varied mosaic—but none of those cultures rose to imperial power. The European colonizers forged empires here, often by

playing one state off against another; the Europeans divided and ruled. Out of this foreign intervention came the modern map of Southeast Asia, as only Thailand (formerly Siam) survived the colonial era as an independent entity. Thailand was useful to two competing powers, the French to the east and the British to the west. It was a convenient buffer, and while the colonists carved pieces off Thailand's domain, the kingdom endured.

Indeed, the Europeans accomplished what local powers could not: the formation of comparatively large, multicultural states that encompassed diverse peoples and societies and welded them together. Were it not for the colonial intervention, it is unlikely that the 17,000 islands of far-flung Indonesia would today constitute the world's fourth largest country in terms of population. Nor would the nine sultanates of Malaysia have been united, let alone with the peoples of northern Borneo across the South China Sea. For good or ill, the colonial intrusion consolidated a realm of few culture cores and numerous ministates into less than a dozen countries.

PHYSICAL GEOGRAPHY

As Figure 10-1 shows, Southeast Asia is a realm in which high relief dominates the physiography. From the Arakan Mountains in western Myanmar (Burma) to the glaciers (yes, glaciers!) of the Indonesian part of New Guinea, elevations rise above 10,000 feet (3300 m) in many locales.

The relief map reminds us that this is not only the Pacific Rim but also the Pacific Ring of Fire, where the crust is unstable, earthquakes are common, and volcanoes are active. Among the islands, Borneo is the sole exception. Borneo is a slab of ancient crust, pushed high above sea level by tectonic forces and eroded into its present mountainous topography.

As Figure 10-1 underscores, rivers rise in the highland backbones of the islands and peninsulas, and deposit their sediments as they wind their way toward the coast; the physiography of Sumatera demonstrates this unmistakably. The volcanic hills, plateaus, and better-drained lowlands are fertile and, in the warmth of tropical climates, can yield multiple crops of rice.

On the peninsular mainland we see a pattern that is already familiar: rivers rising in the Asian interior that create alluvial plains and deltas. The Mekong River is the Chang/Yangzi of Southeast Asia: you can trace it all the way from China via Laos, Thailand, and Cambodia into southern Vietnam, where it forms a massive and populous delta. In the west, Myanmar's key river is the Irrawaddy; Thailand's is the Chao Phraya. In the north, the Red River Basin is the breadbasket of northern Vietnam.

No survey of the physical geography of Southeast Asia would be complete without reference to the realm's seas, gulfs, straits, and bays. Irregular and indented coastlines such as these, with thousands of islands near and far, create difficult problems when it comes to drawing boundaries in the waters offshore (*maritime boundaries*). Southeast Asia has one of the most complex maritime boundary frameworks in the world.

POPULATION GEOGRAPHY

Compared to the huge population numbers and densities in the habitable regions of South Asia and China, demographic totals for the countries of Southeast Asia, with the exception of Indonesia, seem modest. Again, comparisons with Europe come to mind. Three countries—Thailand, the Philippines,

and Vietnam—have populations between 60 and 85 million. Laos, quite a large country territorially (comparable to the United Kingdom), had just 5.8 million inhabitants in 2004. Cambodia, substantially larger than Greece, had 12.7 million. It is noteworthy that of Southeast Asia's 552 million inhabitants, well over half (308 million, 56 percent) live on the islands of Indonesia and the Philippines, leaving the realm's mainland countries with only about 44 percent of the population.

The Ethnic Mosaic

Southeast Asia's peoples come from a common stock just as (Caucasian) Europeans do, but this has not prevented the emergence of regionally or locally discrete ethnic or cultural groups. Figure 10-3 displays the broad distribution of ethnolinguistic groups in the realm, but be aware that this is a generalization. At the scale of this map, numerous small groups cannot be depicted.

The map shows the rough spatial coincidence, on the mainland, between major ethnic group and modern political state. The Burman dominate in the country once known as Burma (Myanmar); the Thai occupy the state once known as Siam (now Thailand); the Khmer form the nation of Cambodia and extend northward into Laos; and the Vietnamese inhabit the long strip of territory facing the South China Sea.

Territorially, by far the largest population is classified in Figure 10-3 as Indonesian, the inhabitants of the great archipelago that extends from Sumatera* west of the Malay Peninsula to the Malukus (Moluccas) in the east and from the lesser Sunda Islands in the south to the Philippines in

the north. Collectively, all these peoples—the Filipinos, Malays, and Indonesians—shown in Figure 10-3 are known as Indonesians, but they have been divided by history and politics. Note, on the map, that the Indonesians in Indonesia itself include Javanese, Madurese, Sundanese, Balinese, and other large groups; hundreds of smaller ones are not shown. In the Philippines, too, island isolation and contrasting ways of life are reflected in the cultural mosaic. Also part of this Indonesian ethnic-cultural complex are the Malays, whose heartland lies on the Malay Peninsula but who form minorities in other areas as well. Like most Indonesians, the Malays are Muslims, but Islam is a more powerful force in Malay society than, in general, in Indonesian culture.

In the northern part of the mainland region, numerous minorities inhabit remote parts of the countries in which the Burman (Burmese), Thai, and Vietnamese dominate. Those minorities, as a comparison of Figures 10-2 and 10-3 indicates, tend to occupy areas on the peripheries of their countries, away from the core areas, where the terrain is mountainous and the forest is dense, and where the governments of the national states do not have complete control. This remoteness and sense of detachment give rise to notions of secession, or at least resistance to governmental efforts to establish authority, often resulting in bitter ethnic conflict.

Immigrants

Figure 10-3 also reminds us that, again like Eastern Europe, Southeast Asia is home to major ethnic minorities from outside the realm. On the Malay Peninsula, note the South Asian (Hindustani) cluster. Hindu communities with Indian ancestries exist in many parts of the peninsula, but in the southwest they form the majority in a small area. In Singapore, too, South Asians form a significant minority. These communities arose during the colonial period, but South Asians arrived in this realm many centuries earlier, propagating Buddhism and

---

*As in Africa, names and spellings have changed with independence. In this chapter, we will use the contemporary spellings, except when we refer to the colonial period. Thus Indonesia's four major islands are Jawa, Sumatera, Kalimantan (the Indonesian part of Borneo), and Sulawesi. The Dutch called them Java, Sumatra, Dutch Borneo, and Celebes, respectively.

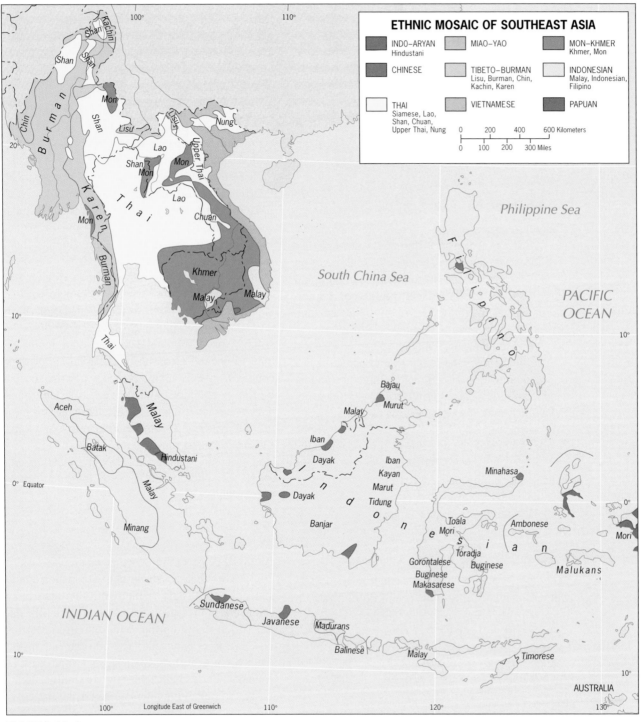

ETHNIC MOSAIC OF SOUTHEAST ASIA

- INDO–ARYAN Hindustani
- MIAO–YAO
- MON–KHMER Khmer, Mon
- CHINESE
- TIBETO–BURMAN Lisu, Burman, Chin, Kachin, Karen
- INDONESIAN Malay, Indonesian, Filipino
- THAI Siamese, Lao, Shan, Chuan, Upper Thai, Nung
- VIETNAMESE
- PAPUAN

0   200   400   600 Kilometers

0   100   200   300 Miles

FIGURE 10-3

leaving architectural and cultural imprints on places as far away as Jawa and Bali.

By far the largest immigrant minority in Southeast Asia, however, is Chinese. The Chinese began arriving here during the Ming and early Qing (Manchu) Dynasties, but the largest exodus occurred during the late colonial period (1870–1940), when as many as 20 million immigrated. The European powers at first encouraged this influx, using the Chinese in administration and trade. But soon these **3 Overseas Chinese** began to move into the major cities, where they established "Chinatowns" and gained control over much of the commerce. By the time the Europeans tried to reduce Chinese immigration, World War II was about to start and the colonial era would soon end.

Today, Southeast Asia is home to as many as 30 million Overseas Chinese, more than half the world total. Their lives have often been difficult. The Japanese relentlessly persecuted those Chinese who lived in Malaya during World War II; during the 1960s, Chinese in Indonesia were accused of communist sympathies, and hundreds of thousands were killed. In the late 1990s, Indonesian mobs again attacked Chinese and their property, this time because of their relative wealth and because many Chinese became Christians during the colonial period, now targeted by Islamic throngs.

Figure 10-4 shows the migration routes and current concentrations of Chinese in Southeast Asia. Most originated in China's Fujian and Guangdong provinces, and many invested much of their wealth in China when it

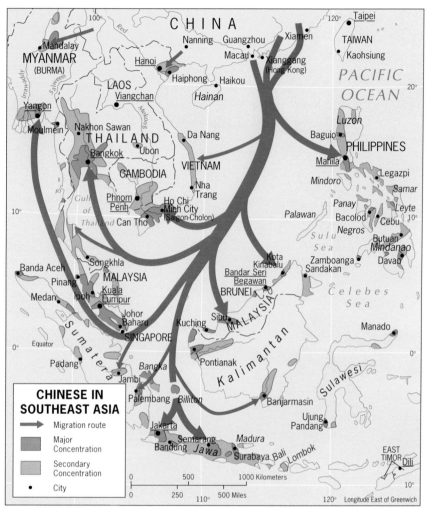

FIGURE 10-4

Peninsula, gained power over the northern part of the island of Borneo, and established themselves in Burma as well. Other colonial powers also gained footholds but not for long. The exception was Portugal, which held on to its eastern half of the island of Timor (Indonesia) until after the Dutch had been ousted from their East Indies.

Figure 10-5 shows the colonial framework in the late nineteenth century, before the United States assumed control over the Philippines in 1898. Note that while Thailand survived as an independent state, it lost territory to the British in Malaya and Burma and to the French in Cambodia and Laos.

The Colonial Imprint

The colonial powers divided their possessions into administrative units as they did in Africa and elsewhere. Some of these political entities became independent states when the colonial powers withdrew. France, one of the mainland's leading colonial powers, divided its Southeast Asian empire into five units. Three of these units lay along the east coast: Tonkin in the north next to China, centered on the basin of the Red River; Cochin China in the south, with the Mekong Delta as its focus; and between these two, Annam. The other two French territories were Cambodia, facing the Gulf of Thailand, and Laos, landlocked in the interior. Out of these five French dependencies there emerged the three states of Indochina. The three east coast territories ultimately became one state, Vietnam; the other two (Cambodia and Laos) each achieved separate independence.

The British ruled two major entities in Southeast Asia (Burma and Malaya) in addition to a large part of northern Borneo and many small islands in the South China Sea. Burma was attached to Britain's Indian empire; from 1886 until 1937 it was governed from distant New Delhi. But when British India became independent in 1947 and split into several countries, Burma was not part of the grand design that created West and East Pakistan (the latter now Bangladesh), Ceylon (now Sri Lanka), and

opened up to foreign businesses. The Overseas Chinese of Southeast Asia played a major role in the economic miracle of the Pacific Rim.

HOW THE POLITICAL MAP EVOLVED

The leading colonial competitors in Southeast Asia were the Dutch, French, British, and Spanish (with

the last replaced by the Americans in their stronghold, the Philippines). The Japanese had colonial objectives here as well, but these came and went during the course of World War II.

The Dutch acquired the greatest prize: control over the vast archipelago now called Indonesia (formerly the Netherlands East Indies). France established itself on the eastern flank of the mainland, controlling all territory east of Thailand and south of China. The British conquered the Malay

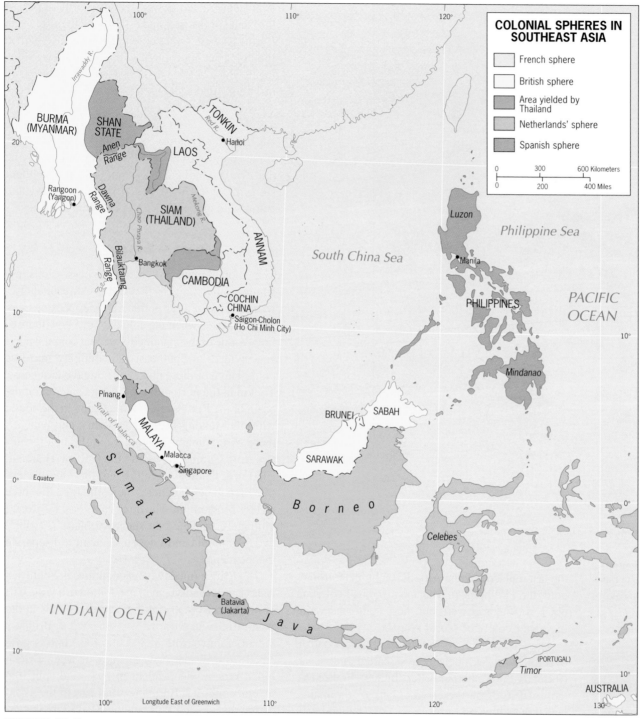

FIGURE 10-5

India. Instead, in 1948 Burma (today called Myanmar) was given the status of a sovereign republic.

In Malaya, the British developed a complicated system of colonies and protectorates that eventually gave rise to the equally complex, far-flung Malaysian Federation. Included were the former Straits Settlements (Singapore was one of these colonies), the nine protectorates on the Malay Peninsula (former sultanates of the Muslim era), the British dependencies of Sarawak and Sabah on the island of Borneo, and numerous islands in the Strait of Malacca and the South China Sea. The original Federation of Malaysia was created in 1963 by the political unification of recently independent mainland Malaya, Singapore, and the former British dependencies on the largely Indonesian island of Borneo. Singapore, however, left the Federation in 1965 to become a sovereign city-state, and the remaining units were later restructured into peninsular Malaysia and, on Borneo, Sarawak and Sabah. Thus the term *Malaya* properly refers to the geographic area of the Malay Peninsula, including Singapore and other nearby islands; the term *Malaysia* identifies the politico-geographical entity of which Kuala Lumpur is the capital city.

The Hollanders took control of the "spice islands" through their Dutch East India Company, and the wealth that they extracted from what is today Indonesia brought the Netherlands its Golden Age. From the mid-seventeenth to the late eighteenth century, the Dutch could develop their East Indies sphere of influence almost without challenge, for the British and French were preoccupied with the Indian subcontinent. By playing the princes of Indonesia's states against one another in the search for economic concessions and political influence, by placing the Chinese in positions of responsibility, and by imposing systems of forced labor in areas directly under its control, the Company had a ruinous effect on the Indonesian societies it subjugated. Java (Jawa), the most populous and productive island, became the focus of Dutch administration; from its capital at Batavia (now Jakarta), the Company extended its sphere of influence into Sumatra (Sumatera), Dutch Borneo (Kalimantan), Celebes (Sulawesi), and the smaller islands of the East Indies. This was not accomplished overnight, and the struggle for territorial control was carried on long after the Dutch East India Company had yielded its administration to the Netherlands government. Dutch colonialism thus threw a girdle around Indonesia's more than 17,000 islands, paving the way for the creation of the realm's largest and most populous nation-state (224 million today).

In the colonial tutelage of Southeast Asia, the Philippines, long under Spanish domination, had a unique experience. As early as 1571, the islands north of Indonesia were under Spain's control (they were named for Spain's King Philip II). Spanish rule began when Islam was reaching the southern Philippines via northern Borneo. The Spaniards spread their Roman Catholic faith with great zeal, and between them the soldiers and priests consolidated Hispanic dominance over the mostly Malay population. Manila, founded in 1571, became a profitable waystation on the route between southern China and western Mexico (Acapulco usually was the trans-Pacific destination for the galleons leaving Manila's port). There was much profit to be made, but the indigenous people shared little in it. Great landholdings were awarded to loyal Spanish civil servants and to men of the church. Oppression eventually yielded revolution, and Spain was confronted with a major uprising in the Philippines

when the Spanish-American War broke out elsewhere in 1898.

As part of the settlement of that war, the United States replaced Spain in Manila. That was not the end of the revolution, however. The Filipinos now took up arms against their new foreign ruler, and not until 1905, after terrible losses of life, did American forces manage to "pacify" their new dominion. Subsequently, U.S. administration in the Philippines was more progressive than Spain's had been. In 1934, Congress passed the Philippine Independence Law, providing for a ten-year transition to sovereignty. But before independence could be arranged, World War II intervened. In 1941, Japan conquered the islands, temporarily ousting the Americans; U.S. forces returned in 1944 and, with strong Filipino support, defeated the Japanese in 1945. The agenda for independence was resumed, and in 1946 the sovereign Republic of the Philippines was proclaimed.

Today, all of Southeast Asia's states are independent, but centuries of colonial rule have left strong cultural imprints. In their urban landscapes, their ed-ucation systems, their civil service, and countless other ways, this realm still carries the marks of its colonial past.

Cultural-Geographic Legacies

The French, who ruled and exploited a crucial quadrant of Southeast Asia, had a name for their empire: *Indochina.*

The *Indo* part of Indochina refers to the cultural imprints from South Asia: the Hindu presence, the importance of Buddhism (which came to Southeast Asia via Sri Lanka [Ceylon] and its seafaring merchants), the influences of Indian architecture and art (especially sculpture), writing and literature, and social structures and patterns.

The *China* in the name Indochina signifies the role of the Chinese here. Chinese emperors coveted Southeast Asian lands, and China's power reached deep into the realm. Social and political upheavals in China, combined with the opportunities created by the European colonists, sent millions of Sinicized people southward. Chinese traders,

FROM THE FIELDNOTES

"Like most major Southeast Asian cities, Bangkok's urban area includes a large and prosperous Chinese sector. No less than 14 percent of Thailand's population of 63.6 million is of Chinese ancestry, and the great majority of Chinese live in the cities. In Thailand, this large non-Thai population is well integrated into local society, and intermarriage is common. Still, Bangkok's 'Chinatown' is a distinct and discrete part of the great city. There is no mistaking Chinatown's limits: Thai commercial signs change to Chinese, goods offered for sale also change (Chinatown contains a large cluster of shops selling gold, for example), and the urban atmosphere, from street markets to bookshops, is dominantly Chinese. This is a boisterous, noisy, energetic part of multicultural Bangkok, a vivid reminder of the Chinese commercial success in Southeast Asia."

pilgrims, seafarers, fishermen, and others sailed from southeastern China to the coasts of Southeast Asia and established settlements there. Over time, those settlements attracted more Chinese emigrants, and Chinese influence in the realm grew (Fig. 10-4). Not surprisingly, relations between the Chinese settlers and the earlier inhabitants of Southeast Asia have at times been strained, even violent. The Chinese presence in Southeast Asia is long-term, but the invasion has continued into modern times.

The name *Indochina* can only refer to a part of mainland Southeast Asia, though. It cannot be used to refer to the realm as a whole. Although the "Indo" segment of this regional name can be taken to refer also to the Buddhist influences that dominate here, it makes no reference to the momentous arrival of Islam, introduced by Arab seafarers in the twelfth and thirteenth centuries and destined to change the cultural geography of the realm.

SOUTHEAST ASIA'S POLITICAL GEOGRAPHY

Southeast Asia is a laboratory for the study of political geography. This realm is a patchwork of nations and states ranging from ancient Buddhist kingdoms to modern Islamic sultanates.

Political geographers for many years have studied the causes of the cyclical rise and decline of states. They have attributed this phenomenon to various factors including climate change (see Chapter 7) and ideological contradiction (see Chapter 2). One historic figure, Friedrich Ratzel (1844–1904), conceptualized the state as a biological organism whose life, from birth through maturation and eventual senility and collapse, mirrors that of any living **4** thing. Ratzel's **organic theory** of state development held that nations, being aggregates of human beings, would over the long term live and die as their citizens did.

To understand the state better, it is useful to learn more about its components. Southeast Asia's states display these in great variety. We focus first on national boundaries on land (deferring maritime boundaries until later), and then we concentrate on the territorial morphology, or shape, of this realm's states.

The Boundaries

Boundaries are sensitive parts of a state's anatomy: just as people are territorial about their individual properties, so nations and states are sensitive about their territories and borders. The saying that "good fences make good neighbors" certainly applies to states, but, as we know, the boundaries between states are not always "good fences."

5 **Boundaries**, in effect, are contracts between states. That contract takes the form of a treaty that contains the *definition* of the boundary in the form of elaborate description. Next, cartographers perform the *delimitation* of the treaty language, drawing the boundary on official, large-scale maps. And throughout human history, states have used those maps to build fences, walls, or other barriers in a process called *demarcation*.

Once established, we can classify boundaries geographically. Some are sinuous, conforming to rivers or mountain crests (*physiographic*) or coinciding with breaks or transitions in the cultural landscape (*anthropogeographic*). As any world political map shows, many boundaries are simply straight lines, delimited without reference to physical or cultural features. These *geometric* boundaries can produce problems when the cultural landscape changes where they exist.

In general, the boundaries of Southeast Asia were better defined than those of several other post-colonial areas of the world, notably Africa, the Arabian Peninsula, and Turkestan. The colonial powers that established the original treaties tried to define boundaries to lie in remote and/or sparsely peopled areas: for example, across

interior Borneo. Nevertheless, certain Southeast Asian boundaries have produced problems, among them the geometric boundary between Papua, the portion of New Guinea ruled by Indonesia, and Papua New Guinea, the eastern component of the island.

Even on a small-scale map of the kind we use in this chapter, we can categorize the boundaries of this realm. A comparison between Figures 10-2 and 10-3 reveals that the boundary between Thailand and Myanmar over long segments is anthropogeographic, that is, ethnic-cultural, notably where the name *Karen*, the Myanmar minority, appears in Figure 10-3. Figure 10-1 shows that a large segment of the Vietnam-Laos boundary is physiographic-political, coinciding with the Annamese Highlands.

Boundaries also can be classified genetically, that is, as their evolution relates to the cultural landscapes they traverse. A leading political geographer, Richard Hartshorne (1899–1992), proposed a four-level *genetic boundary classification*. All four of these boundary types can be observed in Southeast Asia.

Certain boundaries, Hartshorne reasoned, were defined and delimited before the present-day human landscape developed. In Figure 10-6 (upper-left map), the boundary between Malaysia and Indonesia on the island of Borneo is an example of the first **6** boundary type, the **antecedent boundary**. Most of this border passes through sparsely inhabited tropical rainforest, and the break in settlement can even be detected on the small-scale world population map (Fig. G-9).

A second category of boundaries evolved as the cultural landscape of an area took shape, part of the ongoing process of accommodation. These **7** **subsequent boundaries** are represented in Southeast Asia by the map in the upper right of Figure 10-6, which shows in some detail the border between Vietnam and China. This border is the result of a long process of adjustment and modification, the end of which may not yet have come.

The third category involves boundaries drawn forcibly across a unified or at least homogeneous cultural landscape. The colonial powers did this when they divided the island of New Guinea by delimiting a boundary in a nearly straight line (curved in only one place to accommodate a bend in the Fly River), as shown in the lower-left map of **8** Figure 10-6. The **superimposed boundary** they delimited gave the Netherlands the western half of New Guinea. When Indonesia became independent in 1949, the Dutch did not yield their part of New Guinea, which is peopled mostly by ethnic Papuans, not Indonesians. In 1962, the Indonesians invaded the territory by force of arms, and in 1969 the United Nations recognized its authority there. This made the colonial, superimposed boundary the eastern border of Indonesia and had the effect of projecting Indonesia from Southeast Asia into the Pacific Realm. Geographically, all of New Guinea forms part of the Pacific Realm.

The fourth genetic boundary type is the so-called **9** **relict boundary**—a border that has ceased to function but whose imprints (and sometimes influence) are still evident in the cultural landscape. The boundary between the former North and South Vietnam (Fig. 10-6, lower-right map) is a classic example: once demarcated militarily, it has had relict status since 1976 following the reunification of Vietnam in the aftermath of the Indochina War (1964–1975).

Southeast Asia's boundaries have colonial origins, but they have continued to influence the course of events in postcolonial times. Take one instance: the physiographic boundary that separates the main island of Singapore from the rest of the Malay Peninsula, the Johor Strait (see Fig. 10-11). That physiographic-political boundary facilitated, perhaps crucially, Singapore's secession from the state of Malaysia. Without it, Malaysia might have been persuaded to stop the separation process; at the very least, territorial issues would have arisen to slow the sequence of events. As it was, no land boundary needed to be defined. The Johor Strait demarcated Singapore and left no question as to its limits.

State Territorial Morphology

Boundaries define and delimit states; they also create the mosaic of often interlocking territories that give individual countries their shape, also known as their *morphology*. The **10** **territorial morphology** of a state affects its condition, even its survival. Vietnam's extreme elongation has influenced its existence since time immemorial. As we will see, Indonesia has tried to redress its fragmentation into thousands of islands by promoting unity through the "transmigration" of Jawanese

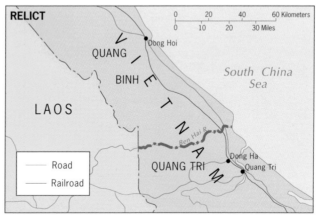

GENETIC POLITICAL BOUNDARY TYPES

FIGURE 10-6

from the most populous island to many of the others.

Political geographers identify five dominant state territorial configurations, all of which we have encountered in our world regional survey but which we have not categorized until now. All but one of these shapes are represented in Southeast Asia, and Figure 10-7 provides the terminology and examples:

- **11** **Compact states** have territories shaped somewhere between round and rectangular, without major indentations. This encloses a maximum amount of territory within a minimum length of boundary. Southeast Asian example: Cambodia.

- **12** **Protruded states** (sometimes called *extended*) have a substantial, usually compact territory from which extends a peninsular corridor that may be landlocked or coastal. Southeast Asian examples: Thailand and Myanmar.

- **13** **Elongated states** (also called *attenuated*) have territorial dimensions in which the length is at least six times the average width, creating a state that lies astride environmental or cultural transitions. Southeast Asian example: Vietnam.

- **14** **Fragmented states** consist of two or more territorial units separated by foreign territory or by water. Subtypes are mainland-mainland, mainland-island, and island-island. Southeast Asian examples: Malaysia, Indonesia, the Philippines, and East Timor.

- **15** **Perforated states** completely surround the territory of other states, so that they have a "hole" in them. No Southeast Asian example; the most illustrative current case is South Africa, perforated by Belgium-sized Lesotho.

In the discussion that follows, we will have frequent occasion to refer to this geographic property of Southeast Asia's states. For so comparatively small a realm with so few states, Southeast Asia displays a considerable variety of state morphologies. When we link these features to other geographic aspects (such as relative location), we obtain useful insights into the regional framework.

One point of caution: states' territorial morphologies do not determine their viability, cohesion, unity, or lack thereof; they can, however, influence these qualities. Cambodia's compactness has not ameliorated its divisive political geography, for example. But as we will find in the pages that follow, shape plays a key role in the still-unfolding political and economic geography of Southeast Asia.

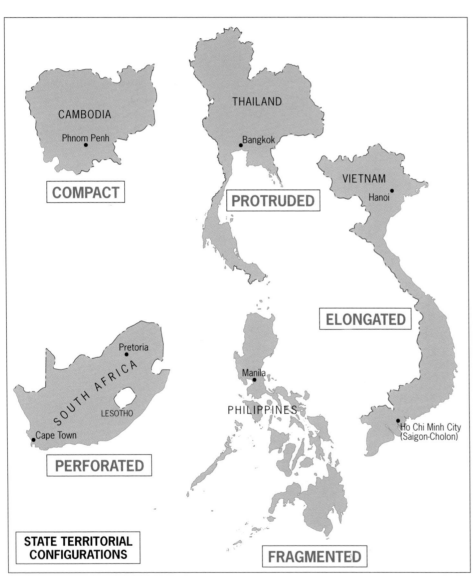

FIGURE 10-7

REGIONS OF THE REALM

Southeast Asia's first-order regionalization must be based on its mainland-island fragmentation. But as we have noted there are physiographic, historical, and cultural reasons to include the Malaysian (southern) part of the Malay Peninsula in the insular region, as shown in Figure 10-2. Using the political framework as our grid, we see that the regions of Southeast Asia are constituted as follows:

Mainland Region Vietnam, Cambodia, Laos, Thailand, Myanmar (Burma)

Insular Region Malaysia, Singapore, Indonesia, East Timor, Brunei, the Philippines

Note, however, that the realm boundary excludes the Indonesian zone of New Guinea (Papua), which is part of the Pacific geographic realm.

MAINLAND SOUTHEAST ASIA

Five countries form the mainland region of Southeast Asia: two of them protruded, one compact, one elongated, and one landlocked. Two colonial powers, buffered by Thailand, shaped its modern historical geography. One religion, Buddhism, dominates cultural landscapes, but this is a multicultural, multiethnic region. Although one of the least urbanized regions in the world, it contains several major cities. And as Figure 10-2 shows, two countries (Vietnam and Myanmar) possess more than one core area each. We approach the region from the east.

COUNTRIES OF INDOCHINA

Former French Indochina gave rise to three modern states: Vietnam, Cambodia, and Laos. Here the United States fought and lost a disastrous war that ended in 1975 but whose impact on America continues to be felt today.

After the Indochina War started formally in 1964 (U.S. involvement actually began earlier), some scholars warned that the conflict might spill over from Vietnam into Laos and Cambodia, and hence into Thailand, Malaysia, and even Myanmar 16 (Burma). This view was based on the **domino theory**, which holds that destabilization and conflict from any cause in one country can result in the collapse of order in one or more neighboring countries, triggering a chain of events that can affect a series of contiguous states in a region.

History proved these scholars wrong; Cambodia and Laos were affected but not the other states. This seemed to invalidate the domino theory, which was grounded in the capitalist-communist struggle of the twentieth century. But communist insurgency was (and is) only one way a country may be destabilized. Ethnic conflict (Equatorial Africa) and cultural strife (former Yugoslavia)—even environmental and economic causes—can set the domino effect in motion.

Elongated **Vietnam** still carries the scars of war, but consider this: about 60 percent of Vietnam's population of 81.9 million is under 21 years of age, so the vast majority of Vietnamese have no

FROM THE FIELDNOTES

"It was a relatively cool day in Hanoi, which was just as well because the drive from Haiphong had taken nearly four hours, much of the time waiting behind ox-carts, and the countryside was hot and humid. On my way to an office along Hang Bai Street I ran into this group of youngsters and their teacher on their way to the Ho Chi Minh memorial. He right away realized the level of my broken French, and explained patiently that he taught history, and took his class on a field trip several times a year. 'They get all dressed up and it is a big moment for them,' he said, 'and there is a lot of history to be visited in Hanoi.' The kids had seen me take this picture and now they crowded around. I told them that I noticed that rouge seemed to be a favorite color in their outfits.' 'But that should not surprise you, monsieur,' said the teacher, smiling. 'This is Vietnam, communist Vietnam, and red is the color that rallies us!' In the city at the opposite end of the country named after Ho Chi Minh (founder of his country's Communist Party, president from 1945 to 1969, victor over the French colonizers, and leader in the war against the United States), my experiences with teachers and students had been quite different, to the point that most called their city Saigon. But here in Hanoi, the ideological flame still burned brightly."

personal memory of that terrible conflict. The more immediate concerns in Vietnam today are to reconnect the country with the outside world and to integrate its 1200-mile (2000-km) strip of attenuated territory through better infrastructure (Fig. 10-8).

The French colonizers recognized that Vietnam, whose average width is under 150 miles (240 km), was not a homogeneous colony, so they divided it into three units: (1) Tonkin, land of the Red River Delta and centered on Hanoi in the north; (2) Cochin China, region of the Mekong Delta and centered on Saigon in the south; and (3) Annam, focused on the ancient city of Hué, in the middle (Fig. 10-5). Today, the Vietnamese prefer to use *Bac Bo*, *Nam Bo*, and *Trung Bo* to designate these areas.

The Vietnamese (or Annamese, also Annamites, after their cultural heartland) speak the same language, although the northerners can easily be distinguished from southerners by their accent. As elsewhere in their colonial empire, the French made their language the *lingua franca* of Indochina, but their tenure was cut short by the Japanese, who invaded Vietnam in 1940. During the Japanese occupation, Vietnamese nationalism became a powerful force, and after the Japanese defeat in 1945, the French could not regain control. In 1954, the French suffered a disastrous final trouncing on the battlefield at Dien Bien Phu in the far northwest and were ousted from the country.

But even after its forces routed the colonizers, Vietnam did not become a unified state. Separate regimes took control: a communist one in Hanoi and a noncommunist counterpart in Saigon. Vietnam's pronounced elongation had made things difficult for the French; now it played its role during the postcolonial period. Note, in Figure 10-8, that Vietnam is widest in the north and south, with a narrow "waist" in its middle zone. North and South Vietnam were worlds apart, and those worlds were represented in Hanoi by communism and in Saigon by anticommunism. For more than a decade the United States tried to prop up the Saigon regime that con-

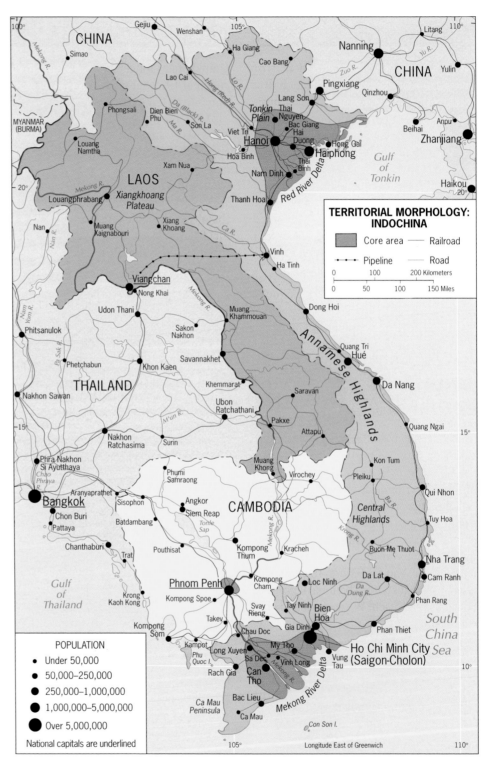

FIGURE 10-8

FROM THE FIELDNOTES

"Although modernization has changed the skylines of Vietnam's major cities (Saigon more so than Hanoi), most of the country's urban areas look and function much as they have for generations. Main arteries throb with commerce and traffic (still mostly bicycles and mopeds); side streets are quieter and more residential. No high-rises here, but the problems of dense urban populations are nevertheless evident. Every time I turned into a side street, the need for street cleaning and refuse collection was obvious, and drains were clogged more often than not. 'We haven't allowed our older neighborhoods to be destroyed like they have in China,' said my colleague from Hanoi University, 'but we also don't have an economy growing so fast that we can afford to provide the services these people need.' Certainly the Pacific Rim contrast between burgeoning coastal China and slower-growing Vietnam—a matter of communist-government policy—is evident in their cities. 'As you see, nothing much has changed here,' she said. 'Not much evidence of globalization. But we're basically self-sufficient, nobody goes hungry, and the gap between the richest and the poorest here in Vietnam is a fraction of what it is now in China. We decided to keep control.' You would find more agreement on these points here in Hanoi than in Saigon, but the Vietnamese state is stable and progressing."

trolled the south, but the communists prevailed, and like China, Vietnam has a communist government today. As many as 2 million Vietnamese refugees set out on often-flimsy boats onto the South China Sea; of those who survived, a majority were settled in the United States.

Today, contrasts between north and south continue. The capital, Hanoi, carries the imprints of its Soviet-era tutelage in the form of Ho Chi Minh's mausoleum and rows of faceless apartment build-ings representative of the Soviet "socialist city." With 4 million residents, Hanoi anchors the northern (Tonkin) core area of Vietnam, the basin of the Red River (its agricultural hinterland). In the paddies, irrigation water is still raised by bucket. On the roads, goods move by human- or animal-drawn cart. As yet little has changed here—except people's expectations.

But Vietnam's communist government has been slow to open the country to foreign investment and involvement. Instead, it has in some ways confirmed the fears of those who resisted communist rule all along: it has harassed Buddhist and Christian groups and detained their leaders, and it has revived old conflicts between minorities in the mountains and ethnic Vietnamese in control. Researchers and visitors report that land is being taken from the hill peoples in large swaths and planted with coffee and other export crops; in 2002 the government announced that Vietnam had become the world's largest exporter of coffee. That claim may be premature, but the deforestation of the interior highlands is already having an environmental effect as annually worsening floods afflict the once-stable countryside.

By 2003, Vietnam's economy lagged behind those of other countries with much potential on the Pacific Rim. Foreign investment has been limited, and many plans to privatize state industries have been shelved. Like China, Vietnam remains a communist country, but unlike China it has not had the leadership of a Deng Xiaoping. Hanoi's avowed decision to mix communist ideology with market economics never resulted in far-reaching change: the emphasis remained on communist statism, not on economic reform. As a result, Vietnam is experiencing neither the turmoil seen in China nor the huge regional disparities arising there. During the mid-1990s, when Saigon prospered and Hanoi stagnated, it appeared that north and south were once again growing apart. Today, Saigon's exuberance is restrained by Hanoi's dogmatism—and with it, Vietnam's economic progress.

Compact **Cambodia** is heir to the ancient Khmer Empire whose capital was Angkor and whose legacy is a vast landscape of imposing monuments including Buddhist-inspired Angkor Wat. Today, 90 percent of Cambodians are Khmers, with small minorities of Vietnamese and Chinese. The present capital, Phnom Penh, lies on the Mekong River (Fig. 10-8), which crosses Cambodia before it enters and forms its great delta in Vietnam.

Geographically, Cambodia enjoys several advantages; compact states enclose a maximum of territory within a minimum of boundary, and cultural homogeneity tends to diminish centrifugal forces. But neither spatial morphology nor homogeneous ethnicity could withstand the impact of the Indochina War, which led to communist revolution and the systematic murder of as many as 2 million Cambodians by the Maoist terror group, the *Khmer Rouge*. Once self-sufficient and a food exporter, Cambodia today must import food. Chronic instability and rural dislocation make this

Southeast Asia's second-poorest country. Its postwar trauma continues.

Landlocked **Laos** has no fewer than five neighbors, one of which is East Asia's giant, China (Fig. 10-8). The Mekong River forms a long stretch of its western boundary, and the important sensitive border with Vietnam to the east lies in mountainous terrain. With 5.8 million people (about half of them ethnic Lao, related to the Thai of Thailand), Laos lies surrounded by comparatively powerful states. The country has no railroads, just a few miles of paved roads, and very little industry; it is only 17 percent urbanized (the capital, Viangchan, lies on the Mekong and has an oil pipeline to Vietnam's coast). Like Vietnam, Laos has a communist government, and like Cambodia, the country continues to be plagued by social unrest as political opposition takes the form of anarchy.

The Mekong River

No discussion of mainland Southeast Asia would be complete without a look at its greatest river. From its source among the snowy peaks of China's Qinghai and Xizang, the Mekong River rushes and flows some 2600 miles (4200 km) to its delta in southernmost Vietnam (Fig. 10-8). This "Danube of Southeast Asia" crosses or borders five of the realm's countries, supporting rice farmers and fishing people, forming a transport route where roads are few, and providing electricity from dams upstream. Tens of millions of people depend on the waters of the Mekong, from subsistence farmers in Laos to apartment dwellers in China. The Mekong Delta in southern Vietnam is one of the realm's most densely populated areas and produces enormous harvests of rice.

But problems loom. China is building a series of dams across the Lancang (as the Mekong is called there) to supply Yunnan Province with electricity. Although such hydroelectric dams should not interfere with water flow, countries downstream worry that a severe dry spell in the interior would impel the Chinese to slow the river's flow to keep

the reservoirs full. Cambodia is concerned over the future of the Tonle Sap, a large natural lake filled by the Mekong. In Vietnam, farmers worry about salt water invading the paddies should the Mekong's level drop. All along the river, fish catches are already falling, and rare species, from the Irrawaddy dolphin to the Siamese crocodile, face extinction. And the Chinese may not be the only dam builders in the future: Thailand has expressed an interest in building a dam on the Thai-Laotian border where it is demarcated by the Mekong.

In such situations, the upstream states have an advantage over those downstream. Several international organizations have been formed to coordinate development in the Mekong Basin, including the Mekong River Commission (MRC) founded half a century ago (China and Myanmar have so far refused to join). China has offered to sell electricity from its dams to Thailand, Laos, and Myanmar. Coordinated efforts to reduce logging in the Mekong's drainage basin have had some effect. After consultations with the MRC, Australia built a bridge linking Laos and Thailand. There is even a plan to make the Mekong navigable from Yunnan to the coast, creating an alternative outlet for interior China.

Sail the Mekong today, however, and you are struck by the slowness of development along this artery. Wooden boats, thatch-roofed villages, and teeming paddies mark a river still crossed by antiquated ferries and flanked by few towns. Of modern infrastructure, one sees little. And yet the Mekong and its basin form the lifeline of mainland Southeast Asia's dominantly rural societies.

PROTRUDED THAILAND AND MYANMAR

In virtually every way, **Thailand** is the leading state of the mainland region. In contrast to its neighbors, Thailand has been a strong participant in the Pacific Rim's economic development. Its capital, Bangkok (population: 8.1 million), is one of the two largest urban centers in the region and one of the world's most prominent primate cities. The country's population, 63.6 million in 2004, is growing at the slowest rate in the entire realm, equivalent to that of fully urbanized Singapore. Over the past few decades, only political instability and uncertainty have inhibited economic progress. Thailand is a constitutional monarchy; its moves toward stable democracy have been halting. Corruption remains a major problem here.

Thailand is the textbook example of a protruded state. From a relatively compact heartland, in which lie the core area, capital, and major areas of productive capacity, a 600-mile (1000-km) corridor of land, in places less than 20 miles (32 km) wide, extends southward to the border with Malaysia (Fig. 10-9). The boundary that defines this protrusion runs down the length of the Malay Peninsula to the Kra Isthmus, where neighboring Myanmar peters out and Thailand fronts the Andaman Sea (an arm of the Indian Ocean) as well as the Gulf of Thailand. In the entire country, no place lies farther from the capital than the southern end of this tenuous protrusion.

Consider this spatial situation in the context of Figure 10-3 (p. 344). Note that the Malay ethnic population group extends from Malaysia more than 200 miles (300 km) into Thai territory. In Thailand's five southernmost provinces, 85 percent of the inhabitants are Muslims (the figure for Thailand as a whole is less than 5 percent). The political border between Thailand and Malaysia is porous; in fact, you can cross it at will almost anywhere along the Kolok River by canoe, and inland it is a matter of walking a forest trail. For more than a century, the southern provinces have had closer ties with Malaysia across the border than with Bangkok 600 miles (1000 km) away, and the Thai government has not tried to restrict movement or impose onerous rules on the Muslim population here. But in the new era of the War on Terror and in view of rising violence in this southern frontier, the south is coming to national attention. The Pattani United Liberation Organization, named after the functional capital of the Muslim south, was active in the 1960s and

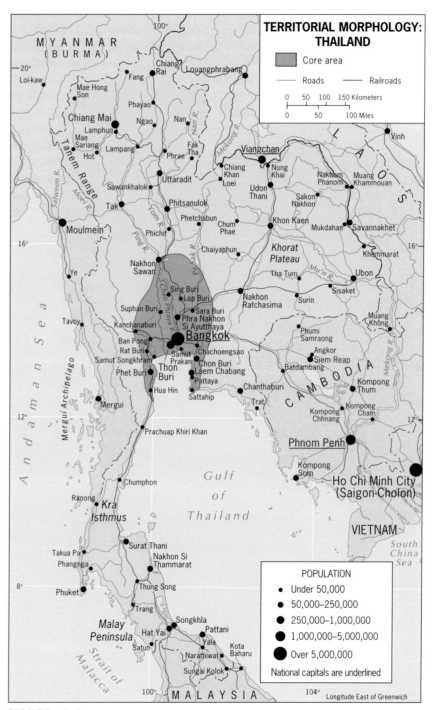

FIGURE 10-9

1970s but has been dormant since. Then a series of bombings at a Pattani hotel, several schools, a Chinese shrine, and a Buddhist temple in 2001 and 2002 raised fears that Muslim-inspired violence was on the rise again. In April 2002, Bangkok newspapers reported that police had found bomb-making materials in the house of a Muslim religious leader. In August 2003, the notorious al-Qaeda terrorist Hambali, who worked to link al-Qaeda with its Southeast Asian counterpart, Jemaah Islamiah, was captured in Thailand. Thailand's pronounced protrusion, its porous border, its dependence on tourism, and the vulnerable location of its Andaman-coast tourist facilities combine to raise concern over this distant part of its national territory.

As Figure 10-2 shows, Thailand occupies the heart of the mainland region of Southeast Asia. While Thailand has no Red, Mekong, or Irrawaddy Delta, its central lowland is watered by a set of streams that flow off the northern highlands and the Khorat Plateau in the east. One of these streams, the Chao Phraya, is the Rhine of Thailand. From the head of the Gulf of Thailand to Nakhon Sawan, this river is a highway of traffic. Barge trains loaded with rice head for the coast, ferry boats head upstream, freighters transport tin and tungsten (of which Thailand is among the world's leading producers). Bangkok sprawls on both sides of the lower Chao Phraya, here flanked by skyscrapers, pagodas, factories, boatsheds, ferry landings, luxury hotels, and modest dwellings in crowded confusion. On the right bank of the Chao Phraya, Bangkok's west side, lie the city's remaining *klong* (canal) neighborhoods, where waterways and boats form the transport system. Bangkok still is known as the Venice of Asia, although many *klongs* have been filled in and paved over to serve as roadways.

Thailand during the 1990s seemed destined to become another Pacific economic tiger. The low-wage Thai workforce, varied economy (oil and fish from the Gulf of Thailand, rice from the interior, tourism), and fast-growing foreign investment combined to raise GDP levels higher than those of

any other mainland Southeast Asian country with the exceptions of Singapore and Malaysia. Since then, Thailand's growth has slowed due to a combination of domestic and foreign causes. Mismanagement and corruption, misguided rural development policies, a refugee influx from unstable neighbors, inadequate control over opium production from the "Golden Triangle" where Myanmar, Laos, and Thailand meet, and a worsening AIDS crisis have compounded Thailand's problems.

Thailand's territorial morphology therefore creates problems as well as opportunities. If the problems in the southernmost reaches of its protrusion can be solved, there will be potential for a three-nation development scheme involving Malaysia and Indonesia. But in 2004, that "if" loomed large among Thailand's concerns.

Thailand's also-protruded neighbor, **Myanmar** (still referred to as Burma in some anti-regime quarters), is one of the world's poorest countries where, it seems, time has stood still for centuries. Long languishing under one of the world's most corrupt military dictatorships, Myanmar shares with Thailand a long stretch of the Kra Isthmus—but look again at Figure 10-9 and note the contrast in surface communications.

Myanmar's territorial morphology is complicated by a shift in the Burmese core area that took place during colonial times. Prior to the colonial period, the focus of embryonic Burma lay in the so-called dry zone between the Arakan Mountains and the Shan Plateau, which covers the country's triangular eastern extension toward the Laotian border (Fig. 10-1). The urban focus of the state was Mandalay, which had a central situation and relative proximity to the non-Burmese highlands all around. Then the British developed the agricultural potential of the Irrawaddy Delta, and Rangoon (now called Yangon) became the hub of the colony. The Irrawaddy waterway links the old and the new core areas, but the center of gravity has shifted to the south.

The political geography of Myanmar constitutes a particularly good example of the role and effect of

territorial morphology on internal state structure. Not only is Myanmar a protruded state: its core area is also surrounded on the west, north, and east by a horseshoe of great mountains—where many of the country's 11 minority peoples had their homelands before the British occupation. The colonial boundaries had the effect of incorporating these peoples into the domain of the Burman people, who constitute about two-thirds of the population (50.3 million today). When the British departed and Burma became independent in 1948, the less numerous peoples traded one master for another. In 1976, nine of these indigenous peoples formed a union to demand the right to self-determination in their homelands.

As Figure 10-3 shows, the peripheral peoples of Myanmar occupy a significant part of the state. The Shan of the northeast and far north, who are related to the neighboring Thai, account for about 7 percent of the population, or 3.4 million. The Karen (just under 10 percent, 4.8 million) live in the neck of Myanmar's protrusion and have proclaimed that they wish to create an autonomous territory within a federal Myanmar. The Mon (about 1.2 million) were in what is today Myanmar long before the Burmans, and introduced Buddhism to the area; they want the return of ancestral lands from which they were ousted. Although the powerful military have dealt these aspirations a series of setbacks, centrifugal forces continue to bedevil the central regime. Its response has been to exert power by all available means rather than to accommodate these forces, even to the point of stifling political discourse (let alone opposition) among the Burman themselves.

INSULAR SOUTHEAST ASIA

On the peninsulas and islands of Southeast Asia's southern and eastern periphery lie six of the realm's 11 states (Fig. 10-2). Few regions in the world contain so diverse a set of countries. Malaysia, the former British colony, consists of two major areas separated by hundreds of miles of South China Sea. The realm's southernmost state, Indonesia, sprawls across thousands of islands from Sumatera in the west to New Guinea in the east. North of the Indonesian archipelago lies the Philippines, a nation that once was a U.S. colony. These are three of the most severely fragmented states on Earth, and each has faced the challenges that such politico-spatial division brings. This insular region of Southeast Asia also contains two small but important sovereign entities: a city-state and a sultanate. The city-state is Singapore, once a part of Malaysia (and one instance in which internal centrifugal forces were too great to be overcome). The sultanate is Brunei, an oil-rich Muslim territory on the island of Borneo that seems transplanted from the Persian Gulf. In addition, a third small entity, East Timor, achieved independence in 2002. Few parts of the world are more varied or interesting geographically.

MAINLAND-ISLAND MALAYSIA

The state of Malaysia represents one of the three types of fragmented states discussed earlier: the mainland-island type, in which one part of the national territory lies on a continent and the other on an island. Malaysia is a colonial political artifice that combines two quite disparate components into a single state: the southern end of the Malay Peninsula and the northern part of the island of Borneo. These are known, respectively, as West Malaysia and East Malaysia (Fig. 10-2). The name *Malaysia* came into use in 1963, when the original Federation of Malaya, on the Malay Peninsula, was expanded to incorporate the areas of Sarawak and Sabah in Borneo. When the name Malaya is used, it refers to the peninsular part of the Federation, whereas Malaysia refers to the total entity.

The Malays of the peninsula, traditionally a rural people, displaced older aboriginal communities there and today make up about 61 percent of the country's population of 25.3 million. They pos-

sess a strong cultural identity expressed in adherence to the Muslim faith, a common language, and a sense of territoriality that arises from their perceived Malayan origins and their collective view of Chinese, Indian, European, and other foreign intruders.

The Chinese came to the Malay Peninsula and to northern Borneo in substantial numbers during the colonial period, and today they constitute about one-fourth of Malaysia's population (they are the largest single group in Sarawak).

Hindu South Asians were in this area long before the Europeans, and for that matter before the Arabs and Islam arrived on these shores. Today they still form a substantial minority of over 7 percent of the population, clustered, like the Chinese, on the western side of the peninsula (Fig. 10-3).

The populous peninsular part of Malaysia remains the country's dominant sector with 11 of its 13 States and nearly 80 percent of its population. Here the Malay-dominated government has strictly controlled economic and social policies while pushing the country's modernization. During the Asian economic boom of the 1990s, Malaysia's planners embraced the notion of symbols: the (then) capital, Kuala Lumpur, was endowed with the world's tallest building; a space-age airport outpaced Malaysia's needs; a high-tech administrative capital was built at Putrajaya, and a nearby development was called Cyberjaya—all part of a so-called *Multimedia Supercorridor* to anchor Malaysia's core area (Fig. 10-10).

The chief architect of this program was Malaysia's long-term and autocratic head of state, Mahathir bin Mohamad, leader of the Malay-dominated party that forms the majority in government. Dr. Mahathir not only had the support of the great majority of the country's Malays, but also that of the important and influential ethnic Chinese minority, which saw him as the only acceptable alternative to the more fundamentalist Islamic party challenging his rule. But Malaysia's headlong rush to modernize caused a backlash among more conservative Muslims which, in 2001, led to Islamist victories

in two States, tin-producing Kelantan and energy-rich but socially-poor Terengganu. As the fundamentalist governments in those two States imposed strict religious laws, Malaysians talked of two "corridors" marking their country: the Multimedia Supercorridor in the west and the "Mecca Corridor" in the east (Fig. 10-10). But after Mahathir's resignation and the appointment of Abdullah Badawi as his more moderate successor, Islamist fervor in the "Mecca Corridor," which had featured calls for a *jihad* in Malaysia, began to wane. In the March 2004 elections, the Islamist party lost the two State legislatures it had gained in 2001, and its leader even lost his seat in the federal parliament. Democracy, flawed as it still may be in this Muslim country, was a victor.

The Malay Peninsula's primacy began long ago, during colonial times, when the British created a substantial economy based on rubber plantations, palm-oil extraction, and mining (tin, bauxite, copper, iron). The Strait of Malacca (Melaka) became one of the world's busiest and most strategic waterways, and Singapore, at the southern end of it, a prized possession (Singapore seceded from the Malaysian Federation in 1965).

Malaysia, despite the loss of Singapore and notwithstanding its recurrent ethnic troubles, became a major player on the burgeoning Pacific Rim. The strong skills and modest wages of the local workforce attracted many companies, and the government capitalized on its opportunities, for example, by encouraging the creation of a high-technology manufacturing complex on the island of Pinang, where Chinese outnumber Malays by two to one and a future Singapore may be in the making.

The decision to combine the eleven Sultanates of Malaya with the States of Sabah and Sarawak on Borneo, creating the country now called Malaysia, had far-reaching consequences. These two States make up 60 percent of Malaysia's territory (although they represent only 20 percent of the population). They endowed Malaysia with major energy resources and huge stands of timber. They also complicated Malaysia's ethnic makeup because each State

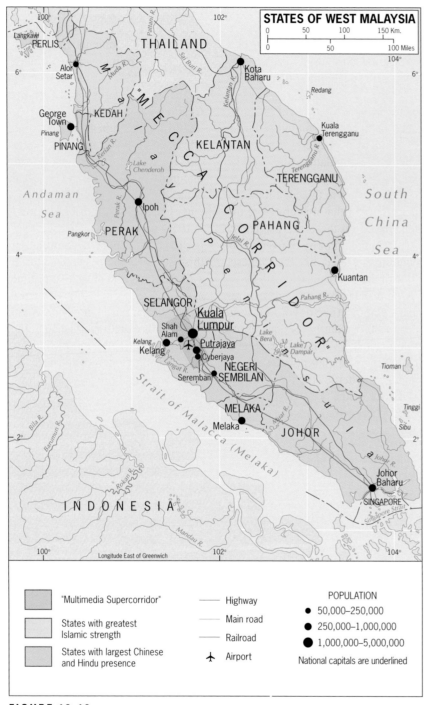

STATES OF WEST MALAYSIA

POPULATION
● 50,000–250,000
● 250,000–1,000,000
● 1,000,000–5,000,000

"Multimedia Supercorridor"

States with greatest Islamic strength

States with largest Chinese and Hindu presence

— Highway
— Main road
— Railroad
✈ Airport

National capitals are underlined

FIGURE 10-10

is home to more than two dozen indigenous groups (in fact, the immigrant Chinese form the largest single group in Sarawak). These locals complain that the federal government in Kuala Lumpur treats East Malaysia as a colony, and politics here are contentious and fractious. It is likely that Malaysia will confront devolutionary forces in East Malaysia as time goes on.

Also located on Borneo—where Sarawak and Sabah meet—is **Brunei**, a rich, oil-exporting Islamic sultanate far from the Persian Gulf. Brunei, the remnant of a former Islamic kingdom that once controlled all of Borneo and areas beyond, came under British control and was granted independence in 1984. Just slightly larger than Delaware and with a population of about 400,000, the sultanate is a mere ministate—except for the discovery of oil in 1929 and natural gas in 1965, which made this Southeast Asia's richest country after Singapore. And there are indications that further discoveries will be made in the offshore zone owned by Brunei. The Sultan of Brunei rules as an absolute monarch; his palace in the capital, Bandar Seri Begawan, is reputed to be the world's largest. He will have no difficulty finding customers for his oil in energy-poor eastern Asia.

SINGAPORE

In 1965, a fateful event occurred in Southeast Asia. Singapore, crown jewel of British colonialism in this realm, seceded from the recently independent (1963) Malaysian Federation and became a sovereign state, albeit a ministate (Fig. 10-11). With its magnificent relative location, its human resources, and firm government, Singapore then overcame the limitations of space and the absence of raw materials to become one of the economic tigers on the Pacific Rim.

With a mere 240 square miles (600 sq km) of territory, space is at a premium in Singapore, and this is a constant worry for the government. Singapore's only local spatial advantage over Hong Kong is that its small territory is less fragmented (there are just a few small islands in addition to the compact main

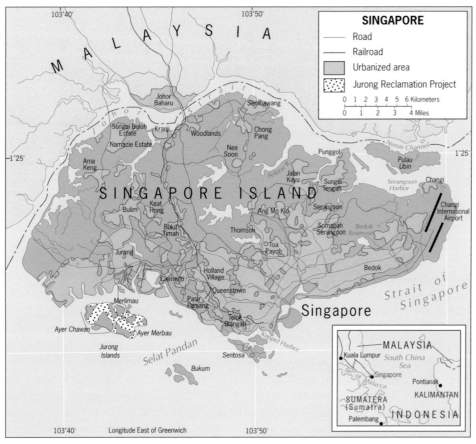

FIGURE 10-11

more than 15—to almost U.S. $25,000—that of neighboring Malaysia reached just U.S. $8330. Among other things, Singapore became (and for many years remained) the world's largest producer of disk drives for small computers.

To accomplish its revival, Singapore has moved in several directions. First, it will focus on three growth areas: information technology, automation, and biotechnology. Second, there are notions of a "Growth Triangle" involving Singapore's developing neighbors, Malaysia and Indonesia; those two countries would supply the raw materials and cheap labor, and Singapore the capital and technical know-how. Third, Singapore opened its doors to capitalists of Chinese ancestry who left Hong Kong when China took it over and who wanted to relocate their enterprises here. Singapore's population is 77 percent Chinese, 14 percent Malay, and 8 percent South Asian. The government is Chinese-dominated, and its policies have served to sustain Chinese control. Indeed, Singapore's combination of authoritarianism and economic success often is cited in China itself as proving that communism and market economies can coexist.

INDONESIA

The fourth-largest country in the world in terms of human numbers is also the globe's most expansive **18 archipelago**. Spread across more than 17,000 islands, Indonesia's 224 million people live both separated and clustered—separated by water and clustered on islands large and small.

The complicated map of Indonesia requires close attention (Fig. 10-12). Five large islands dominate the archipelago territorially, but one of these, New Guinea in the east, is not part of the Indonesian culture sphere, although its western half is under Indonesian control. The other four major islands are collectively known as the Greater Sunda islands: *Jawa* (Java), smallest but by far the most populous and important; *Sumatera* (Sumatra) in the west, directly across the Strait of Malacca from

island). With a population of 4.3 million and an expanding economy, Singapore must develop space-conserving, high-tech and service industries.

Benefiting from its relative location, the old port of Singapore had become one of the world's busiest (by numbers of ships served) even before **17** independence. It thrived as an ***entrepôt*** between the Malay Peninsula, Southeast Asia, Japan, and other emerging economic powers on the Pacific Rim and beyond. Crude oil from Southeast Asia still is unloaded and refined at Singapore, then shipped to Asian destinations. Raw rubber from the adjacent peninsula and from Indonesia's island of Sumatera is shipped to Japan, the United States, China, and other countries. Timber from

Malaysia, rice, spices, and other foodstuffs are processed and forwarded via Singapore. In return, automobiles, machinery, and equipment are imported into Southeast Asia through Singapore.

But that is the old pattern. Singapore's leaders want to redirect the city-state's economy and move toward high-tech industries for the future. In Singapore the government tightly controls business as well as other aspects of life. (Some newspapers and magazines have been banned for criticizing the regime, and there are even fines for such things as eating on the subway and failing to flush a public toilet.) Its overall success after secession has tended to keep the critics quiet: while GNP per capita from 1965 to 2002 multiplied by a factor of

FROM THE FIELDNOTES

"My first visit to Singapore was by air, and I got to know the central city, where I was based, rather well. But the second was by freighter from Hong Kong, and this afforded quite a different perspective. Here was one reason for Singapore's success: I have simply never seen a neater, cleaner, better organized, or more modern port facility. Even piles of loose items were carefully stacked. Loading and offloading went quietly and orderly. Singapore is one of the world's leading *entrepôts*, where goods are brought in, stored, and transshipped. Nobody does it better, and Singapore carefully guards its reputation for dependability, on-schedule loading, and lack of corruption."

Malaysia; *Kalimantan*, the Indonesian sector of large, compact, minicontinent Borneo; and wishbone-shaped, distended *Sulawesi* (Celebes) to the east. Extending eastward from Jawa are the Lesser Sunda Islands, including Bali and, near the eastern end, Timor. Another important island chain within Indonesia is the Maluku (Molucca) Islands, between Sulawesi and New Guinea. The central water body of Indonesia is the Java Sea.

Indonesia is a Dutch colonial creation, and the Dutch chose Jawa as their colonial headquarters, making Batavia (now Jakarta) their capital. Today, **Jawa** remains the core of Indonesia. With about 125 million inhabitants, Jawa is one of the world's most densely peopled places and one of the most agriculturally productive. Jawa also is the most highly urbanized part of a country in which more than 60 percent of the people still live on the land, and the Pacific Rim boom of the 1990s had strong impact here. The city of Jakarta (13.1 million), on the northwestern coast, became the heart of a larger conurbation now known as *Jabotabek*, consisting of the capital as well as Bogor, Tangerang, and Bekasi. During the 1990s the population of this megalopolis grew

from 15 to 20 million, and it is predicted to reach 30 million by 2010. Already, Jabotabek is home to over 10 percent of Indonesia's entire population and 25 percent of its urban population. Thousands of factories, their owners taking advantage of low prevailing wages, were built in this area, straining its infrastructure and overburdening the port of Jakarta. On an average day, hundreds of ships lie at anchor, awaiting docking space to offload raw materials and take on finished products.

Always in Indonesia, Jawa is where the power lies. As a cultural group (though itself heterogeneous), the Jawanese constitute about 56 percent of the country's population. In the center of the island, a politically powerful sultanate centers on Yogyakarta; Jawa also is the main base for the two national Islamic movements, one comparatively moderate and the other increasingly revivalist.

Sumatera, Indonesia's westernmost island, forms the western shore of the busy Strait of Malacca; Singapore lies across the Strait from approximately the middle of the island. Although much larger than Jawa, Sumatera has only about one-third as many people (45 million). In colonial

times the island became a base for rubber and palm-oil plantations; its high relief makes possible the cultivation of a wide range of crops, and neighboring Bangka and Belitung yield petroleum and natural gas. Palembang is the key urban center in the south, but current attention focuses on the north. There, the Batak people accommodated colonialism and Westernization and made Medan one of Indonesia's Pacific Rim boom cities. Farther north the Aceh fought the Dutch into the twentieth century and now demand greater autonomy, if not outright independence, for their province from the Indonesian government. The central Province of Riau, too, is politically restive, and Sumatera presents Jakarta with economic assets and political liabilities.

Kalimantan is the Indonesian part of the island of Borneo, a slab of the Earth's crystalline crust whose backbone of tall mountains is of erosional, not volcanic, origin. Larger than Texas, Borneo has a deep, densely forested interior that is a last refuge for some 30,000 orangutans. Elephants, rhinoceroses, and tigers survive even as loggers attack their habitat. Borneo's Pleistocene heritage survives because its human population still is small (13 million

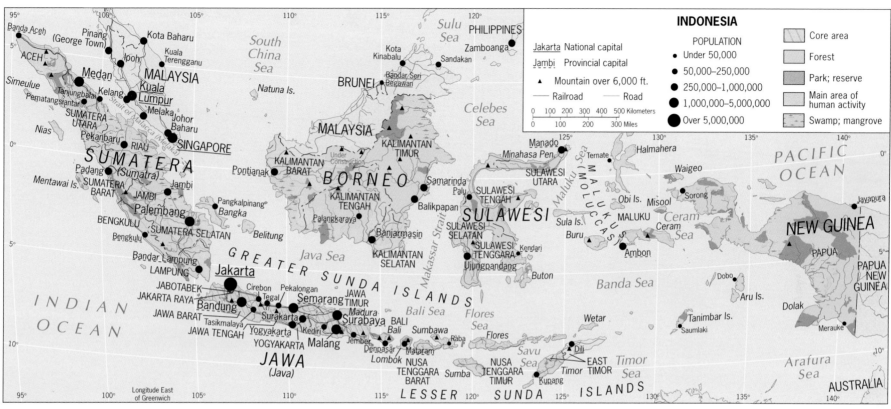

FIGURE 10-12

on the Indonesian side, 5 million on the Malaysian) and because indigenous peoples, principally the Dayak clans, had less impact on the natural environment than the Indonesian and Chinese immigrants who log the forests and clear land for farms. As Figure 10-12 shows, the only towns of any size in Kalimantan lie on or near the coast; routes into the interior still are few and far between.

Sulawesi consists of a set of intersecting, volcanic mountain ranges rising above sea level; the 500-mile (800-km) Minahasa Peninsula is still growing by volcanic action into the Philippine Sea. This northern peninsula, a favorite of the Dutch colonizers, remains the most developed part of an otherwise rugged and remote island, with Manado its relatively prosperous focus. Seven major ethnic groups inhabit the valleys and basins between the mountains, but the population of 16 million also

includes immigrants from Jawa, especially in and around the southern center of Ujungpandang. Subsistence farming is the leading mode of life, although logging, some mining, and fishing augment the economy.

Papua, the Indonesian name for the western part of New Guinea, has become an issue in Indonesian politics. Bounded on the east by a classic superimposed geometric boundary (Fig. 10-12), it was taken over by Indonesia from the Dutch in 1969. Papua constitutes about 22 percent of Indonesia's territory, but its population is barely more than 2 million—just 1 percent of the nation. The indigenous inhabitants of this province, which is in effect a colony, are Papuan, most living in remote reaches of this mountainous and densely forested island. Papua is economically important to Indonesia, for it contains what is reputed to be

the world's richest gold mine and its second-largest open-pit copper mine, but political consciousness has reached the Papuans. The Free Papua Movement has held small rallies in the capital, Jayapura, displaying a Papuan flag and demanding recognition.

Diversity in Unity

Indonesia's survival as a unified state is as remarkable as India's and Nigeria's. With more than 300 discrete ethnic clusters, over 250 languages, and just about every religion practiced on Earth (although Islam dominates), actual and potential centrifugal forces are powerful here. Wide waters and high mountains perpetuate cultural distinctions and differences. Indonesia's national motto is *bhinneka tunggal ika*: diversity in unity.

"I drove from Manado on the Minahasa Peninsula in northeastern Sulawesi to see the ecological crisis at Lake Tondano, where a fast-growing water hyacinth is clogging the water and endangering the local fishing industry. On the way, in the town of Tomolon, I noticed this side street lined with prefabricated stilt houses in various stages of completion. These, I was told, were not primarily for local sale. They were assembled from wood taken from the forests of Sulawesi's northern peninsula, then taken apart again and shipped from Manado to Japan. 'It's a very profitable business for us,' the foreman told me. 'The wood is nearby, the labor is cheap, and the market in Japan is insatiable. We sell as many as we can build, and we haven't even begun to try marketing these houses in Taiwan or China.' At least, I thought, this wood was being converted into a finished product, unlike the mounds of logs and planks I had seen piled up in the ports of Borneo awaiting shipment to East Asia."

What Indonesia has achieved is etched against the country's continuing cultural complexity. There are dozens of distinct aboriginal cultures; virtually every coastal community has its own roots and traditions. And the majority, the rice-growing Indonesians, include not only the numerous Jawanese—who have their own cultural identity—but also the Sundanese (who constitute 14 percent of Indonesia's population), the Madurese (8 percent), and others. Perhaps the best impression of the cultural mosaic comes from the string of islands that extends eastward from Jawa to Timor (Fig. 10-12). The rice-growers of Bali adhere to a modified version of Hinduism, giving the island a unique cultural atmosphere; the population of Lombok is mainly Muslim, with some Balinese Hinduism; Sumbawa is a Muslim community; Flores is mostly Roman Catholic. In western Timor, Protestant groups dominate; in the now-independent east, where the Portuguese ruled, Roman Catholicism prevails. Nevertheless, Indonesia nominally is the world's largest Muslim country: overall, 88 percent of the people adhere to Islam, and in the cities the silver domes of neighborhood mosques rise above the townscape. But Islam is not (perhaps not yet) the issue it is in Malaysia, where observance generally is stricter and where minorities fear Islamization as Malay power grows.

Transmigration and the Outer Islands

As noted earlier, Indonesia's population of 224 million makes it the world's fourth most populous country, but Jawa, we also noted, contains approximately 56 percent of it. With about 125 million people on an island the size of Louisiana, population pressure is enormous here. Moreover, Indonesia's annual rate of population growth remains at 1.6 percent, resulting in a doubling time of 44 years. To deal with this problem, and at the same time to strengthen the core's power over outlying areas, the Indonesian government long pursued a **19** policy known as **transmigration**, inducing Jawanese (and Madurese from the adjacent island of Madura) to relocate to other islands. Several million Jawanese have moved to locales as distant as the Malukus, Sulawesi, and Sumatera; many Madurese were resettled in Kalimantan.

In fact, the Dutch colonialists started the Transmigration Program, but then President Sukarno abandoned it and it was not revived until 1974, when President Suharto saw economic and political opportunities in it. Over the next 25 years, nearly 8 million people were relocated; the World Bank gave Indonesia nearly U.S. $1 billion to facilitate the program before it pulled out amid reports of forced migration.

The results were mixed. World Bank project reviews suggested that the vast majority of the transmigrants were happy with the land, schools, medical services, and other amenities with which they were provided, but independent surveys painted a different picture. About half of the migrants never managed to rise above a life of subsistence; many never received even the most basic needs for survival in their new environment; land was simply taken away from indigenous inhabitants, notably in Kalimantan; the newcomers did much damage to the traditional cultures they invaded; and the program led to massive forest and wetland destruction. And thousands of migrants died in clashes with local inhabitants.

In March 2001, the Transmigration Program was canceled as one of the final acts of the short-lived government of President Abdurrahman Wahid, but the damage it did will trouble Indonesia for generations. From northern Sumatera to western Kalimantan to central Sulawesi to the islands of the Malukus to Papua, cultural strife and its polarizing effect will challenge Indonesian governments of the future.

■ EAST TIMOR

The easternmost of the lessor Sunda Islands is Timor, the eastern part of which was a Portuguese colony, overrun by Indonesia in 1975 and annexed in 1976, which became the scene of a bitter struggle for independence. When the people of East Timor in 1999 were allowed to express their views on independence, this Connecticut-sized territory with fewer than 800,000 inhabitants took the first steps toward nationhood.

East Timor became an independent state in May 2002. As the map shows, its sovereignty creates several politico-geographical complications (Fig. 10-13). The political entity of East Timor consists of a main territory, where the capital named Dili is located, and a small exclave on the north coast of (Indonesian) West Timor called Ocussi (spellings vary, and it is sometimes mapped as Ocussi-Ambeno or Ambeno Province). Although there is a road from the exclave via Kefamenanu to the main territory, relations between East Timor and Indonesia may not make this link feasible, so that the two parts of the new state have to be connected by boat traffic. Indeed, the sea may come to mean more to East Timor than mere connectivity. Beneath the waters offshore lie valuable oil and gas reserves (Fig. 10-13) whose ownership will depend on how East Timor's maritime boundaries (a concept to be discussed in Chapter 12) will be drawn. If Australia and Indonesia obey the longstanding United Nations rules governing this process, East Timor will find itself with the resources to ensure its development into a self-sustaining state.

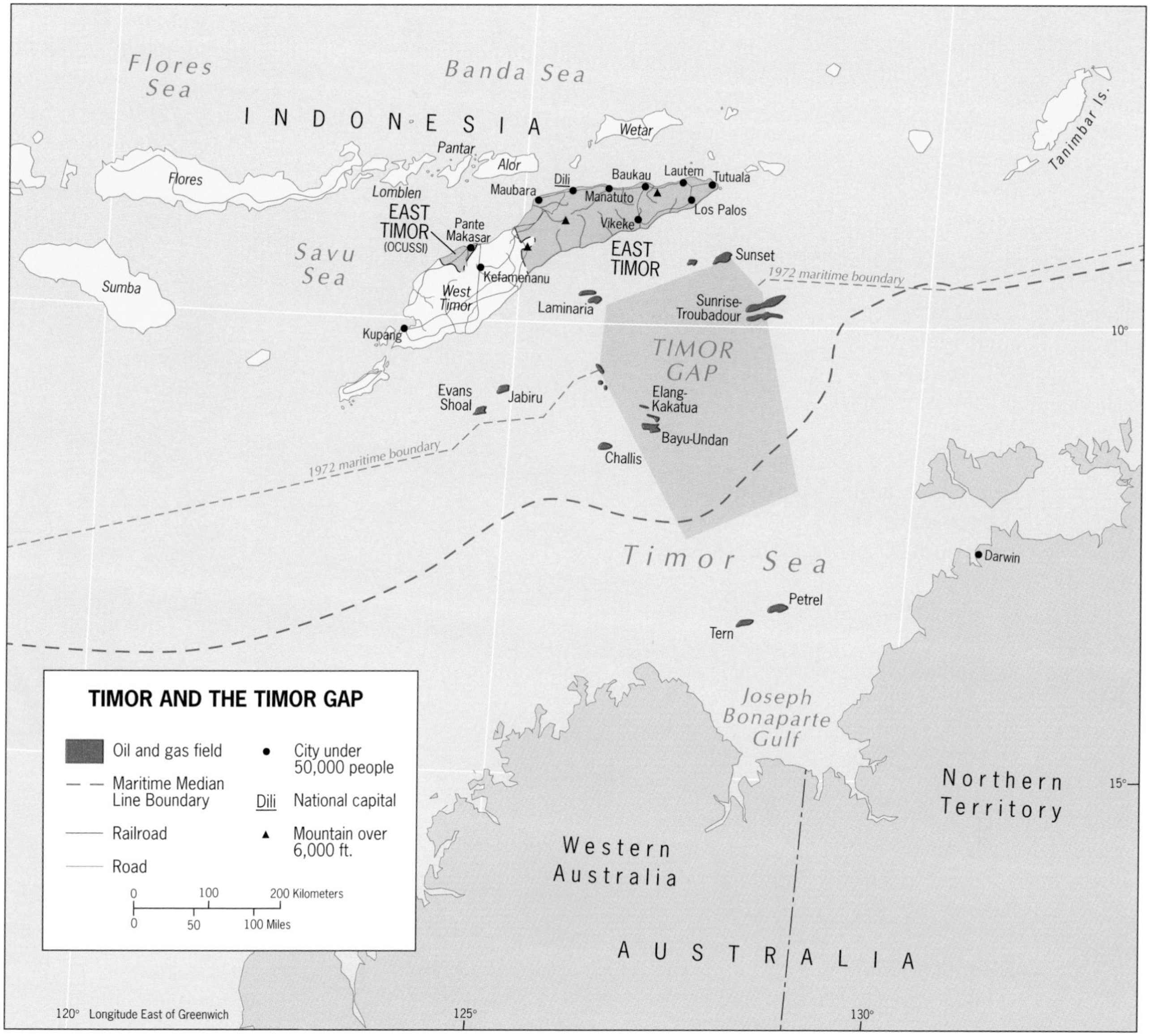

THE PHILIPPINES

North of Indonesia, across the South China Sea from Vietnam, and south of Taiwan lies an archipelago of more than 7000 islands (only about 460 of them larger than one square mile in area) inhabited by 83.9 million people. The inhabited islands of the Philippines can be viewed as three groups: (1) Luzon, largest of all, and Mindoro in the north, (2) the Visayan group in the center, and (3) Mindanao, second largest, in the south (Fig. 10-14). Southwest of Mindanao lies a small group of islands, the Sulu Archipelago, nearest to Indone-

FIGURE 10-13

sia, where Muslim-based insurgencies have kept the area in turmoil.

Few of the generalizations we have been able to make for Southeast Asia could apply in the Philippines without qualification. The country's location relative to the mainstream of change in this part of the world has had much to do with this situation.

The islands, inhabited by peoples of Malay ancestry with Indonesian strains, shared with much of the rest of Southeast Asia an early period of Hindu cultural influence, which was strongest in the south and southwest and diminished northward. Next came a Chinese invasion, felt more strongly on the largest island of Luzon in the northern part of the Philippine archipelago. Islam's arrival was delayed somewhat by the position of the Philippines well to the east of the mainland and to the north of the Indonesian islands. The few southern Muslim beachheads were soon overwhelmed by the Spanish invasion during the sixteenth century. Today the Philippines, adjacent to the world's largest Muslim state (Indonesia), is 83 percent Roman Catholic, 7 percent Protestant, and only 5 percent Muslim.

The Philippines' small Muslim population, concentrated in the southeastern flank of the archipelago (Fig. 10-14), has long decried its marginalization in this dominantly Christian country. Over the past 30 years a half-dozen Muslim organizations, including the Moro ("Moor") National Liberation Front and the Moro Islamic Liberation Front, have promoted the Muslim cause with tactics ranging from peaceful negotiation with the government to violent insurgency. Densely forested Basilan Island became the base for an especially extreme group, Abu Sayyaf, which received support through the al-Qaeda network and brought the War on Terror to the Philippines when American troops joined Filipino forces in pursuit. Deathly bombings in southern towns underscore a Muslim challenge that is pirating a disproportionate share of Manila's operating budget. Nor is this the Philippines' only security problem. In 2004, a dormant communist insurrection revived, requiring a response in difficult, remote mountain terrain where military operations are especially costly.

People and Culture

Out of the Philippines melting pot, where Mongoloid-Malay, Arab, Chinese, Japanese, Spanish, and American elements have met and mixed, has emerged the distinctive Filipino culture. It is not a homogeneous or unified culture, as is reflected by the nearly 90 Malay languages in use in the islands, but it is in many ways unique. At independence in 1946, the largest of the Malay languages, Tagalog (also called Pilipino), became the country's official language. But English is widely learned as a second language, and a Tagalog-English hybrid, "Taglish," is increasingly heard today. As in other

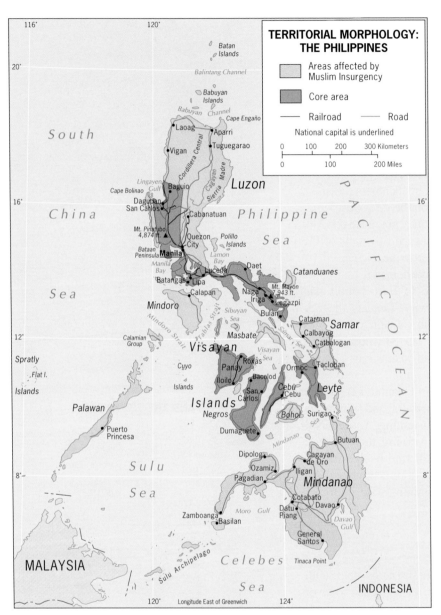

FIGURE 10-14

countries of Southeast Asia, the Chinese component of the population is small (less than 3 percent) but dominant in local business.

The Philippines' population, concentrated where the good farmlands lie, is densest in three general areas (Fig. G-9): (1) the northwestern and south-

central part of Luzon, (2) the southeastern extension of Luzon, and (3) the islands of the Visayan Sea between Luzon and Mindanao. Luzon is the site of the capital, Manila-Quezon City (10.7 million, more than one-eighth of the entire national population), a major metropolis facing the South China Sea. Alluvial as well as volcanic soils, together with ample moisture in this tropical environment, produce self-sufficiency in rice and other staples and make the Philippines a net exporter of farm products despite a high population growth rate of 2.4 percent.

In recent years, the population issue has divided this dominantly Roman Catholic society, with the government promoting family planning and the clergy opposing it. But behind this debate lies another of the Philippines' assets: in a realm of mostly undemocratic regimes, the Philippines has come out of its period of authoritarian rule a rejuvenated, if not yet robust, democracy.

The Philippines seems to get little mention in discussions of developments on the Asian Pacific Rim, and yet it would seem to be well positioned to share in the Pacific Rim's economic growth. Governmental mismanagement and political instability have slowed the country's participation, but during the 1990s the situation improved. Despite a series of jarring events—the ouster of U.S. military bases, the damaging eruption of a volcano near the capital, the violence of Muslim insurgents, and a dispute over the nearby Spratly Islands in the South China Sea—the Philippines made substantial economic progress during the decade. Its electronics and tex-

tile industries (mostly in the Manila hinterland) expanded continuously, and more foreign investment arrived. But agriculture continues to dominate the Philippines' economy, unemployment remains high, further land reform is badly needed, and social restructuring (reducing the controlling influence over national affairs by a comparatively small group of families) must occur. However, progress is being made. The country now is a lower-middle-income economy, and given a longer period of stability and success in reducing the population growth rate, it will rise to the next level and finally take its place among Pacific Rim growth poles.

DISCOVERY TOOLS   www.wiley.com/college/deblij

The *Concepts and Regions* Website, featuring *GeoDiscoveries*, offers many additional resources to enhance your understanding and experience of this chapter. Be sure to explore the following:

GeoDiscoveries Video Clips

Southeast Asia for Virtual Tourists
Ho Chi Minh City, A City of Growth
Urban and Rural Thailand
Cultural Attractions in Thailand
Singapore is Rich in Development

GeoDiscoveries Animations

Boundary Types
Territorial Configurations

GeoDiscoveries Interactivities

The Intentional Tourist: Travels in Southeast Asia
Empires of the Sons and Daughters
The Ethnic Mosaic
Touring the Cities of Southeast Asia
Know Boundaries

Photo Galleries

Nha Trang, Vietnam
Bangkok, Thailand
Mekong River, Laos
Hanoi, Vietnam
Ho Chi Minh City, Vietnam
Siem Rap, Cambodia
Xiangkhoang Province, Laos
Kuala Lumpur, Malaysia
Ambon, Indonesia
Port Kelang, Malaysia
Singapore

Virtual Field Trip

Fortress Corregidor: A Philippines World War II Relict

More To Explore

Systematic Essay: Political Geography
Issues in Geography: Overseas Chinese; Domino Theory; The Mighty Mekong; Pinang: A Future Singapore?; Rich and Broken Brunei
Major Cities: Saigon, Bangkok, Jakarta, Manila

Expanded Regional Coverage

Mainland Southeast Asia, Insular Southeast Asia

The Austral Realm

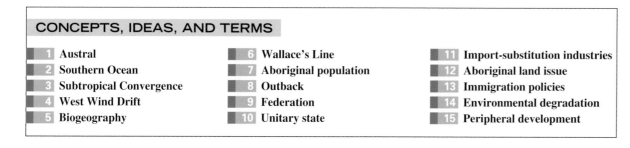

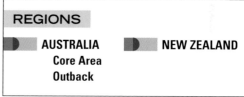

THE AUSTRAL REALM is geographically unique. It is the only geographic realm that lies entirely in the Southern Hemisphere. It is also the only realm that has no land link of any kind to a neighboring realm and is thus completely surrounded by ocean and sea. It is second only to the Pacific as the world's least populous realm. Appropriately, its name refers to its location **1** (**Austral** means south)—a location far from the sources of its dominant cultural heritage but close to its newfound economic partners on the western Pacific Rim.

DEFINING THE REALM

Two countries constitute the Austral Realm: Australia, in every way the dominant one, and New Zealand, physiographically more varied than its giant partner. Between them lies the Tasman Sea.

To the west lies the Indian Ocean, to the east the Pacific, and to the south the frigid Southern Ocean.

This southern realm is at a crossroads. On the doorstep of populous Asia, its Anglo-European legacies are now infused by other cultural strains. Polynesian Maori in New Zealand and Aboriginal communities in Australia are demanding better terms of life. Pacific Rim markets are buying huge

MAJOR GEOGRAPHIC QUALITIES OF AUSTRAL REALM

1. Australia and New Zealand constitute a geographic realm by virtue of territorial dimensions, relative location, and dominant cultural landscape.

2. Despite their inclusion in a single geographic realm, Australia and New Zealand differ physiographically. Australia has a vast, dry, low-relief interior; New Zealand is mountainous.

3. Australia and New Zealand are marked by peripheral development—Australia because of its aridity, New Zealand because of its topography.

4. The populations of Australia and New Zealand are not only peripherally distributed but also highly clustered in urban centers.

5. The realm's human geography is changing—in Australia because of Aboriginal activism and Asian immigration, and in New Zealand because of Maori activism and Pacific-islander immigration.

6. The economic geography of Australia and New Zealand is dominated by the export of livestock products (and in Australia also by wheat production and mining).

7. Australia and New Zealand are being integrated into the economic framework of the western Pacific Rim, principally as suppliers of raw materials.

quantities of raw materials. Japanese and other Asian tourists fill hotels and resorts. Queensland's tropical Gold Coast resembles Honolulu's Waikiki. The streets of Sydney and Melbourne display a multicultural panorama unimagined just two generations ago. All these changes have stirred political debate. Issues ranging from immigration quotas to indigenous land rights dominate, exposing social fault lines (city versus Outback in Australia, North and South in New Zealand). Aborigines and Maori were here first, and the Europeans came next. Now Asia looms in Australia's doorway.

LAND AND ENVIRONMENT

Physiographic contrasts between massive, compact Australia and elongated, fragmented New Zealand are related to their locations with respect to the Earth's tectonic plates (consult Fig. G-4). Australia, with some of the geologically most ancient rocks on the planet, lies at the center of its own plate, the Australian Plate. New Zealand, younger and less stable, lies at the convulsive convergence of the Australian and Pacific Plates. Earthquakes are rare in Australia and volcanic eruptions are unknown; New Zealand has plenty of both. This locational contrast is also reflected by differences in relief (Fig. 11-1). Australia's highest relief occurs in what Australians call the Great Dividing Range, the mountains that line the east coast from the Cape York Peninsula to southern Victoria, with an outlier in Tasmania. The highest point along these old, now eroding mountains is Mount Kosciusko, 7310 feet (2228 m) tall. In New Zealand, entire ranges are higher than this, and Mount Cook reaches 12,315 feet (3764 m).

West of Australia's Great Dividing Range, the physical landscape generally has low relief, with some local exceptions such as the Macdonnell Ranges near the center; plateaus and plains dominate (Fig. 11-2). The Great Artesian Basin is a key physiographic region, providing underground water sources in what would otherwise be desert country; to the south lies the continent's predominant river system, the Murray-Darling. The area mapped as *Western Plateau and Margins* in Figure 11-2 contains much of Australia's mineral wealth.

Figure G-8 reveals the effects of latitudinal location and interior isolation on Australia's climatology. In this respect, Australia is far more varied than New Zealand, its climates ranging from tropical in the far north, where rainforests flourish, to Mediterranean in parts of the south. The interior is dominated by desert and steppe conditions, the steppes providing the grasslands that sustain tens of millions of livestock. Only in the east does Australia have an area of humid temperate climate, and here lies most of the country's economic core area. New Zealand, by contrast, is totally under the influence of the Southern and Pacific oceans, creating moderate, moist conditions, temperate in the north and colder in the south.

2 Twice now we have referred to the **Southern Ocean**, but try to find this ocean on maps and globes published by famous cartographic organizations such as the National Geographic Society and Rand McNally. From their maps you would conclude that the Atlantic, Pacific, and Indian oceans reach all the way to the shores of Antarctica. Australians and New Zealanders know better. They experience the frigid waters and persistent winds of this great weathermaker on a daily basis.

For us geographers, it is a good exercise to turn the globe upside down now and then. After all, the usual orientation is quite arbitrary. Modern map-making started in the Northern Hemisphere, and the cartographers put their hemisphere on top and the other at the bottom. That is now the norm, and it can distort our view of the world. In bookstores in the Southern Hemisphere, you sometimes see tongue-in-cheek maps showing Australia and Argentina at the top, and Europe and Canada at the bottom. But this matter has a serious side. A reverse view of the globe shows us how vast the ocean encircling Antarctica is. The Southern Ocean may be remote, but its existence is real.

Where do the northward limits of the Southern Ocean lie? This ocean is bounded not by land but **3** by a marine transition called the **Subtropical Convergence**. Here the cold, extremely dense waters

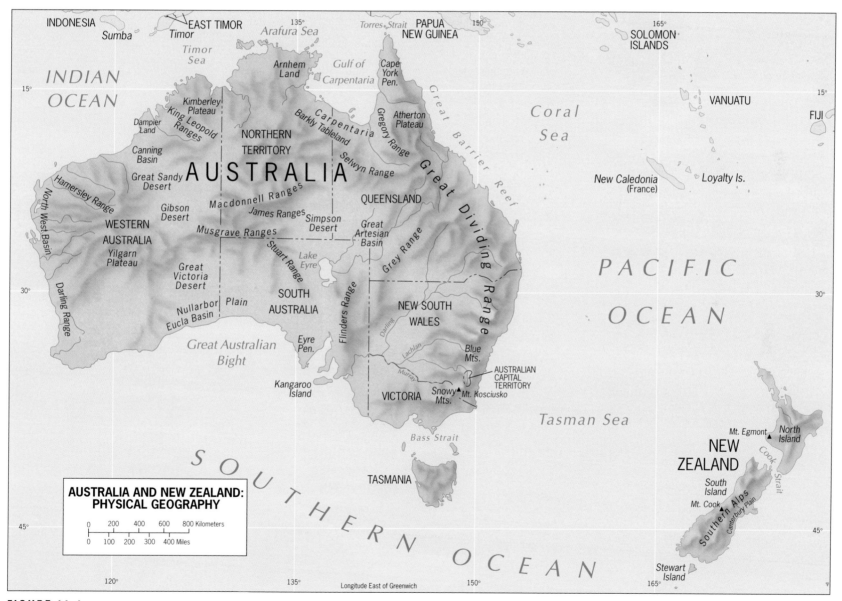

AUSTRALIA AND NEW ZEALAND: PHYSICAL GEOGRAPHY

FIGURE 11-1

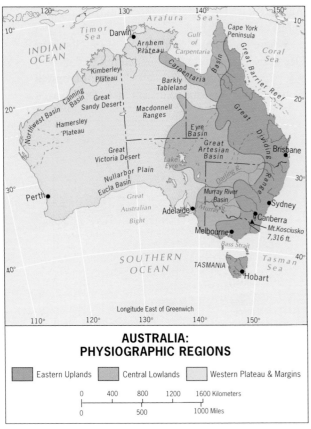

**AUSTRALIA:
PHYSIOGRAPHIC REGIONS**

■ Eastern Uplands ■ Central Lowlands □ Western Plateau & Margins

0   400   800   1200   1600 Kilometers

0   500   1000 Miles

FIGURE 11-2

Biogeography

One of this realm's defining characteristics is its wildlife. Australia is the land of kangaroos and koalas, wallabies and wombats, possums and platypuses. These and numerous other *marsupials* (animals whose young are born very early in their development and then carried in an abdominal pouch) owe their survival to Australia's early isolation during the breakup of Gondwana (see Fig. 6-2). Before more advanced mammals could enter Australia and replace the marsupials, as happened in other parts of the world, the landmass was separated from Antarctica and India, and today it contains the world's largest assemblage of marsupial fauna.

Australia's vegetation also has distinctive qualities, notably the hundreds of species of eucalyptus trees native to this geographic realm. Many other plants form part of Australia's unique flora, some with unusual adaptation to the high temperatures and low humidity that characterize much of the continent.

The study of fauna and flora in spatial perspective combines the disciplines of biology and geography **5** in a field known as **biogeography**, and Australia is a giant laboratory for biogeographers. In the In-

troduction (pp. 11-13), we noted that several of the world's climatic zones are named after the vegetation that marks them: tropical savanna, steppe, tundra. When climate, soil, vegetation, and animal life reach a long-term, stable adjustment, vegetation forms the most visible element of this ecosystem.

Biogeographers are especially interested in the distribution of plant and animal species, and in the relationships between plant and animal communities and their natural environments. (The study of plant life is called *phytogeography*; the study of animal life is called *zoogeography*.) These scholars seek to explain the distributions the map reveals. In 1876 one of the founders of biogeography, Alfred Russel Wallace, published a book entitled *The Geographical Distribution of Animals* in which he fired the first shot in a long debate: where does the zoogeographic boundary of Australia's fauna lie? Wallace's fieldwork in the area revealed that Australian forms exist not only in Australia itself but also in New Guinea and in some islands to the west. So Wallace proposed that the faunal boundary should lie between Borneo and Sulawesi, and just east of Bali (Fig. 11-3).

6 **Wallace's Line** soon was challenged by other researchers, who found species Wallace had missed

of the Southern Ocean meet the warmer waters of the Atlantic, Pacific, and Indian oceans. It is quite sharply defined by changes in temperature, chemistry, salinity, and marine fauna. Flying over it, you can actually observe it in the changing colors of the water: the Antarctic side is a deep gray, the northern side a greenish blue.

Although the Subtropical Convergence moves seasonally, its position does not vary far from latitude 40° South, which also is the approximate northern limit of Antarctic icebergs. Defined this way, the great Southern Ocean is a huge body of water that moves clockwise (from west to east) around Antarctica, which is why we also call it the **4** **West Wind Drift**.

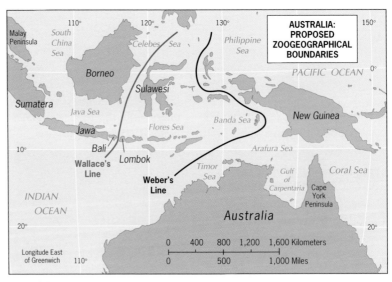

**AUSTRALIA:
PROPOSED
ZOOGEOGRAPHICAL
BOUNDARIES**

0   400   800   1,200   1,600 Kilometers

0   500   1,000 Miles

FIGURE 11-3

and who visited islands Wallace had not. There was no question that Australia's zoogeographic realm ended somewhere in the Indonesian archipelago, but where? Western Indonesia was the habitat of non-marsupial animals such as tigers, rhinoceroses, and elephants, as well as primates; New Guinea clearly was part of the realm of the marsupials. How far had the more advanced mammals progressed eastward along the island stepping stones toward New Guinea? The zoogeographer Max Weber found evi-

dence that led him to postulate his own *Weber's Line*, which, as Figure 11-3 shows, lay very close to New Guinea.

Not all research in zoogeography or phytogeography deals with such large questions. Much of it focuses on the relationships between particular species and their habitats, that is, the environment they normally occupy and of which they form a part. Such environments change, and the changes can spell disaster for the species. In Australia, the

7 arrival of the **Aboriginal population** (between 50,000 and 60,000 years ago) had comparatively limited effect on the habitats of the extant fauna. But the invasion of the European colonizers and the introduction of their livestock led to the destruction of habitats and the extinction of many native species. Whether in East Africa or in western Australia, the key to the conservation of what remains of natural flora and fauna lies in the knowledge embodied by the field of biogeography.

REGIONS OF THE REALM

Australia is the dominant component of the Austral Realm, a continent-scale country in a size category that also includes China, Canada, the United States, and Brazil. For two reasons, however, Australia has fewer regional divisions than do the aforementioned countries: Australia's relatively uncomplicated physiography and its diminutive human numbers. Our discussion, therefore, uses the core-periphery concept as a basis for investigating Australia and focuses on New Zealand as a region by itself.

AUSTRALIA

On January 1, 2001, Australia celebrated its 100th birthday as a state, the Commonwealth of Australia, recognizing (still) the British monarch as the head of state and entering its second century with a strong economy, stable political framework, high standard of living for most of its people, and favorable prospects ahead. Positioned on the Pacific Rim, nine-tenths as large as the 48 contiguous U.S. States, well endowed with farmlands and vast pastures, major rivers, ample underground water, minerals, and energy resources, served by good natural harbors, and populated by 20 million mostly-well-educated people, Australia is one of the most fortunate countries on Earth.

Not everyone in Australia shares adequately in all this good fortune, however, and the less advantaged made their voices heard during the celebrations. The country's indigenous (Aboriginal) population, though a small minority today of about 300,000, remains disproportionately disadvantaged in almost every way, from lower life expectancies to higher unemployment than average, from lower high school graduation rates to higher imprisonment ratios. But the nation is now embarked on a campaign to address these ills, with actions ranging from public demonstrations supporting reconciliation to official expressions of regret for past mistreatment and from enhanced social services to favorable court decisions involving Aboriginal land claims.

When Australia was born as a federal state, its per-capita GNP, as reckoned by economic geographers, was the highest in the world. Australia's bounty fueled a huge flow of exports to Europe, and Australians prospered. That golden age could not last forever, and eventually the country's share of world trade declined. Still, Australia today ranks among the top 15 countries in the world in terms of GNP, and for the vast majority of Australians life is comfortable.

The statistics bear witness to Australia's current good fortune. In terms of the indicators of development discussed in Chapter 9, Australia is far ahead

of all its western Pacific Rim competitors except Japan (which has dropped in the rankings because of its own economic stagnation). As Australians celebrated their first century, they were, on average, earning far more than Thais, Malaysians, or Koreans. In terms of consumption of energy per person, the number of automobiles and miles of roads, levels of health, and literacy, Australia had all the properties of a developed country. Australian cities, where 85 percent of all Australians live, are not encircled by crowded shantytowns. Nor is the Australian countryside inhabited by a poverty-stricken peasantry.

Distance

Australians often talk about distance. One of their leading historians, Geoffrey Blainey, labeled it a "tyranny"—an imposed remoteness from without and a divisive part of life within. Even today, Australia is far from nearly everywhere on Earth. A jet flight from Los Angeles to Sydney takes 14 hours nonstop and is correspondingly expensive. Freighters carrying products to European markets take ten days to two weeks to get there. Inside Australia, distances also are of continental proportions, and Australians pay the price—literally. Until some upstart private airlines started a price war, Australians paid more per mile for their domestic flights than air passengers anywhere else in the world.

But distance also was an ally, permitting Australians to ignore the obvious. Australia was a British progeny, a European outpost. Once you had arrived as an immigrant from Britain or Ireland, there were a wide range of environments, magnificent scenery, vast open spaces, and seemingly limitless opportunities. When the Japanese Empire expanded, Australia's remoteness saved the day. When immigration became an issue, Australia in its comfortable isolation could adopt an all-white admission policy that was not officially terminated until 1976. When boat people by the hundreds of thousands fled Vietnam in the aftermath of the Indochina War, almost none reached Australian shores.

Today Australia is changing and rapidly so. Immigration policy now focuses on the would-be immigrants' qualifications, skills, financial status, age, and facility with the English language. With regard to skills, high-technology specialists, financial experts, and medical personnel are especially welcome. Relatives of earlier immigrants, as well as a quota of genuine asylum-seekers, also are readily admitted. In recent years, total immigration has been limited to about 80,000 annually, but the number may have to increase if only to keep the country's population growing. According to recent data, Australia's natural rate of increase now barely hovers above zero.

Already, Australia's changed immigration policies have dramatically altered cultural landscapes, especially in the urban areas. The country is fast becoming a truly multicultural society; in Sydney, for example, one in seven residents is of Asian ancestry, a ratio that will rise to one in five by 2010. During the 1990s, when for a time Japan was Australia's leading trade partner, Australian schools began to teach Japanese to tens of thousands of children. The Austral Realm is clearly in transition.

Core and Periphery

Australia is a large landmass, but its population is heavily concentrated in a core area that lies in the east and southeast, most of which faces the Pacific Ocean (here named the Tasman Sea between Australia and New Zealand). As Figure 11-4 shows, this crescent-like Australian heartland extends from north of the city of Brisbane to the vicinity of Adelaide and includes the largest city, Sydney, the capital, Canberra, and the second-largest city, Melbourne. A secondary core area has developed in the far southwest, centered on Perth and its outport, Fremantle. Beyond lies the vast periphery, **8** which the Australians call the **Outback**.

To better understand the evolution of this spatial arrangement, it helps to refer again to the map of world climates (Fig. G-8). Environmentally, Australia's most favored strips face the Pacific and Southern oceans, and they are not large. We can describe the country as a coastal rimland with cities, towns, farms, and forested slopes giving way to the vast, arid, interior Outback. On the western flanks of the Great Dividing Range lie the extensive grassland pastures that catapulted Australia into its first commercial age—and on which still graze one of the largest sheep herds on Earth (over 160 million sheep, producing more than one-fifth of all the wool sold in the world). Where it is moister, to the north and east, cattle by the millions graze on ranchlands. This is frontier Australia, over which livestock have ranged for nearly two centuries.

Aboriginal Australians reached this landmass as long as 50,000 to 60,000 years ago, crossed the Bass Strait into Tasmania, and had developed a patchwork of indigenous cultures when Captain Arthur Phillip sailed into what is today Sydney Harbor (1788) to establish the beginnings of modern Australia. The Europeanization of Australia doomed the continent's Aboriginal societies. The first to suffer were those situated in the path of British settlement on the coasts, where penal colonies and free towns were founded. Distance protected the Aboriginal communities of the northern interior longer than elsewhere; in Tasmania, the indigenous Australians were exterminated in just decades after having lived there for perhaps 45,000 years.

Eventually, the major coastal settlements became the centers of seven different colonies, each with its own hinterland; by 1861, Australia was delimited by its now-familiar pattern of straight-line boundaries (Fig. 11-4). Sydney was the focus for New South Wales; Melbourne, Sydney's rival, anchored Victoria. Adelaide was the heart of South Australia, and Perth lay at the core of Western Australia. Brisbane was the nucleus of Queensland, and Hobart was the seat of government in Tasmania. The largest clusters of surviving Aboriginal people were in the so-called Northern Territory, with Darwin, on Australia's tropical north coast, its colonial city. Notwithstanding their shared cultural heritage, the Australian colonies were at odds not only with London over colonial policies but also with each other over economic and political issues. The building of an Australian nation during the late nineteenth century was a slow and difficult process.

A Federal State

On January 1, 1901, following years of difficult negotiations, the Australia we know today finally emerged: the Commonwealth of Australia, consisting of six States and two Federal Territories (Fig. 11-4). The two Federal Territories are the *Northern Territory*, assigned to protect the interests of the large Aboriginal population concentrated there, and the *Australian Capital Territory*, carved from southern New South Wales to accommodate the federal capital of Canberra, inaugurated in 1927.

Australia's six States are New South Wales (capital Sydney), at 6.8 million the most populous and politically powerful; Queensland (Brisbane), with the Great Barrier Reef offshore and tropical rainforests in its north; Victoria (Melbourne), small but populous by Australian standards with nearly 5 million inhabitants; South Australia (Adelaide), where the Murray-Darling river system reaches the sea; Western Australia (Perth) with barely 2 million people in an area of nearly 1 mil-

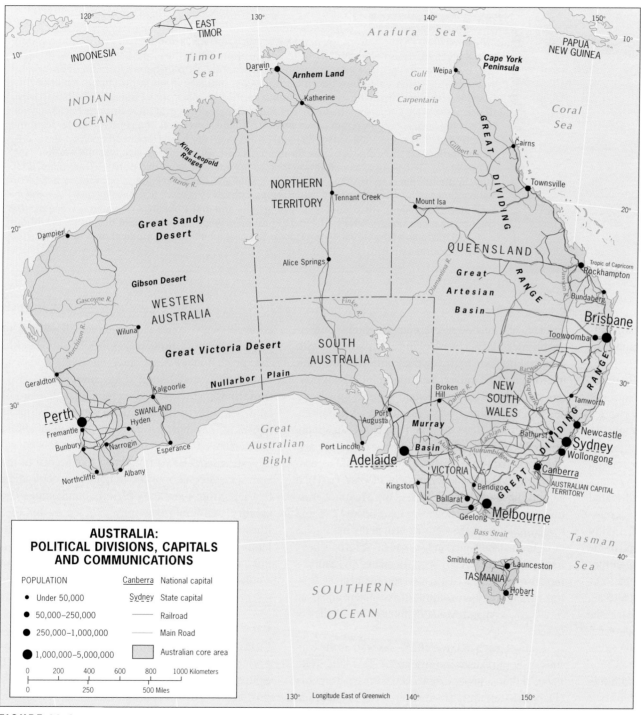

**AUSTRALIA:
POLITICAL DIVISIONS, CAPITALS
AND COMMUNICATIONS**

POPULATION

Canberra — National capital
Sydney — State capital

• Under 50,000
• 50,000–250,000
● 250,000–1,000,000
● 1,000,000–5,000,000

——— Railroad
——— Main Road
▢ Australian core area

0   200   400   600   800   1000 Kilometers
0   250   500 Miles

FIGURE 11-4

lion square miles; and Tasmania (Hobart), the island across the Bass Strait from the mainland and in the path of the storms of the Southern Ocean.

In earlier chapters, we have referred to the concept of federalism, an idea with ancient Greek and Roman roots familiar to Americans, Canadians, South Asians, and, more recently, Russians. It is a notion of communal association whose name comes from the Latin *foederis*, implying alliance and coexistence, a union of consensus **9** and common interest—a **federation**. It stands in contrast to the idea that states should be centralized, or unitary. For this, too, the ancient Romans had a term: *unitas*, meaning unity. Most European countries are **10** **unitary states**, including the United Kingdom of Great Britain and Northern Ireland. Although the majority of Australians came from that tradition (a kingdom, no less), they managed to overcome their differences and establish a Commonwealth that was, in effect, a federation of States with different viewpoints, economies, and objectives, separated by vast distances along the rim of an island continent. And yet, the experiment succeeded.

An Urban Culture

During this century of federal association, the Australians developed an urban culture. Despite those vast open spaces and romantic notions of frontier and Outback, 85 percent of all Australians live in cities and towns. On the map, Australia's areal functional organization is similar to Japan's: large cities lie along the coast, the centers of manufacturing complexes as well as

"My most vivid memory from my first visit to Alice Springs is spotting vineyards and a winery in this parched, desert environment as the plane approached the airport. I asked a taxi driver to take me there, and got a lesson in economic geography. Drip irrigation from an underground water supply made viticulture possible; the tourist industry made it profitable. None of this, however, is evident from the view seen here: a spur of the Macdonnell Ranges overlooks a town of bare essentials under the hot sun of the Australian desert. What Alice Springs has is centrality: it is the largest settlement in a vast area Australians often call 'the centre.' Not far from the midpoint on the nearly 2000-mile (3200-km) Stuart Highway from Darwin on the Northern Territory's north coast to Adelaide on the Southern Ocean, Alice Springs also was the northern terminus of the Central Australian Railway [before the line was extended north to Darwin in 2003], seen in the middle distance. The shipping of cattle and minerals is a major industry here. You need a sense of humor to live here, and the locals have it: the town actually lies on a river, the intermittent Todd River. An annual boat race is held, and in the absence of water the racers carry their boats along the dry river bed. No exploration of Alice Springs would be complete without a visit to the base of the Royal Flying Doctor Service, which brings medical help to outlying villages and homesteads."

the foci of agricultural areas. Contributing to this situation in Japan was mountainous topography; in Australia, it was an arid interior. There, however, the similarity ends. Australia's territory is 20 times larger than Japan's, and Japan's population is more than six times that of Australia's. Japan's port cities are built to receive raw materials and to export finished products. Australia's cities forward minerals and farm products from the Outback to foreign markets and import manufactures from overseas. Distances in Australia are much greater, and spatial interaction (which tends to decrease with increasing distance) is less. In comparatively small, tightly organized Japan, you can travel from one end of the country to the other along highways, through tunnels, and over bridges with utmost speed and efficiency. In Australia, the overland trip from Sydney to Perth, or from Darwin to Adelaide using the newly completed rail link via Alice Springs, is time-consuming and slow. Nothing in Australia compares to Japan's high-speed bullet trains.

For all its vastness and youth, Australia nonetheless developed a remarkable cultural identity, a sameness of urban and rural landscapes that persists from one end of the continent to the other. Sydney, often called the New York of Australia, lies on a spectacular estuarine site, its compact, high-rise central business district overlooking a port bustling with ferry and freighter traffic. Sydney is a vast, sprawling metropolis of 4.1 million, with multiple outlying centers studding its far-flung suburbs; brash modernity and reserved British ways blend here. Melbourne, sometimes regarded as the Boston of Australia, prides itself on its more interesting architecture and more cultured ways. Brisbane, the capital of Queensland, which also anchors Australia's Gold Coast and adjoins the Great Barrier Reef, is the Miami of Australia; unlike Miami, however, its residents can find nearby relief from the summer heat in the mountains of its immediate hinterland (as well as at its beaches). Perth, Australia's San Diego, is one of the world's most isolated cities, separated from its nearest Australian neighbor by two-thirds of a continent and from Southeast Asia and Africa by thousands of miles of ocean.

And yet, each of these cities—as well as the capitals of South Australia (Adelaide), Tasmania (Hobart), and, to a lesser extent, the Northern Territory (Darwin)—exhibits an Australian character of unmistakable quality. Life is orderly and unhurried. Streets are clean, slums are few, graffiti rarely seen. By American and even European standards, violent crime (though rising) is uncommon. Standards of public transportation, city schools, and healthcare provision are high. Spacious parks, pleasing waterfronts, and plentiful sunshine make Australia's urban life more acceptable than that almost anywhere else in the world. Critics of Australia's way of life say that this very pleasant state of affairs has persuaded Australians that hard work is not really necessary. But a few days' experience in the commercial centers of the major cities contradicts that assertion: the pace of life is quickening. The country's cultural geography evolved as that of a European outpost, prosperous and secure in its isolation. Now Australia must reinvent itself as a major link in an Australo-Asian chain, a Pacific partner in a transformed regional economic geography.

Economic Geography

From the very beginning, however, goods imported from Britain (and later from the United States) were expensive, largely because of transport costs. This encouraged local entrepreneurs to set up their own industries in and near the developing cities—activities economic geographers call **11** **import-substitution industries**.

When the prices of foreign goods became lower because transportation was more efficient and therefore cheaper, local businesses demanded protection from the colonial governments, and high tariffs were erected against imported goods. Local products now could continue to be made inefficiently because their market was guaranteed. If Japan could not afford this, how could Australia? We can see the answer on the map (Fig. 11-5). Even before federation in 1901, all the colonies could export valuable minerals whose earnings shored up those inefficient, uncompetitive local industries. By the time the colonies unified, pastoral industries were also contributing income. So the miners and the farmers paid for those imports Australians could not produce themselves, plus the products made in the cities. No wonder the cities grew: here were secure manufacturing jobs, jobs in state-run service enterprises, and jobs in the growing government bureaucracy. When we noted earlier that Australians once had the highest per-capita GNP in the world, this was achieved in the mines and on the farms, not in the cities.

But the good times had to come to an end. The prices of farm products fluctuated, and international market competition increased. The cost of mining, transporting, and shipping ores and minerals also rose. Australians like to drive (there are more road miles per person in Australia than in the United States), and expensive petroleum imports were needed. Meanwhile, the government-protected industries had been further fortified by strong labor unions. Not surprisingly, as the economy declined, the

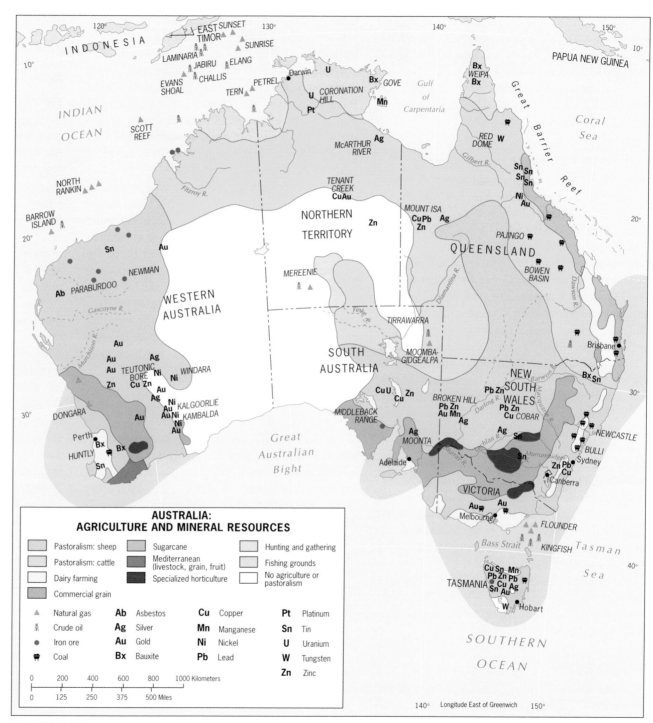

FIGURE 11-5

AUSTRALIA:
AGRICULTURE AND MINERAL RESOURCES

Pastoralism: sheep
Pastoralism: cattle
Dairy farming
Commercial grain
Sugarcane
Mediterranean (livestock, grain, fruit)
Specialized horticulture
Hunting and gathering
Fishing grounds
No agriculture or pastoralism

Natural gas
Crude oil
Iron ore
Coal

Ab Asbestos
Ag Silver
Au Gold
Bx Bauxite

Cu Copper
Mn Manganese
Ni Nickel
Pb Lead

Pt Platinum
Sn Tin
U Uranium
W Tungsten
Zn Zinc

0   200   400   600   800   1000 Kilometers
0   125   250   375   500 Miles

national debt rose, inflation grew, and unemployment crept upward. In 2002, only 21 percent by value of Australian exports were the kind of high-tech goods that have made East Asia's economic tigers so successful. That was more than double the 1985 figure but still far below what the country needed to produce.

Agricultural Abundance

And yet, Australia has material assets of which other countries on the Pacific Rim can only dream. In agriculture, sheep-raising was the earliest commercial venture, but it was the technology of refrigeration that brought world markets within reach of Australian beef producers. Wool, meat, and wheat have long been the country's big three income earners; Figure 11-5 displays the vast pastures in the east, north, and west that constitute the ranges of Australia's huge herds. The zone of commercial grain farming forms a broad crescent extending from northeastern New South Wales through Victoria into South Australia, and covers much of the hinterland of Perth. Keep in mind the scale of this map: Australia is only slightly smaller than the 48 contiguous States of the United States! Commercial grain farming in Australia is big business. As the climatic map would suggest, sugarcane grows along most of the warm, humid coastal strip of Queensland, and Mediterranean crops (including grapes for Australia's highly successful wines) cluster in the hinterlands of Adelaide and Perth. Mixed horticulture concentrates in the basin of the Murray River system, including rice, grapes, and citrus fruits, all under irrigation. And, as elsewhere in the world, dairying has developed near the large urban areas. With its considerable range of environments, Australia yields a diversity of crops.

Mineral Wealth

Australia's mineral resources, as Figure 11-5 shows, also are diverse. Major gold discoveries in Victoria and New South Wales produced a ten-year gold rush starting in 1851 and ushered in a new eco-

nomic era. By the middle of that decade, Australia was producing 40 percent of the world's gold. Subsequently, the search for more gold led to the discoveries of other minerals. New finds are still being made today, and even oil and natural gas have been found both inland and offshore (see the symbols in Fig. 11-5 in the Bass Strait between Tasmania and the mainland, and off the northwestern coast of Western Australia). Coal is mined at many locations, notably in the east near Sydney and Brisbane but also in Western Australia and even in Tasmania; before coal prices fell, this was a valuable export. Major deposits of metallic and nonmetallic minerals abound—from the complex at Broken Hill and the mix of minerals at Mount Isa to the huge nickel deposits at Kalgoorlie and Kambalda, the copper of Tasmania, the tungsten and bauxite of northern Queensland, and the asbestos of Western Australia. A glance at the map reveals the wide distribution of iron ore (the red dots), and for this raw material as for many others, Japan has been Australia's best customer in recent years.

Manufacturing's Limits

Australian manufacturing, as we noted earlier, remains oriented to domestic markets. One cannot expect to find Australian automobiles, electronic equipment, or cameras challenging the Pacific Rim's economic tigers for a place on world markets—not yet, at any rate. Australian manufacturing is diversified, producing some machinery and equipment made of locally produced steel as well as textiles, chemicals, paper, and many other items. These industries cluster in and near the major urban areas where the markets are. The domestic market in Australia is not large, but it remains relatively affluent. This makes it attractive to foreign producers, and Australia's shops are full of high-priced goods from Japan, South Korea, Taiwan, and Hong Kong. Indeed, despite its long-term protectionist practices, Australia still does not produce many goods that could be manufactured at home. Overall, the economy continues to display symptoms

of a still-developing rather than a fully developed country.

Australia's Future

The Commonwealth of Australia is changing, but its neighbors in Southeast and South Asia are changing even faster. Australia's European bonds are weakening and its Asian ties are strengthening, but Australia will be buffeted by the winds of change on the Pacific Rim. There is substantial risk for any economy that depends strongly on the export of raw materials and agricultural staples. But Australia faces challenges at home as well. These include: (1) Aboriginal issues, (2) immigration issues, (3) environmental issues, and (4) issues involving Australia's status and role.

The Aboriginal issues currently focus on two questions: the formal admission by government and majority of mistreatment of the Aboriginal minority with official apologies and reparations, and the ownership of land. Obviously, the second question has major geographic implications. The Aboriginal population of 300,000 (including many of mixed ancestry) has been gaining influence in national affairs, and in the 1980s Aboriginal leaders began a campaign to obstruct exploration on what they designated as ancestral and sacred lands. Until 1992, Australians had taken it for granted that Aborigines had no right to land ownership, but in that year the Australian High Court made the first of a series of rulings in favor of Aboriginal claimants. A subsequent court decision implied that vast areas (probably as much as 78 percent of the whole continent) could potentially be subject to Aboriginal claims 12 (Fig. 11-6). Today the **Aboriginal land issue** remains mainly (though not exclusively) an Outback issue, but it has the potential to overwhelm Australia's court system and to inhibit economic growth.

The immigration issue is older than Australia itself. Fifty years ago, when Australia had less than half the population it has today, 95 percent of the people were of European ancestry, and more than

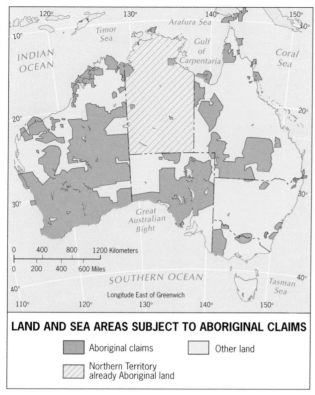

FIGURE 11-6

LAND AND SEA AREAS SUBJECT TO ABORIGINAL CLAIMS

- ■ Aboriginal claims
- ▨ Northern Territory already Aboriginal land
- ☐ Other land

FROM THE FIELDNOTES

"While in Darwin I had to visit a government office, and waiting in line with me was this Aboriginal girl, her brother, and her father. They lived about two hours from the city, she said (which I gathered was the length of their bus ride), and they had to come here from time to time for filing forms and visiting the doctor. When I asked whether they enjoyed coming to Darwin, all three shook their heads vigorously. 'It is not a friendly place,' she said. 'But we must come here.' In fact I had been surprised at the comparatively small number of Aboriginal people I had seen in Darwin, far fewer than in Alice Springs. Certainly laid-back Alice Springs is a very different place from larger, busier, and more modern Darwin."

three-quarters of them came from the British Isles. **13** Eugenic (race-specific) **immigration policies** maintained this situation until the 1970s. Today, the picture is dramatically different: of 19.9 million Australians, only about one-third have British-Irish origins, and Asian immigrants outnumber both European immigrants and the natural increase each year. During the early 1990s, nearly 150,000 legal immigrants arrived in Australia annually, most from Hong Kong, Vietnam, China, the Philippines, India, and Sri Lanka. Immigration quotas have since been reduced, most recently to 80,000, but Asian immigrants continue to outnumber those from Western sources. This situation roils Australian society, and restrictions on foreign ownership of Australian real estate reflect fears of what extremist groups call the "Asianization" of the country. But the success of

many Asian settlers in Australia evinces the opportunities still available in this free and open society. Sydney, the main recipient of the Asian influx, has become a mosaic of ethnic districts, some of which have gone through periods of gang violence and drug dealing but have stabilized and even prospered over time. Still, multiculturalism will remain a long-term challenge for Australia in the twenty-first century. **14** **Environmental degradation** is almost synonymous with Australia. To both Aborigines and European settlers, its ecology paid a heavy price. Great stands of magnificent forest were destroyed. In Western Australia, centuries-old trees were simply "ringed" and left to die, so that the sun could penetrate through their leafless crowns to nurture the grass below. Then the sheep could be driven into these new pastures. In Tasmania, where Aus-

tralia's native eucalyptus tree reaches its greatest dimensions (comparable to North American redwood stands), tens of thousands of acres of this irreplaceable treasure have been lost to chain saws and pulp mills. Many of Australia's unique marsupial species have been destroyed, and many more are endangered or threatened. "Never have so few people wreaked so much havoc on the ecology of so large an area in so short a time," observed a geographer in Australia recently. But awareness of this environmental degradation is growing. In Tasmania, the "Green" environmentalist political party has become a force in State affairs, and its activism has slowed deforestation, dam-building, and other "development" projects. Still, many Australians fear the environmentalist movement as an obstacle to economic growth at a time when the economy needs stimulation. This, too, is an issue for the future.

Status and Role

Several issues involving Australia's status at home, relations with neighbors, and position in the world are stirring up national debate as well. A domestic question is whether Australia should become a republic, ending the status of the British monarch as the head of state, or continue its status quo in the British Commonwealth. A 1999 referendum proved that a majority of Australians were not prepared to abandon the monarchy—at least not in favor of "a president appointed by two-thirds majority of the membership of the Commonwealth Parliament" as the ballot put it. Although polls had predicted a republican victory, the language on the ballot probably changed the outcome; most Australians seem to favor a republic, but more of them distrust their politicians. If the ballot had said "in favor of a popularly elected president," chances are that Australia would be on its way to becoming a federal republic. This issue will undoubtedly arise again.

Relations with neighboring Indonesia and East Timor have been complicated. For many years, Australia had what may be called a special relationship with Indonesia, whose help it needs in curbing illegal seaborne immigration. It was also profitable for Australia to counter international (UN) opinion and recognize Indonesia's 1976 annexation of Portuguese East Timor, for in doing so Australia could deal directly with Jakarta for the oil reserves under the Timor Sea (see Fig. 10-13). Thus Australia gave neither recognition nor support to the rebel movement that fought for independence in East Timor. But this story had a relatively happy ending. When

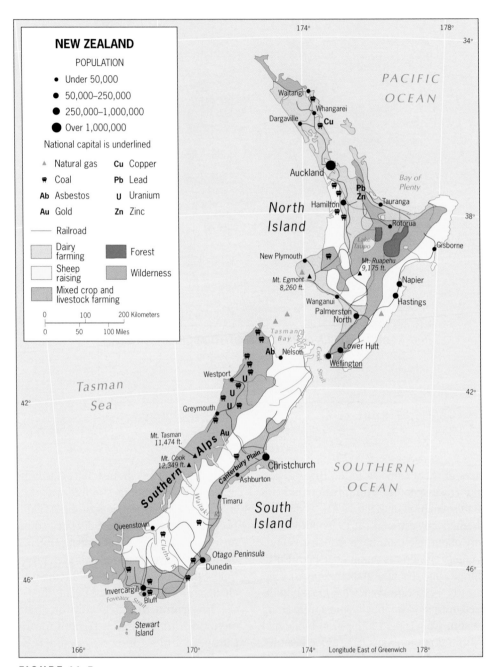

FIGURE 11-7

the East Timorese campaign for independence succeeded in 1999 and Indonesian troops began an orgy of murder and destruction, Australia sent an effective peacekeeping force and spearheaded the United Nations effort to stabilize the situation. Today, a new chapter has opened in Australia's relations with these northern neighbors.

Australia also has a long-term relationship with Papua New Guinea (PNG), as we will note in Chapter 12. This association, too, has gone through difficult times. In recent years, the inhabitants of Papua New Guinea have strongly resisted privatization, World Bank involvement, and globalization generally. Australia assists PNG in several spheres, but its motives are sometimes questioned. In 2001, the construction of a projected gas pipeline from PNG to the Australian State of Queensland precipitated fighting among tribespeople over land rights, resulting in dozens of casualties, and Australian public opinion reflected doubts over the appropriateness of this venture.

Immediately after the September 2001 attack on U.S. targets in New York and Washington, Australian leaders expressed strong support for the American campaign against terrorism, but they made it a point also to assure the Indonesian government that the War on Terror was not a war on Islam. Although the great majority of Australians supported this stance, their will was severely tested in late 2002 when a terrorist attack on a nightclub in Kuta Beach on the island of Bali killed 88 vacationing Australians.

When political violence and chaos overwhelmed the Solomon Islands east of Papua New Guinea in 2003, Australian forces intervened: a failing state in Australia's neighborhood could become a base for terrorist activity.

Territorial dimensions, relative location, and raw-material wealth have helped determine Australia's place in the world and, more specifically, on the Pacific Rim. Australia's population, still just 20 million in the first decade of our new century, is smaller than Malaysia's and barely larger than that of the Caribbean island of Hispaniola. But Aus-

tralia's importance in the international community far exceeds its human numbers.

NEW ZEALAND

Fifteen hundred miles east-southeast of Australia, in the Pacific Ocean across the Tasman Sea, lies New Zealand. In an earlier age, New Zealand would have been part of the Pacific geographic realm because its population was Maori, a people with Polynesian roots. But New Zealand, like Australia, was invaded and occupied by Europeans.

Today, its population of 4 million is almost 80 percent European, and the Maori form a minority of less than 400,000, with many of mixed Euro-Polynesian ancestry.

New Zealand consists of two large mountainous islands and many scattered smaller islands (Fig. 11-7). The two large islands, with the South Island somewhat larger than the North Island, look diminutive in the great Pacific Ocean, but together they are larger than Britain. In contrast to Australia, the two main islands are mostly mountainous or hilly, with several peaks rising far higher than any on the Australian landmass. The South Island has a

FROM THE FIELDNOTES

"The drive from Christchurch to Arthur's Pass on the South Island of New Zealand was a lesson in physiography and biogeography. Here, on the east side of the Southern Alps, you leave the Canterbury Plain and its agriculture and climb into the rugged topography of the glacier-cut, snowcapped mountains. A last pasture lies on a patch of flatland in the foreground; in the background is the unmistakable wall of a U-shaped valley sculpted by ice. Natural vegetation ranges from pines to ferns, becoming even more luxuriant as you approach the moister western side of the island."

spectacular snowcapped range appropriately called the Southern Alps, with numerous peaks reaching beyond 10,000 feet (3300 m). The smaller North Island has proportionately more land under low relief, but it also has an area of central highlands along whose lower slopes lie the pastures of New Zealand's chief dairying district. Hence, while Australia's land lies relatively low in elevation and exhibits much low relief, New Zealand's is on the average high and is dominated by rugged relief.

Human Spatial Organization

Thus the most promising areas for habitation are the lower-lying slopes and lowland fringes on both islands. On the North Island, the largest urban area, Auckland, occupies a comparatively low-lying peninsula. On the South Island, the largest lowland is the agricultural Canterbury Plain, centered on Christchurch. What makes these lower areas so attractive, apart from their availability as cropland, is their magnificent pastures. The range of soils and pasture plants allows both summer and winter grazing. Moreover, the Canterbury Plain, the chief farming region, also produces a wide variety of vegetables, cereals, and fruits. About half of all New Zealand is pasture land, and much of the farming provides fodder for the pastoral industry. Sixty million sheep and eight million cattle dominate these livestock-raising activities, with wool, meat, and dairy products providing nearly two-thirds of the islands' export revenues.

Despite their contrasts in size, shape, physiography, and history, New Zealand and Australia have much in common. Apart from their joint British heritage, they share a sizeable pastoral economy, a small local market, the problem of great distances to world markets, and a desire to stimulate (through protection) domestic manufacturing. The high degree of urbanization in New Zealand (77 percent of the total population) again resembles Australia: substantial employment in city-based industries, mostly the processing and packing of livestock and farm products, and government jobs.

Spatially, New Zealand shares with Australia its **15** pattern of **peripheral development** (Fig. G-9), imposed not by desert but by high rugged mountains. The country's major cities—Auckland and the capital of Wellington (together with its satellite of Hutt) on the North Island, and Christchurch and Dunedin on the South Island—are all located on the coast, and the entire rail and road system is peripheral in its configuration (Fig. 11-7). This is more pronounced on the South Island than in the north because the Southern Alps are New Zealand's most formidable barrier to surface communications.

The Maori Factor and New Zealand's Future

Like Australia, New Zealand has a history of difficult relations with its indigenous population. The Maori, who account for about 10 percent of the country's population today, appear to have reached these islands during the tenth century AD. By the time the European colonists arrived, they had had a tremendous impact on the islands' ecosystems, especially on the North Island where most Maori lived. In 1840, the Maori and the British signed a treaty at Waitangi that granted the colonists sovereignty over New Zealand but guaranteed Maori rights over established tribal lands. Although the British abrogated parts of the treaty in 1862, the Maori had reason to believe that vast reaches of New Zealand, as well as offshore waters, were theirs.

As in Australia, judicial rulings during the 1990s supported the Maori position, which led to expanded claims and growing demands; one of the Maori's persistent complaints is the slow pace of integration of the minority into modern New Zealand society. Although Maori claims encompass much of the South Island, they also cover prominent sites in the

"It was Sunday morning in Christchurch on New Zealand's South Island, and the city center was quiet. As I walked along Linwood Street I heard a familiar sound, but on an unfamiliar instrument: the Bach sonata for unaccompanied violin in G minor—played magnificently on a guitar. I followed the sound to the artist, a Maori musician of such technical and interpretive capacity that there was something new in every phrase, every line, every tempo. I was his only listener; there were a few coins in his open guitar case. Shouldn't he be playing before thousands, in schools, maybe abroad? No, he said, he was happy here, he did alright. A world-class talent, a street musician playing Bach on a Christchurch side street, where tourists from around the world were his main source of income. Talk about globalization."

major cities. Today, the Maori question is the leading national issue in the country.

Dominant cultural heritage and prevailing cultural landscape form two criteria on which the delimitation of the Austral Realm is based. But in both Australia and New Zealand, the cultural mosaic is changing, and the convergence with neighboring realms is proceeding.

DISCOVERY TOOLS

www.wiley.com/college/deblij

The *Concepts and Regions* Website, featuring *GeoDiscoveries*, offers many additional resources to enhance your understanding and experience of this chapter. Be sure to explore the following:

Photo Galleries

Australia's Great Interior Desert
Sydney, Australia
Newcastle, Australia
Auckland, New Zealand
Christchurch, New Zealand
Alice Springs, Australia
Darwin, Australia
Arthur's Pass, New Zealand

More To Explore

Systematic Essay: Biogeography
Issues in Geography: Is There a Southern Ocean?
Major Cities: Sydney

Expanded Regional Coverage

Core Area, Outback, New Zealand

Chapter 12

The Pacific Realm

CONCEPTS, IDEAS, AND TERMS

1. Marine geography
2. Territorial sea
3. High seas
4. Continental shelf
5. Exclusive Economic Zone (EEZ)
6. Maritime boundary
7. Median-line boundary
8. High-island cultures
9. Low-island cultures
10. Antarctic Treaty

REGIONS

MELANESIA

MICRONESIA

POLYNESIA

ETWEEN THE AMERICAS to the east and the western Pacific Rim to the west lies the vast Pacific Ocean, larger than all the world's land areas combined. In this greatest of all oceans lie tens of thousands of islands, some large (New Guinea is by far the largest), most small (many are uninhabited). Together, the land area of these islands is a mere 376,000 square miles (974,000 sq km), about the size of Texas plus New Mexico, and over 90 percent of this lies in New Guinea.*

DEFINING THE REALM

The Pacific geographic realm—land and water—covers nearly an entire hemisphere of this world, the one commonly called the Sea Hemisphere (Fig. 12-1). This Sea Hemisphere meets the Russian and North American realms in the far north and merges into the Southern Ocean in the south. Despite the preponderance of water, this fragmented, culturally complex realm does possess regional identities. It includes the Hawaiian Islands, Tahiti, Tonga, and Samoa—fabled names in a world apart.

In terms of modern cultural and political geography, Indonesia and the Philippines are not part of the Pacific Realm, although Indonesia's political system reaches into it; nor are Australia and New Zealand part of it. Before the European invasion and colonization, Australia would have been included because of its Aboriginal population and New Zealand because of its Maori population's

*The figures in Table G-1 do not match these totals because only the political entity of Papua New Guinea is listed, not the Indonesian province of Papua that occupies the western part of the island. Here, as Figure 10-2 showed, the political and the realm boundaries do not coincide.

STUDY TOOLS

3D GLOBE
AREA AND DEMOGRAPHIC DATA

FLASHCARDS
MAP QUIZZES

AUDIO PRONUNCIATION GUIDE
CHAPTER QUIZ

ANNOTATED WEBLINKS
LONELY PLANET WEBLINKS

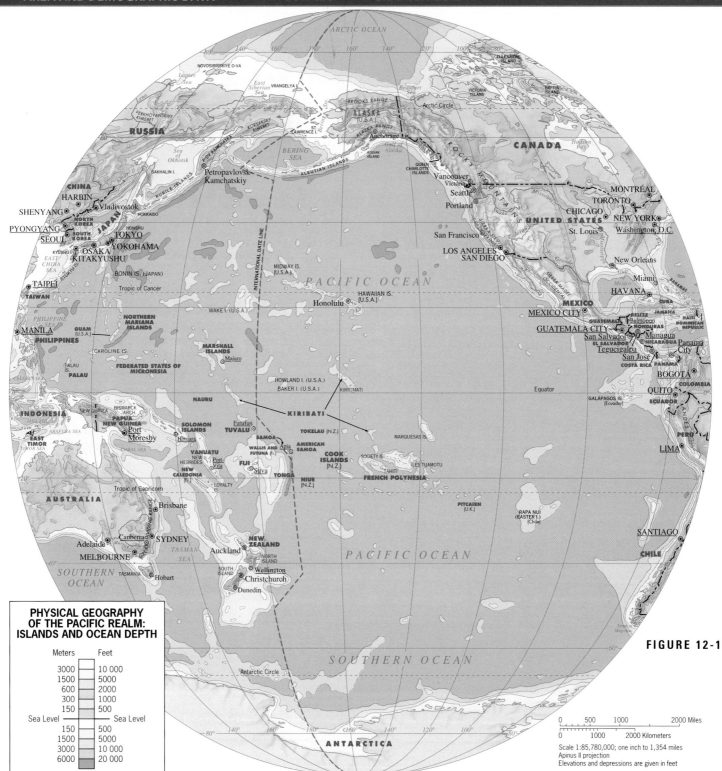

PHYSICAL GEOGRAPHY
OF THE PACIFIC REALM:
ISLANDS AND OCEAN DEPTH

Meters	Feet
3000	10 000
1500	5000
600	2000
300	1000
150	500
Sea Level	Sea Level
150	500
1500	5000
3000	10 000
6000	20 000

FIGURE 12-1

Scale 1:85,780,000; one inch to 1,354 miles
Apinus II projection
Elevations and depressions are given in feet

0 500 1000 2000 Miles
0 1000 2000 Kilometers

MAJOR GEOGRAPHIC QUALITIES OF THE PACIFIC REALM

1. The Pacific Realm's total area is the largest of all geographic realms. Its land area, however, is the smallest, as is its population.

2. The island of New Guinea, with 7.7 million people, alone contains over 80 percent of the Pacific Realm's population.

3. The Pacific Realm, with its wide expanses of water and numerous islands, has been strongly affected by United Nations Law of the Sea provisions regarding states' rights over economic assets in their adjacent waters.

4. The highly fragmented Pacific Realm consists of three regions: Melanesia (including New Guinea), Micronesia, and Polynesia.

5. Melanesia forms the link between Papuan and Melanesian cultures in the Pacific.

6. The Pacific Realm's islands and cultures may be divided into volcanic *high-island* cultures and coral-based *low-island* cultures.

7. In Micronesia, U.S. influence has been particularly strong and continues to affect local societies.

8. In Polynesia, local cultures are nearly everywhere severely strained by external influences. In Hawai'i, as in New Zealand, indigenous culture has been largely submerged by Westernization.

9. Indigenous Polynesian culture continues to exhibit a remarkable consistency and uniformity throughout the Polynesian region, its enormous dimensions and dispersal notwithstanding.

Polynesian affinities. But the Europeanization of their countries has engulfed black Australians and Maori New Zealanders, and the regional geography of Australia and New Zealand today is decidedly not Pacific. In New Guinea, on the other hand, Pacific peoples remain numerically and culturally the dominant element.

The Pacific islands were colonized by the French, British, and Americans; an indigenous Polynesian kingdom in the Hawaiian Islands was annexed by the United States and is now the fiftieth State. Still, today, the map is an assemblage of independent and colonial territories (Fig. 12-1). Paris controls New Caledonia and French Polynesia. The United States administers Guam and American Samoa, the Line Islands, Wake Island, Midway Islands, and several smaller islands; the United States also has special relationships with other territories, former dependencies that are now nominally independent. The British, through New Zealand, have responsibility for the Pitcairn group of islands, and New Zealand administers and supports the Cook, Tokelau, and Niue Islands. Easter Island, the storied speck of land in the southeastern Pacific, is part of Chile. Indonesia rules Papua, the western half of the island of New Guinea.

Other island groups have become independent states. The largest are Fiji, once a British dependency, the Solomon Islands (also formerly British), and Vanuatu (until 1980 ruled jointly by France and Britain). Also on the current map, however, are such microstates as Tuvalu, Kiribati, Nauru, and Palau. Foreign aid is crucial to the survival of most of these countries. Tuvalu, for example, has a total area of around 10 square miles, a population of some 10,000, and a per-capita GNP of about $600, derived from fishing, copra sales, and some tourism. But what really keeps Tuvalu going is an international trust fund set up by Australia, New Zealand, the United Kingdom, Japan, and South Korea. Annual grants from that fund, as well as money sent back to families by workers who have left for New Zealand and elsewhere, allow Tuvalu to survive.

THE PACIFIC REALM AND ITS MARINE GEOGRAPHY

Certain land areas may not be part of the Pacific Realm (we cited the Philippines and New Zealand), but the Pacific Ocean extends from the shores of North and South America to mainland East and Southeast Asia and from the Bering Sea to the Subtropical Convergence. This means that several seas, including the Sea of Japan (East Sea), the East China Sea, and the South China Sea, are part of the Pacific Ocean. As we will see below, this relationship matters. Pacific coastal countries, large and small, mainland and island, compete for jurisdiction over the waters that bound them.

The Pacific Realm and its ocean, therefore, form ▮ an ideal place to focus on **marine geography**. This field encompasses a variety of approaches to the study of oceans and seas; some marine geographers focus on the biogeography of coral reefs, others on the geomorphology of beaches, and still others on the movement of currents and drifts. A particularly interesting branch of marine geography has to do with the definition and delimitation of political boundaries at sea. Here geography meets political science and maritime law.

Littoral (coastal) states do not end where atlas maps suggest they do. States have claimed various forms of jurisdiction over coastal waters for centuries, closing off bays and estuaries and ordering foreign fishing fleets to stay away from nearby fishing grounds. Thus arose the notion of the ▮ **territorial sea**, where all the rights of a coastal ▮ state would prevail. Beyond lay the **high seas**, free, open, and unfettered by national interests.

It was in the interest of colonizing, mercantile states to keep territorial seas narrow and high seas wide, thus interfering as little as possible with their commercial fleets. In the seventeenth and eighteenth centuries the territorial sea was 3, 4, or at most 6 nautical miles wide, and the colonizing powers claimed the same widths for their colonies (1 nautical mile = 1.15 statute miles [1.85 km]).

In the twentieth century these constraints weakened. States without trading fleets saw no reason to limit their territorial seas. States with nearby fishing grounds traditionally exploited by their own fleets wanted to keep the increasing number of foreign ▮ trawlers away. States with shallow **continental shelves**, offshore continuations of coastal plains, wished to control the resources on and below the seafloor, made more accessible by improving technology. States disagreed on the methods by which offshore boundaries, whatever their width, should be defined. Early efforts by the League of Nations in the 1920s to resolve these issues had only partial success, mainly in the technical area of boundary delimitation.

In 1945, the United States helped precipitate what has become known as the "scramble for the oceans." President Harry S Truman issued a proclamation that claimed U.S. jurisdiction and control over all the resources "in and on" the continental shelf down to its margin, around 100 fathoms (600 feet) deep. In some areas, the shallow continental shelf of the United States extends more than 200 miles offshore, and Washington did not want foreign countries drilling for oil just beyond the 3-mile territorial sea.

Few observers foresaw the impact the Truman Proclamation would have, not only on U.S. waters but on the oceans everywhere, including the Pacific. It set off a rush of other claims. In 1952, a group of South American countries, some with little continental shelf to claim, issued the Declaration of Santiago, claiming exclusive fishing rights up to a distance of 200 nautical miles (230 statute miles) off their coasts. Meanwhile, as part of the Cold War competition, the Soviet Union urged its allies to claim a 12-mile territorial sea.

UNCLOS Intervention

Now the United Nations intervened, and a series of UNCLOS (United Nations Conference on the Law of the Sea) meetings began. These meetings addressed issues ranging from the closure of bays to the width and delimitation of the territorial sea, and after three decades of negotiations they achieved a convention that changed the political and economic geography of the oceans forever. Among its key provisions were the authorization of a 12-mile territorial sea for all countries and the establishment ▮ of a 200-mile (230-statute-mile) **Exclusive Economic Zone (EEZ)** over which a coastal state would have total economic rights. Resources in and under this EEZ (fish, oil, minerals) belong to the coastal state, which could either exploit them or lease, sell, or share them as it saw fit.

These provisions had a far-reaching impact on the world's oceans and seas (Fig. 12-2), and especially so on the Pacific. Unlike the Atlantic Ocean, the Pacific is studded with islands large and small, and a microstate consisting of one small island suddenly acquired an EEZ covering 166,000 square nautical miles. European colonial powers still holding minor Pacific possessions (notably France) saw their maritime jurisdictions vastly expanded. Small low-income archipelagos could now bargain with large, rich fishing nations over fishing rights in their EEZs. And for all the UNCLOS Convention's provisions for the "right of innocent passage" of shipping through EEZs and via narrow straits, the world's high seas have obviously been diminished.

The extension of the territorial sea to 12 nautical miles and the EEZ to an additional 188 nautical ▮ miles created new **maritime boundary** problems. Waters less than 24 miles wide separate many ▮ countries all over the world, so that **median lines**, equidistant from opposite shores, have been delimited to establish their territorial seas. And even more countries lie closer than 400 nautical miles apart, requiring further maritime-boundary delimitation to determine their EEZs. In such maritime regions as the North Sea, the Caribbean Sea, and the Japan, East China, and South China Seas, a maze of maritime boundaries emerged, some of them subject to dispute. Political changes on land can lead to significant modifications at sea.

A case in point involves newly independent East Timor and its neighbor across the Timor Sea, Australia. As we noted in Chapter 10 (and Fig. 10-13), Australia had divided the waters and seafloor of the Timor Sea with Indonesia while recognizing Indonesia's annexation of East Timor. When East Timor achieved independence, the so-called *Timor Gap* became an issue: where was the median line that would divide Timorese and Australian claims to the oil and gas reserves in this area? The Australians argued that since the line defined with Indonesia predated Timor's independence, it should continue in effect, giving Australia the bulk of the energy resources. But UNCLOS regulations,

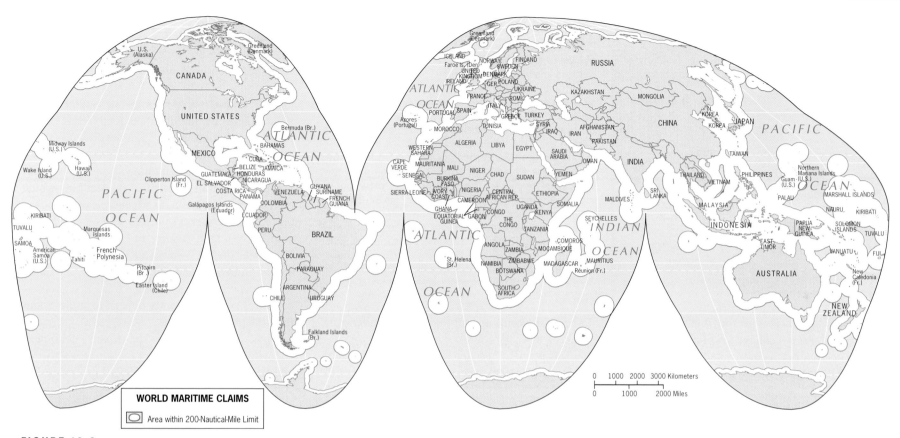

WORLD MARITIME CLAIMS

⬭ Area within 200-Nautical-Mile Limit

FIGURE 12-2

to which Australia subscribed, required a redelimitation, giving East Timor a much larger share. This led Australia to withdraw from UNCLOS in 2002 so that it would not be bound by its rules. Australia offered East Timor most of the revenues from one gasfield but nothing from the other reserves. Thus East Timor faced a choice between accepting a disadvantageous deal that would net it tens of millions of dollars or challenging Australia over a prize that may amount to as much as U.S. $6 billion. Small, poor countries do not have it easy when they confront large, rich ones.

The UNCLOS provisions created opportunities for some states to expand their spheres of influence. Wider territorial-sea and EEZ allocations raised the stakes: claiming an island now entailed potential control over a huge maritime area. In Chapters 9 and 10, we referred to several island disputes off mainland East Asia, which involve Japan and Russia, Japan and South Korea, Japan and China, China and Vietnam, and China and the Philippines. Ownership of many islands there is uncertain, and small specks of island territory have become large stakes in the scramble for the oceans. For example, in the case of the Spratly Islands (Fig. 10-2), six countries claim ownership, including both Taiwan and China. China's island claims in the South China Sea support Beijing's contention that this body of water is part and parcel of the Chinese state—a position that worries other states with coasts facing it.

Figure 12-2 reveals what EEZ regulations have meant to Pacific Realm countries such as Tuvalu, Kiribati, and Fiji, with nearly circular EEZs now surrounding the clusters of islands in this vast ocean space. Japan, Taiwan, and other fishing nations have purchased fishing rights in these EEZs from the island governments. Nonetheless, violations of EEZ rights do occur; recently, Vanuatu and the Philippines were at odds over unauthorized Filipino fishing in Vanuatu's EEZ.

The process of boundary delimitation continues. In an earlier edition of this book, we included a map of the South China Sea and nearby waters off East and Southeast Asia, showing the median-line boundaries delimited according to

"Having arrived late at night, I did not have a perspective yet of the setting of Pago Pago—but even before I got up I knew what the leading industry here had to be. An intense, overpowering smell of fish filled the air. Sure enough, daylight revealed a huge tuna processing plant right across the water, with fishing boats arriving and departing continuously. American Samoa has a huge Pacific Ocean EEZ, one of the richest fishing grounds in the whole realm, and seafood exports rank ahead of tourism as the leading source of external income for the territory, an industry that is mostly American-owned. As the photograph shows, American Samoa is a group of 'high-islands' of volcanic origin, with considerable relief and environmental variety. Tuvalu, the country I had just left, was the complete opposite—a 'low-island' territory whose EEZ has been leased to Japanese fishing fleets, which do their processing on board and employ no locals."

sions in Article 76 of the 1982 UNCLOS Convention were re-examined. Where a continental shelf extends beyond the 200-mile limit of the EEZ, the Convention apparently allows a coastal state to claim that extension as a "natural prolongation" of its landmass. Although there is as yet no regulation that would also allow the state to extend its EEZ to the edge of the continental shelf, it is not difficult to foresee such an amendment to the Convention. As it stands, states are now delimiting their proposed "natural prolongations" under a 2009 deadline, which puts poorer states at a disadvantage since the required marine surveys are very expensive. In any case, an article in *Science* in December 2002 reported that the big winners are likely to include the United States, Canada, Australia, New Zealand, Russia, and India—although a total of 60 coastal countries could ultimately benefit to some extent from the provision.

UNCLOS specifications. But that map changed when coastal states engaged in bilateral and multilateral negotiations—and argued over island ownership with major boundary implications. The Pacific (and world) map of maritime boundaries remains a work in progress.

This point is illustrated by a development that occurred late in 1999, when the implications of provi-

Thus the "scramble for the oceans" continues, and with it the constriction of the world's high seas and open waters.

REGIONS OF THE REALM

Sail across the Pacific Ocean, and one spectacular vista follows another. Dormant and extinct volcanoes, sculpted by erosion into basalt spires draped by luxuriant tropical vegetation and encircled by reefs and lagoons, tower over azure waters. Low atolls with nearly snow-white beaches, crowned by stands of palm trees, seem to float in the water. Pacific islanders, where foreign influences have not overtaken them, appear to take life with enviable ease.

More intensive investigations reveal that such Pacific sameness is more apparent than real. Even the Pacific Realm, with its long sailing traditions, its still-diffusing populations, and its historic migrations, has a durable regional framework. Figure 12-3 outlines the three regions that constitute the Pacific Realm:

Melanesia Papua (Indonesia), Papua New Guinea, Solomon Islands, Vanuatu, New Caledonia (France), Fiji

Micronesia Palau, Federated States of Micronesia, Northern Mariana Islands, Republic of the Marshall Islands, Nauru, western Kiribati, Guam (U.S.)

Polynesia Hawaiian Islands (U.S.), Samoa, American Samoa, Tuvalu, Tonga, eastern Kiribati, Cook and other New Zealand-administered islands, French Polynesia

Ethnic, linguistic, and physiographic criteria are among the bases for this regionalization of the Pacific Realm, but we should not lose sight of the dimensions. Not only is the land area small, but also the population of this entire realm (including Indonesia's Papua) in 2004 was only about 11 million—8.5 million without Papua—about the same as one very

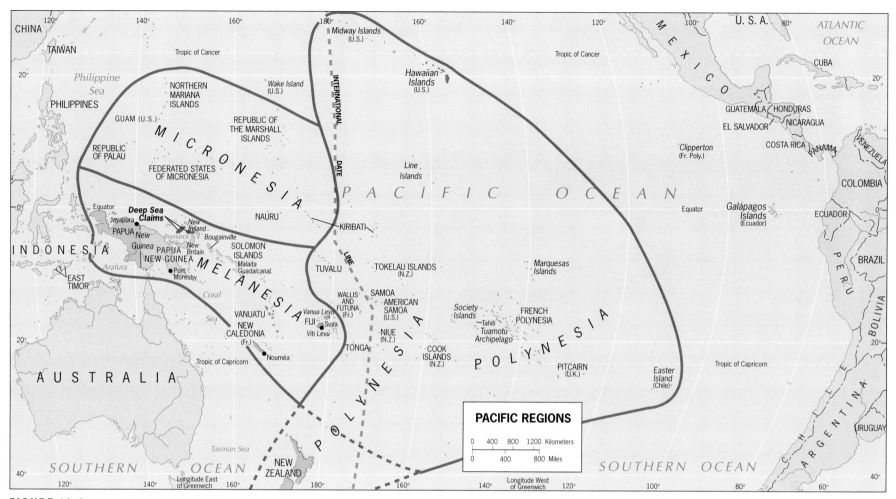

FIGURE 12-3

large city. Fewer people live in this realm even than in another vast area of far-flung settlements—the oases of North Africa's Sahara.

MELANESIA

The large island of New Guinea lies at the western end of a Pacific region that extends eastward to Fiji and includes the Solomon Islands, Vanuatu, and New Caledonia (Fig. 12-3). The human mosaic here is complex, both ethnically and culturally. Most of the 7.7 million people of New Guinea (including the Indonesian part, the province of Papua, and the independent state of Papua New Guinea) are Papuans, and a large minority is Melanesian. Altogether there are as many as 700 communities speaking different languages; the Papuans are most numerous in the densely forested highland interior and in the south, while the Melanesians inhabit the north and east.

The region as a whole has more than 8 million inhabitants, making this the most populous Pacific region by far.

With 5.2 million people today, **Papua New Guinea (PNG)** became a sovereign state in 1975 after nearly a century of British and Australian administration. Almost all of PNG's limited development is taking place along the coasts, while most of the interior remains hardly touched by the changes that transformed neighboring Australia. Perhaps

FROM THE FIELDNOTES

"Arriving in the capital of New Caledonia, Nouméa, in 1996 was an experience reminiscent of French Africa 40 years earlier. The French tricolor was much in evidence, as were uniformed French soldiers. European French residents occupied hillside villas overlooking palm-lined beaches, giving the place a Mediterranean cultural landscape. And New Caledonia, like Africa, is a source of valuable minerals. It is one of the world's largest nickel producers, and from this vantage point you could see the huge treatment plants, complete with concentrate ready to be shipped (left, under conveyor). What you cannot see here is how southern New Caledonia has been ravaged by the mining operations, which have denuded whole mountainsides. Working in the mines and in this facility are the local Kanaks, Melanesians who make up about 45 percent of the population of about 240,000. Violent clashes between Kanaks and French have obstructed government efforts to change New Caledonia's political status in such a way as to accommodate pressures for independence as well as continued French administration."

second, followed by copper, silver, timber, and several agricultural products including coffee and cocoa, reflecting the country's resource and environmental diversity. Pacific Rim developments have affected even PNG: most exports go to the nearest neighbor, Australia, but Japan ranks close behind.

Turning eastward, it is a measure of Melanesia's cultural fragmentation that as many as 120 languages are spoken in the approximately 1000 islands that make up the **Solomon Islands** (about 80 of these islands support almost all the people, numbering about 475,000). Inter-island, historic animosities among the islanders were worsened by the events of World War II, when thousands of Malaitans were moved by U.S. forces to Guadalcanal. This started a postwar cycle of violence that led Australia to intervene in 2003.

New Caledonia, still under French rule, is in a very different situation. Only about 45 percent of the population of about 240,000 are Melanesian; 37 percent are of French ancestry, many of them descended from the inhabitants of the penal colony France established here in the nineteenth century. Nickel mines, based on reserves that rank among the world's largest, dominate New Caledonia's export economy. The mining industry attracted additional French settlers, and social problems arose. Most of the French population lives in or near the capital city of Nouméa, steeped in French cultural landscapes, in the southeastern quadrant of the island. Melanesian demands for an end to colonial rule have led to violence, and the two communities are still in the process of coming to terms.

On its eastern margins, Melanesia includes one of the Pacific Realm's most interesting countries, **Fiji**. On two larger and over 100 smaller islands live nearly 1 million Fijians, of whom 51 percent are Melanesians and 44 percent South Asians, the latter brought to Fiji from India during the British colonial occupation to work on the sugar plantations. When Fiji achieved independence in 1970, the native Fijians owned most of the land and held political control, while the Indians were concentrated in the towns (chiefly Suva, the cap-

four-fifths of the population lives in what we may describe as a self-sufficient subsistence economy, growing root crops and hunting wildlife, raising pigs, and gathering forest products. Old traditions of the kind lost in Australia persist here, protected by remoteness and the rugged terrain.

Welding this disparate population into a nation is a task hardly begun, and PNG faces numerous obstacles in addition to its cultural complexity. Not only are hundreds of languages in use, but more than half the population is illiterate. English, the official

language, is used by the educated minority but is of little use beyond the coastal zone and its towns. The capital, Port Moresby, has about 380,000 inhabitants, reflecting the low level of urbanization (15 percent) in this developing economy.

Yet Papua New Guinea is not without economic opportunities. As Figure G-10 shows, it now ranks in the lower-middle-income group of states (in the 1980s it was one of the world's poorest). Oil was discovered in the 1980s, and by the late 1990s crude oil was PNG's largest export by value. Gold now ranks

ital) and dominated commercial life. It was a recipe for trouble, and it was not long in coming when, in a later election, the politically active Indians outvoted the Fijians for seats in the parliament. A coup by the Fijian military was followed by a revision of the constitution, which awarded a majority of seats to ethnic Fijians. Before long, however, the Fijian majority splintered, but a coalition government for some time proved the constitution workable—until 2000. In May 1999, Fiji's first prime minister of Indian ancestry had taken office, angering some ethnic Fijians to the point of staging another coup. The prime minister and members of the government were taken hostage at the parliament building in Suva, and the perpetrators demanded that Fijians would henceforth govern the country.

This action, and Fiji's inability to counter it, had a devastating impact on the country. Foreign trading partners stopped buying Fijian products. The tourist industry suffered severely. And in the end, although the coup leaders were ousted and arrested, they secured a deal that gave ethnic Fijians the control they had sought. Fiji's future thus remains clouded.

Melanesia, the most populous region in the Pacific Realm, also is bedeviled by centrifugal forces of many kinds. No two countries present the same form of multiculturalism; each has its own challenges to confront, and some of these challenges spill over into neighboring (or more distant foreign) islands.

◆ MICRONESIA

North of Melanesia and east of the Philippines lie the islands that constitute the region known as Micronesia (Fig. 12-3). The name (*micro* means small) refers to the size of the islands: the 2000-plus islands of Micronesia are not only tiny (many of them no larger than 1 square mile), but they are also much lower-lying, on an average, than those of Melanesia. **8** Some are volcanic islands (**high islands**, as the people call them), but they are outnumbered by **9** islands composed of coral, the **low islands** that

FROM THE FIELDNOTES

"Back in Suva, Fiji more than 20 years after I had done a study of its CBD, I noted that comparatively little had changed, although somehow the city seemed more orderly and prosperous than it was in 1978. At the Central Market I watched a Fijian woman bargain with the seller over a batch of taro, the staple starch-provider in local diets. I asked her how she would serve it. 'Well, it's like your potato,' she said. 'I can make a porridge, I can cut it up and put it in a stew, and I can even fry pieces of it and make them look like the French fries they give you at the McDonald's down the street.' Next I asked the seller where his taro came from. 'It grows over all the islands,' he said. 'Sometimes, when there's too much rain, it may rot—but this year the harvest is very good.' . . . Next I walked into the crowded Indian part of the CBD. On a side street I got a reminder of the geographic concept of agglomeration. Here was a colonial-period building, once a hotel, that had been converted into a business center. I counted 15 enterprises, ranging from a shoestore to a photographer and including a shop where rubber stamps were made and another selling diverse tobacco products. Of course (this being the Indian sector of downtown Suva) tailors outnumbered all other establishments."

barely lie above sea level. Guam, with 210 square miles (550 sq km), is Micronesia's largest island, and no island elevation anywhere in Micronesia reaches 3300 feet (1000 m).

The high-island/low-island dichotomy is useful not only in Micronesia, but also throughout the realm. Both the physiographies of these islands and the economies they support differ in crucial ways. High islands wrest substantial moisture from the

ocean air; they tend to be well watered and have good volcanic soils. As a result, agricultural products show some diversity, and life is reasonably secure. Populations tend to be larger on these high islands than on the low islands, where drought is the rule and fishing and the coconut palm are the mainstays of life. Small communities cluster on the low islands, and over time, many of these have died out. The major migrations, which sent fleets to populate islands from Hawai'i to New Zealand, tended to originate in the high islands.

Until the mid-1980s, Micronesia was largely a United States Trust Territory (the last of the post-World War II trusteeships supervised by the United Nations), but that status has now changed. As Figure 12-3 shows, today Micronesia is divided into countries bearing the names of independent states. The **Marshall Islands**, where the United States tested nuclear weapons (giving prominence to the name *Bikini*), now is a republic in "free association" with the United States, having the same status as the Federated States of Micronesia and (since 1994) Palau. The **Northern Mariana Islands** are a commonwealth "in political union" with the United States. In effect, the United States provides billions of dollars in assistance to these countries, in return for which they commit themselves to avoid foreign policy actions that are contrary to U.S. interests. There are other conditions: **Palau**, for example, granted the United States rights to existing military bases for 50 years following independence.

Also part of Micronesia is the U.S. territory of **Guam**, where independence is not in sight and where U.S. military installations and tourism provide the bulk of income, and the remarkable Republic of **Nauru**. With a population of barely 12,000 and 8 square miles of land, this microstate got rich by selling its phosphate deposits to Australia and New Zealand, where they are used as fertilizer. Per-capita incomes rose to U.S. $10,000, making Nauru one of the Pacific's high-income societies. But the phosphate deposits ran out, and an island scraped bare faced an economic crisis.

In this region of tiny islands, most people subsist on farming or fishing, and virtually all the countries need infusions of foreign aid to survive. The natural economic complementarity between the high-island farming cultures and the low-island fishing communities all too often is negated by distance, spatial as well as cultural. Life here may seem idyllic to the casual visitor, but for the Micronesians it often is a daily challenge.

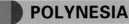

POLYNESIA

To the east of Micronesia and Melanesia lies the heart of the Pacific, enclosed by a great triangle stretching from the Hawaiian Islands to Chile's Easter Island to New Zealand. This is Polynesia (Fig. 12-3), a region of numerous islands (*poly* means many), ranging from volcanic mountains rising above the Pacific's waters (Mauna Kea on Hawai'i reaches nearly 13,800 feet [over 4200 m]), clothed by luxuriant tropical forests and drenched by well over 100 inches of rainfall each year, to low coral atolls where a few palm trees form the only vegetation and where drought is a persistent problem. The Polynesians have somewhat lighter-colored skin and wavier hair than do the other peoples of the Pacific Realm; they are often also described as having an excellent physique. Anthropologists differentiate between these original Polynesians and a second group, the Neo-Hawaiians, who are a blend of Polynesian, European, and Asian ancestries. In the U.S. State of Hawai'i—actually an archipelago of more than 130 islands—Polynesian culture has been both Europeanized and Orientalized.

Its vastness and the diversity of its natural environments notwithstanding, Polynesia clearly constitutes a geographic region within the Pacific Realm. Polynesian culture, though spatially fragmented, exhibits a remarkable consistency and uniformity from one island to the next, from one end of this widely dispersed region to the other. This consistency is particularly expressed in vocabularies, technologies, housing, and art forms. The Polynesians are uniquely adapted to their maritime environment, and long before European sailing ships began to arrive in their waters, Polynesian seafarers had learned to navigate their wide expanses of ocean in huge double canoes as long as 150 feet (45 m). They traveled hundreds of miles to favorite fishing zones and engaged in inter-island barter trade, using maps constructed from bamboo sticks and cowrie shells and navigating by the stars. However, modern descrip-

tions of a Pacific Polynesian paradise of emerald seas, lush landscapes, and gentle people distort harsh realities. Polynesian society was forced to get used to much loss of life at sea when storms claimed their boats; families were ripped apart by accident as well as migration; hunger and starvation afflicted the inhabitants of smaller islands; and the island communities were often embroiled in violent conflicts and cruel retributions.

The political geography of Polynesia is complex. In 1959, the Hawaiian Islands became the fiftieth State to join the United States. The State's population is now 1.3 million, with over 80 percent living on the island of Oahu. There, the superimposition of cultures is symbolized by the panorama of Honolulu's skyscrapers against the famous extinct volcano at nearby Diamond Head. The Kingdom of **Tonga** became an independent country in 1970 after seven decades as a British protectorate; the British-administered Ellice Islands were renamed **Tuvalu**, and along with the Gilbert Islands to the north (now renamed **Kiribati**), they received independence from Britain in 1978. Other islands continued under French control (including the Marquesas Islands and Tahiti), under New Zealand's administration (Rarotonga), and under British, U.S., and Chilean flags.

In the process of politico-geographical fragmentation, Polynesian culture has suffered severe blows. Land developers, hotel builders, and tourist dollars have set **Tahiti** on a course along which Hawai'i has already traveled far. The Americanization of eastern **Samoa** has created a new society different from the old. Polynesia has lost much of its ancient cultural consistency; today, the region is a patchwork of new and old—the new often bleak and barren, with the old under intensifying pressure.

The countries and cultures of the Pacific Realm lie in an ocean on whose rim a great drama of economic and political transformation will play itself out during the twenty-first century. Already, the realm's own former margins—in Hawai'i in the north and in New Zealand in the south—have been so recast by foreign intervention that little remains of the kingdoms and cultures that once prevailed. Now the Pacific world faces changes far greater even than those brought here by European colonizers. Once upon a time the shores and waters of the Mediterranean Sea formed an arena of regional transformation that changed the world. Then it was the Atlantic, avenue of the Industrial Revolution and stage of fateful war. Now it seems to be the turn of the Pacific as the world's largest country and next superpower (China) faces the richest and most powerful (the United States). Giants will jostle for advantage in the Pacific; how will the weak societies of the Pacific Realm fare?

A FINAL CAVEAT: PACIFIC AND ANTARCTIC

South of the Pacific Realm lies Antarctica and its encircling Southern Ocean. The combined area of these two geographic expanses constitutes 40 percent of the entire planet—two-fifths of the Earth's surface containing one one-thousandth of the world's population.

Is Antarctica a geographic realm? In physiographic terms, yes, but not on the basis of the criteria we use in this book. Antarctica is a continent, nearly twice as large as Australia, but virtually all of it is covered by a dome-shaped ice sheet nearly 2 miles (3.2 km) thick near its center. The continent often is referred to as the "white desert" because, despite all of its ice and snow, annual precipitation is low (less than 6 inches [15 cm] per year). Temperatures are frigid, with winds so strong that Antarctica also is called the "home of the blizzard." For all its size, no functional regions have developed here, no towns, no transport networks except the supply lines of research stations. And Antarctica still is a frontier, even a scientific frontier still slowly giving up its secrets. Underneath all that ice lie some 70 lakes of which Lake Vostok is the largest with over 5400 square miles (14,000

FROM THE FIELDNOTES

"The Antarctic Peninsula is geologically an extension of South America's Andes Mountains, and this vantage point in the Gerlach Strait leaves you in no doubt; the mountains rise straight out of the frigid waters of the Southern Ocean. We have been passing large, flat-topped icebergs, but these do not come from the peninsula's shores; rather, they form on the leading edges of ice shelves or on the margins of the mainland where continental glaciers slide into the sea. Here along the peninsula, the high-relief topography tends to produce jagged, irregular icebergs. The mountain range that forms the Antarctic Peninsula continues across Antarctica under thousands of feet of ice and is known as the Transantarctic Mountains. Looking at this place you are reminded that beneath all this ice lies an entire continent, still little-known, with fossils, minerals, even lakes yet to be discovered and studied."

sq km). It may be as much as 2000 feet (600 m) deep. No one has yet seen a sample of its water.

Like virtually all frontiers, Antarctica has always attracted pioneers and explorers. Whale and seal hunters destroyed huge populations of Southern Ocean fauna during the eighteenth and nineteenth centuries, and explorers planted the flags of their countries on Antarctic shores. Between 1895 and 1914, the quest for the South Pole became an international obsession; Roald Amundsen, the Norwegian, reached it first in 1911. All this led to national claims in Antarctica during the interwar period (1918–1939).

The geographic effect was the partitioning of Antarctica into pie-shaped sectors centered on the South Pole (Fig. 12-4). In the least frigid area of the continent, the Antarctic Peninsula, British, Argentinian, and Chilean claims overlapped—and still do.

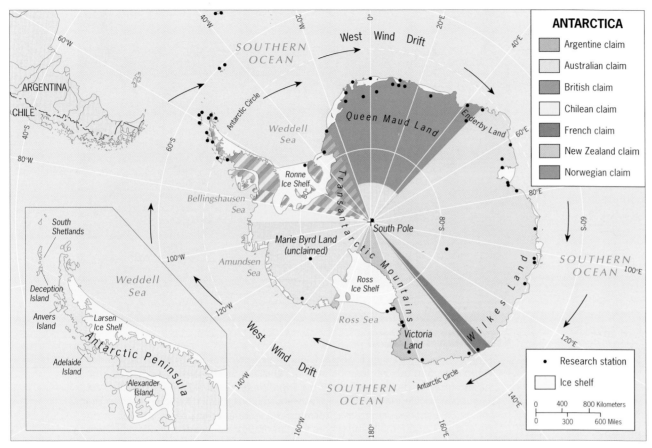

FIGURE 12-4

is nearly as large as the North and South Atlantic. However distant actual exploitation may be, countries want to keep their stakes here.

But the claimant states (those with territorial claims) recognize the need for cooperation. During the late 1950s, they joined in the International Geophysical Year (IGY) that launched major research programs and established a number of permanent research stations throughout the continent. This spirit of cooperation led to the 1961 ⑩ signing of the **Antarctic Treaty**, which ensures continued scientific collaboration, prohibits military activities, protects the environment, and holds national claims in abeyance. In 1991, when the treaty was extended under the terms of the Wellington Agreement, concerns were raised that it does not do enough to control future resource exploitation.

In an age of growing national self-interest and increasing raw-material consumption, the possibility exists that Antarctica and its offshore waters may yet become an arena for international rivalry. Until now, its remoteness and its forbidding environments have saved it from that fate. The entire world benefits from this because evidence is mounting that Antarctica plays a critical role in the global environmental system, so that human modifications may have worldwide (and unpredictable) consequences.

One sector, Marie Byrd Land (shown in neutral beige on the main map), was never formally claimed by any country.

Why should states be interested in territorial claims in so remote and difficult an area? Both land and sea contain raw materials that may some day become crucial: proteins in the waters, and fuels and minerals beneath the ice. Antarctica (5.5 million sq mi/14.2 million sq km) is almost twice as large as Australia, and the Southern Ocean (see pp. 368-370)

DISCOVERY TOOLS www.wiley.com/college/deblij

The *Concepts and Regions* Website, featuring *GeoDiscoveries*, offers many additional resources to enhance your understanding and experience of this chapter. Be sure to explore the following:

Photo Galleries

Papeete, Tahiti
Nouméa, New Caledonia
Suva, Fiji
Bora Bora, Pago Pago, American Samoa
Tuvalu
Gerlach Strait, Antarctica

More To Explore

Systematic Essay: Marine Geography

Virtual Field Trip

Hawai'i's Cultural Diversity

Expanded Regional Coverage

Melanesia, Micronesia, Polynesia

Using the Maps

If, as is often said, a picture is worth a thousand words, then a map is worth a million. This book contains more than 200 maps, locational and thematic, to help you understand the text. Please make it a point to study each map when it is referred to and do so as carefully as you would a paragraph of text.

Maps display locations and patterns in geographic space. They are a kind of shorthand through which encoded spatial messages are transmitted to the reader. Such shorthand is necessary because the real world is so complex that a great deal of geographic information must be compressed into the small confines of maps that can fit onto the pages of this book. Cartographers (mapmakers) must carefully choose which information to include; these decisions force them to omit many other things in order to prevent cluttering a map with less relevant information. For instance, Figure A shows several city blocks in central London but avoids mapping individual buildings because they would interfere with the key information being presented—the spatial distribution of cholera deaths in that part of London during an epidemic.

MAP READING

Deciphering the coded messages contained in our maps—*map reading*—is not difficult, but a few rules

of the road do help. The need to miniaturize portions of the world on small maps is discussed in the section on map scale (pp. 7-8), and Figures A and B offer two additional and contrasting examples.

Orientation, or direction, on maps can usually be discerned by reference to the geographic grid of latitude and longitude. *Latitude* is measured from 0° to 90° north and south of the equator (parallels of latitude are always drawn in an east-west direction), with the equator being 0° and the North and South Poles being 90°N and 90°S, respectively. Meridians of *longitude* (always drawn north-south) are measured 180° east and west of the *prime meridian* (0°), which passes through the Greenwich Observatory next to London; the 180th meridian, for the most part, serves as the *international date line* that lies in the middle of the Pacific Ocean (see Fig. 12-1). Inspection of Figure B shows that north is not automatically at the top of a map; instead, the direction of north curves along every meridian, with all such lines of longitude converging at the North Pole. The many minor directional distortions in this map are unavoidable: it is geometrically impossible to transfer the grid of a three-dimensional sphere (globe) onto a two-dimensional flat map. Therefore, compromises in the form of *map projections* must be devised in which, for example, properties such as areal size and distance are preserved but directional constancy is sacrificed.

MAP SYMBOLS

Once the background mechanics of scale and orientation are understood, the main task of decoding the map's content can proceed. The content of most maps in this book is organized within the framework of point, line, and area symbols, which are made especially clear through the use of color. These symbols are usually identified in the map's *legend*, but sometimes accompanying text must tell the map reader more. In Figure A, for example, we are informed that the dots reflect cholera deaths, but we are not informed that each P symbol represents a municipal water pump at the time of this epidemic, 1854. *Point symbols*, shown as dots on maps, can tell us two things: the location of each phenomenon, and, sometimes, its quantity. Look, for example, at Figure 1-14 (p. 51). Not only does this map show the locations of Western European cities, but also their relative population size.

Line symbols connect places or areas between which some sort of movement or flow has occurred or is occurring. Figure B, taken from the chapter on Southeast Asia, shows the migration of Chinese into this realm, a mass movement that originated mainly in southeastern China and produced coastal Chinese communities on the islands and peninsulas of this multicultural part of the world. Note that we make no attempt to rank the sizes of Southeast Asian cities, so

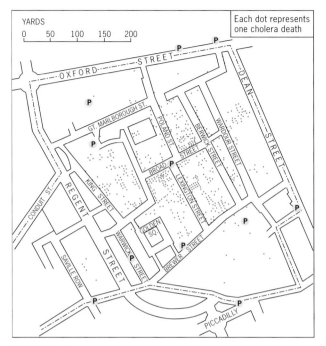

FIGURE A

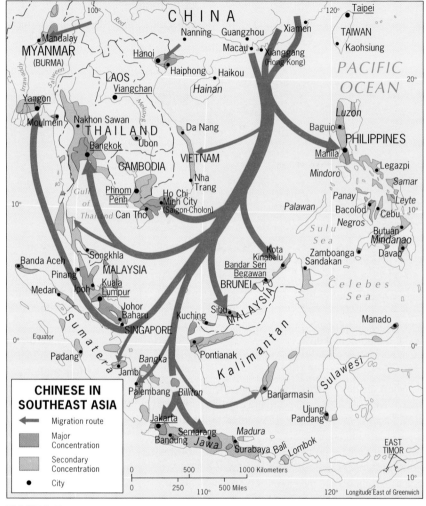

FIGURE B

as not to obscure the key theme. *Area symbols* are used to classify two-dimensional spaces. Primary and secondary concentrations of Chinese residents in Southeast Asia are differentiated by color in Figure B. Area symbols can also be used to communicate quantitative information: for example, the light-tan-colored zones in Figure G-7 (p. 11) delimit semiarid areas within which annual precipitation averages from 12 to 20 inches (30 to 50 cm).

MAP INTERPRETATION

The explanation of cartographic patterns is one of the geographer's most important tasks. Although that task is performed for you throughout this book, readers should be aware that today's practitioners use many sophisticated techniques and cutting-edge computer technology to analyze vast quantities of areal data. These modern methods notwithstanding,

geographic inquiry still focuses on the search for meaningful *spatial relationships*. This longstanding concern of the discipline is classically demonstrated in Figure A. By showing on his map that cholera fatalities clustered around municipal water pumps, Dr. John Snow was able to persuade city authorities to shut them off; almost immediately, the number of new disease victims dwindled to zero, thereby confirming Snow's theory that contaminated drinking water was crucial in the spread of cholera.

Sometimes maps can reveal information or insights beyond those anticipated or intended by the cartographer. Take a look at the ethnolinguistic map on page 260, Fig. 7-20. That map was drawn before the events of September 11, 2001 and subsequent developments in and around Afghanistan. Its relevance and meaning now exceed its original, illustrative intent. So study maps such as this in detail, and you will vastly increase what you get out of this book and this course.

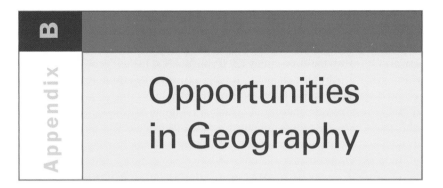

Opportunities in Geography

*T*he chapters of this book give an idea of the wide range of topics and interests that geographers pursue, particularly in the many discussions of concepts (regional and otherwise) and frequent references to the content of geography's systematic fields. There are specializations within each of those topical fields as well as in regional studies—but an introductory book such as this lacks enough space to discuss all of them. This appendix, therefore, is designed to help you, should you decide to major or minor in geography and/or to consider it as a career option.

AREAS OF SPECIALIZATION

As in all disciplines, areas of concentration or specialization change over time. In North American geography early in the twentieth century, there was a period when most geographers were physical geographers, and the natural landscape was the main object of geographic analysis. Then the pendulum swung toward human (cultural) geography, and students everywhere focused on the imprints of human activity on the surface of the Earth. Still later the analysis of spatial organization became a major concern. In the meantime, geography's attraction for some students lay in technical areas: in cartography, in aerial and satellite remote sensing, in computer-assisted spatial data analysis, and, most recently, in geographic information systems (GIS).

All this meant that geography posed (and continues to pose) a challenge to its professionals. New developments require that we keep up to date, but we must also continue to build on established foundations.

Regional Geography

One of these established foundations, of course, is regional geography, which encompasses a large group of specializations. Some geographers specialize in the theory of regions: how they should be defined, how they are structured, and how their internal components work. This leads in the direction of *regional science*, and some geographers have preferred to call themselves regional scientists. But make no mistake: regional science is regional geography.

Another, older approach to regional geography involves specialization in an area of the world ranging in size from a geographic realm to a single region or even a State or part of a State. There was a time when regional geographers, because of the interdisciplinary nature of their knowledge, were sought after by government agencies. Courses in regional geography abounded in universities' geography departments; regional geographers played key roles in international studies and research programs. But then the drive to make geography a more rigorous science and to search for universal (rather than regional) truths contributed to a decline in regional geography. The results were not long in coming, and

in recent years you have probably seen the issue of "geographic illiteracy" discussed in newspapers and magazines. Now the pendulum is swinging back again, and regional geography and regional specialization are reviving. This is a propitious time to consider regional geography as a professional field.

Your Personal Interests

Geography, as we have pointed out, is united by several bonds, of which regional geography is but one. Regional geography exemplifies the spatial view that all geographers hold; the spatial approach to study and research binds physical and human geographers, regionalists, and topical specialists. Another unifying theme is an abiding interest in the relationships between human societies and natural environments. We have referred to that topic frequently in this book; as an area of specialization it has gone through difficult times. Perhaps more than anything, geography remains a field of *synthesis*, of understanding interrelationships.

Geography also is a field science, using "field" in another context. In the past, almost all major geography departments required a student's participation in a "field camp" as part of a master's degree program; thus many undergraduate programs included field experience. It was one of those bonding practices in which students and faculty with diverse interests met, worked together, and learned from one another.

Today, few field camps of this sort are offered, but that does not change what geography is all about. If you see an opportunity for field experience with professional geographers—even just a one-day reconnaissance—take it. But realize this: a few days in the field with geographic instruction may hook you for life.

Some geographers, in fact, are far better field-data gatherers than analysts or writers. In this respect they are not alone: this also happens in archeology, geology, and biology (among other field disciplines). This does not mean that these field workers do not contribute significantly to knowledge. Often on a research team some of the members are better in the field, and others excel in subsequent analysis. From all points of view, however, fieldwork is important.

Geography, then, is practiced in the field and in the office, in physical and human contexts, in generality and detail. Small wonder that so many areas of specialization have developed! If you check the undergraduate catalogue of your college or university, you will see some of these specializations listed as semester-length courses. But no geography department, no matter how large, could offer them all.

How does an area of specialization develop, and how can one become a part of it? The way in which geographic specializations have developed tells us much about the entire discipline. Some major areas, now old and established, began as research and theory-building by one scholar and his or her students. These graduate students dispersed to the faculties of other universities and began teaching what they had learned. Thus, for example, did Carl Sauer's cultural geography spread from the University of California, Berkeley during the middle decades of the twentieth century.

It is one of the joys of geography that the basics and methods, once learned, are applicable to so many features of the human and physical world. Geographers have specialized in areas as disparate as shopping-center location and glacier movement, tourism and coastal erosion, real estate and wildlife, retirement communities and sports. Many of these specializations began with the interests and energies of a single scholar. Some thrived and grew into major geographic pursuits; others remained one-person shows, but with potential. When you discuss your own interests with a faculty advisor, you may refer to a university where you would like to do graduate work. "Oh yes," the answer may be, "Professor X is in their geography department, working on just that." Or perhaps your advisor will suggest another university where a member of the faculty is known to be working on the topic in which you are interested. That is the time to write a letter or e-mail of inquiry. What is the professor working on now? Are graduate students involved? Are research funds available? What are the career prospects after graduation?

The Association of American Geographers, or **AAG** (1710 16th Street, N.W., Washington, D.C. 20009-3198 [**www.aag.org**]), recognizes no less than 54 so-called Specialty Groups. In academic year 2004-2005, the Specialty Group roster included the following branches of geography:

Africa
Applied Geography
Asian Geography
Bible
Biogeography
Canadian Studies
Cartography
China
Climate
Coastal and Marine Geography
Communication Geography
Cryosphere (Earth's Ice System)
Cultural and Political Ecology
Cultural Geography
Developing Areas
Disability
Economic Geography
Energy and Environment
Environmental Perception and Behavioral
  Geography
Ethics, Justice, and Human Rights
Ethnic Geography
European

Geographic Information Systems (GIS)
Geographic Perspectives on Women
Geography Education
Geography of Religions and Belief Systems
Geomorphology (Landform Analysis)
Hazards
Historical Geography
History of Geography
Human Dimensions of Global Change
Indigenous Peoples
Latin America
Medical Geography
Microcomputers
Middle East
Military Geography
Mountain Geography
Political Geography
Population
Qualitative Research
Recreation, Tourism, and Sport
Regional Development and Planning
Remote Sensing
Rural Geography
Russian, Central Eurasian, and East European
Sexuality and Space
Socialist and Critical Geography
Spatial Analysis and Modeling
Transportation Geography
Urban Geography
Viticulture and Oenology
Water Resources
Worldwide Web

If you contact the AAG at the Internet address given above, click on the green "Membership" tab and then on "Specialty Groups" in the drop-box; then scroll down the alphabetical listing to obtain the e-mail address of the current chairperson of any group. We encourage you to contact the professional geographer who chairs the group(s) you are interested in. These elected leaders are ready, enthusiastic, and willing to provide you with the information you are looking for.

All this may seem far in the future. Still, the time to start planning for graduate school is now. Applications for admission and financial assistance must be made shortly after the *beginning* of your senior year. That makes your junior year a year of decision.

AN UNDERGRADUATE PROGRAM

The most important concern for any geography major or minor is basic education and training in the field. An undergraduate curriculum contains all or several of the following courses (titles may vary):

1. Physical Geography of the Global Environment (weather and climate, natural landscapes, landforms, soils, moisture and surface water flows, elementary biogeography)
2. Introduction to Human Geography (basic principles of cultural and economic geography)
3. World Regional Geography (world geographic realms)

These beginning courses are followed by more specialized courses, including both methodological and substantive ones:

4. Geographic Information Technology
5. Cartographic Theory and Techniques
6. Introduction to Quantitative Methods of Analysis
7. Introduction to Geographic Information Systems (GIS)
8. Analysis of Remotely Sensed Data
9. Cultural Geography
10. Political Geography
11. Urban Geography
12. Economic Geography
13. Historical Geography
14. Geomorphology
15. Geography of United States-Canada, Europe, and/or other world geographic realms

You can see how Physical Geography of the Global Environment would be followed by the more specialized analysis of landforms covered in Geomorphology, and how Human Geography now divides into such areas as intermediate and/or advanced cultural and economic geography. As you progress, the focus becomes even more specialized. Thus Economic Geography may be followed by:

16. Industrial Geography
17. Transportation Geography
18. Geography of Development

At the same time, regional concentrations may come into sharper focus:

19. Geography of Western Europe (or other major regions)

In these more advanced courses, you will use the technical knowledge acquired from courses numbered **4** through **8** (and perhaps others). Now you can avail yourself of the opportunity to develop these skills further. Many departments offer such courses as these:

20. Advanced Quantitative Methods
21. Computer Cartography
22. Intermediate GIS
23. Advanced GIS
24. Advanced Satellite Imagery Interpretation

From this list (which represents only part of a comprehensive curriculum), it is evident that you cannot, even in four years of undergraduate study, register for all courses. The geography major in many universities requires a minimum of only 30 (semester-hour) credits—just 10 courses, fewer than half those listed here. This is another reason to begin thinking about specialization at an early stage.

Because of the number and variety of possible geography courses, most departments require that their majors complete a core program that includes courses in substantive areas as well as theory and methods. That core program is important, and you should not be tempted to put off these courses until your last semesters. What you learn in the core program will make what follows (or should follow) much more meaningful.

You should also be aware of the flexibility of many undergraduate programs, something that can be especially important to geographers. Imagine that you are majoring in geography and develop an interest in Southeast Asia. But the regional specialization of the geographers in your department may be focused somewhere else—say, Middle and South America. However, courses on Southeast Asia are indeed offered by other departments, such as anthropology, history, or political science. If you are going to be a regional specialist, those courses will be very useful and should be part of your curriculum—but you are not able to receive geography credit for them. After successfully completing those courses, however, you may be able to register for an independent study or reading course in the Geography of Southeast Asia if a faculty member is willing to guide you. Always discuss such matters with the undergraduate advisor or chairperson of your geography department.

LOOKING AHEAD

By now, as a geography major, you will be thinking of the future—either in terms of graduate school or a salaried job. In this connection, if there is one important lesson to keep in mind, it is to *plan ahead* (a redundancy for emphasis). The choice of graduate school is one of the most important you will make in your life. The professional preparation you acquire as a graduate student will affect your competitiveness in the job market for years to come.

Choosing a Graduate School

Your choice of a graduate school hinges on several factors, and the geography program it offers is one of them. Possibly, you are constrained by residency factors, and your choice involves the schools of only one State. Your undergraduate record and grade point average affect the options. As a geographer, you may have strong feelings in favor of or against particular parts of the country. And although some schools may offer you financial support, others may not.

Certainly the programs and specializations of the prospective graduate department are extremely important. If you have settled on your own area of interest, it is wise to find a department that offers opportunities in that direction. If you have yet to decide, it is best to select a large department with several options. Some students are so impressed by the work and writings of a particular geographer that they go to his or her university solely to learn from and work with that scholar. In every case, information and preparation are crucial. Many a prospective graduate student has arrived on campus eager to begin work with a favorite professor, only to find that the professor is away on a sabbatical leave!

Fortunately, information can be acquired with little difficulty. One of the most useful publications of the AAG is its *Guide to Geography Programs in North America*, published annually. A copy of this annual directory should be available in the office of your geography department, but if you plan to enter graduate school, a personal copy would be an asset. Not only does the *Guide* describe the programs, requirements, financial aid, and other aspects of geography departments in North America, but it also lists all faculty members and their current research and teaching specializations. Moreover, it also contains a complete listing of the Association's approximately 8,000 members and their specializations. And Internet addresses are now provided as well for accessing the Web sites of geography departments, which they increasingly utilize to supplement their entries in the *Guide* and provide even more detailed information about their programs. Although you will find the *Guide* to be very useful in your decision-making, we strongly recommend that you also explore the Web sites of departments you wish to consider.

You may discover one particular department that stands out as the most interesting and most appropriate for you. But do not limit yourself to one school. Consult the *Guide*'s 21-page table (entitled *Program Specialties*) that links 35 geographic specializations to every department listed, and then explore each appropriate match-up. After careful investigation, it is best to rank a half-dozen schools (or more), write or e-mail all of them for admission application forms, and apply to several. Multiple applications are a bit costly, but the investment is worth it.

Assistantships and Scholarships

One reason to apply to several universities has to do with financial (and other) support for which you may be eligible. If you have a reasonably well-rounded undergraduate program behind you and a good record of achievement, you are eligible to become a teaching assistant (TA) in a graduate department. Such assistantships usually offer full or partial tuition plus a monthly stipend (during the nine-month academic year). Conditions vary, but this position may make it possible for you to attend a university that would otherwise be out of reach. A tuition waiver alone can be well over $10,000 annually. Application for an assistantship is made directly to the department; you should write or e-mail the department contact listed in the latest AAG *Guide*, who will either respond directly to you or forward your inquiry to the departmental committee that evaluates applications.

What does a TA do? The responsibilities vary, but often teaching assistants are expected to lead discussion sections (of a larger class taught by a professor), laboratories, or other classes. They prepare and grade examinations and help undergraduates deal with problems arising from their courses. This is an excellent way to determine your own ability and interest in a teaching career.

In some instances, especially in larger geography departments, research assistantships are available. When a member of the faculty is awarded a large grant (e.g., by the National Science Foundation) for a research project, that grant may make possible the appointment of one or more research assistants (RAs). These individuals perform tasks generated by the project and are rewarded by a modest salary (usually comparable to that of a TA). Normally, RAs do not receive tuition waivers, but sometimes the department and the graduate school can arrange a waiver to make the research assistantship more attractive to better students. Usually, RAs are chosen from among graduate students already on campus who have proven their interest and ability. Sometimes, however, an incoming student is appointed. Always ask about opportunities.

Geography students also are among those eligible for many scholarships and fellowships offered by universities and off-campus organizations. When you write your introductory letter, be sure to inquire about all forms of financial aid.

JOBS FOR GEOGRAPHERS

Upon completing your bachelor's degree, you may decide to take a job rather than going on to graduate school. Again, this is a decision best made early in your junior year, for two main reasons: (1) so that you can tailor your curriculum for a vocational objective, and (2) so that you can start searching for a job well before graduation.

Internships

A very good way to enter the job market—and to become familiar with the working environment—is by taking an internship in an agency, office, or firm. Many organizations find it useful to have interns. Internships help organizations train beginning professionals and give companies an opportunity to observe the performance of trainees. Many an intern has ultimately been employed by his or her organization. Some employers have even suggested what courses the intern should take in the next academic year to improve future performance. For example, an urban or regional planning agency that employs an intern might suggest that the intern add a relevant GIS course or urban-planning course to his or her program of study.

Some organizations will appoint interns on a continuing basis, say, two afternoons a week around the year; others make full-time, summer-only appointments available. Occasionally, an internship can be linked to a departmental curriculum, yielding academic credit as well as vocational experience. Your department undergraduate advisor or the chairperson will be the best source of assistance.

One of the most interesting internship programs is offered by the National Geographic Society in Washington, D.C. Every year, the Society invites three groups of about eight interns each to work with the permanent staff of its various departments. Application forms are available in every geography department in the United States, and competition is strong. The application itself is a useful exercise, as it tells you what the Society (and other organizations) look for in your qualifications.

Professional Opportunities

A term you will sometimes see in connection with jobs is *applied geography*. This, one supposes, distinguishes geography teaching from practical geography. In fact, however, all professional geography in education, business, government, and elsewhere is "applied." In the past a large majority of geography graduates became teachers—in elementary and high schools, colleges, and universities. More recently, geographers have entered other arenas in increasing numbers. In part, this is related to the decline in geographic education in schools, but it also reflects the growing recognition of geographic skills by employers in business and government.

Nevertheless, what you as a geographer can contribute to a business is not yet as clear to the managers of many companies as it should be. The anybody-can-do-geography attitude is a form of ignorance you will undoubtedly confront. (This is so even in precollegiate education, where geography was submerged in social studies—and often taught by teachers who had never taken a course in the discipline!)

So, once employed, you may have to prove not only yourself but also the usefulness of the skills and capacities you bring to your job. There is a positive side to this. Many employers, once dubious about hiring a geographer, learn how geography can contribute—and become enthusiastic users of geographic talent. Where one geographer is employed, whether in a travel firm, publishing house, or planning office, you will soon find more.

Geographers are employed in business, government, and education. Planning is a profession that employs many geographers. Employment in business has grown in recent years (see the following section). Governments at national, State, and local levels have always been major employers of geographers. And in education, where geography was long in decline, the demand for geography teachers will grow again.

For more detailed information about employment, you should contact the AAG (at the address given earlier) for the inexpensive booklet entitled *Careers in Geography*; or check with your geography department, whose office is very likely to have a copy.

Business and Geography

With their education in global and international affairs, their knowledge of specialized areas of interest to business, and their training in cartography, GIS, methods of quantitative analysis, and writing, geographers with a bachelor's degree are attractive business-employment prospects. Some undergraduate students have already chosen the type of business they will enter; for them, there are departments that offer concentrations in their areas of interest. Several geography departments, for instance, offer a curriculum that concentrates on tourism and travel—a business in which geographic skills are especially useful.

These days, many companies want graduates with a strong knowledge of international affairs and fluency in at least one foreign language in addition to their skills in other areas. The business world is quite different from academia, and the transition is not always easy. Your employer will want to use your abilities to enhance profits. You may at first be placed in a job where your geographic skills are not immediately applicable, and it will be up to you to look for opportunities to do so.

One of our students was in such a situation some years ago. She was one of more than a dozen new employees doing what was essentially clerical work. (Some companies will use this type of work to determine a new employee's punctuality, work habits, adaptability, and productivity.) One day she heard that the company was considering establishing a branch to sell its product in East Africa. The student had a regional interest in East Africa as an undergraduate and had even taken a year of Swahili-language training. On her own time she wrote a carefully documented memorandum to the company's president and vice presidents, describing factors that should be taken into consideration in the projected expansion. That report

evinced her regional skills, locational insights, and knowledge of the local market and transport problems, along with the probable cultural reaction to the company's product, the country's political circumstances, and (last but significantly) this employee's ability to present such issues effectively. She supported her report with good maps and several illustrations. Soon she received a special assignment to participate in the planning process, and her rise in the company's ranks had begun. She had seized the opportunity and demonstrated the utility of her geographic skills.

Geography graduates have established themselves in businesses of all kinds: banking, international trade, manufacturing, retailing, and many more. Should you join a large firm, you may be pleasantly surprised by the number of other geographers who hold jobs there—not under the title of geographer but under countless other titles ranging from analyst and cartographer to market researcher and program manager. These are positions for which the appointees have competed with other graduates, including business graduates. As we noted earlier, once an employer sees the assets a geographer can bring, the role of geographers in the company is assured.

Government and Geography

Government has long been a major employer of geographers at the national and State as well as local level. *Careers in Geography* (the AAG booklet) estimates that at least 2,500 geographers are working for governments, about half of them for the federal government. In the U.S. State Department, for instance, there is an Office of the Geographer staffed by professional geographers. Other agencies where geographers are employed include the Defense Mapping Agency, the Bureau of the Census, the U.S. Geological Survey, the Central Intelligence Agency, and the Army Corps of Engineers. Still other employers are the Library of Congress, the National Science Foundation, and the Smithsonian Institution. Many other branches of government also have positions for which geographers are eligible.

Opportunities also exist at State and local levels. Several States now have their own Office of the State Geographer; all States have agencies engaged in planning, resource analysis, environmental protection, and transportation policy-making. All these agencies need geographers who have skills in cartography, remote sensing, spatial database analysis, and the operation of geographic information systems.

Securing a position in government requires early action. If you want a job with the federal government, start at the beginning of your senior year. Every State capital and many other large cities have a Federal Job Information Center (FJIC). (The Washington office is at 1900 E Street, N.W., Washington, D.C. 20415.) You may request information about a particular agency and its job opportunities; it is appropriate to write or e-mail directly to the personnel office of the agency or agencies in which you are interested. You may also write to the Office of Personnel Management (OPM), Washington, D.C. 20415, which has offices in most large cities.

Planning and Geography

Planning has become one of geography's allied professions. The planning process is a complex one comprising people trained in many fields. Geographers, with their cartographic, locational, regional, and analytical skills, are sought after by planning agencies. Many an undergraduate student gets that first professional opportunity as an intern in a planning office.

Planning is done by many agencies and offices at levels ranging from the federal to the municipal. Cities have planning offices, as do regional authorities. Working in a planning office can be a very rewarding experience because it involves the solving of social and economic problems, the conservation and protection of the environment, the weighing of diverse and often conflicting arguments and viewpoints, and much interaction with workers trained in other fields. Planning is a superb learning experience.

A career in planning can be much enhanced by a background in geography, but you will have to adjust your undergraduate curriculum to include courses in such areas as public administration, public finance, and other related fields. Thus a career in planning itself requires early planning on your part. At many universities, the geography department is closely as-

sociated with the planning department, and your faculty advisor can inform you about course requirements. But if you have your eye on a particular office or agency, you should also request information from its director about desired and required skills.

Planning is by no means a monopoly of government. Government-related organizations such as the Agency for International Development (AID), the World Bank, and the International Monetary Fund (IMF) have planning offices, as do nongovernmental organizations (NGOs), banks, airline companies, industrial firms, multinational corporations, and research institutes. Private-sector opportunities for planners have been expanding, and you may wish to explore them. An important organization is the planners' equivalent of the geographers' AAG: the American Planning Association, 1776 Massachusetts Avenue, N.W., Washington, D.C. 20006. The AAG's *Careers in Geography* provides additional information on this expanding field and where you might pursue its study.

Teaching Geography

If you are presently a freshman or sophomore, your graduation may coincide with the end of a long decline in the geography-teaching profession. Just a few decades ago, teaching geography in elementary or high school was the goal of thousands of undergraduate geography majors. But then began the merging of geography into the hybrid field called "social studies," and prospective teachers no longer needed to have any training in geography in many States. In Florida, for instance, teachers were formerly required to take courses in regional geography and conservation (as taught in geography departments), but in the 1970s those requirements were dropped. Education planners fell victim to the myth that the teaching of geography does not require any training. What was left of geography often was taught by teachers whose own fields were history, civics, or even basketball coaching!

If you read the daily newspapers, you have seen reports of the predictable results. As mentioned earlier, "geographic illiteracy" has become a common

complaint (often made by the same education planners who pushed geography into the social studies program and eliminated teacher-education requirements). Now States are returning to the education requirement so that teachers will learn some geography. And geography is returning to elementary and high school curricula—assisted in particular by the National Geographic Society, which supports State-level Geographic Alliances of educators and academic geographers. The need for geography teachers will soon be on the upswing again.

So this may be a good time to consider teaching geography as a career. You should do some research, however, because States vary in their progressiveness in this arena. You should also visit the School of Education in your college or university and ask questions about instructional opportunities in geography. In addition, contact not only the AAG but also the *National Council for Geographic Education (NCGE)* at 206-A Martin Hall, Jacksonville State University, Jacksonville, AL 36265-1602 (**www.ncge.org**). Your State geographic society or Geographic Alliance also may be helpful; ask your department advisor or chairperson for details.

SOME FINAL THOUGHTS

The opportunities in geography are many, but often they are not as obvious as those in other fields. You will find that good and timely preparation produces results and, frequently, unexpected rewards.

The discussion in this appendix has provided a comprehensive answer to the oft-asked question, "What can one do with geography?" If you wish to explore this question further, along with the AAG's *Careers in Geography* we recommend *On Becoming a Professional Geographer*, a book edited by Martin S. Kenzer (Merrill/Macmillan, 1989).

We wish you every success in all your future endeavors. And, speaking for the entire community of professional geographers, we would be delighted to have you join our ranks should you choose a geography-related career.

Glossary

Aboriginal population   See **indigenous peoples**.*

Absolute location   The position or place of a certain item on the surface of the Earth as expressed in degrees, minutes, and seconds of **latitude**, 0° to 90° north or south of the equator, and **longitude**, 0° to 180° east or west of the *prime meridian* passing through Greenwich, England (a suburb of London).

Accessibility   The degree of ease with which it is possible to reach a certain location from other locations. *Inaccessibility* is the opposite of this concept.

Acculturation   Cultural modification resulting from intercultural borrowing. In **cultural geography**, the term refers to the change that occurs in the culture of **indigenous peoples** when contact is made with a society that is technologically superior.

Agglomeration   Process involving the clustering or concentrating of people or activities.

Agrarian   Relating to the use of land in rural communities, or to **agricultural** societies in general.

Agriculture   The purposeful tending of crops and livestock in order to produce food and fiber.

Alluvial   Refers to the mud, silt, and sand (collectively *alluvium*) deposited by rivers and streams. *Alluvial plains* adjoin many larger rivers; they consist of these renewable deposits that are laid down during floods, creating fertile and productive soils. Alluvial **deltas** mark the mouths of rivers such as the Nile and the Ganges.

Altiplano   High-elevation plateau, basin, or valley between even higher mountain ranges, especially in the Andes of South America.

Altitudinal zonation   Vertical regions defined by physical-environmental zones at various elevations (see Fig. 4–13), particularly in the highlands of South and Middle America. See *tierra caliente*, *tierra templada*, *tierra fría*, *tierra helada*, and *tierra nevada*.

American Manufacturing Belt   North America's near-rectangular Core Region, whose corners are Boston, Milwaukee, St. Louis, and Baltimore. Dominated the industrial geography of the U.S. and Canada during the industrial age; still a formidable economic powerhouse that remains the realm's geographic heart.

*Words in boldface type are defined elsewhere in this Glossary.

Antecedent boundary   A political boundary that existed before the **cultural landscape** emerged and stayed in place while people moved in to occupy the surrounding area.

Anthracite coal   Hardest and highest carbon-content coal, and therefore of the highest quality.

Apartheid   Literally, *apartness*. The Afrikaans term for South Africa's pre-1994 policies of racial separation, a system that produced highly segregated socio-geographical patterns.

Aquaculture   The use of a river segment or an artificial pond for the raising and harvesting of food products, including fish, shellfish, and even seaweed (particularly in Japan).

Aquifer   An underground reservoir of water contained within a porous, water-bearing rock layer.

Arable   Land fit for cultivation by one farming method or another. See **physiologic density**.

Archipelago   A set of islands grouped closely together, usually elongated into a *chain*.

Area   A term that refers to a part of the Earth's surface with less specificity than **region**. For example, *urban area* alludes generally to a place where urban development has occurred, whereas *urban region* requires certain specific criteria upon which such a designation is based (e.g., the spatial extent of commuting or the built townscape).

Areal functional organization   A geographic principle for understanding the evolution of regional organization, whose five interrelated tenets are applied to the spatial development of Japan on p. 332.

Areal interdependence   A term related to **functional specialization**. When one area produces certain goods or has certain raw materials and another area has a different set of raw materials and produces different goods, their needs may be *complementary*; by exchanging raw materials and products, they can satisfy each other's requirements.

Arithmetic density   A country's population, expressed as an average per unit area, without regard for its **distribution** or the limits of **arable** land—see also **physiologic density**.

Aryan   From the Sanskrit *Arya* ("noble"), a name applied to an ancient people who spoke an Indo-European language and who moved into northern India from the northwest.

Atmosphere   The Earth's envelope of gases that rests on the oceans and land surface and penetrates open spaces

within soils. This layer of nitrogen (78 percent), oxygen (21 percent), and traces of other gases is densest at the Earth's surface and thins with altitude.

Autocratic   A government that holds absolute power, often ruled by one person or a small group of persons who control the country by despotic means.

Balkanization   The fragmentation of a **region** into smaller, often hostile political units.

Barrio   Term meaning "neighborhood" in Spanish. Usually refers to an urban community in a Middle or South American city; also applied to low-income, inner-city concentrations of Hispanics in such western U.S. cities as Los Angeles.

Bauxite   Aluminum ore; usually deposited at shallow depths in the wet tropics.

Biogeography   The study of *flora* (plant life) and *fauna* (animal life) in spatial perspective.

Birth rate   The *crude birth rate* is expressed as the annual number of births per 1000 individuals within a given population.

Bituminous coal   Softer coal of lesser quality than **anthracite**, but of higher grade than **lignite**. When heated and converted to coking coal or *coke*, it is used to make steel.

Break-of-bulk point   A location along a transport route where goods must be transferred from one carrier to another. In a port, the cargoes of oceangoing ships are unloaded and put on trains, trucks, or perhaps smaller river boats for inland distribution.

Buffer state   See **buffer zone**.

Buffer zone   A country or set of countries separating ideological or political adversaries. In southern Asia, Afghanistan, Nepal, and Bhutan were parts of a buffer zone between British and Russian-Chinese imperial spheres. Thailand was a *buffer state* between British and French colonial domains in mainland Southeast Asia.

Caliente   See **tierra caliente**.

Cartogram   A specially transformed map not based on traditional representations of **scale** or area.

Cartography   The art and science of making maps, including data compilation, layout, and design. Also concerned with the interpretation of mapped patterns.

Caste system   The strict **social stratification** and segregation of people—specifically in India's Hindu society—on the basis of ancestry and occupation.

Cay A low-lying small island usually composed of coral and sand. Pronounced *kee*, and often spelled "key."

Central business district (CBD) The downtown heart of a central city, the CBD is marked by high land values, a concentration of business and commerce, and the clustering of the tallest buildings.

Centrality The strength of an urban center in its capacity to attract producers and consumers to its facilities; a city's "reach" into the surrounding region.

Centrifugal forces A term employed to designate forces that tend to divide a country—such as internal religious, linguistic, ethnic, or ideological differences.

Centripetal forces Forces that unite and bind a country together—such as a strong national culture, shared ideological objectives, and a common faith.

Chaebol Giant corporation controlling numerous companies and benefiting from government connections and favors, dominant in South Korea's **economic geography**; key to the country's **development** into an **economic tiger**, but more recently a barrier to free-market growth.

Charismatic Personal qualities of certain leaders that enable them to capture and hold the popular imagination, to secure the allegiance and even the devotion of the masses. Gandhi, Mao Zedong, and Franklin D. Roosevelt were good examples in the 20th century.

China Proper The eastern and northeastern portions of China that contain most of the country's huge population; mapped in Figure 9–3.

Choke point A narrowing of an international waterway causing marine traffic congestion, requiring reduced speeds and/or sharp turns, and increasing the risk of collision as well as vulnerability to attack. When the waterway narrows to a distance of less than 24 miles (38 km), this necessitates the drawing of a **median line (maritime) boundary**. Examples are the Hormuz Strait between Oman and Iran at the entrance to the Persian Gulf, and the Strait of Malacca between Malaysia and Indonesia.

City-state An independent political entity consisting of a single city with (and sometimes without) an immediate **hinterland**. The ancient city-states of Greece have their modern equivalent in Singapore.

Climate The long-term conditions (over at least 30 years) of aggregate **weather** over a region, summarized by averages and measures of variability; a synthesis of the succession of weather events we have learned to expect at any given location.

Climate change theory An alternative to the **hydraulic civilization theory**; holds that changing **climate** (rather than a monopoly over **irrigation** methods) could have provided certain cities in the ancient **Fertile Crescent** with advantages over others.

Climatology The geographic study of **climates**. Includes not only the classification of climates and the analysis of their regional distribution, but also broader environmental questions that concern climate change, interrelationships with soil and vegetation, and human-climate interaction.

Coal See **anthracite coal**, **bituminous coal**, and **lignite**.

Collectivization The reorganization of a country's **agriculture** under communism that involves the expropriation of private holdings and their incorporation into relatively large-scale units, which are farmed and administered cooperatively by those who live there.

Colonialism See **imperialism**.

Commercial agriculture For-profit **agriculture**.

Common market A **free-trade area** that not only has created a **customs union** (a set of common tariffs on all imports from outside the area) but also has eliminated restrictions on the movement of capital, labor, and enterprise among its member countries.

Compact state A politico-geographical term to describe a **state** that possesses a roughly circular, oval, or rectangular territory in which the distance from the geometric center to any point on the boundary exhibits little variance. Poland and Cambodia are examples of this shape category.

Complementarity Exists when two regions, through an exchange of raw materials and/or finished products, can specifically satisfy each other's demands.

Confucianism A philosophy of ethics, education, and public service based on the writings of Confucius (*Kongfuzi*); traditionally regarded as one of the cornerstones of Chinese **culture**.

Congo See the footnote on p. 198, which makes the distinction between the two Equatorial African countries, Congo and The Congo.

Coniferous forest A forest of cone-bearing, needleleaf evergreen trees with straight trunks and short branches, including spruce, fir, and pine.

Contagious diffusion The distance-controlled spreading of an idea, innovation, or some other item through a local population by contact from person to person—analogous to the communication of a contagious illness.

Conterminous United States The 48 **contiguous** or adjacent States that occupy the southern half of the North American realm. Alaska is not contiguous to these States because western Canada lies in between; neither is Hawai'i, separated from the mainland by over 2000 miles of ocean.

Contiguous Adjoining; adjacent.

Continental drift The slow movement of continents controlled by the **processes** associated with **plate tectonics**.

Continental shelf Beyond the coastlines of many landmasses, the ocean floor declines very gently until the depth of about 660 feet (200 m). Beyond the 660-foot line the sea bottom usually drops off sharply, along the *continental slope*, toward the much deeper mid-oceanic basin. The submerged continental margin is called the continental shelf, and it extends from the shoreline to the upper edge of the continental slope.

Continentality The variation of the continental effect on air temperatures in the interior portions of the world's landmasses. The greater the distance from the moderating influence of an ocean, the greater the extreme in summer and winter temperatures. Continental interiors also tend to be dry when the distance from oceanic moisture sources becomes considerable.

Conurbation General term used to identify a large multi-metropolitan complex formed by the coalescence of two or more major **urban areas**. The Atlantic Seaboard **Megalopolis**, extending along the northeastern U.S. coast from southern Maine to Virginia, is a classic example.

Copra The dried-out, fleshy interior of a coconut that is used to produce coconut oil.

Cordillera Mountain chain consisting of sets of parallel ranges, especially the Andes in northwestern South America.

Core See **core area**; **core-periphery relationships**.

Core area In geography, a term with several connotations. *Core* refers to the center, heart, or focus. The core area of a **nation-state** is constituted by the national heartland, the largest population cluster, the most productive region, and the part of the country with the greatest **centrality** and **accessibility**—probably containing the capital city as well.

Core-periphery relationships The contrasting spatial characteristics of, and linkages between, the *have* (core) and *have-not* (periphery) components of a national or regional **system**.

Corridor In general, refers to a spatial entity in which human activity is organized in a linear manner, as along a major transport route or in a valley confined by highlands. More specifically, the politico-geographical term for a land extension that connects an otherwise **landlocked state** to the sea.

Cultural diffusion The **process** of spreading and adoption of a cultural element, from its place of origin across a wider area.

Cultural ecology The multiple interactions and relationships between a **culture** and its **natural environment**.

Cultural environment See **cultural ecology**.

Cultural geography The wide-ranging and comprehensive field of geography that studies spatial aspects of human **cultures**.

Cultural landscape The forms and artifacts sequentially placed on the **natural landscape** by the activities of various human occupants. By this progressive imprinting of the human presence, the physical (natural) landscape is modified into the cultural landscape, forming an interacting unity between the two.

Cultural pluralism   See **plural(istic) society**.

Cultural revival   The regeneration of a long-dormant **culture** through internal renewal and external infusion.

Culture   The sum total of the knowledge, attitudes, and habitual behavior patterns shared and transmitted by the members of a society. This is anthropologist Ralph Linton's definition; hundreds of others exist.

Culture area   See **culture region**.

Culture hearth   Heartland, source area, innovation center; place of origin of a major **culture**.

Culture region   A distinct, culturally discrete spatial unit; a **region** within which certain cultural norms prevail.

Customs union   A **free-trade area** in which member countries set common tariff rates on imports from outside the area.

Death rate   The *crude death rate* is expressed as the annual number of deaths per 1000 individuals within a given population.

Deciduous   A deciduous tree loses its leaves at the beginning of winter or the onset of the dry season.

Definition   In political geography, the written legal description (in a treaty-like document) of a boundary between two countries or territories—see **delimitation**.

Deforestation   The clearing and destruction of forests (especially tropical rainforests) to make way for expanding settlement frontiers and the exploitation of new economic opportunities.

Deglomeration   Deconcentration.

Delimitation   In political geography, the translation of the written terms of a boundary treaty (the **definition**) into an official cartographic representation (map).

Delta   **Alluvial** lowland at the mouth of a river, formed when the river deposits its alluvial load on reaching the sea. Often triangular in shape, hence the use of the Greek letter whose symbol is Δ.

Demarcation   In political geography, the actual placing of a political boundary on the **cultural landscape** by means of barriers, fences, walls, or other markers.

Demographic transition model   Multi-stage model, based on Western Europe's experience, of changes in population growth exhibited by countries undergoing industrialization. High **birth rates** and **death rates** are followed by plunging death rates, producing a huge net population gain; this is followed by the convergence of birth and death rates at a low overall level. See Figure 8–6.

Demography   The interdisciplinary study of population—especially **birth rates** and **death rates**, growth patterns, longevity, **migration**, and related characteristics.

Desert   An arid area supporting sparse vegetation, receiving less than 10 inches (25 cm) of precipitation per year. Usually exhibits extremes of heat and cold because the moderating influence of moisture is absent.

Desertification   The **process** of **desert** expansion into neighboring **steppelands** as a result of human degradation of fragile semiarid environments.

Development   The economic, social, and institutional growth of national **states**.

Devolution   The **process** whereby regions within a **state** demand and gain political strength and growing autonomy at the expense of the central government.

Dhows   Wooden boats with characteristic triangular sails, plying the seas between Arabian and East African coasts.

Dialect   Regional or local variation in the use of a major language, such as the distinctive accents of many residents of the U.S. South or New England.

Diffusion   The spatial spreading or dissemination of a **culture** element (such as a technological innovation) or some other phenomenon (e.g., a disease outbreak). For the various channels of outward geographic spread from a source area, see **contagious**, **expansion**, **hierarchical**, and **relocation diffusion**.

Distance decay   The various degenerative effects of distance on human spatial structures and interactions.

Diurnal   Daily.

Divided capital   In political geography, a country whose central administrative functions are carried on in more than one city is said to have divided capitals. The Netherlands and South Africa are examples.

Domestication   The transformation of a wild animal or wild plant into a domesticated animal or a cultivated crop to gain control over food production. A necessary evolutionary step in the development of humankind: the invention of **agriculture**.

Domino theory   The belief that political destabilization in one **state** can result in the collapse of order in a neighboring state, triggering a chain of events that, in turn, can affect a series of **contiguous** states.

Double cropping   The planting, cultivation, and harvesting of two crops successively within a single year on the same plot of farmland.

Doubling time   The time required for a population to double in size.

Dry canal   An overland rail and/or road **corridor** across an **isthmus** dedicated to performing the transit functions of a canalized waterway. Best adapted to the movement of containerized cargo, there must be a port at each end to handle the necessary **break-of-bulk** unloading and reloading.

Ecology   The study of the many interrelationships between all forms of life and the natural environments in which they have evolved and continue to develop. The study of *ecosystems* focuses on the interactions between specific organisms and their environments. See also **cultural ecology**.

Economic geography   The field of geography that focuses on the diverse ways in which people earn a living, and how the goods and services they produce are expressed and organized spatially.

Economic tiger   One of the burgeoning beehive countries of the western **Pacific Rim**. Following Japan's route since 1945, these countries have experienced significant modernization, industrialization, and Western-style economic growth since 1980. Three leading economic tigers today are South Korea, Taiwan, and Singapore. Term is increasingly used more generally to describe any fast-developing economy.

Economies of scale   The savings that accrue from large-scale production wherein the unit cost of manufacturing decreases as the level of operation enlarges. Supermarkets operate on this principle and are able to charge lower prices than small grocery stores.

Ecosystem   See **ecology**.

Ecumene   The habitable portions of the Earth's surface where permanent human settlements have arisen.

Elite   A small but influential upper-echelon social class whose power and privilege give it control over a country's political, economic, and cultural life.

El Niño-Southern Oscillation (ENSO)   A periodic, large-scale, abnormal warming of the sea surface in the low latitudes of the eastern Pacific Ocean that has global implications, disturbing normal **weather** patterns in many parts of the world, especially South America.

Elongated state   A **state** whose territory is decidedly long and narrow in that its length is at least six times greater than its average width. Chile and Vietnam are two classic examples.

Emigrant   A person **migrating** away from a country or area; an out-migrant.

Empirical   Relating to the real world, as opposed to theoretical abstraction.

Enclave   A piece of territory that is surrounded by another political unit of which it is not a part.

Endemism   Refers to a disease in a host population that affects many people in a kind of equilibrium without causing rapid and widespread deaths.

Entrepôt   A place, usually a port city, where goods are imported, stored, and transshipped; a **break-of-bulk point**.

Environmental degradation   The accumulated human abuse of a region's **natural landscape** that, among other things, can involve air and water pollution, threats to plant and animal ecosystems, misuse of **natural resources**, and generally upsetting the balance between people and their habitat.

Epidemic   A local or regional outbreak of a disease.

Escarpment   A cliff or very steep slope; frequently marks the edge of a plateau.

Estuary   The widening mouth of a river as it reaches the sea; land subsidence or a rise in sea level has overcome the tendency to form a **delta**.

Ethnic cleansing   The slaughter and/or forced removal of one **ethnic** group from its homes and lands by another, more powerful ethnic group bent on taking that territory.

Ethnicity   The combination of a people's **culture** (traditions, customs, language, and religion) and racial ancestry.

European Union (EU)   **Supranational** organization constituted by 25 European countries to further their common economic interests. These countries are: Germany, France, Italy, the United Kingdom, Belgium, the Netherlands, Luxembourg, Denmark, Finland, Sweden, Austria, Ireland, Portugal, Spain, and Greece. In 2004, 10 additional countries were admitted, raising the EU total to 25: Estonia, Latvia, Lithuania, the Czech Republic, Slovakia, Poland, Hungary, Slovenia, Malta, and Cyprus.

European state model   A **state** consisting of a legally defined territory inhabited by a population governed from a capital city by a representative government.

Exclave   A bounded (non-island) piece of territory that is part of a particular **state** but lies separated from it by the territory of another state. Alaska is an exclave of the United Sates.

Exclusive Economic Zone (EEZ)   An oceanic zone extending up to 200 **nautical miles** from a shoreline, within which the coastal **state** can control fishing, mineral exploration, and additional activities by all other countries.

Expansion diffusion   The spreading of an innovation or an idea through a fixed population in such a way that the number of those adopting grows continuously larger, resulting in an expanding area of dissemination.

Extraterritoriality   Politico-geographical concept suggesting that the property of one **state** lying within the boundaries of another actually forms an extension of the first state.

Fatwa   Literally, a legal opinion or proclamation issued by an Islamic cleric, based on the holy texts of Islam, long applicable only in the *Umma*, the realm ruled by the laws of Islam. In 1989 the Iranian ayatollah Khomeini extended the reach of the *fatwa* by condemning to death a British citizen and author living in the United Kingdom.

Favela   Shantytown on the outskirts or even well within an urban area in Brazil.

Fazenda   Coffee plantation in Brazil.

Federal state   A political framework wherein a central government represents the various subnational entities within a **nation-state** where they have common interests—defense, foreign affairs, and the like—yet allows these various entities to retain their own identities and to have their own laws, policies, and customs in certain spheres.

Federation   See **federal state**.

Fertile Crescent   Crescent-shaped zone of productive lands extending from near the southeastern Mediterranean coast through Lebanon and Syria to the **alluvial** lowlands of Meso-potamia (in Iraq). Once more fertile than today, this is one of the world's great source areas of **agricultural** and other innovations.

First Nations   Canada's indigenous peoples of American descent, whose U.S. counterparts are called Native Americans.

Fjord   Narrow, steep-sided, elongated, and inundated coastal valley deepened by glacier ice that has since melted away, leaving the sea to penetrate.

Floodplain   Low-lying area adjacent to a mature river, often covered by **alluvial** deposits and subject to the river's floods.

Forced migration   Human **migration** flows in which the movers have no choice but to relocate.

Formal region   A type of **region** marked by a certain degree of homogeneity in one or more phenomena; also called *uniform region* or *homogeneous region*.

Forward capital   Capital city positioned in actually or potentially contested territory, usually near an international border; it confirms the **state's** determination to maintain its presence in the region in contention.

Fossil fuels   The energy resources of **coal**, natural gas, and petroleum (oil), so named collectively because they were formed by the geologic compression and transformation of tiny plant and animal organisms.

Four Motors of Europe   Rhône-Alpes (France), Baden-Württemberg (Germany), Catalonia (Spain), and Lombardy (Italy). Each is a high-technology-driven region marked by exceptional industrial vitality and economic success not only within Europe but on the global scene as well.

Fragmented state   A **state** whose territory consists of several separated parts, not a **contiguous** whole. The individual parts may be isolated from each other by the land area of other states or by international waters. The United States and Indonesia are examples.

Francophone   French-speaking. Quebec constitutes the heart of Francophone Canada.

Free-trade area   A form of economic integration, usually consisting of two or more **states**, in which members agree to remove tariffs on trade among themselves. Usually accompanied by a **customs union** that establishes common tariffs on imports from outside the trade area, and sometimes by a **common market** that also removes internal restrictions on the movement of capital, labor, and enterprise.

Fría   See **tierra fría**.

Frontier   Zone of advance penetration, usually of contention; an area not yet fully integrated into a national **state**.

FTAA (Free Trade Area of the Americas)   The ultimate goal of **supranational** economic integration in North, Middle, and South America: the creation of a single-market trading bloc (perhaps as soon as 2005) that would involve every country in the Western Hemisphere between the Arctic shore of Canada and Cape Horn at the southern tip of Chile.

Functional region   A **region** marked less by its sameness than its dynamic internal structure; because it usually focuses on a central node, also called *nodal region* or *focal region*.

Functional specialization   The production of particular goods or services as a dominant activity in a particular location.

Fundamentalism   See **revivalism (religious)**.

Gentrification   The upgrading of an older residential area through private reinvestment, usually in the downtown area of a central city. Frequently this involves the displacement of established lower-income residents, who cannot afford the heightened costs of living, and conflicts are not uncommon as such neighborhood change takes place.

Geographic realm   The basic spatial unit in our world regionalization scheme. Each realm is defined in terms of a synthesis of its total human geography—a composite of its leading cultural, economic, historical, political, and appropriate environmental features.

Geography of development   The field of geography concerned with spatial aspects and regional expressions of **development**.

Geometric boundaries   Political boundaries **defined** and **delimited** (and occasionally **demarcated**) as straight lines or arcs.

Geomorphology   The geographic study of the configuration of the Earth's solid surface—the world's landscapes and their constituent landforms.

Ghetto   An intraurban region marked by a particular **ethnic** character. Often an inner-city poverty zone, such as the black ghetto in U.S. central cities. Ghetto residents are involuntarily segregated from other income and racial groups.

Glaciation   See **Pleistocene Epoch**.

Globalization   The gradual reduction of regional contrasts at the world scale, resulting from increasing international cultural, economic, and political exchanges.

Green Revolution   The successful recent development of higher-yield, fast-growing varieties of rice and other cereals in certain developing countries.

Gross domestic product (GDP)   The total value of all goods and services produced in a country by that state's economy during a given year.

Gross national product (GNP)   The total value of all goods and services produced in a country by that state's economy during a given year, plus all citizens' income from foreign investment and other external sources.

Growth pole   An urban center with certain attributes that, if augmented by a measure of investment support, will stimulate regional economic development in its **hinterland**.

Hacienda   Literally, a large estate in a Spanish-speaking country. Sometimes equated with the **plantation**, but there are important differences between these two types of agricultural enterprise (see pp. 143–144).

Heartland theory   The hypothesis, proposed by British geographer Halford Mackinder during the early twentieth century, that any political power based in the heart of Eurasia could gain sufficient strength to eventually dominate the world. Further, since Eastern Europe controlled access to the Eurasian interior, its ruler would command the vast "heartland" to the east.

Hegemony   The political dominance of a country (or even a region) by another country. The former Soviet Union's postwar grip on Eastern Europe, which lasted from 1945 to 1990, was a classic example.

Helada   See **tierra helada**.

Hierarchical diffusion   A form of **diffusion** in which an idea or innovation spreads by trickling down from larger to smaller adoption units. An urban **hierarchy** is usually involved, encouraging the leapfrogging of innovations over wide areas, with geographic distance a less important influence.

Hierarchy   An order or gradation of phenomena, with each level or rank subordinate to the one above it and superior to the one below. The levels in a national urban hierarchy are constituted by hamlets, villages, towns, cities, and (frequently) the **primate city**.

High island   Volcanic islands of the Pacific Realm that are high enough in elevation to wrest substantial moisture from the tropical ocean air (see **orographic precipitation**). They tend to be well watered, their volcanic soils enable productive agriculture, and they support larger populations than **low islands**—which possess none of these advantages and must rely on fishing and the coconut palm for survival.

High sea   Areas of the oceans away from land, beyond national jurisdiction, open and free for all to use.

Highveld   A term used in Southern Africa to identify the high, grass-covered plateau that dominates much of the region. The lowest-lying areas in South Africa are called *lowveld*; areas that lie at intermediate elevations are the *middleveld*.

Hinterland   Literally, "country behind," a term that applies to a surrounding area served by an urban center. That center is the focus of goods and services produced for its hinterland and is its dominant urban influence as well. In the case of a port city, the hinterland also includes the inland area whose trade flows through that port.

Historical inertia   A term from manufacturing geography that refers to the need to continue using the factories, machinery, and equipment of heavy industries for their full, multiple-decade lifetimes to cover major initial investments—even though these facilities may be increasingly obsolete.

Holocene   The current *interglaciation* epoch (the warm period of glacial contraction between the glacial expansions of an **ice age**); extends from 10,000 years ago to the present. Also known as the *Recent Epoch*.

Humus   Dark-colored upper layer of a soil that consists of decomposed and decaying organic matter such as leaves and branches, nutrient-rich and giving the soil a high fertility.

Hydraulic civilization theory   The theory that cities able to control **irrigated** farming over large **hinterlands** held political power over other cities. Particularly applies to early Asian civilizations based in such river valleys as the Chang (Yangzi), the Indus, and those of Mesopotamia.

Hydrologic cycle   The **system** of exchange involving water in its various forms as it continually circulates among the **atmosphere**, the oceans, and above and below the land surface.

Ice age   A stretch of geologic time during which the Earth's average atmospheric temperature is lowered; causes the equatorward expansion of continental ice sheets in the higher latitudes and the growth of mountain glaciers in and around the highlands of the lower latitudes.

Iconography   The identity of a region as expressed through its cherished symbols; its particular **cultural landscape** and personality.

Immigrant   A person **migrating** into a particular country or area; an in-migrant.

Imperialism   The drive toward the creation and expansion of a colonial empire and, once established, its perpetuation.

Import-substitution industries   The industries local entrepreneurs establish to serve populations of remote areas when transport costs from distant sources make these goods too expensive to import.

Inaccessibility   See **accessibility**.

Indentured workers   Contract laborers who sell their services for a stipulated period of time.

Indigenous peoples   Native or *aboriginal* peoples; often used to designate the inhabitants of areas that were conquered and subsequently colonized by the **imperial** powers of Europe.

Industrial Revolution   The term applied to the social and economic changes in agriculture, commerce, and especially manufacturing and urbanization that resulted from technological innovations and specialization in late-eighteenth-century Europe.

Informal sector   Dominated by unlicensed sellers of homemade goods and services, the primitive form of capitalism found in many developing countries that takes place beyond the control of government.

Infrastructure   The foundations of a society: urban centers, transport networks, communications, energy distribution systems, farms, factories, mines, and such facilities as schools, hospitals, postal services, and police and armed forces.

Insular   Having the qualities and properties of an island. Real islands are not alone in possessing such properties of **isolation** : an **oasis** in the middle of a **desert** also has qualities of insularity.

Insurgent state   Territorial embodiment of a successful guerrilla movement. The establishment by anti-government insurgents of a territorial base in which they exercise full control; thus a state within a **state**.

Intercropping   The planting of several types of crops in the same field; commonly used by **shifting cultivators**.

Interglaciation   See **Pleistocene Epoch**.

Intermontane   Literally, between mountains. The location can bestow certain qualities of natural protection or **isolation** to a community.

Internal migration   **Migration** flow within a country, such as ongoing westward and southward movements toward the **Sunbelt** in the United States.

International migration   **Migration** flow involving movement across an international boundary.

Intervening opportunity   In trade or **migration** flows, the presence of a nearer opportunity that greatly diminishes the attractiveness of sites farther away.

Inuit   **Indigenous** peoples of North America's Arctic zone, formerly known as Eskimos.

Irredentism   A policy of cultural extension and potential political expansion by a **state** aimed at a community of its nationals living in a neighboring state.

Irrigation   The artificial watering of croplands.

Isohyet   A line connecting points of equal rainfall total.

Isolation   The condition of being geographically cut off or far removed from mainstreams of thought and action. It also denotes a lack of receptivity to outside influences, caused at least partially by poor **accessibility**.

Isotherm   A line connecting points of equal temperature.

Isthmus   A **land bridge**; a comparatively narrow link between larger bodies of land. Central America forms such a link between Mexico and South America.

Jakota Triangle   The easternmost region of the East Asian realm, consisting of *Ja*pan, (South) *Ko*rea, and *Tai*wan.

Juxtaposition Contrasting places in close proximity to one another.

Karst The distinctive natural landscape associated with the chemical erosion of soluble limestone rock.

Land alienation One society or culture group taking land from another. In Subsaharan Africa, for example, European **colonialists** took land from **indigenous** Africans and put it to new uses.

Land bridge A narrow **isthmian** link between two large landmasses. They are temporary features—at least in terms of geologic time—subject to appearance and disappearance as the land or sea level rises and falls.

Land reform The spatial reorganization of **agriculture** through the allocation of farmland (often expropriated from landlords) to **peasants** and tenants who never owned land.

Land tenure The way people own, occupy, and use land.

Landlocked An interior **state** surrounded by land. Without coasts, such a country is disadvantaged in terms of **accessibility** to international trade routes, and in the scramble for possession of areas of the **continental shelf** and control of the **exclusive economic zone** beyond.

Latitude Lines of latitude are **parallels** that are aligned east-west across the globe, from 0° latitude at the equator to 90° North and South latitude at the poles.

Leached soil Infertile, reddish-appearing, tropical soil whose surface consists of oxides of iron and aluminum; all other soil nutrients have been dissolved and transported downward into the subsoil by percolating water associated with heavy rainfall.

Leeward The protected or downwind side of a **topographic** barrier with respect to the winds that flow across it.

Lignite Low-grade, brown-colored variety of coal.

Lingua franca A "common language" prevalent in a given area; a second language that can be spoken and understood by many peoples, although they speak other languages at home.

Littoral Coastal or coastland.

Llanos The interspersed **savanna** grasslands and scrub woodlands of the Orinoco River's wide basin that covers much of interior Colombia and Venezuela.

Location theory A logical attempt to explain the locational pattern of an economic activity and the manner in which its producing areas are interrelated. The agricultural location theory that underlies the **von Thünen model** is a leading example.

Loess Deposit of very fine silt or dust that is laid down after having been windborne for a considerable distance. Notable for its fertility under **irrigation** and its ability to stand in steep vertical walls.

Longitude Angular distance (0° to 180°) east or west as measured from the *prime meridian* (0°) that passes through the Greenwich Observatory in suburban London, England. For much of its length across the mid-Pacific Ocean, the 180th meridian functions as the *international date line*.

Low island Low-lying coral islands of the Pacific Realm that—unlike **high islands**—cannot wrest sufficient moisture from the tropical ocean air to avoid chronic drought. Thus productive agriculture is impossible and their modest populations must rely on fishing and the coconut palm for survival.

Madrassa Revivalist (**fundamentalist**) religious school where the curriculum focuses on Islamic religion and law and requires rote memorization of the Qu'ran (Koran), Islam's holy book. Founded in former British India, these schools were most numerous in present-day Pakistan but have diffused to Turkey in the west and Indonesia in the east.

Maghreb The region occupying the northwestern corner of Africa, consisting of Morocco, Algeria, and Tunisia.

Main Street Canada's dominant **conurbation** that is home to nearly two-thirds of the country's inhabitants; extends southwestward from Quebec City in the mid-St. Lawrence Valley to Windsor on the Detroit River.

Mainland-Rimland framework Twofold regionalization of the Middle American realm based on its modern cultural history. The Euro-Amerindian *Mainland*, stretching from Mexico to Panama (minus the Caribbean coastal strip), was a self-sufficient zone dominated by **hacienda land tenure**. The Euro-African *Rimland*, consisting of that Caribbean coastal zone plus all of the Caribbean islands to the east, was the zone of the **plantation** that heavily relied on trade with Europe.

Maquiladora The term given to modern industrial plants in Mexico's U.S. border zone. These foreign-owned factories assemble imported components and/or raw materials, and then export finished manufactures, mainly to the United States. Import duties are disappearing under **NAFTA**, bringing jobs to Mexico and the advantages of low wage rates to the foreign entrepreneurs.

Marchland An area or **frontier** of uncertain boundaries that is subject to various national claims and an unstable political history. Refers specifically to the movement of various armies across such zones.

Marine geography The geographic study of oceans and seas. Its practitioners investigate both the physical (e.g., coral-reef **biogeography**, ocean-**atmosphere** interactions, coastal **geomorphology**) and human (e.g., **maritime boundary-making**, fisheries, beachside development) aspects of oceanic environments.

Maritime boundary An international boundary that lies in the ocean. Like all boundaries, it is a vertical plane, extending from the seafloor to the upper limit of the air space in the atmosphere above the water.

Median line boundary An international **maritime boundary** drawn where the width of a sea is less than 400 **nautical miles**. Because the **states** on either side of that sea claim **exclusive economic zones** of 200 nautical miles, it is necessary to reduce those claims to a (median) distance equidistant from each shoreline. **Delimitation** on the map almost always appears as a set of straight-line segments that reflect the configurations of the coastlines involved.

Medical geography The study of health and disease within a geographic context and from a spatial perspective. Among other things, this field of geography examines the sources, **diffusion** routes, and distributions of diseases.

Megacity Informal term referring to the world's most heavily populated cities; in this book, the term refers to a **metropolis** containing a population of greater than 10 million.

Megalopolis When spelled with a lower-case *m*, a synonym for **conurbation**, one of the large coalescing supercities forming in diverse parts of the world. When capitalized, refers specifically to the multi-metropolitan corridor that extends along the northeastern U.S. seaboard from north of Boston to south of Washington, D.C. (Fig. 3–9).

Mercantilism Protectionist policy of European **states** during the sixteenth to the eighteenth centuries that promoted a state's economic position in the contest with rival powers. The acquisition of gold and silver and maintaining a favorable trade balance (more exports than imports) were central to the policy.

Meridian Line of **longitude**, aligned north-south across the globe, that together with **parallels** of **latitude** forms the global grid system. All meridians converge at both poles and are at their maximum distances from each other at the equator.

Mestizo Derived from the Latin word for *mixed*, refers to a person of mixed white and Amerindian ancestry.

Metropolis Urban **agglomeration** consisting of a (central) city and its suburban ring. See **urban (metropolitan) area**.

Metropolitan area See **urban (metropolitan) area**.

Migration A change in residence intended to be permanent. See also **forced**, **internal**, **international**, and **voluntary migration**.

Migratory movement Human relocation movement from a source to a destination without a return journey, as opposed to cyclical movement (see **nomadism**).

Model An idealized representation of reality built to demonstrate its most important properties. A **spatial** model focuses on a geographical dimension of the real world, such as the **von Thünen model** that explains agricultural location patterns in a commercial economy.

Monsoon   Refers to the seasonal reversal of wind and moisture flows in certain parts of the subtropics and lower-middle latitudes. The *dry monsoon* occurs during the cool season when dry offshore winds prevail. The *wet monsoon* occurs in the hot summer months, which produce onshore winds that bring large amounts of rainfall. The air-pressure differential over land and sea is the triggering mechanism, with windflows always moving from areas of relatively higher pressure toward areas of relatively lower pressure. Monsoons make their greatest regional impact in the coastal and near-coastal zones of South Asia, Southeast Asia, and East Asia.

Mosaic culture   The emerging cultural-geographic framework of the United States, dominated by the fragmentation of specialized social groups into homogeneous communities of interest marked not only by income, race, and ethnicity but also by age, occupational status, and lifestyle. The result is an increasingly heterogeneous socio-spatial complex, which resembles an intricate mosaic composed of myriad uniform—but separate—tiles.

Mulatto   A person of mixed African (black) and European (white) ancestry.

Multinationals   Internationally active corporations that can strongly influence the economic and political affairs of many countries they operate in.

Muslim   An adherent of the Islamic faith.

Muslim Front   A term used by certain scholars for the African Transition Zone of northern Africa, which is primarily regarded as a still-expanding frontier of Islam that affects countries from Guinea in the west to the African Horn in the east (see Fig. 6–17).

NAFTA (North American Free Trade Agreement) The **free-trade area** launched in 1994 involving the United States, Canada, and Mexico.

Nation   Legally a term encompassing all the citizens of a **state**, it also has other connotations. Most definitions now tend to refer to a group of tightly-knit people possessing bonds of language, **ethnicity**, religion, and other shared **cultural** attributes. Such homogeneity actually prevails within very few states.

Nation-state   A country whose population possesses a substantial degree of **cultural** homogeneity and unity. The ideal form to which most **nations** and **states** aspire—a political unit wherein the territorial state coincides with the area settled by a certain national group or people.

NATO (North Atlantic Treaty Organization)   Established in 1950 at the height of the Cold War as a U.S.-led

supranational defense pact to shield postwar Europe against the Soviet military threat. NATO is now in transition, expanding its membership while modifying its objectives in the post-Soviet era.

Natural hazard   A natural event that endangers human life and/or the contents of a **cultural landscape**.

Natural increase rate   Population growth measured as the excess of live births over deaths per 1000 individuals per year. Natural increase of a population does not reflect either **emigrant** or **immigrant** movements.

Natural landscape   The array of landforms that constitutes the Earth's surface (mountains, hills, plains, and plateaus) and the physical features that mark them (such as water bodies, soils, and vegetation). Each **geographic realm** has its distinctive combination of natural landscapes.

Natural resource   Any valued element of (or means to an end using) the environment; includes minerals, water, vegetation, and soil.

Nautical mile   By international agreement, the nautical mile—the standard measure at sea—is 6076.12 feet in length, equivalent to approximately 1.15 statute miles (1.85 km).

Neocolonialism   The term used by developing countries to underscore that the entrenched colonial system of international exchange and capital flow has not changed in the postcolonial era—thereby perpetuating the huge economic advantages of the developed world.

Network (transport)   The entire regional **system** of transportation connections and nodes through which movement can occur.

Nevada   See **tierra nevada**.

Nomadism   Cyclical movement among a definite set of places. Nomadic peoples mostly are **pastoralists**.

Nucleation   Cluster; **agglomeration**.

Oasis   An area, small or large, where the supply of water (from an **aquifer** or a major river such as the Nile) permits the transformation of the adjacent **desert** into productive cropland.

Occidental   Western. Also see ***Oriental***.

Offshore banking   Term referring to financial havens for foreign companies and individuals, who channel their earnings to accounts in such a country (usually an "offshore" island-state) to avoid paying taxes in their home countries.

OPEC (Organization of Petroleum Exporting Countries)   The international oil *cartel* or syndicate formed by a number of producing countries to promote their common economic interests through the formulation of joint pricing policies and the limitation of market options for consumers. The 11 member-states (as of mid-2004) are: Algeria, Indonesia, Iran, Iraq, Kuwait, Libya, Nigeria, Qatar, Saudi Arabia, United Arab Emirates (UAE), and Venezuela.

Organic theory   Friedrich Ratzel's theory of **state** development that conceptualized the state as a biological organism whose life—from birth through maturation to eventual senility and collapse—mirrors that of any living thing.

Oriental   The root of the word *oriental* is from the Latin for *rise*. Thus it has to do with the direction in which one sees the sun "rise"—the east; *oriental* therefore means Eastern. ***Occidental*** originates from the Latin for *fall*, or the "setting" of the sun in the west; *occidental* therefore means Western.

Orographic precipitation   Mountain-induced precipitation, especially where air masses are forced to cross **topographic** barriers. Downwind areas beyond such a mountain range experience the relative dryness known as the **rain shadow effect**.

Outback   The name given by Australians to the vast, peripheral, sparsely-settled interior of their country.

Outer city   The non-central-city portion of the American **metropolis**; no longer "sub" to the "urb," this outer ring was transformed into a full-fledged city during the late twentieth century.

Pacific Rim   A far-flung group of countries and parts of countries (extending clockwise on the map from New Zealand to Chile) sharing the following criteria: they face the Pacific Ocean; they evince relatively high levels of economic development, industrialization, and urbanization; their imports and exports mainly move across Pacific waters.

Pacific Ring of Fire   Zone of crustal instability along tectonic **plate** boundaries, marked by earthquakes and volcanic activity, that ring the Pacific Ocean basin.

Paddies (paddyfields)   Ricefields.

Pandemic   An outbreak of a disease that spreads worldwide.

Pangaea   A vast, singular landmass consisting of most of the areas of the present-day continents. This supercontinent began to break up more than 200 million years ago when still-ongoing **plate** divergence and **continental drift** became dominant processes (see Fig. 6–2).

Parallel   An east-west line of **latitude** that is intersected at right angles by **meridians** of **longitude**.

Pastoralism   A form of **agricultural** activity that involves the raising of livestock.

Peasants   In a **stratified** society, peasants are the lowest class of people who depend on **agriculture** for a living. But they often own no land at all and must survive as tenants or day workers.

Peninsula   A comparatively narrow, finger-like stretch of land extending from the main landmass into the sea. Florida and Korea are examples.

Peon (*peone*)   Term used in Middle and South America to identify people who often live in serfdom to a wealthy landowner; landless **peasants** in continuous indebtedness.

Per capita   Capita means *individual*. Income, production, or some other measure is often given per individual.

Perforated state   A **state** whose territory completely surrounds that of another state. South Africa, which encloses Lesotho and is perforated by it, is a classic example.

Periodic market   Village market that opens every third day or at some other regular interval. Part of a regional network of similar markets in a preindustrial, rural setting where goods are brought to market on foot and barter remains a major mode of exchange.

Periphery   See **core-periphery relationships**.

Permafrost   Permanently frozen water in the near-surface soil and bedrock of cold environments, producing the effect of completely frozen ground. Surface can thaw during brief warm season.

Physical geography   The study of the geography of the physical (natural) world. Its subfields include **climatology**, **geomorphology**, **biogeography**, soil geography, **marine geography**, and water **resources**.

Physical landscape   Synonym for **natural landscape**.

Physiographic political boundaries   Political boundaries that coincide with prominent physical features in the **natural landscape**—such as rivers or the crest ridges of mountain ranges.

Physiographic region (province)   A **region** within which there prevails substantial **natural-landscape** homogeneity, expressed by a certain degree of uniformity in surface **relief**, **climate**, vegetation, and soils.

Physiography   Literally means *landscape description*, but commonly refers to the total **physical geography** of a place; includes all of the natural features on the Earth's surface, including landforms, **climate**, soils, vegetation, and water bodies.

Physiologic density   The number of people per unit area of **arable** land.

Pilgrimage   A journey to a place of great religious significance by an individual or by a group of people (such as a pilgrimage to Mecca for **Muslims**).

Plantation   A large estate owned by an individual, family, or corporation and organized to produce a cash crop. Almost all plantations were established within the tropics; in recent decades, many have been divided into smaller holdings or reorganized as cooperatives.

Plate tectonics   Plates are bonded portions of the Earth's mantle and crust, averaging 60 miles (100 km) in thickness. More than a dozen such plates exist (see Fig. G–4), most of continental proportions, and they are in mo-tion. Where they meet one slides under the other, crumpling the surface crust and producing significant volcanic and earthquake activity; a major mountain-building force.

Pleistocene Epoch   Recent period of geologic time that spans the rise of humankind, beginning about 2 million years ago. Marked by *glaciations* (repeated advances of continental ice sheets) and milder *interglaciations* (ice sheet contractions). Although the last 10,000 years are known as the **Holocene** Epoch, Pleistocene-like conditions seem to be continuing and we are most probably now living through another Pleistocene interglaciation; thus the glaciers likely will return.

Plural(istic) society   A society in which two or more population groups, each practicing its own **culture**, live adjacent to one another without mixing inside a single **state**.

Polder   Land reclaimed from the sea adjacent to the shore of the Netherlands by constructing dikes and then pumping out the water trapped behind them.

Political geography   The study of the interaction of geographical area and political **process**; the spatial analysis of political phenomena and processes.

Pollution   The release of a substance, through human activity, which chemically, physically, or biologically alters the air or water it is discharged into. Such a discharge negatively impacts the environment, with possible harmful effects on living organisms—including humans.

Population density   The number of people per unit area. Also see **arithmetic density** and **physiologic density** measures.

Population distribution   The way people have arranged themselves in geographic space. One of human geography's most essential expressions because it represents the sum total of the adjustments that a population has made to its natural, cultural, and economic environments.

Population explosion   The rapid growth of the world's human population during the past century, attended by ever-shorter **doubling times** and accelerating *rates* of increase.

Population geography   The field of geography that focuses on the spatial aspects of **demography** and the influences of demographic change on particular places.

Population implosion   The opposite of the **population explosion**, refers to the declining populations of many European countries and Russia in which the **death rate** exceeds the **birth rate** and **immigration** rate.

Population movement   See **migration**; **migratory movement**.

Population projection   The future population total that demographers forecast for a particular country. For example, in Table G–1 such projections are given for all the world's countries for 2025.

Population (age-sex) structure   Graphic representation (*profile*) of a population according to age and gender.

Postindustrial economy   Emerging economy, in the United States and a handful of other highly advanced countries, as traditional industry is increasingly eclipsed by a higher-technology productive complex dominated by services, information-related, and managerial activities.

Primary economic activity   Activities engaged in the direct extraction of **natural resources** from the environment such as mining, fishing, lumbering, and especially **agriculture**.

Primate city   A country's largest city—ranking atop the urban **hierarchy**—most expressive of the national culture and usually (but not always) the capital city as well.

Process   Causal force that shapes a spatial pattern as it unfolds over time.

Protruded state   Territorial shape of a **state** that exhibits a narrow, elongated land extension (or *protrusion*) leading away from the main body of territory. Thailand is a leading example.

Push-pull concept   The idea that **migration** flows are simultaneously stimulated by conditions in the source area, which tend to drive people away, and by the perceived attractiveness of the destination.

Qanat   In **desert** zones, particularly in Iran and western China, an underground tunnel built to carry **irrigation** water by gravity flow from nearby mountains (where **orographic precipitation** occurs) to the arid flatlands below.

Quaternary economic activity   Activities engaged in the collection, processing, and manipulation of *information*.

Quinary economic activity   Managerial or control-function activity associated with decision-making in large organizations.

Rain shadow effect   The relative dryness in areas downwind of mountain ranges caused by **orographic precipitation**, wherein moist air masses are forced to deposit most of their water content as they cross the highlands.

Realm   See **geographic realm**.

Region   A commonly used term and a geographic concept of central importance. An **area** on the Earth's surface marked by specific criteria.

Regional boundary   In theory, the line that circumscribes a **region**. But razor-sharp lines are seldom encountered, even in nature (e.g., a coastline constantly changes depending upon the tide). In the **cultural landscape**, not only are regional boundaries rarely self-evident, but when they are ascertained by geographers they most often turn out to be **transitional** borderlands.

Regional complementarity   See **complementarity**.

Regional disparity The spatial unevenness in standard of living that occurs within a country, whose "average," overall income statistics invariably mask the differences that exist between the extremes of the wealthy **core** and the poorer **periphery**.

Regional state A "natural economic zone" that defies political boundaries, and is shaped by the global economy of which it is a part; its leaders deal directly with foreign partners and negotiate the best terms they can with the national governments under which they operate.

Relative location The regional position or **situation** of a place relative to the position of other places. Distance, **accessibility**, and connectivity affect relative location.

Relict boundary A political boundary that has ceased to function, but the imprint of which can still be detected on the **cultural landscape**.

Relief Vertical difference between the highest and lowest elevations within a particular area.

Relocation diffusion Sequential **diffusion process** in which the items being diffused are transmitted by their carrier agents as they relocate to new areas. The most common form of relocation diffusion involves the spreading of innovations by a **migrating** population.

Revivalism (religious) Religious movement whose objectives are to return to the foundations of that faith and to influence **state** policy. Often called **religious fundamentalism**; but in the case of Islam, **Muslims** prefer the term revivalism.

Rift valley The trough or trench that forms when a strip of the Earth's crust sinks between two parallel faults (surface fractures).

Sahel Semiarid **steppeland** zone extending across most of Africa between the southern margins of the arid Sahara and the moister tropical **savanna** and forest zone to the south. Chronic drought, **desertification**, and overgrazing have contributed to severe famines in this area since 1970.

Savanna Tropical grassland containing widely spaced trees; also the name given to the tropical wet-and-dry climate (*Aw*).

Scale Representation of a real-world phenomenon at a certain level of reduction or generalization. In **cartography**, the ratio of map distance to ground distance; indicated on a map as a bar graph, representative fraction, and/or verbal statement. *Macroscale* refers to a large area of national proportions; *microscale* refers to a local area no bigger than a county.

Scale economies See **economies of scale**.

Secondary economic activity Activities that process raw materials and transform them into finished industrial products. The *manufacturing* sector.

Sedentary Permanently attached to a particular area; a population fixed in its location. The opposite of **nomadic**.

Separate development The spatial expression of South Africa's "grand" **apartheid** scheme, whereby non-white groups were required to settle in segregated "homelands." The policy was dismantled when white-minority rule collapsed in the early 1990s.

Sequent occupance The notion that successive societies leave their cultural imprints on a place, each contributing to the cumulative **cultural landscape**.

Shantytown Unplanned slum development on the margins of cities in developing countries, dominated by crude dwellings and shelters mostly made of scrap wood, iron, and even pieces of cardboard.

Sharecropping Relationship between a large landowner and farmers on the land whereby the farmers pay rent for the land they farm by giving the landlord a share of the annual harvest.

Sharia The criminal code based in Islamic law that prescribes corporal punishment, amputations, stonings, and lashing for both major and minor offenses. Its occurrence today is associated with the spread of **religious revivalism** in **Muslim** societies.

Shatter belt Region caught between stronger, colliding external cultural-political forces, under persistent stress, and often fragmented by aggressive rivals. Eastern Europe and Southeast Asia are classic examples.

Shifting agriculture Cultivation of crops in recently cut and burned tropical-forest clearings, soon to be abandoned in favor of newly cleared nearby forest land. Also known as *slash-and-burn agriculture*.

Sinicization Giving a Chinese cultural imprint; Chinese **acculturation**.

Site The internal locational attributes of an urban center, including its local spatial organization and physical setting.

Situation The external locational attributes of an urban center; its **relative location** or regional position with reference to other non-local places.

Social stratification See **stratification (social)**.

Spatial Pertaining to space on the Earth's surface. Synonym for *geographic(al)*.

Spatial diffusion See **diffusion**.

Spatial interaction See **complementarity**, **transferability**, and **intervening opportunity**.

Spatial model See **model**.

Spatial process See **process**.

Spatial system The components and interactions of a **functional region**, which is defined by the areal extent of those interactions. Also see **system**.

Special Economic Zone (SEZ) Manufacturing and export center within China, created in the 1980s to attract foreign investment and technology transfers. Six SEZs—all located on southern China's Pacific coast—currently operate: Shenzhen, adjacent to Hong Kong; Zhuhai; Shantou; Xiamen; Hainan Island, in the far south; and still-building Pudong, across the river from Shanghai.

Squatter settlement See **shantytown**.

State A politically organized territory that is administered by a sovereign government and is recognized by a significant portion of the international community. A state must also contain a permanent resident population, an organized economy, and a functioning internal circulation system.

State capitalism Government-controlled corporations competing under free-market conditions, usually in a tightly regimented society. South Korea is a leading example. Also see *chaebol*.

State planning Involves highly centralized control of the national planning process, a hallmark of communist economic systems. Soviet central planners mainly pursued a grand political design in assigning production to particular places; their frequent disregard of **economic geography** contributed to the eventual collapse of the U.S.S.R.

State territorial morphology A **state's** geographical shape, which can have a decisive impact on its spatial cohesion and political viability. A **compact** shape is most desirable; among the less efficient shapes are those exhibited by **elongated**, **fragmented**, **perforated**, and **protruded** states.

Stateless nation A **national** group that aspires to become a **nation-state** but lacks the territorial means to do so; the Palestinians and Kurds of Southwest Asia are classic examples.

Steppe Semiarid grassland; short-grass prairie. Also the name given to the semiarid climate type (*BS*).

Stratification (social) In a layered or stratified society, the population is divided into a **hierarchy** of social classes. In an industrialized society, the working class is at the lower end; **elites** that possess capital and control the means of production are at the upper level. In the traditional **caste system** of Hindu India, the "untouchables" form the lowest class or caste, whereas the still-wealthy remnants of the princely class are at the top.

Subduction In **plate tectonics**, the **process** that occurs when an oceanic plate converges head-on with a plate carrying a continental landmass at its leading edge. The lighter continental plate overrides the denser oceanic plate and pushes it downward.

Subsequent boundary A political boundary that developed contemporaneously with the evolution of the major elements of the **cultural landscape** through which it passes.

Subsistence Existing on the minimum necessities to sustain life; spending most of one's time in pursuit of survival.

Subtropical Convergence A narrow marine **transition zone**, girdling the globe at approximately latitude 40°S, that marks the equatorward limit of the frigid Southern Ocean and the poleward limits of the warmer Atlantic, Pacific, and Indian Oceans to the north.

Suburban downtown In the United States (and increasingly in other advantaged countries), a significant concentration of major urban activities around a highly **accessible** suburban location, including retailing, light industry, and a variety of leading corporate and commercial operations. The largest are now coequal to the American central city's **central business district (CBD)**.

Sunbelt The popular name given to the southern tier of the United States, which is anchored by the mega-States of California, Texas, and Florida. Its warmer climate, superior recreational opportunities, and other amenities have been attracting large numbers of relocating people and activities since the 1960s; broader definitions of the Sunbelt also include much of the western U.S., particularly Colorado and the coastal Pacific Northwest.

Superimposed boundary A political boundary emplaced by powerful outsiders on a developed human landscape. Usually ignores pre-existing cultural-spatial patterns, such as the border that still divides North and South Korea.

Supranational A venture involving three or more **states**—political, economic, and/or cultural cooperation to promote shared objectives. The **European Union** is one such organization.

System Any group of objects or institutions and their mutual interactions. Geography treats systems that are expressed **spatially**, such as in **functional regions**.

Systematic geography Topical geography: **cultural**, **political**, **economic geography**, and the like.

Taiga The subarctic, mostly **coniferous** snowforest that blankets northern Russia and Canada south of the **tundra** that lines the Arctic shore.

Takeoff Economic concept to identify a stage in a country's **development** when conditions are set for a domestic Industrial Revolution.

Taxonomy A **system** of scientific classification.

Technopole A planned techno-industrial complex (such as California's Silicon Valley) that innovates, promotes, and manufactures the products of the **postindustrial** informational economy.

Tectonics See **plate tectonics**.

Templada See **tierra templada**.

Terracing The transformation of a hillside or mountain slope into a step-like sequence of horizontal fields for intensive cultivation.

Territoriality A country's or more local community's sense of property and attachment toward its territory, as expressed by its determination to keep it inviolable and strongly defended.

Territorial sea Zone of seawater adjacent to a country's coast, held to be part of the national territory and treated as a segment of the sovereign **state**.

Tertiary economic activity Activities that engage in *services*—such as transportation, banking, retailing, education, and routine office-based jobs.

Tierra caliente The lowest of the **altitudinal zones** into which the human settlement of Middle and South America is classified according to elevation. The *caliente* is the hot humid coastal plain and adjacent slopes up to 2,500 feet (750 m) above sea level. The natural vegetation is the dense and luxuriant tropical rainforest; the crops include sugar and bananas in the lower areas, and coffee, tobacco, and corn along the higher slopes.

Tierra fría The cold, high-lying **altitudinal zone** of settlement in Andean South America, extending from about 6,000 feet (1,800 m) in elevation up to nearly 12,000 feet (3,600 m). **Coniferous** trees stand here; upward they change into scrub and grassland. There are also important pastures within the *fría*, and wheat can be cultivated.

Tierra helada In Andean South America, the highest-lying habitable **altitudinal zone**—ca. 12,000 to 15,000 feet (3,600 to 4,500 m)—between the tree line (upper limit of the *tierra fría*) and the snow line (lower limit of the *tierra nevada*). Too cold and barren to support anything but the grazing of sheep and other hardy livestock.

Tierra nevada The highest and coldest **altitudinal zone** in Andean South America (lying above 15,000 feet [4,500 m]), an uninhabitable environment of permanent snow and ice that extends upward to the Andes' highest peaks of more than 20,000 feet (6,000 m).

Tierra templada The intermediate **altitudinal zone** of settlement in Middle and South America, lying between 2,500 feet (750 m) and 6,000 feet (1,800 m) in elevation. This is the "temperate" zone, with moderate temperatures compared to the *tierra caliente* below. Crops include coffee, tobacco, corn, and some wheat.

Topography The surface configuration of any segment of **natural landscape**.

Toponym Place name.

Transculturation Cultural borrowing and two-way exchanges that occur when different **cultures** of approximately equal complexity and technological level come into close contact.

Transferability The capacity to move a good from one place to another at a bearable cost; the ease with which a commodity may be transported.

Transhumance Seasonal movement of people and their livestock in search of pastures. Movement may be vertical (into highlands during the summer and back to lower elevations in winter) or horizontal, in pursuit of seasonal rainfall.

Transition zone An area of **spatial** change where the peripheries of two adjacent **realms** or **regions** join; marked by a gradual shift (rather than a sharp break) in the characteristics that distinguish these neighboring geographic entities from one another.

Transmigration The policy of the Indonesian government to induce residents of the overcrowded, **core-area** island of Jawa to move to the country's other islands.

Treaty ports **Extraterritorial enclaves** in China's coastal cities, established by European colonial invaders under unequal treaties enforced by gunboat diplomacy.

Tropical deforestation See **deforestation**.

Tropical savanna See **savanna**.

Tsunami A seismic (earthquake-generated) sea wave that can attain gigantic proportions and cause coastal devastation.

Tundra The treeless plain that lies along the Arctic shore in northernmost Russia and Canada, whose vegetation consists of mosses, lichens, and certain hardy grasses.

Turkestan Northeasternmost region of the North Africa/Southwest Asia realm. Known as Soviet Central Asia before 1992, its five (dominantly Islamic) former S.S.R.'s have become the independent countries of Kazakhstan, Uzbekistan, Turkmenistan, Kyrgyzstan, and Tajikistan. Today Turkestan has expanded to include a sixth state, Afghanistan.

Unitary state A **nation-state** that has a centralized government and administration that exercises power equally over all parts of the state.

Urbanization A term with several connotations. The proportion of a country's population living in urban places is its level of urbanization. The **process** of urbanization involves the movement to, and the clustering of, people in towns and cities—a major force in every geographic realm today. Another kind of urbanization occurs when an expanding city absorbs rural countryside and transforms it into suburbs; in the case of cities in disadvantaged countries, this also generates peripheral **shantytowns**.

Urban (metropolitan) area The entire built-up, non-rural area and its population, including the most recently constructed suburban appendages. Provides a better picture of the dimensions and population of such an area than the delimited municipality (central city) that forms its heart.

Urban realms model A spatial generalization of the contemporary large American city. It is shown to be a widely dispersed, multi-centered **metropolis** consisting of increasingly independent zones or *realms*, each focused on its own **suburban downtown**; the only exception is the shrunken central realm, which is focused on the **central business district** (see Figs. 3–11 and 3–12).

Veld See **highveld**.

Voluntary migration Population movement in which people relocate in response to perceived opportunity, not because they are forced to migrate.

Von Thünen's Isolated State model Explains the location of agricultural activities in a commercial economy. A **process** of spatial competition allocates various farming activities into concentric rings around a central market city, with profit-earning capability the determining force in how far a crop locates from the market. The original (1826) Isolated State model now applies to the continental scale (see Fig. 1–6).

Weather The immediate and short-term conditions of the **atmosphere** that impinge on daily human activities.

Wet monsoon See **monsoon**.

Windward The exposed, upwind side of a **topographic** barrier that faces the winds that flow across it.

World geographic realm See **geographic realm**.

List of Maps

Photo Credits

Page 101: George W. Moore. Page 119: Andy Ryan. All other photos: H. J. de Blij

Index

A LEGACY OF SERVICE TO GEOGRAPHY

John Wiley & Sons is the worldwide leader in geography publishing. We began our partnership with geography education in 1911 and continue to publish college texts, professional books, journals, and technology products that help teachers teach and students learn. We are committed to making it easier for students to visualize spatial relationships, think critically about their interactions with the environment, and appreciate the earth's dynamic landscapes and diverse cultures.

To serve our customers we have partnered with the AAG, the AGS, and the NCGE while building extraordinary relationships with Microsoft and Rand McNally that allow us to bundle discounted copies of Encarta and the Goode's Atlas with all of our textbooks.

Wiley Geography continues this legacy of service during each academic year with with outstanding first editions and revisions.

Regional Geography

▶ de Blij/Muller   *Geography: Realms, Region, and Concepts 11e* (0-471-15224-2)

▶ de Blij/Muller   *Concepts and Regions in Geography 2e* (0-471-64991-0)   **new**

▶ Blouet/Blouet   *Latin America and the Caribbean 4e Update* (0-471-48052-5)   **new**

▶ Weightman   *Dragons and Tigers: Geography of South, East and Southeast Asia 1e Update* (0-471-48476-8)   **new**

Physical Geography

▶ Strahler/Strahler   *Physical Geography: Science and Systems of the Human Environment 3e* (0-471-48053-3)   **new**

▶ Strahler/Strahler   *Introducing Physical Geography 3e Media Version* (0-471-66969-5)   **new**

▶ Marsh/Grossa   *Environmental Geography 3e* (0-471-48280-3)   **new**

▶ MacDonald   *Biogeography: Introduction to Space, Time and Life* (0-471-24193-8)

Human Geography

▶ de Blij/Murphy   *Human Geography 7e* (0-471-44107-4)

▶ Kuby/Harner/Gober   *Human Geography in Action 3e* (0-471-43055-2)

GIS and Remote Sensing

▶ Chrisman   *Exploring GIS 2e* (0-471-31425-0)

▶ DeMers   *Fundamentals of GIS 3e* (0-471-20491-9)   **new**

▶ Lillesand/Kiefer/ Chipman   *Remote Sensing and Image Interpretation 5e* (0-471-45152-5)

Political Geography

▶ Glassner/Fahrer   *Political Geography 3e* (0-471-35266-7)

Natural Resources

▶ Cutter/Renwick   *Exploitation, Conservation, Preservation 4e* (0-471-15225-0)

▶ Cech   *Principles of Water Resources 2e* (0-471-48475-X)   **new**